ADVANCES IN CATALYSIS

VOLUME 45

ADVANCES IN CATALYSIS

Impact of Surface Science on Catalysis

VOLUME 45

Edited by

BRUCE C. GATES
University of California
Davis, California

HELMUT KNÖZINGER
University of Munich
Munich, Germany

ACADEMIC PRESS

A Harcourt Science and Technology Company

San Diego San Francisco New York
Boston London Sydney Tokyo

This book is printed on acid-free paper. ♾

Academic Press
A Harcourt Science and Technology Company
525 B Street, Suite 1900, San Diego, California 92101-4495, USA
http://www.academicpress.com

Academic Press
Harcourt Place, 32 Jamestown Road, London NW1 7BY, UK
http://www.hbuk.co.uk/ap/

International Standard Book Number: 0-12-007845-7

PRINTED IN THE UNITED STATES OF AMERICA
00 01 02 03 04 SB 9 8 7 6 5 4 3 2 1

Contents

Adsorption Energetics and Bonding from Femtomole Calorimetry and from First Principles Theory

Qingfeng Ge, Rickmer Kose, and David A. King

Active Sites on Oxides: From Single Crystals to Catalysts

Hicham Idriss and Mark A. Barteau

Catalysis and Surface Science: What Do We Learn from Studies of Oxide-Supported Cluster Model Systems?

H.-J. Freund, M. Bäumer, and H. Kuhlenbeck

Sum Frequency Generation: Surface Vibrational Spectroscopy Studies of Catalytic Reactions on Metal Single-Crystal Surfaces

Gabor A. Somorjai and Keith R. McCrea

Contributors

Numbers in parentheses indicate the pages on which the authors' contributions begin.

M. BÄUMER, *Fritz-Haber-Institut der Max-Planck-Gesellschaft, D-14195 Berlin, Germany* (333)

MARK A. BARTEAU, *Center for Catalytic Science and Technology, Department of Chemical Engineering, University of Delaware Newark, Delaware 19716* (261)

GERHARD ERTL, *Fitz-Haber-Institut der Max-Planck-Gesellschaft, D-14195 Berlin, Germany* (1)

H.-J. FREUND, *Fritz-Haber-Institut der Max-Planck-Gesellschaft, D-14195 Berlin, Germany* (333)

QINGFENG GE, *Department of Chemistry, University of Cambridge, Cambridge CB2 1EW, United Kingdom* (207)

B. HAMMER, *Institute of Physics, Aalborg University, DK-9220 Aalborg, Denmark* (71)

HICHAM IDRISS, *Department of Chemistry, University of Auckland, Auckland, New Zealand* (261)

DAVID A. KING, *Department of Chemistry, University of Cambridge, Cambridge CB2 1EW, United Kingdom* (207)

RICKMER KOSE, *Sandia National Laboratories, Livermore, California 94550* (207)

H. KUHLENBECK, *Fritz-Haber-Institut der Max-Planck-Gesellschaft, D-14195 Berling, Germany* (333)

KEITH R. MCCREA, *Department of Chemistry, University of California, Berkeley, California 94720* (385)

J. K. NØRSKOV, *Center for Atomic-Scale Materials Physics, Department of Physics, Technical University of Denmark, DK-2800 Lyngby, Denmark* (71)

GABOR A. SOMORJAI, *Department of Chemistry, University of California, Berkeley, California 94720* (385)

JOOST WINTTERLIN, *Fritz-Haber-Institut der Max-Planck-Gesellschaft, D-14195 Berlin, Germany* (131)

Preface

This is the first volume of *Advances in Catalysis* dedicated to a single theme, the impact of surface science on catalysis. Surface science emerged in the 1960s with the development of reliable ultrahigh vacuum apparatus, providing exact structures of surfaces of metal single crystals (determined by low-energy electron diffraction) as well as information about their compositions (determined by electron spectroscopies) and the relationships between surface structure and composition and catalytic reaction rates (the latter measured with single-crystal samples transferred without contamination between an ambient-pressure catalytic reactor chamber and an ultrahigh vacuum chamber). Catalysis provided much of the driving force for the early development of surface science.

This volume of the *Advances* begins with a chapter by Ertl about dynamics of surface reactions. The chapter integrates dynamics over a wide range of time scales. The quantum level concerns energy exchange between different degrees of freedom of surface species. If these are equilibrated, then transition state theory is applied to formulate the rates of elementary steps; combinations of elementary steps constituting a catalytic cycle are described by macroscopic kinetics. The chapter illustrates the concepts for reactions on single crystals of metal.

Nørskov and Hammer summarize advances resulting from the use of density functional theory to describe reactions at surfaces, mostly those of transition metals. The predictions of adsorption and surface reactions account well for variations in reactivity from one metal to another. This chapter illustrates the dramatic advances in the application of theory to reactivity and catalysis on surfaces.

Wintterlin focuses on scanning tunneling microscopy (STM) for characterization of reactions on metal surfaces. The technique has even brought forth moving pictures of individual atoms reacting on surfaces during catalysis. It has allowed visualization of catalytic sites in action, for example, atomic steps on ruthenium where NO is dissociated. It also illustrates some of the limitations of commonly assumed kinetics of surface reactions based on Langmuir adsorption; for example, CO oxidation is shown to take place at boundaries between islands of O and islands of CO on Pt(111).

King, Ge, and Kose summarize work integrating calorimetry into surface science. Energetics of reactions on some of the best-defined surface structures (single crystals of metal) have emerged, with both the initial and final states being well characterized. The data provide fruitful ground for the merging of experiment and theory (which is also addressed in depth in this chapter) and improved understanding of surface chemical bonding.

Barteau and Idriss assess the surface science of oxides. Oxides received much less attention than metals in the early period of surface science, but now the techniques for characterization of oxide surfaces in ultrahigh vacuum are well developed, and many new insights into surface reactivity and catalysis by oxides have emerged. The data are providing connections between the chemistry of metal oxide powder surfaces and the surfaces of single-crystal oxides. Data characterizing oxides present as ordered thin films on metal single crystals have provided excellent insights into the surface chemistry of oxides because some thin-film oxide surfaces have reactivities nearly matching those of the oxide itself—and the electrical conductivity of the sample nearly matches that of the metal and allows electron spectroscopy experiments without the buildup of substantial charge on the sample.

The supported thin films have been used to disperse metals in the form of clusters or particles, and these are models of supported metal catalysts, as described by Freund, Bäumer, and Kuhlenbeck. These samples offer the advantages of planar solids used in many surface science experiments by allowing interrogation by electron and particle beam spectroscopies, and they also offer some dazzling STM images of metal clusters that could not have been obtained with the clusters dispersed on porous powder supports.

Some of the most insightful investigations of solid catalysts have been those providing evidence of the surface species during catalysis, but the ultrahigh vacuum techniques of classic surface science fall short in this regard because of the pressure limitation. Now a new method, sum frequency generation spectroscopy, described by McCrea and Somorjai, bridges the gap and allows measurement of vibrational spectra of species on working single-crystal catalyst surfaces; it has been useful for distinguishing spectator species from reactive intermediates in alkene hydrogenation on platinum, for example.

As surface science continues its rapid development, this volume illustrates how it is still driven by the challenges of catalysis and how both theory and scanning tunneling microscopy have forcefully emerged as essential tools. It is also evident how surface science continues to shore up the foundation of catalytic science.

B. C. Gates
H. Knözinger

Dynamics of Reactions at Surfaces

GERHARD ERTL
Fritz-Haber-Institut der Max-Planck-Gesellschaft
D-14195 Berlin, Germany

The rate of a catalytic reaction is determined by the dynamics of the individual steps involved, which may be classified according to a rough hierarchy: The quantum level concerns the energy exchange between the different degrees of freedom. If these are in thermal equilibrium at all stages, the concepts of transition state theory may be applied to formulate the rates of the elementary processes constituting the reaction mechanism. Combination of these steps eventually enables rationalization of the macroscopic kinetics. The contributions of surface science to the current knowledge of the various levels of these complex phenomena are illustrated by selected examples. © 2000 Academic Press.

I. Introduction

Catalysis concerns the *rate* of a chemical reaction; catalysis is a dynamic phenomenon. A heterogeneously catalyzed reaction consists of a sequence of elementary steps such as adsorption, surface diffusion, chemical transformation of adsorbed species, and desorption, the identification and characterization of which comprise the reaction mechanism and the temporal behavior of which determines, together with the relevant transport processes, the overall rate. The contributions of surface science to catalysis consist mainly of studies of these phenomena and, from them, elucidation of the general concepts underlying catalysis, rather than contributions to the solution of specific problems in technological processes. Nonetheless,

Abbreviations: a_o, lattice constant; E_A, energy of lowest empty electronic state of gas-phase molecule (affinity level); E_F, Fermi energy; $\langle E_{trans} \rangle$, mean translational energy of desorbing molecule; $\langle E_{\perp} \rangle$, mean translational energy of incident particle along surface normal; ER, Eley–Rideal; L, length of domain boundary; LH, Langmuir–Hinshelwood; ML, monolayers; N, number of adsorbed atoms/molecules; P, relative partial pressure; r, reaction rate; s_0, sticking coefficicent at zero coverage; t_{relax}, energy relaxation time; t_{site}, mean residence time of a particle on specific surface site; T_{ad}, adsorbate temperature; T_{el}, electron temperature; T_{ph}, phonon temperature; T_{rot}, rotational temperature; T_s, surface temperature; v_f, mean velocity; $\delta = \theta / \theta_{max}$, relative coverage; ε_A, energy of lowest empty electronic state of neutral adsorbed particle; θ, surface coverage; Θ, angle of desorbing particle relative to surface normal; ϕ, work function.

0360-0564/00 $35.00

there are ample examples demonstrating the positive impact of surface science even in this latter sense.

The various aspects of the dynamics of surface reactions and catalysis may be classified in a hierarchical scheme in terms of (qualitative) time and length scales (Fig. 1):

1. The efficiency of a catalyst (or a catalytic system comprising even the reactor design) will be determined by the *macroscopic kinetics* of the overall reaction, including all chemical transformations plus the relevant physical processes of energy and mass transport. The resulting yield depends on external parameters, such as temperature, flow rates, and concentrations (or partial pressures) of the species participating in the reaction. Modeling of the macroscopic kinetics is frequently achieved by chemical engineers fitting empirical equations (e.g., power law kinetics) for the concentration dependences and rate constants that depend exponentially on temperature with apparent activation energies as parameters. As a next step, assumptions about the underlying reaction mechanism provide correlations between the concentrations (= coverages) of the surface intermediates and the external variables. Such an approach, for example, underlies the famous Temkin equation for modeling the kinetics of ammonia synthesis.

2. The development of sophisticated surface physical methods enabled detailed insights into the atomic processes on surfaces as well as identification and characterization of the properties of the surface species. Descrip-

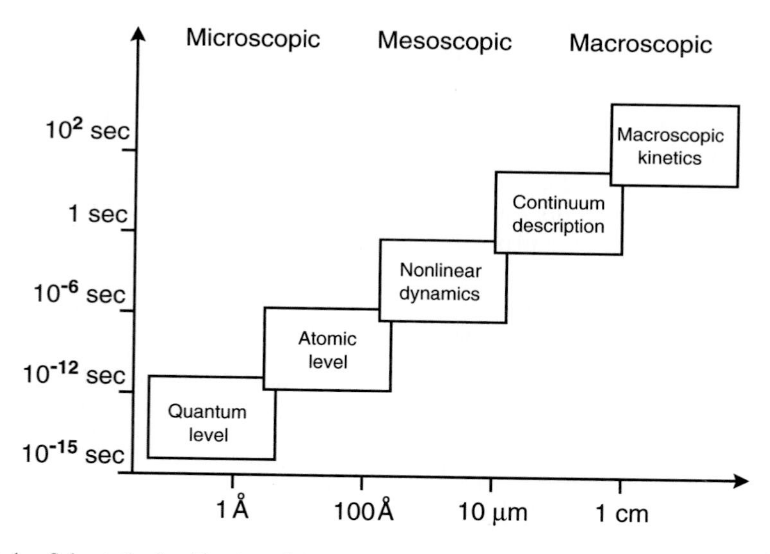

FIG. 1. Schematic classification of the various aspects of the dynamics of surface reactions.

tion of the progress of a catalytic reaction in terms of an approach called microkinetics is achieved by modeling the macroscopic kinetics through correlating the atomic processes with the macroscopic parameters in the framework of a suitable *continuum model.* By analogy to kinetics of homogeneous reactions, the individual particles are described by continuum variables for their partial coverages, which to a first approximation are correlated with the external parameters (partial pressures and temperature) through Langmuir-type relationships. The latter may eventually be properly modified by taking into account effects of surface heterogeneity or variations of the kinetics parameters with coverage.

3. Even if the assumptions underlying a specific continuum model are not completely fulfilled, the model may nevertheless provide a good approximation over a limited range of external parameters (which will frequently also be the case for technological catalysis). The formulation of the rate laws for the full sequence of elementary reactions will then usually lead to a set of nonlinear coupled (ordinary) differential equations for the concentrations (coverages) of the various surface species involved. The resulting temporal behavior under constant continuous-flow conditions will not necessarily be stationary; instead, for certain parameter ranges, it may be oscillatory or even chaotic. The spatial distributions may not be uniform either, and the existence of local variations in surface coverage causes coupling of the reaction with adsorbate diffusion or other transport (e.g., heat transfer) processes. As a consequence, the formation of spatiotemporal concentration patterns on a *mesoscopic scale* may occur. Phenomena of this type belong to the general field of nonlinear dynamics, with widespread manifestations in chemistry, physics, biology, etc. Only the development of suitable experimental tools together with theoretical progress has made this subject accessible in the field of heterogeneous catalysis.

4. Formulation of rate equations in terms of a continuum model of course requires detailed knowledge of the properties of the surface species on an atomic level. Even then, the usual "ansatz" in terms of the partial coverages assumes a random distribution of noninteracting particles, and this assumption is usually not fulfilled. The occurrence of interactions between the adsorbed species and their occupation of nonequivalent adsorption sites complicate an appropriate description, apart from the fact that (i) the surface is nonuniform even when it consists of single-crystal planes (because of the existence of active sites, etc.) and (ii) the surface may undergo structural transformations under the influence of the adsorbates. Detailed investigation of these effects on an *atomic scale* represents the classical domain of surface science.

5. The basic idea of transition state theory (TST), forming the framework of chemical kinetics, consists of the assumption that at all stages along

the reaction coordinate thermal equilibrium is established, leading to the temperature (T) as the only essential (macroscopic) parameter. This assumption requires that energy exchange between the degrees of freedom of the particles interacting with the surface and the heat bath of solid occur much faster than the elementary step initiating nuclear motion. Processes of energy transfer between the various degrees of freedom on the *quantum level* form the ultimate basis for chemical transformations.

This review is an attempt to illustrate some of the contributions of surface science to the various cited levels with the goal of a better understanding of the dynamics of heterogeneous catalysis. The main emphasis is thus placed on the microscopic aspects, and the other categories are discussed more briefly. The discussion is restricted to metal surfaces, the author's area of expertise.

II. The Quantum Level

A. General

The dynamics of elementary processes of surface reactions is governed by the exchange of energy between the various degrees of freedom of the adsorbate and the substrate surface. For the former, these comprise the (quantized) states of vibration and (frustrated) rotation and translation, whereas, for a metallic surface virtually continuous distributions of electronic excitations and vibronic excitations (= phonons) have to be taken into consideration. The existence of thermal equilibrium between the populations of these degrees of freedom is the basis for the validity of TST and hence the prerequisite for any description of the rate of a chemical reaction by a continuum model, i.e., in terms of temperature-dependent rate constants and concentration variables. Such a situation may, of course, be expected only if the nuclear motion characterizing the elementary process under consideration occurs on a time scale much longer than the relaxation time to reach thermal equilibrium.

An example of a typical reaction time is the mean residence time of a particle on a specific adsorption site, t_{site}, before it moves to a neighboring site. Scanning tunneling microscope (STM) investigations of the system O/Ru(0001) indicated a value of $t_{site} \approx 6 \times 10^{-2}$ s at 300 K, and this value is changed by about one order of magnitude as a result of interactions with neighboring adparticles (*1*). From the activation energy for surface diffusion, one may estimate that this residence time decreases to the order of 10^{-6} s at 500 K; with more weakly held adsorbates, it of course becomes much smaller still, to a lower limit of about 10^{-13} s given by the period of a single vibration.

An estimate of a typical energy relaxation time, t_{relax}, can again be determined on the basis of STM observations (*2, 3*). Thermally activated dissociation of O_2 molecules chemisorbed as peroxo-like species on a Pt(111) surface leads to the formation of pairs of chemisorbed O atoms. If the temperature is low enough ($\lesssim$180 K), these atoms are not mobile once they are in thermal equilibrium with the solid, i.e., they do not change their positions over longer periods of time. Interestingly, neighboring adatoms are not located on adjacent sites; rather, in the pairs, the atoms are separated from each other by 5–8 Å. This effect is a consequence of the finite time needed for damping the adsorption energy into the heat bath of the solid. As illustrated by the schematic potential diagram of Fig. 2, dissociative chemisorption along the reaction coordinate comprises mutual separation of the two atoms parallel to the surface plane. Obviously, this does not occur with the atoms in the ground state (Fig. 2, solid line) but instead

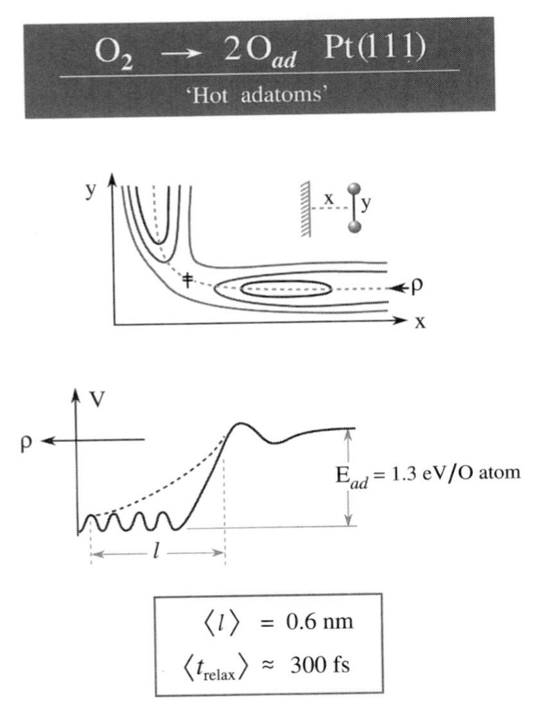

FIG. 2. Schematic potential diagram (two-dimensional and one-dimensional along the reaction coordinate ρ) for dissociative chemisorption of O_2 on Pt(111).

along the dotted line in Fig. 2 (involving "hot" adatoms). From the distance ($\langle l \rangle$) traversed until thermal accommodation is reached and from the mean velocity determined by the chemisorption energy, an average relaxation time on the order $t_{relax} \approx 3 \times 10^{-13}$ s can be estimated. It thus must be inferred that surface reactions occurring on a time scale $<10^{-12}$ s do not take place in thermal equilibrium so that the concept of TST also breaks down. Examples of the experimental manifestation of such processes are presented later.

B. "Thermal" Surface Reactions: Electron vs. Phonon Coupling

First, however, consider the question of whether coupling to phonon or electronic excitations is the dominant mechanism for energy exchange in "thermal" surface reactions. For particles incident on a surface, this problem has been discussed extensively in the early literature. Electronic excitation had been favored in theoretical treatments, and clear experimental manifestation of this effect in reactions between partners with pronounced differences in electronegativity (such as oxygen interacting with alkali metal surfaces) is discussed later. In "neutral" adsorption and scattering events, however, it is now generally believed that energy exchange with the lattice vibrations (phonons) is dominant; this is supported by data characterizing inelastic scattering of H_2 and D_2 molecules at Ni surfaces (*4*). With reactions occurring on the surface from the adsorbed state, the experimental situation seems to be rather hopeless because of the strong coupling between electronic and phonon excitations. However, insight was recently gained by separating the relevant time scales through the application of ultrafast (femtosecond) laser techniques (*5*). The experiments were performed with a Ru(0001) surface onto which O + CO were coadsorbed. When the temperature of the sample was continuously increased by heating, only desorption of CO (but no formation of CO_2) was observed. This situation changed when the sample was irradiated with short (120-fs) intense pulses of infrared radiation (800 nm); then, both CO and CO_2 with an intensity ratio of about 35 : 1 came off the surface. Both yields exhibited a strongly nonlinear dependence on laser fluence, signaling that single-particle excitations were not responsible for the reactions. Two-pulse correlation measurements with varying delay times between two subsequent pulses revealed that the relaxation time for the decay of the excitation responsible for CO desorption was on the order of 20 ps, and it was much shorter (only about 3 ps) for CO_2 formation, suggesting the operation of a different activation mechanism for formation of CO_2. Detailed analysis indicated the following: The infrared radiation becomes absorbed within a depth of about 10 nm by the valence electrons of the metal, which very rapidly (in ~10 fs) thermalize among each other to an electron temperature (T_{el}), which couples on a time scale of about 1 ps to the heat bath of the lattice vibrations (T_{ph}) until T_{el} = T_{ph}; the near-surface region cools continuously by heat transport to the bulk, as shown in Fig. 3a. As depicted in Fig. 3b, CO_2 formation is initiated during these first few picoseconds, when T_{el} may reach typical values as high as 6000 K, whereas CO desorption starts only later with the increase of T_{ph}. The latter process is associated with an activation energy of 0.8 eV, and that for CO_2 formation is considerably higher, namely, 1.8 eV, which explains why with "normal" heating only CO desorption occurs. The Fermi–Dirac distribution of electrons above the Fermi level associated with the high T_{el} during the laser shot, however, causes substantial population of an antibonding O 2p-derived level. Coupling to the nuclear motion in the Ru–O bond, as characterized by an adsorbate temperature (T_{ad}), then initiates CO_2 formation. Although the reaction is triggered by electronic excitation through T_{el}, it can thus still be regarded as a thermal reaction.

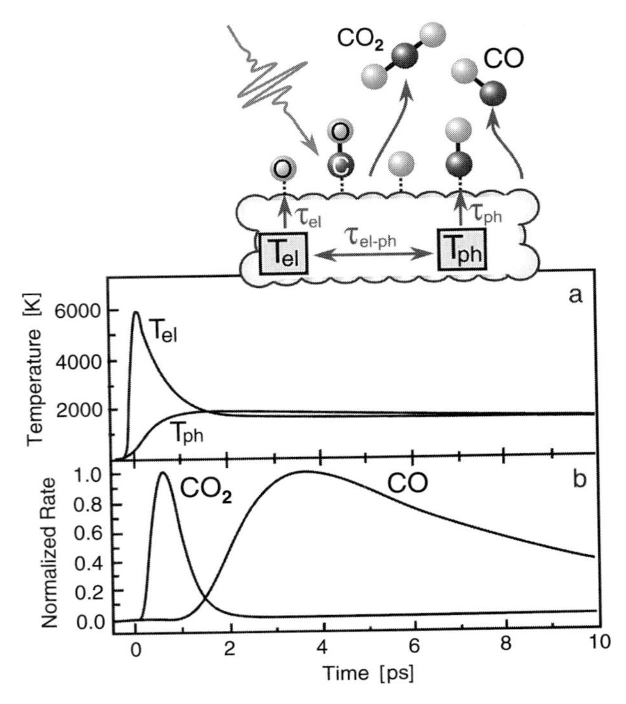

FIG. 3. Surface chemistry on an O + CO/Ru(0001) surface resulting from a femtosecond laser pulse: variation of the electron temperature T_{el} and the phonon temperature T_{ph} with time, leading to CO_2 formation and CO desorption, respectively.

Under normal thermal conditions, T_{el} is always equal to T_{ph}, but this example suggests that even then electronic excitation through partial population of low-lying antibonding orbitals might be the decisive primary step of activation.

In the current case, however, desorption of CO will be driven exclusively by phonon excitations since the lowest lying empty level (derived from the CO–$2\pi^*$ orbital) is centered about 5 eV above E_F and is thus much too high in energy to become substantially populated by electronic heating. How the degrees of freedom of the adsorbate couple to lattice heating in this case is being investigated in experiments in which temporally resolved sum-frequency-generation spectra from the C–O stretch vibration following absorption of a femtosecond laser pulse are recorded. It is to be expected that with this strategy, future experiments will indeed allow the transition state of a surface reaction to be spectroscopically probed.

Next, however, I return to the question of whether there is experimental evidence for the nonvalidity of the assumptions underlying TST for reactions occurring at solid surfaces.

C. Manifestation of Nonthermal Processes

1. *Influence of the Energy Content of the Incident Particles on Adsorption*

The Langmuir isotherm that usually underlies models of catalytic reactions on the continuum level is based on a "hit and stick" model: If a particle from the gas phase hits an empty adsorption site, it sticks (becomes adsorbed); otherwise, it is reflected back. This discussion primarily concentrates on interactions with the bare surface (i.e., in the limit of zero coverage), and effects associated with the presence of other adsorbed species are shifted to the next level of the hierarchy (the atomic level). Figure 4a shows that the sticking coefficient at zero coverage (s_0) for dissociative chemisorption of H_2 impinging with constant kinetic energy (64 meV) onto a Ni(111) surface is independent of the surface temperature (T_s), and s_0 increases continuously with the normal component of the translational energy ($\langle E_\perp \rangle$) of the incident molecules (Fig. 4b) (*4*). In view of the schematic potential of Fig. 2, these results suggest that there is a finite activation barrier of about 0.1 eV for dissociation that, however, cannot be overcome by coupling to the heat bath of the solid but instead requires accumulation of energy to a certain degree of freedom of the impinging molecule. With Ni(110), obviously such a barrier is negligible, since the sticking coefficient is close to unity, independent of kinetic energy and surface temperature. The data for Ni(111) clearly show that the reaction does not occur in thermal equilibrium, and TST does not provide an appropriate description.

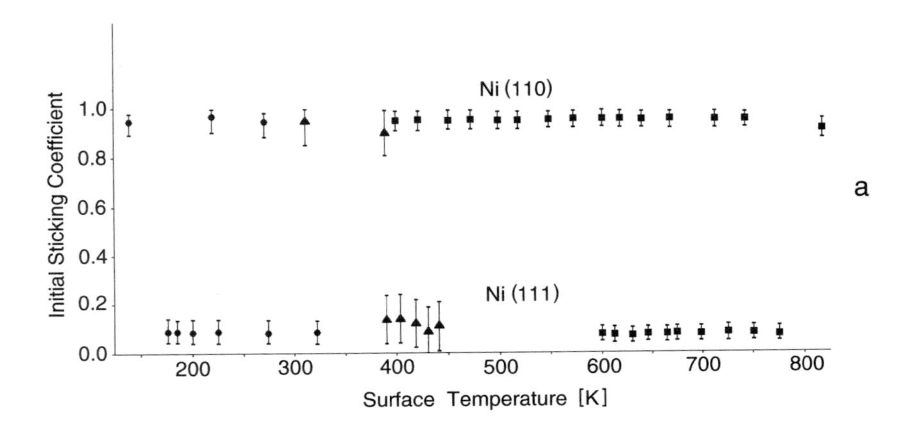

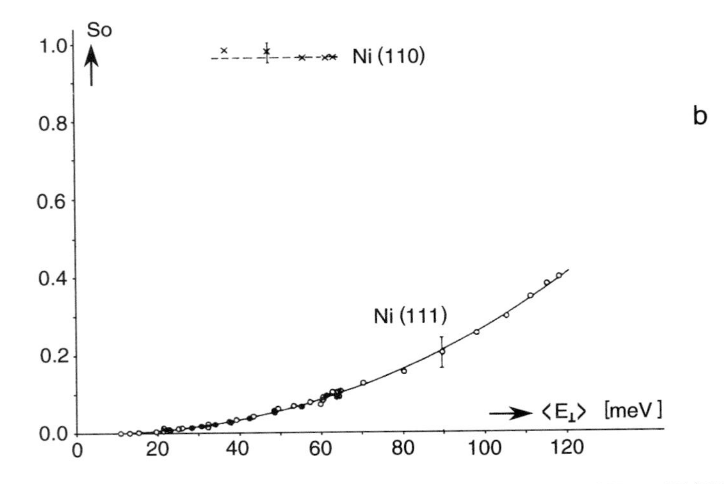

FIG. 4. The initial sticking coefficient for dissociative chemisorption of H_2 on Ni(110) and Ni(111) surfaces (*4*): (a) as a function of surface temperature; (b) as a function of incident kinetic energy.

This conclusion appears *a posteriori* to be rather plausible; since at the beginning the two partners are well separated from each other, coupling between their various energy levels takes place only during the course of the reaction and is hence expected to be incomplete.

Effects of this kind were first reported for dissociative adsorption of hydrogen on copper surfaces, and this system then served as a benchmark for detailed experimental and theoretical studies (*6*). Molecular beam experiments were carried out that involved systematic variation of the translational energy and also probing of the populations of vibrational and rotational levels, the latter even with selection of the polarization ("cartwheel" vs. "helicopter") (*7–10*). The detailed experimental findings prompted the evaluation of sophisticated (up to six-dimensional) potential energy surfaces and theoretical modeling of the dynamics of this prototype and related systems (*11*).

The effects of incomplete energy exchange on collision with the surface are by no means restricted to bond breaking within the lightest molecule (for which the large difference in mass with respect to the surface atoms is particularly unfavorable for momentum transfer). An interesting and widely investigated example in which bond breaking during collision with the surface occurs is the dissociative chemisorption of methane (*12*). Figure 5 shows the dissociation probability for CH_4 and CD_4 on a Ni(111) surface as a function of the translational energy normal to the surface, whereby

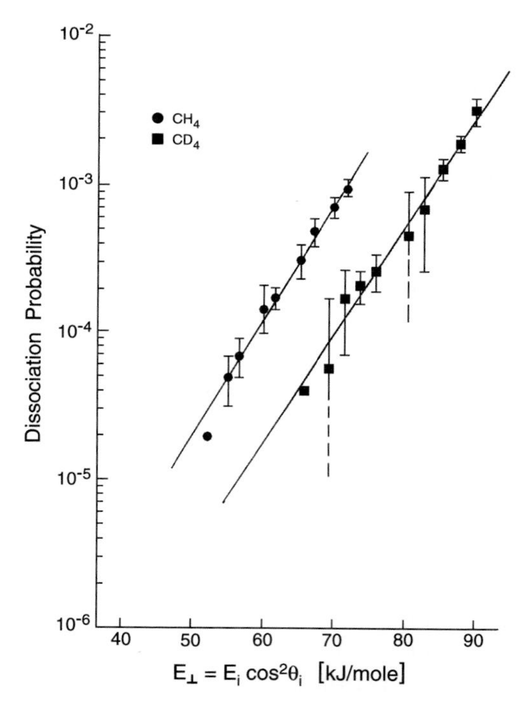

FIG. 5. The dissociation probabilities of CH_4 and CD_4 impinging onto a Ni(111) surface as a function of the normal component of translational energy (*13*).

the surface temperature had no influence (*13*). Additional experiments in which the population of the vibrational levels was varied showed that excitation of the bending and umbrella modes was characterized by an efficiency comparable to that of an increase in the translational energy. These findings suggest that it is in fact the deformation of the methane molecule on impact with the surface that initiates its dissociation. Activation presumably involves the transition into a pyramidal configuration in which three hydrogen atoms together with the carbon atom are within a plane, with each of them interacting with the surface. The pronounced isotope effect is attributed to tunneling of the hydrogen atom along the C–H coordinate. This model of deformation preceding dissociation also explains why previous attempts to promote dissociative chemisorption by direct optical excitations of the vibrational modes were unsuccessful (*14*). In contrast, recent experiments with NO excited to high vibrational states revealed a much higher probability for its dissociative adsorption on Cu(111) than for such adsorption of molecules in the vibrational ground state (*15*).

Another example relevant to an important technological process serves to illustrate how complex the actual situation may be: It is now well established that for catalytic ammonia synthesis, dissociative nitrogen adsorption is the rate-limiting step, with the Fe(111) surface exhibiting the highest activity (*16*). The sticking coefficient for this step under thermal conditions was found to be very low and to depend on surface temperature, from which a sequential reaction mechanism proceeding through two molecular states in thermal equilibrium with the surface was concluded: $N_2 \rightarrow N_{2,ad}(\gamma) \rightarrow N_{2,ad}(\beta) \rightarrow 2N_{ad}$ (*17*). Molecular beam experiments, on the other hand, indicated a pronounced increase of the sticking coefficient with translational energy, pointing to the operation of a "direct" collision-induced mechanism (*18*). Recent detailed theoretical analysis resolved this apparent discrepancy (*19*): Although at higher kinetic energies the direct mechanism indeed prevails, it plays only an insignificant role at the normal temperatures applied in the technological process. The original mechanism proceeding via accommodated molecular precursors dominates under these conditions so that the application of TST is well justified.

Even molecular (nondissociative) adsorption may be associated with an activation energy, for example, if the process proceeds via a physisorbed state into a chemisorbed state (trapping-mediated adsorption). Such a case is presented, for example, by the system O_2/Pt(111), whereby the molecule may be chemisorbed either in a superoxo-like or peroxo-like state, as distinguished by vibrational stretch frequencies of 870 and 690 cm^{-1}, respectively (*20*). Recent experiments with molecular beams of varying kinetic energy showed that with low kinetic energies of the incident molecules both superoxo-like and peroxo-like species are formed, whereas at high kinetic energies only the more strongly bound peroxolike species is populated, reflecting a close correlation between incident translational energy and the preferred trajectories for adsorption (*21*).

Generally, even for nondissociative adsorption without an activation barrier in the entrance channel, the sticking coefficient deviates from unity and decreases with increasing kinetic energy as well as with increasing surface temperature because of less efficient energy exchange (*22*). State-resolved experiments, together with classical trajectory calculations, shed much light on the details of the elementary processes of energy exchange involved (*23*). For example, Fig. 6a shows time-of-flight (TOF) distributions of NO molecules in various rotational states (J'') coming off an oxidized Ge surface upon scattering of a rotationally cold molecular beam with a narrow kinetic energy distribution of approximately 730 meV (*24*). The TOF data exhibit two maxima with high and low mean velocities (v_f) corresponding to molecules being directly scattered or being subject to adsorption and subsequent desorption, respectively. As shown in Fig. 6b, the latter exhibit a translational energy of about 45 meV, independent of their rotational state (curve b in Fig. 6b), and this value is equal to $2kT_s$ (where T_s is the surface temperature). The population of the various rotational levels of this fraction follows a Boltzmann distribution with T_{rot} = 190 K again being equal to T_s so that those molecules originating from trapping/desorption retain the properties of thermal equilibration with the surface after returning to the gas phase. (That this must not necessarily be the case is explained in the next section.) The fast molecules originating from direct inelastic scattering do not, in contrast, exhibit a Boltzmann distribution of their rotational population; rather, they exhibit the effect of a "rotational rainbow" with overpopulation of high rotational quantum levels J''. Their kinetic energy decreases linearly with increasing rotational energy (curve a in Fig. 6b) but to a lesser extent as if the rotational energy originated solely from the translational energy (curve c in Fig. 6b). A qualitative interpretation of these effects is presented in the theoretical work by Muhlhausen *et al.* (*25*).

Under the high-pressure conditions of practical catalysis, both the gas phase and catalyst are usually at the same temperature so that the effects due to restricted energy exchange are masked. Nevertheless, these phenomena are decisive for the numerical values of the (thermally averaged) net kinetics parameters, so that even a detailed analysis of the state-resolved dynamics is of relevance for a better understanding of catalytic activity.

2. *Energy Distributions of Particles Released from Surfaces*

Since a surface in contact with a gas phase will eventually reach the state of thermodynamic equilibrium, the principle of microscopic reversibility

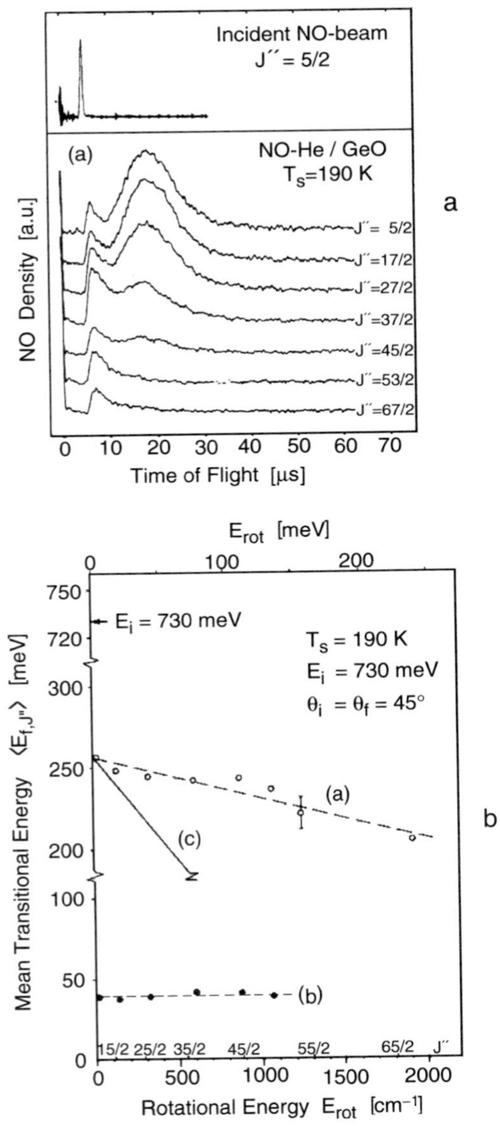

FIG. 6. State-resolved molecular beam experiments characterizing the scattering of NO from an oxidized Ge surface (*24*): (a) Time-of-flight distributions of the incident and reflected beams in various rotational states *J*″; (b) correlation between mean translational energy and rotational energy for the fraction of molecules undergoing direct-inelastic scattering (a) and for those leaving the surface after thermal accommodation (b).

requires that the effects of the quantum states of the incident molecules on sticking also manifest themselves in the reverse process of desorption. Following the arguments of the preceding section, according to which the sticking coefficient decreases with increasing surface temperature and increasing translational energy, it must be concluded that hot particles should be missing in the desorbing flux and hence their mean translation energy should become lower than $2kT_s$, the value corresponding to thermal equilibrium at the surface temperature T_s. This suggestion is confirmed by the results of classical trajectory calculations by Tully (*26*) for the desorption of argon and of xenon from Pt(111). Considerable deviations from Arrhenius behavior for the rates of desorption were found, and Fig. 7 demonstrates the predicted "translational cooling in desorption" effect. For a more strongly chemisorbed system [CO/Ru(0001)], this effect was recently verified experimentally (*27*). The measured TOF distributions after surface irradiation with a fs infrared laser pulse were converted into the mean translational energy of the desorbing molecules ($\langle E_{trans} \rangle$). As shown in Fig. 8a, $\langle E_{trans} \rangle/2k$ as a function of the laser fluence is always far below the actual temperature

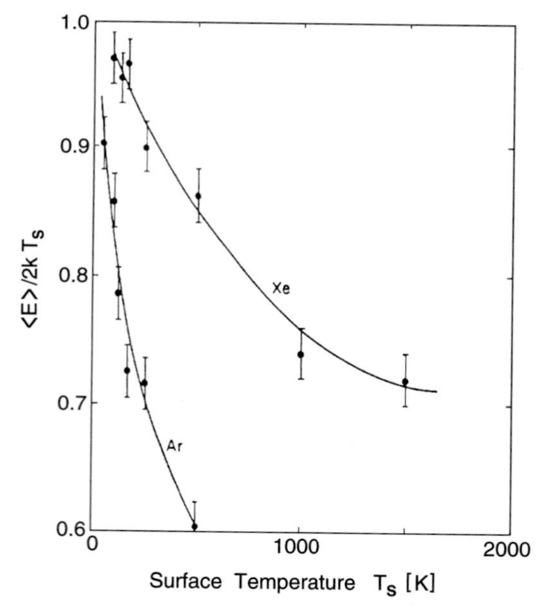

FIG. 7. The mean translational energies of argon and xenon atoms (in units of $2kT_s$) desorbing from Pt(111) as a function of surface temperature T_s as indicated by computer simulations by Tully (*26*).

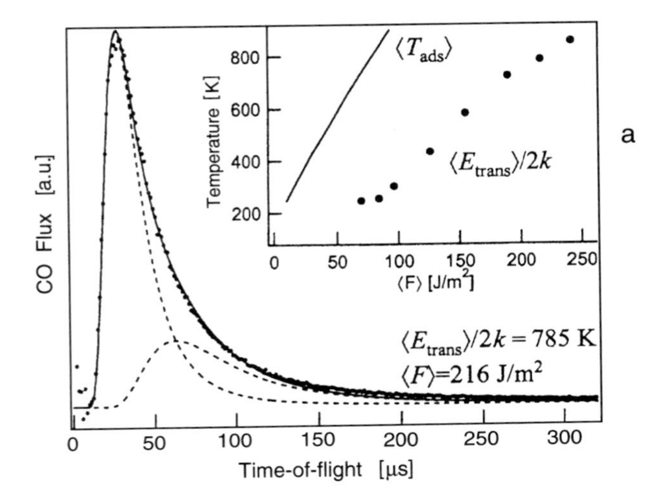

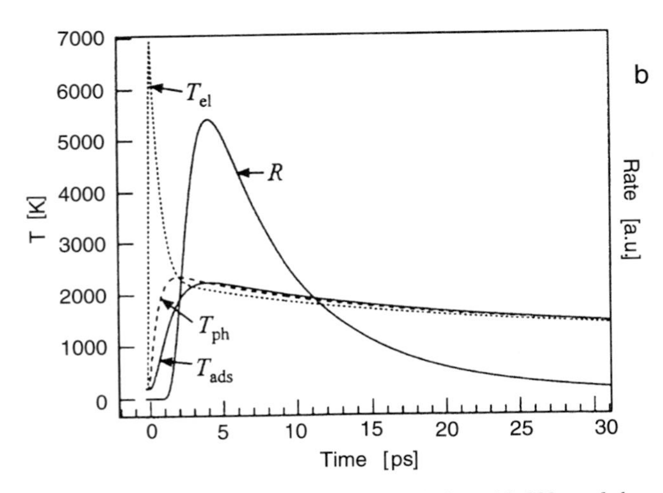

FIG. 8. Desorption of CO from Ru(0001) after irradiation with 800-nm fs-laser pulses (*27*). (a) Time-of-flight distribution for a typical laser fluence (⟨F⟩). The inset shows the variation of the mean translational energy (⟨E_{trans}⟩/$2k$) and of the adsorbate temperature (T_{ads}) with laser fluence (⟨F⟩). (b) Typical temperature transients following a laser shot; temperature of the electrons (T_{el}), the phonons (T_{ph}), the adsorbate (T_{ads}), as well as resulting desorption rate (R) as a function of time.

($\langle T_{ads} \rangle$): the variation of $\langle T_{ads} \rangle$ with time is reproduced together with the desorption rate (R) after a laser shot in Fig. 8b (*27*).

Cooling effects in desorption also occur for the other degrees of freedom of desorbing molecules. For example, Fig. 9 shows that the mean rotational temperature of NO molecules desorbing from a Pt(111) surface is equal to the surface temperature only up to about 400 K; at higher temperatures, T_{rot} levels off (*28*) because of incomplete excitation of the rotational motion during the desorption event. This means that rapidly rotating molecules exhibit a lowered sticking probability.

If adsorption proceeds across a barrier and is promoted by direct collision, then the desorbing particles are expected to exhibit a mean translation energy which is *higher* than that corresponding to the surface temperature. Furthermore, the angular distribution of the desorbing particles deviates from the thermal cosine law and is peaked along the surface normal, as is usually described by a $\cos^n\theta$ dependence ($n > 1$). The previously discussed

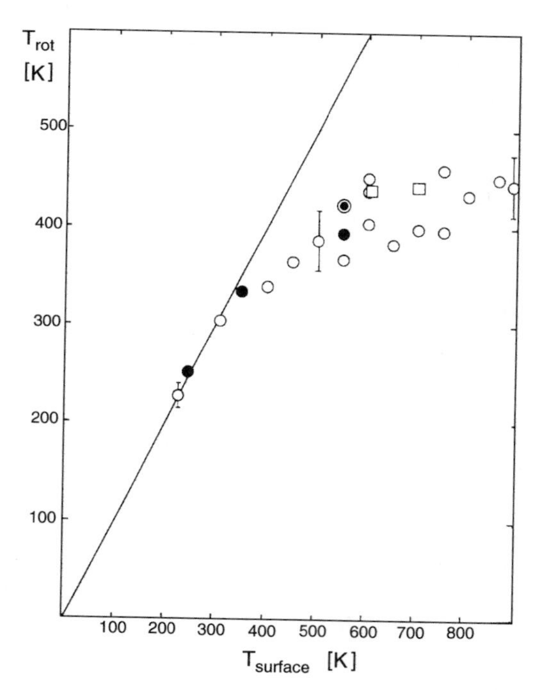

FIG. 9. Variation of the rotational temperature (T_{rot}) of NO molecules desorbing from a Pt(111) surface as a function of surface temperature. The different symbols mark varying kinetic energies of the primary molecules and signal that complete accommodation has occurred (*28*).

benchmark system H_2/Cu has also been investigated, and the predicted effects were confirmed (*29*). Concerning the population of internal degrees of freedom, Kubiak *et al.* (*30*) found, for example, that for H_2 associatively desorbing from Cu(111) the $\nu'' = 1/\nu'' = 0$ vibrational population ratio was about 100 times larger than that for equilibrium at T_s. Further detailed experiments led to the conclusion of a preferred steric orientation of the molecular axis during desorption (*31, 32*). More specifically, six-dimensional density functional theory (DFT) calculations of the potential surface for H_2/Pd(100) predicted substantial differences in sticking coefficient of molecules with their rotational axis parallel (helicopter) or perpendicular (cartwheel) to the surface (*33*). Experimental evidence for this "dynamic filtering" effect in desorption (*34*) was again obtained with the H_2/Cu(111) system (*35*): Molecules desorbing in the vibrational ground state were found to prefer helicopter motion for low rotational quantum numbers but to be isotropic for $\nu'' = 1$ as well as for $J'' > 6$ in the $\nu'' = 0$ state.

An interesting effect was observed for the decomposition of NO_2 on a Ge surface (*36*): The rotational state distribution of the NO molecules formed was found to be non-Boltzmann and essentially independent of surface temperature. It was concluded that NO that was formed through the reaction $NO_2 \rightarrow NO + O_{ad}$ was released directly into the gas phase without intermediate thermal equilibration with the surface, which is attributed to the large exothermicity of this reaction (about 220 kJ/mol).

Recombinative desorption may also be the last step in a catalytic reaction (such as $CO_{ad} + O_{ad} \rightarrow CO_2$ in carbon monoxide oxidation), and the energy content of the product molecules from this process shows some interesting nonthermal features.

The schematic (one-dimensional) energy diagram for this reaction on Pt(111) at low coverages proceeding in the thermal ground state through the Langmuir–Hinshelwood mechanism is depicted in Fig. 10 (*37*). (Modifications of the energetics resulting from interactions between adsorbed particles are discussed later). Since the product molecule interacts only very weakly with the surface, the nascent CO_2 experiences a repulsive potential after passing through the transition state and presumably does not undergo full accommodation before it is released into the gas phase. This suggestion is confirmed by the sharper than cosine angular distributions and excess translational energy (*38*). Depending on the actual adsorption state, the mean translational energy can increase to 330 meV (about 10 times $2kT_s$), and the angular distribution follows a $\cos^9\theta$ relation. (The situation for an additional channel caused by hot O adatoms is discussed later.) The transition state is thought to exhibit a bent configuration, whereas the final product molecule is linear. Hence, one might further expect noticeable excitation of internal degrees of freedom, and this has

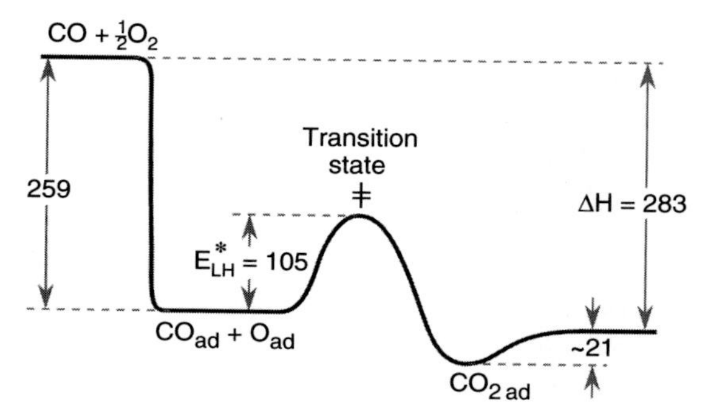

FIG. 10. Schematic potential diagram for the progress of CO oxidation on Pt(111) at low coverages (*37*) (enthalpies in kJ/mol).

indeed been demonstrated by infrared chemiluminescence (*39–43*). For example, CO_2 formed from chemisorbed O on a Pt foil at 800 K was found to exhibit populations of the three vibrational states corresponding to temperatures of about 2000 K, whereas reaction with hot adatoms caused even higher degrees of excitation. A rough estimate showed that the CO_2 molecules carry off about 80% of excess energy from the transition state, with about equal shares in translational and internal degrees of freedom, whereas only about 20% of this energy is released into the heat bath of the solid.

This situation even suggests some conclusions about the nature of the transition state from analysis of the gaseous product molecules. Matsushima (*44*) analyzed in detail the angular distributions of CO_2 formed on different surfaces and under various conditions. For example, Fig. 11 shows some data for Pt(110) (*45*). This surface is reconstructed into a 1×2 missing row phase, as depicted in the inset of Fig. 11a. The angular distribution of CO_2 along the trough, i.e., the [110] direction (Fig. 11a) is peaked along the surface normal, whereas that along the [001] direction (Fig. 11b) exhibits two maxima at ±35°. Their relative intensities vary with CO coverage, and deconvolution suggests an additional peak along the surface normal which grows with increasing CO coverage. The suggested interpretation is sketched in Fig. 12: The inclined direction of CO_2 production arises from reaction between O atoms in the threefold sites of the facets on the reconstructed Pt(110) surface (Fig. 12A), whereas the normal component is formed either by reaction with O atoms at the bottom of the troughs (Fig. 12B) or, as appears to be more likely, on the flat portions of the surface where the 1×2 reconstruction is lifted to 1×1 patches if the local CO coverage is sufficiently high.

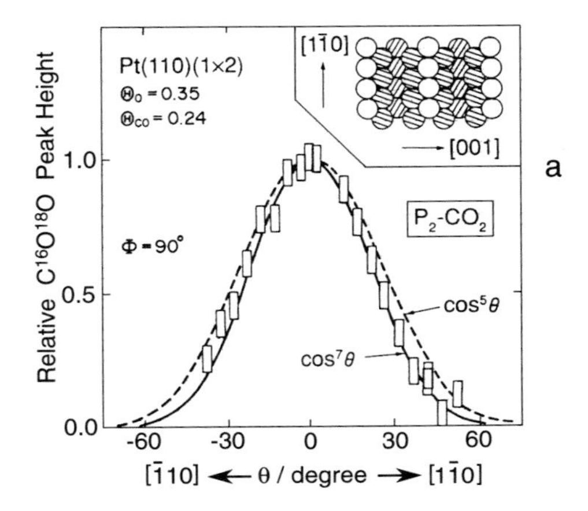

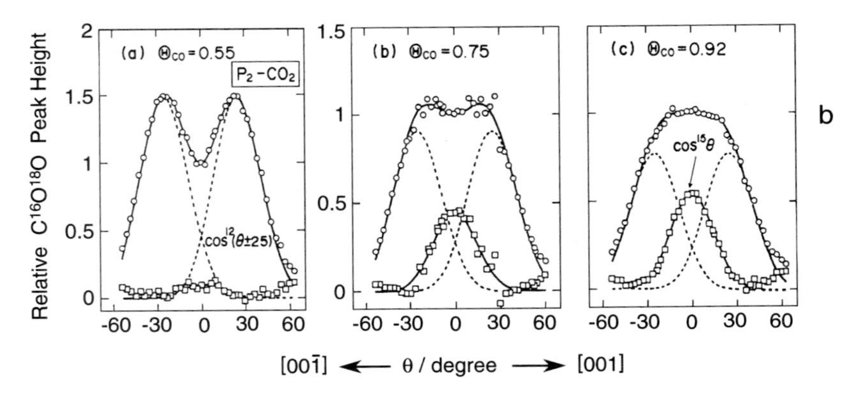

FIG. 11. Polar angle distributions of CO_2 molecules formed by CO oxidation on Pt(110) (*45*): (a) along the [110] direction; (b) along the [001] direction for different CO coverages.

3. *"Intrinsic" Precursors and "Hot" Adatoms*

The term "precursor" has different meanings in the literature, and therefore some clarification is needed:

1. The production of a stable surface species [such as O_{ad} in the case of O_2/Pt(111)] may be preceded by the formation of a surface intermediate (here, molecularly adsorbed O_2, even existing in two different states) which is frequently denoted as precursor. If this "precursor" is a physisorbed species formed by energy dissipation of the impinging particle from which

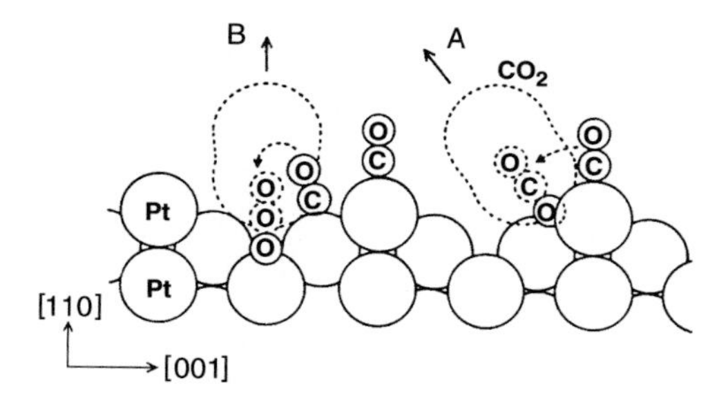

FIG. 12. Suggested geometries for CO_2 formation to account for the experimental results of Fig. 11 (*45*).

a transition into the chemisorbed state occurs, this process is also called "trapping-mediated chemisorption."

2. Langmuir (*46*) observed for the first time that the sticking coefficient does not necessarily decrease linearly with coverage, as would follow from the "hit and stick" model underlying the derivation of his famous isotherm. He attributed this effect to the existence of a transient mobile surface intermediate. The simplest model (*47, 48*) is based on the assumption that an incident particle can also be trapped above an already occupied site. There, it may be accommodated and exhibit a high mobility so that it eventually reaches a free chemisorption site before again desorbing. With increasing coverage, the mean diffusion length to reach such a site also increases, and hence the probability for chemisorption during its surface lifetime decreases—but less than linearly with coverage as in the simple Langmuir model. This "extrinsic" precursor comes into play only at nonzero coverages, but it is in thermal equilibrium with the surface.

3. This is no longer the case with the "intrinsic precursor species," a term that denotes particles trapped on the surface before they attain thermal equilibrium. This state is of relevance for various kinetics phenomena in surface reactions and is discussed in the current context.

Tully (*26*) realized in the context of his trajectory calculations of the interactions of noble gases with a platinum surface that long after equilibration of the normal component of energy, particles may continue to slide long distances across the surface before the tangential momentum is equilibrated. Such an effect had already been suggested by Weinberg and Merrill (*49*), and Harris and Kasemo (*50*) had introduced the concept of a hot precursor in order to rationalize a series of experimental observations of surface reactions. The principle of this concept is illustrated schematically in Fig.

13, in contrast to the strong coupling limit. In the latter case, a particle impinging on a surface instantaneously attains thermal equilibrium and sticks where it hits the surface (consistent with the Langmuir adsorption model). At a distance far from the surface, however, coupling will still be weak, so that the particle hits first the repulsive part of the potential and exchanges its adsorption energy only stepwise. Since the corrugation of the potential parallel to the surface is generally smoother than that perpendicular to it, the particle is expected to travel an appreciable distance across the surface before it comes to rest. If along the way, however, it hits another particle with the same mass, energy transfer will be quite efficient and both particles are expected to remain attached to each other, provided that

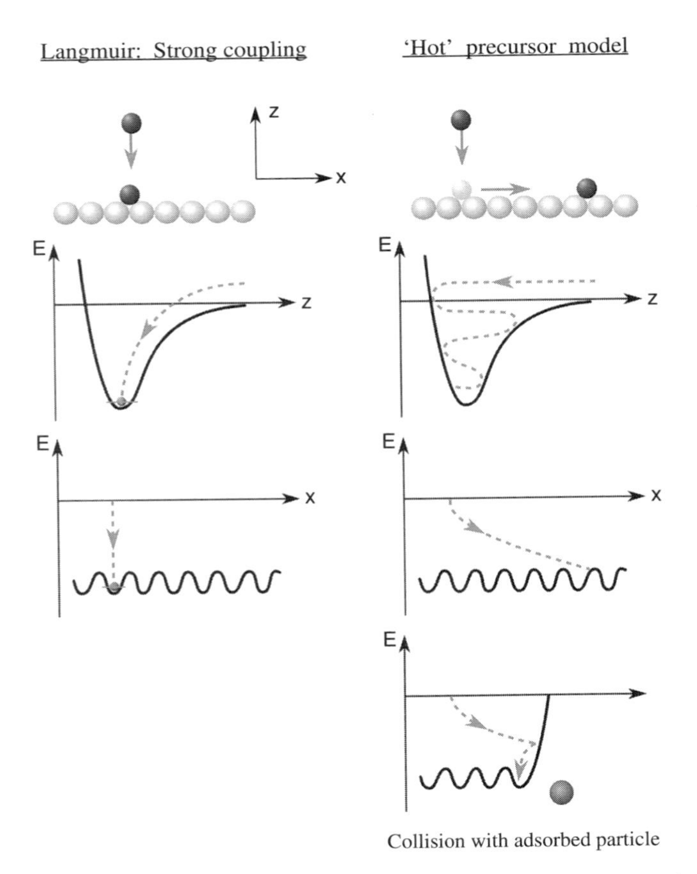

FIG. 13. Schematic dynamics for adsorption following the Langmuir concept (strong coupling) and according to the "hot" precursor model.

the surface temperature is low enough to suppress regular (i.e., thermally activated) surface diffusion.

On the basis of these considerations, a direct experimental demonstration of the existence of this mechanism was achieved (*51*). Figure 14 shows an STM image from an Ag(110) surface onto which 0.02 monolayers (MLs) of O_2 molecules were adsorbed at 65 K. If the randomly arriving molecules had been accommodated at their points of impact, by far the majority of them would have been present as single adparticles, which was by no means the case. The inset in Fig. 14a shows a high-resolution section of the image

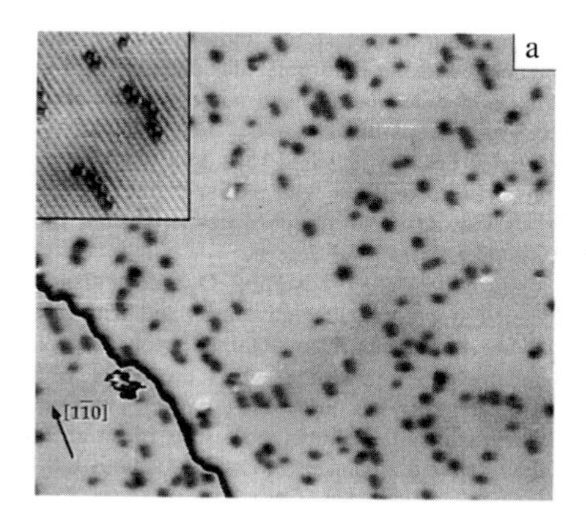

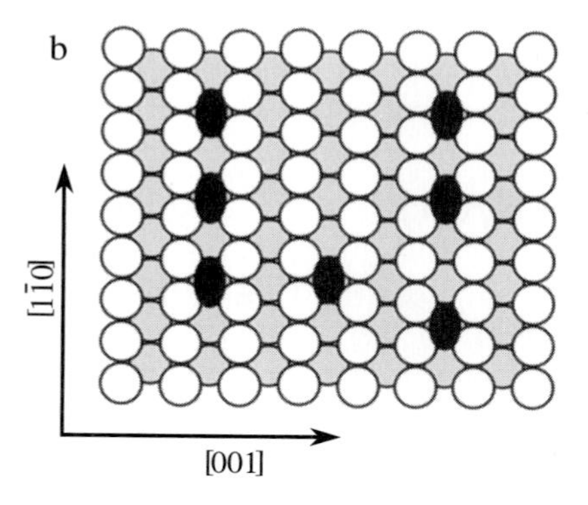

FIG. 14. STM data from adsorption of O_2 molecules on Ag(110) demonstrating the operation of the hot precursor mechanism (*51*): (a) STM image 39 × 23 nm^2 recorded after adsorption of 0.02 monolayers of O_2 at 65 K. Inset: atomic-resolution image of the substrate lattice with O_2 molecules as dark dots; (b) model with adsorbed O_2 molecules (black) and Ag atoms from the topmost (white) and second layer (gray).

in which both the adparticles and the substrate structure are resolved. Only few single adsorbates are discernible, together with ensembles of two to four molecules, as shown in Fig. 14b. Their mutual distance of 2a along the O_2^- direction is the energetically favorable configuration. Even when the total coverage was only 0.3% of a monolayer, about 40% of the adsorbed particles were observed to form ensembles with two or more molecules, whereas the probability for the formation of pairs resulting from random distribution would have been only 0.6%.

The intrinsic precursor may result not only from transient incomplete accommodation of a particle trapped on the surface from the gas phase but also may be the result of a reaction occurring at the surface. The potential diagram for dissociative chemisorption shown in Fig. 2 indicates that the adsorption energy is transferred to the solid while the two atoms are separating from each other. For O_2/Pt(111), this process occurs over two or three lattice constants (and requires about 300 fs), i.e., the system follows the dashed line in Fig. 2. In the case of dissociative chemisorption of O_2 on Al(111) (for which no chemisorbed molecular state exists), the O atoms formed are at least 10 nm apart from each other (*52*), a result for which no satisfactory theoretical explanation exists. A proposed speculation (*53, 54*) is that one of the O atoms is ejected into the gas phase and recaptured by the surface.

These hot adatoms are more energetic than they would be if accommodated with the surface and are hence also expected to be more reactive. Such effects were suggested by Au and Roberts (*55*) for reactions between molecular oxygen and ammonia on MgO surfaces [and later for Cu single-crystal surfaces (*56*)]. The most direct evidence, however, is found in several studies of the oxidation of CO on Pt (*38, 44, 57, 58*). Figure 15 shows temperature-programmed reaction spectroscopy (TPRS) data for oxygen and carbon monoxide that had been coadsorbed on Pt(111) (*57*). O_2 molecules chemisorbed at low temperature dissociated upon heating to 150 K. When this (thermally accommodated) species was afterwards exposed to CO, the TPRS data showed a single peak (β) of CO_2 evolution at about 300 K, arising from the standard Langmuir–Hinshelwood reaction $O_{ad} + CO_{ad} \rightarrow CO_2$. The solid line in Fig. 16 represents the one-dimensional potential for this reaction channel. When, instead, O_2 and CO were coadsorbed at low temperature, subsequent heating gave rise to another peak of CO_2 evolution (α) exactly at the temperature (150 K) at which $O_{2,ad}$ dissociates. Since we know from the previous discussion that hot adatoms are then formed as intermediates, it is concluded that an alternative reaction path (dashed line in Fig. 16) is opened. There is a finite chance that the hot O adatom thermalizes before it hits a neighboring CO molecule, and then CO_2 formation requires a higher activation energy, and hence CO_2

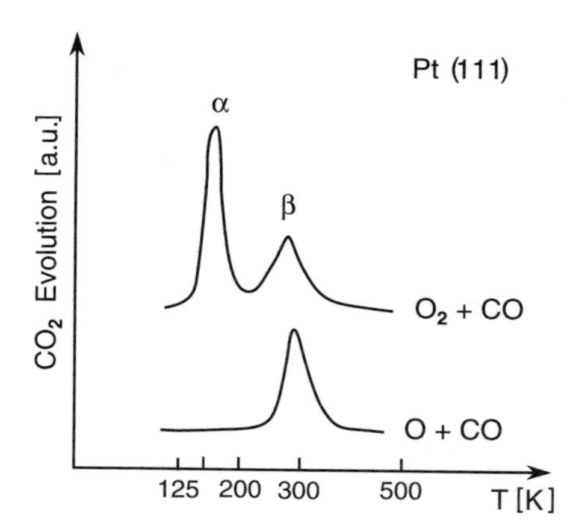

FIG. 15. Temperature-programmed reaction spectroscopy characterizing CO_2 formation on Pt(111), demonstrating the participation of hot O adatoms (*57*).

comes off the surface only at higher temperature (β). This model is supported by the observation that the relative proportions of the α and β peaks may be controlled by the O_2 and CO coverages. As summarized later, quite similar effects are observed when the chemisorbed oxygen molecules are dissociated photochemically (or by tunneling electrons) rather than thermally; these results indicate that the nascent O atoms have comparable energy content in all cases.

4. *Langmuir–Hinshelwood vs. Eley–Rideal Kinetics*

The preceding discussion of reactions involving hot adatoms leads immediately to a "classical" problem in the kinetics of heterogeneous catalysis:

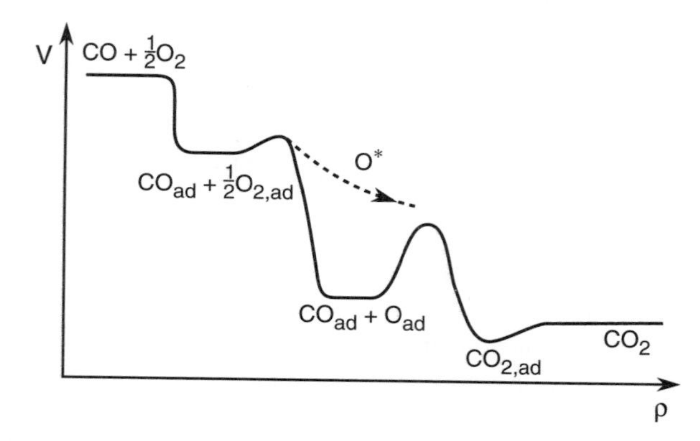

FIG. 16. Schematic potential energy (V) diagram for CO oxidation illustrating the possible participation of hot adatoms O* (ρ is the reaction coordinate).

distinguishing between a reaction of A + B proceeding with both species from the adsorbed state [i.e., thermally equilibrated with the surface, a Langmuir–Hinshelwood (LH) mechanism] and that proceeding via interaction of an adsorbed particle with another directly colliding from the gas phase [an Eley–Rideal (ER) mechanism].

Generally, most surface catalytic reactions are believed to proceed via the LH mechanism. Direct experimental verification has been obtained in experiments in which, after adsorption of the reactants, the gas phase is removed and the sample temperature continuously increased until the product molecules come off. Specifically, the LH mechanism is concluded to operate if the time lag between adsorption and product formation is much longer (i.e., exceeding about 10 ps) than the relaxation time for reaching thermal equilibrium. In the carbon monoxide reaction on Pd(111), for example, the mean reaction time for CO molecules interacting with an O_{ad}-precovered surface was determined by means of a modulated molecular beam technique to vary to about 10^{-4} s (and from these data, the activation energy for the LH reaction $O_{ad} + CO_{ad} \rightarrow CO_2$ was determined) (*59*).

Reactions involving hot adatoms as discussed in the preceding section obviously represent an intermediate situation between LH and ER insofar as the reactants are not in complete thermal equilibrium with the surface (LH), but the reaction does not take place by direct collision without (at least partial) previous accommodation (ER).

Clear identification of the limiting case of an ER mechanism has been found to be rather difficult and is subject to intense research. There seems to be consensus that effects indicating the operation of ER type mechanisms are so far restricted to the impingement of *atoms* (i.e., energetically excited reactants) onto the surface. Concerning the oxidation of CO on Pt(111), the first evidence for non-LH behavior was reported for beams of O *atoms* and CO molecules impinging onto the surface (*60*). In contrast, with O_2 *molecules* and CO under similar experimental conditions, clear evidence for a LH mechanism was again found (*61*); the sharp angular distribution that was observed resembled the data reported by Matsushima (*57*) with hot adatoms. Probing the internal energy content of the product molecules by infrared chemiluminescence (*41*) showed that exposure of the CO-covered surface to O atoms led to the formation of CO_2 with a higher degree of internal excitation than when O_2 molecules were used, indicating that the product had some kind of "memory" of the initial state of one of the reactants. In another work (*62*), this reaction was explicitly claimed to proceed via the ER mechanism, and this suggestion was confirmed by subsequent classical trajectory calculations (*63*). In this latter work it was found that most reactive events occurred within the first few picoseconds after collision, when the incident atom was in only a weakly held state. Apart from these ER events, however, in a significant fraction of reactive encounters the product molecule is only formed a longer time ($>$ 10 ps) after the initial impact, and these events are no longer of the classical ER type but rather have to be classified as hot adatom processes.

The most detailed investigations of the operation of a true ER mechanism were performed with reactions involving the collision of an incident hydrogen atom onto an adsorbed species, such as $H + D_{ad} \rightarrow HD$. The most direct indication was obtained from measurements of the energy content of the product molecules released into the gas phase. The energy balance for this latter process is $E_{HD} = E_{diss}(HD) + E_{kin}(H) - E_{ad}(D)$, which amounts to about 2.3 eV for Cu(111); state-resolved experiments in this case indeed showed such an amount of energy carried away by HD in the translational, rotational, and vibrational degrees of freedom (*64, 65*). Similarly, the reaction $D + D_{ad} \rightarrow D_2$ on Ni(110) was observed to produce rotationally and vibrationally excited molecules. The observations were considered as evidence for an ER mechanism (*66*), and the conclusion was in turn supported by theoretical results (*67–69*). The formation of CH_4 via a collision of H with $CH_{3,ad}$ on Cu(111) was also demonstrated to be associated with significant excess translational energy and a very narrow angular distribution, and again the operation of an ER mechanism was inferred (*70*).

However, classical trajectory calculations for the $H + D_{ad}$ reaction on Si(001) (*68*) indicated that apart from an ER process, reactions involving hot adatoms (i.e., partially accommodated species) as well as collision-induced reactions might also occur. Remarkably, experimental investigation of the $H + D_{ad}$ reaction on Ni(110) revealed not only HD as product molecules but also (to a smaller extent) D_2 (*71*). Furthermore, it was found that the rate of HD production was not simply proportional to the coverage of the surface with D, as expected for a pure ER mechanism (*72, 73*). Thus, a more appropriate classification of the essential steps was proposed (*72*), which is shown in Fig. 17. In addition to the pure ER reaction (Fig. 17a), a fraction of the incident H atoms may also become adsorbed, leading to a modified ER mechanism (Fig. 17b). In addition, the incoming H atom may become trapped (but not completely accommodated) as a hot adatom which either reacts with D_{ad} (Fig. 17c) or transfers its energy to an adsorbed D atom, which then is excited and may react further even to D_2 (Fig. 17d). Simple random walk simulations revealed that indeed all these experimental observations may be qualitatively accounted for within the framework of this concept (*74*).

Similarly, reactions between gaseous D atoms and CH_3I adsorbed on various surfaces were found to follow composite hot adatom and ER mecha-

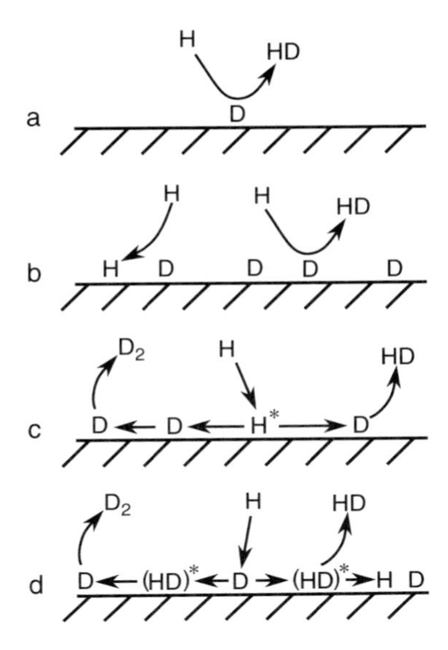

FIG. 17. Schematic classification of the possible reaction types in the H + D_{ad} → HD reaction (*72*). See text for explanation.

nisms (*75*). The formation of methane from a beam of hydrogen atoms impinging onto a carbon-covered Ni(100) surface was concluded to be dominated by ER steps (*76*); hydrogenation of graphite island edges is also believed to involve hot adatoms (*77*).

The general conclusion is that the original ER mechanism as proposed 60 years ago (*78*) hardly operates in its pure form, even in reactions involving atomic primary particles. Instead, hot adatom processes appear to be more common (even with molecular adsorbates).

5. *Collision-Induced Surface Reactions*

The activation barrier for initiating a surface reaction can be overcome either by the energy content of the reacting species or by the kinetic energy of (chemically inert) particles impinging onto an adsorbed layer from the gas phase. Such a collision-induced surface reaction ("chemistry with a hammer") was observed with the dissociation of CH_4 adsorbed on Ni(111) induced by the impact of argon atoms (*79, 80*). Figure 18 shows the variation of the cross section for dissociation of CH_4 as a function of the kinetic energy of the argon atoms in the direction of the surface normal for various values of the total kinetic energy. In contrast to the scaling of the dissocia-

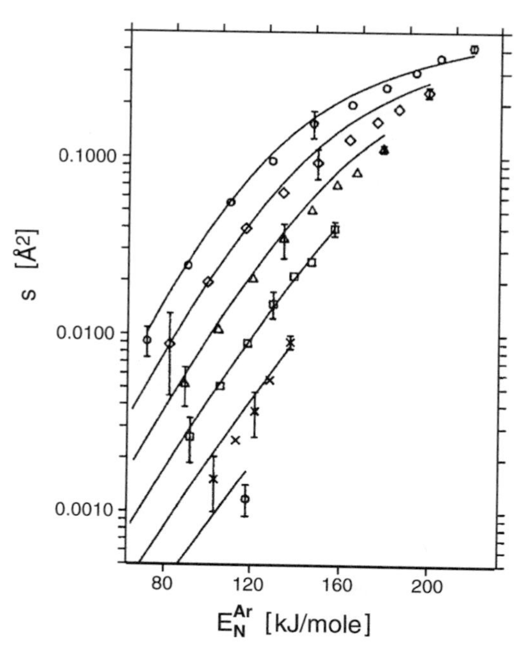

FIG. 18. "Chemistry with a hammer:" Variation of the cross section *s* for dissociation of CH_4 adsorbed on Ni(111) at 47 K induced by impact of argon atoms as a function of the normal component of their kinetic energy (E_N^{Ar}). The different curves correspond to varying amounts of total kinetic energy, ranging (from top to bottom) from 217 to 117 kJ/mol (*80*).

tion probability of an incident CH_4 beam with its normal kinetic energy (Fig. 5), such a simple relationship is no longer found in the current case. This effect could be rationalized by resolving the reaction into two steps. In the first, a fraction of the kinetic energy of the incident argon atom is transferred to the adsorbed molecule, and the amount of energy transferred depends on the point of impact (or impact parameter), which can be evaluated in the framework of a hard-sphere model. The energy accumulated by this process in the normal direction may then be used in the second step to cause dissociation in the same way as if the CH_4 molecule strikes the surface with sufficiently high kinetic energy (as discussed in Section II.C.1). The data represented by the solid lines in Fig. 18 were derived by use of the sketched model, the validity of which is demonstrated by the excellent agreement with the experimental results (*80*). In competition with dissociation, collision-induced desorption also occurs, and with a considerably higher probability (*79*). Once the argon atom transfers energy to the adsorbate, the CH_4 molecule collides with the surface and may rebound into the gas phase if the geometry is not favorable for dissociation.

The effects of collision-induced desorption and dissociation are not restricted to weakly held adlayers; they have also been observed with chemisorbed systems (*81–83*). For O_2 chemisorbed on Ag(110), under the impact of xenon atoms a common threshold of 0.9 eV for both processes was found; the corresponding value is about 1.2 eV for O_2/Pt(111) (*83*). These values are considerably higher than the thermal activation energies. The common thresholds for dissociation and desorption might be a consequence of rapid energy exchange between the various modes of nuclear motion (viz., the O–O vibration leading to dissociation and the M–O_2 vibration causing desorption), but the possible alternative involving the creation of hot adatoms in collision-induced dissociation initiating desorption of neighboring adparticles in a secondary step cannot be ruled out.

A mechanism of the latter kind was discussed previously for the formation of CO_2 from a CO + O_2 adlayer in which thermal dissociation of the adsorbed O_2 molecules creates hot adatoms causing CO_2 formation. By analogy, impingement of xenon atoms on a coadsorbed CO + O_2 adlayer on Pt(111) also caused CO_2 formation in addition to collision-induced desorption of O_2 (*84*).

Finally, in this context, another (in my opinion still rather puzzling) effect needs to be discussed. There are several reports in the literature according to which the rate of CO desorption from metal surfaces is enhanced in the presence of CO in the gas phase (*85–88*). By monitoring the exchange process between $^{13}C^{18}O$ preadsorbed on Ni(100) and postdosed $^{12}C^{16}O$ with infrared spectroscopy, an activation energy for the CO flux-induced desorption of only 25 kJ/mol was determined, which is only about one-fifth of the value for desorption into vacuum (*89*). This effect was attributed to repulsive interactions between preadsorbed particles and molecules arriving from the gas phase onto nearby sites. Strikingly, it was claimed to operate even at coverages below 0.3 (ML) and CO fluxes as low as 10^{-3} ML s^{-1}. Since the incident CO molecules exhibit only thermal kinetic energies, this phenomenon cannot be attributed to the collision-induced processes as discussed previously; it awaits clarification.

In general it is concluded that the continuous bombardment of adsorbate-covered surfaces with particles from the gas phase (even if their mean kinetic energy is low) under the high-pressure conditions of practical catalysis might be of nonnegligible influence on the kinetics beyond the simple concepts of independent adsorption and desorption processes. There is possibly another cause of the difference between high- and low-pressure studies, and early reports (*90*) according to which reaction rates may be affected by the presence of inert gases surely deserve reexamination.

D. Electronic Excitations Caused by Surface Reactions

Langmuir and Kingdon (*91*) reported in 1923 that thermal desorption of cesium from tungsten occurs in the form of Cs^+ ions, and on the basis of this observation they concluded, "Hence cesium atoms leaving a tungsten surface are robbed of their valence electrons by the tungsten." This effect forms the basis for thermionic energy conversion and is readily explained by the fact that it costs less energy to ionize a Cs atom (3.9 eV) than one

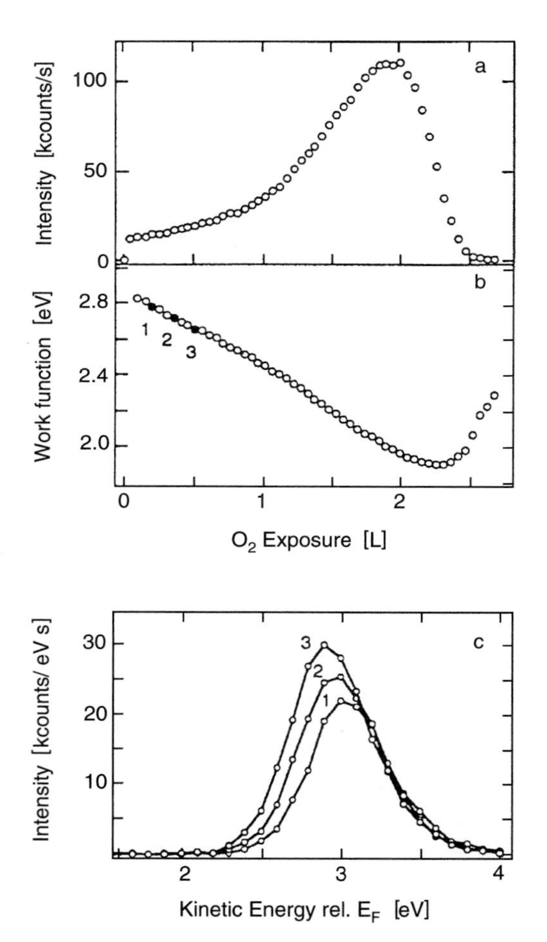

Fig. 19. Exoelectron emission in the reaction of O_2 with a Li surface (*95*): (a) intensity of emitted electrons; (b) work function, each as a function of O_2 exposure; (c) kinetic energy distributions of emitted electrons at the three stages shown in b.

gains by transferring this electron to the Fermi level of W (work function $\gtrsim 5$ eV). It clearly demonstrates the possibility for adiabatic charge transfer in a gas–surface reaction. Less obvious, however, is the interpretation of other earlier observations, according to which interaction of oxygen (and other electronegative molecules) with alkali metal surfaces may cause the emission of electrons (*92*), the origin of which must be attributed to the energy gain associated with the chemical transformation of the surface. This effect of "exoelectron" emission (sometimes referred to as "exoemission") (*93*) has been the subject of intense research in recent years and clearly demonstrates the possibility for electronic excitations by surface reactions starting from the electronic ground state. Closely related are other observations indicating that the same type of reactions may even be associated with the ejection of negative ions (*94*).

For example, Fig. 19 shows the intensity of electrons (Fig. 19a) and the variation of the work function ϕ (Fig. 19b) as a function of O_2 exposure for a Li surface reacting with O_2 (*95*). The electron yield increases continuously up to a maximum, from which it decreases sharply, and the work function simultaneously decreases continuously with progressing oxidation. The energy distributions of the emitted electrons with respect to the Fermi level E_F at the three stages shown in Fig. 19b are plotted in Fig. 19c. These data demonstrate that the low-energy cutoff of emission is determined by ϕ, which explains the increase in the yield with decreasing work function until the surface layer becomes saturated. These phenomena are rationalized in the framework of the theoretical model proposed by Kasemo *et al.* (*96*) and illustrated in Fig. 20: The lowest empty electronic state (affinity level) E_A (which is 0.4 eV lower than the vacuum level for O_2) is continuously lowered on the approach of O_2 to the surface. When it crosses the Fermi level, there is a high probability for its being occupied by an electron tunneling from E_F ("harpooning"), and then the negative ion is further accelerated to the surface, where bond formation occurs. There is, however, a small probability that the neutral particle will be stopped at the surface prior to ionization. The empty state will then be at ε_A, which will be filled by an electron from the metal, whereby the energy released excites another metal electron via the Auger effect. Obviously, this exoelectron attains its maximum kinetic energy if both electrons originate from the Fermi level (and $E_{kin,max} = -\varepsilon_A$), and the minimum kinetic energy is just given by the work function, $E_{kin,min} = \phi$. This model explains *inter alia* why the energy distributions in Fig. 19c have a common high-energy edge (ε_A), whereas their low-energy cutoff is equal to the (varying) work function.

Such a nonadiabatic reaction pathway appears at first glance to be rather improbable since the quenching of electronic excitations at metal surfaces (in this case, this is a hole state) is usually much faster than the time scale

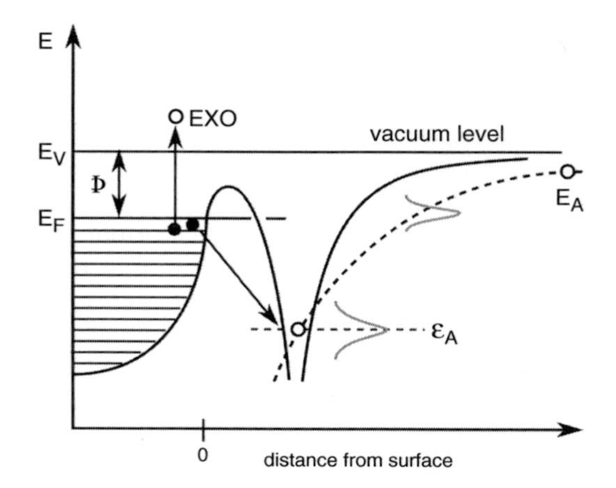

FIG. 20. The mechanism for exoelectron emission: electronic energy diagram for an electronegative particle approaching a surface (*96*).

for nuclear motion, and indeed in the O_2/Li system the probability for exoelectron emission is $<10^{-6}$ e/incident O_2 molecule. The competition between nuclear and electronic motion is directly reflected by the exponential increase of the electron yield with the velocity of the incident molecules, as shown in Fig. 21 for the system O_2 + Cs (*97*). Note that a velocity of

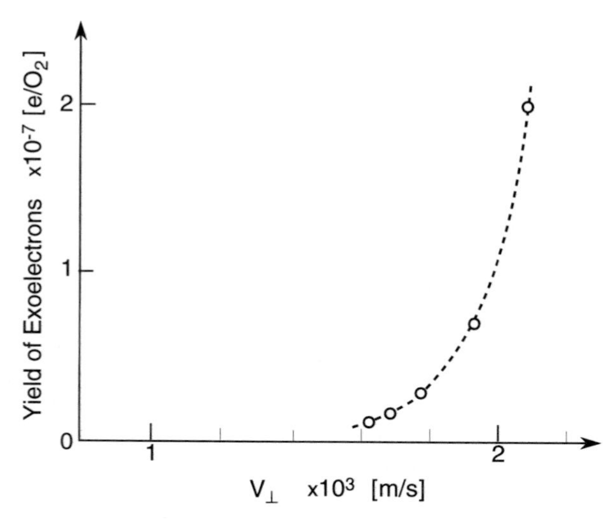

FIG. 21. Variation of the initial yield of exoelectrons in the system O_2 + Cs with the normal velocity of the impinging O_2 molecules (*97*).

2×10^{-3} m/s is equivalent to a distance of 2 Å in 100 fs, which provides an indication of the time scale on which these elementary processes occur.

More detailed discussion of the O_2 + Li system (*95*) revealed that the whole reaction is more complex than stated previously and takes place via the following sequence: (i) The incoming O_2 molecule is resonance ionized by harpooning to O_2^-, which picks up a second electron if it comes closer to the surface; (ii) the O_2^{2-} readily dissociates into two O^- species. The O^- ion at an alkali metal surface offers an excited hole at ε_A which is then filled by Auger deexcitation associated with exoelectron emission.

The intermediate formation of O^- in front of the surface is supported by the detection of ejected O^- ions (albeit with very small probability; $\lesssim 10^{-8}$ O^-/incident O_2) (*98*). Similar effects were reported previously for Cl^- and Br^- ions ejected from alkali metal surfaces exposed to halogens (*99*). Then the yields were considerably higher, and the effects may be correlated with the energetics of the affinity level E_A with respect to E_F (*99, 100*).

The energy scheme of Fig. 20 suggests that apart from Auger decay, deexcitation might also occur via light emission. Fluorescence was indeed reported for Cl_2 interacting with K surfaces, but with considerably lower yield than exoelectron emission (*101*); in the reaction with O_2, the light intensity was always below the detection limit. This result is in agreement with general experience indicating that at metal surfaces fluorescence is completely suppressed by Auger deexcitation for energies up to about 500 eV (*102*).

Direct manifestation of electronic excitations caused by surface reactions through the emission of exoelectrons will be restricted to interactions between electronegative molecules (such as O_2, NO_2, or halogens) and low-work-function (e.g., alkali metal) surfaces since, according to Fig. 20, $\phi < |\varepsilon_A|$ is a necessary prerequisite. Even then, the emitted electrons represent only the tip of the iceberg since many combinations of lower-lying electrons will cause excitations only to energies below the vacuum level. The creation of such hot electrons by a surface reaction, however, was recently demonstrated for the adsorption of H and D atoms on Ag and on Cu surfaces (*103*). The experimental arrangement consisted of a thin metal film evaporated onto a Si(111) surface, whereby a Schottky diode with a barrier height of about 0.6 eV at the interface was formed. Hot electrons (or holes) formed by adsorption of hydrogen atoms at the outer surface can pass through the metal film and cross the Schottky barrier, after which they are recorded as a "chemicurrent." In this way, electron excitation even during adsorption of neutral species was demonstrated, and these results suggest that nonadiabatic energy dissipation through creation of electron–hole pairs at metal surfaces might indeed be more common than has been assumed.

E. Surface Reactions Caused by Electronic Excitations

The reaction between coadsorbed O and CO to give CO_2 on a Ru(0001) surface initiated by femtosecond infrared laser pulses has already been referred to as an example of a reaction caused by heating the electron gas, leading to partial occupation of an antibonding adsorbate level and triggering the transformation. However, this was still considered a thermal reaction in a wider sense since the electron distribution could be described by an electronic temperature T_e that differed considerably from the lattice temperature, T_{ph}. Irradiation of adsorbate-covered surfaces with higher-energy photons (typically with energies up to 6.4 eV) opens up the field of surface photochemistry. Although not directly of relevance for (thermal) heterogeneous catalysis, studies of these phenomena nevertheless provide some important insights into the dynamics of elementary processes involved in surface reactions.

Since the lifetimes of electronic excitations at metal surfaces are several orders of magnitude shorter than the time scales for nuclear motions, photochemical reactions appear to be improbable. Surprisingly, however, the cross sections found are comparable to those determined for reactions with free molecules, and hence surface photochemistry (in particular with inclusion of ultrafast laser techniques) is currently an intense field of research (*104–107*), primarily because the short lifetime of the excitation is compensated by a much larger cross section for absorption of the incident light. The latter process occurs mostly in the near-surface region of the metal (within about 10 nm), where relaxation of the nascent photoexcited electrons through scattering events leads to the rapid establishment of a (time-dependent) energy distribution. As depicted in Fig. 22, these hot electrons may scatter at the surface or be resonantly attached to an empty orbital of the adsorbate complex. This is equivalent to a transition from the ground state potential to that of an excited state, as depicted on the right-hand side of Fig. 22. If the latter is repulsive, the nuclear motion becomes accelerated and eventually takes up sufficient kinetic energy to cause bond breaking after a return to the ground state (*108*).

The efficiency of such processes is governed primarily by the lifetimes of electrons excited to energies above the Fermi level; these recently became accessible to direct experimental determination by means of two-photon photoelectron spectroscopy using femtosecond laser techniques. Measurements of this type have been performed with noble and transition metals (*109–113*); for example, Fig. 23 shows data for Cu, Ag, and Ta as a function

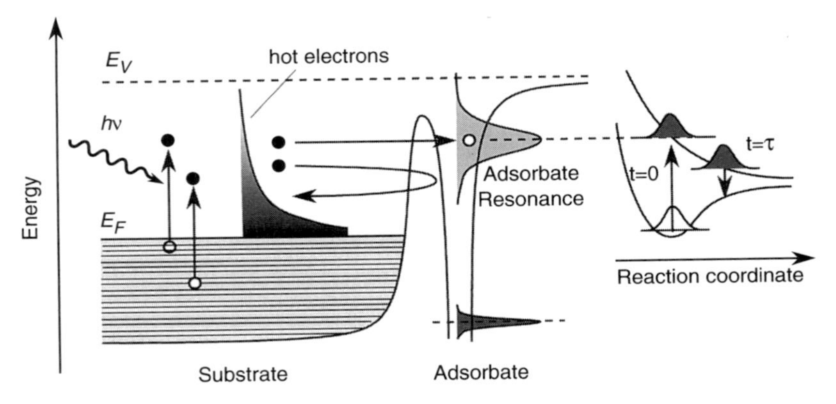

FIG. 22. Energy diagram for an adsorbate-covered metal surface under the influence of light absorption.

of the energy above the Fermi level (*113*). Generally, the relaxation times decay rapidly with increasing energy and decrease well below ~10 fs for $(E - E_F) \gtrsim 2$ eV. The fairly simple Fermi liquid theory provides a surprisingly good description; more advanced theoretical *ab initio* calculations are currently in progress (*114*).

From the various systems being examined with respect to photodesorption (*104–107, 115, 116*), some results for ammonia desorbing from Cu(111)

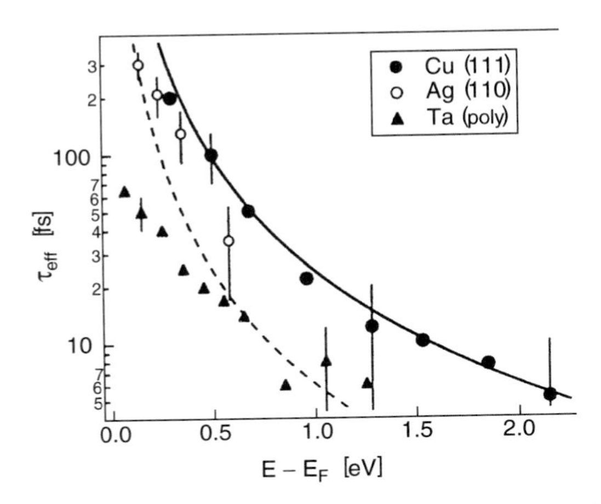

FIG. 23. The effective relaxation times of hot electrons in Cu, Ag, and Ta as a function of their energy above the Fermi level E_F. The solid line is the prediction from Fermi–liquid theory, and the dashed line represents a corresponding estimate for Ta (*113*).

upon irradiation with 6.4-eV photons are presented (*117*). Up to a laser fluence of 8 mJ/cm^2, the desorption yield increases linearly with fluence, indicating that this process is initiated by single-photon absorption rather than by heating of the electron gas. Polarization measurements demonstrate that the primary excitation consists of the creation of hot electrons, as shown in Fig. 22. Ammonia molecules come off the surface strongly peaked along the surface normal and with a mean translational energy of about 0.1 eV, which is far in excess of the thermal energy corresponding to the surface temperature of $\sim$100 K. Surprisingly, there is a pronounced isotope effect: The cross section for desorption of NH_3 is 3.1×10^{-21} cm^2, and that for ND_3 is only 0.7×10^{-21} cm^2 (i.e., the ratio is 4.4). This result suggests that the energy required to break the molecule–surface bond is acquired in an intramolecular (i.e., N–H) coordinate during the short-lived electronic excitation.

The mechanism of desorption induced by electronic transition (DIET) is usually discussed (*107, 108, 116*) in terms of the so-called MGR model (*118, 119*). As shown on the right-hand side of Fig. 22, intermediate formation of a negative ion resonance by capture of a hot electron by the adsorbate causes a Franck–Condon transition from the ground state potential to a repulsive excited state. The situation for the current system is shown in Fig. 24 in the form of potential surfaces as a function of two coordinates, namely, the separation between the plane formed by the three H atoms and the N atom (x) and the distance of the N atom from the surface (z). In the ground state the molecule is bound to the surface through the N atom, and the H atoms are directed away from the surface. The excited state is characterized by a planar geometry, i.e., the minimum of its potential is at $x = 0$. Upon excitation from the ground state minimum (Fig. 24, dashed line with the up arrow), the system ends at a repulsive section of the excited state potential surface, the gradient of which causes the plane of H atoms to bend down, i.e., the umbrella vibration mode of the ammonia molecule becomes excited. After a short time (a few femtoseconds), the system returns to the ground state, but it might have picked up enough energy to desorb. One of the resulting trajectories is plotted in the lower part of Fig. 24 and indicates that the molecule comes off the surface preferentially with the H atoms pointed down. Since the excitation of the N–H umbrella mode is the decisive step, the large isotope effect becomes plausible. Further theoretical studies (*120, 121*) essentially confirmed this concept, and determination of the vibrational state distributions of desorbing molecules provided evidence for additional quantum effects, namely, the nonthermal population of states of different symmetry, which are sensitively influenced by the potential for the inverted geometry (*122*).

In addition to desorption, photoexcitation may also cause bond breaking

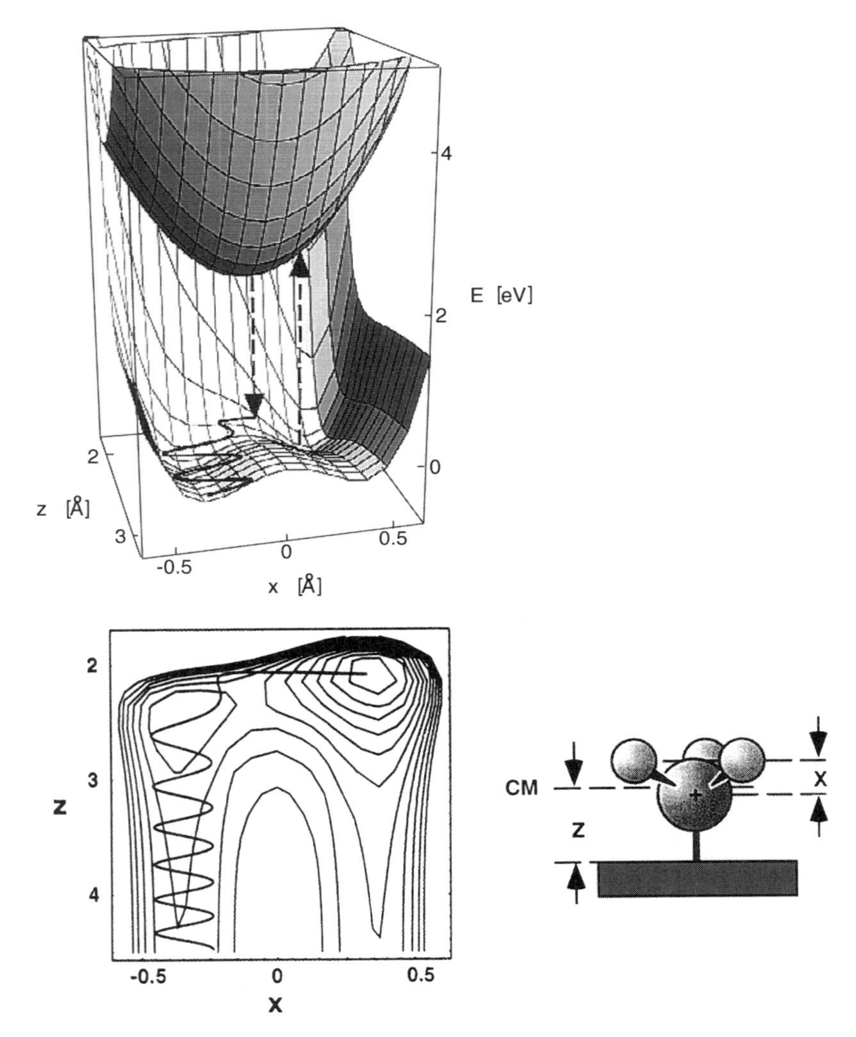

FIG. 24. The mechanism of photoinduced desorption of NH_3 from Cu(111) (*117*). See text for explanation.

in adsorbed molecules. A good example is represented by the system N_2O/Pt(111) (*123*). Irradiation with 6.4-eV photons was found to cause desorption of N_2, N_2O, and O atoms. The internal and translational energy distributions of the analyzed N_2 molecules suggest their complete thermalization before they leave the surface, but the desorbed N_2O molecules were detected with excess translational and vibrational energies, which shows that they retain some "memory" of the excitation step. Interestingly, the O atoms are released with prompt axial recoil with no collisions with neighboring adsorbates along the tilted N_2O molecular bond axis. These ballistic atoms leave the surface in either the ground (^3P) or the first excited (^1D) state with identical kinetic energies, suggesting a common dissociative intermediate.

Instead of being ejected into the gas phase, photogenerated particles may travel along the surface and exhibit enhanced reactivity with other species prior to thermalization, and it was indeed the formation of CO_2 upon irradiation of a $CO + O_2$ adlayer on Pt(111) that prompted the first hint of a hot adatom mechanism in this reaction (*124*). Later detailed investigations of the same system (*125*) showed that CO_2 formed by photochemical reaction between adsorbed O_2 and CO at 25 K substrate temperature exhibited an angular distribution strongly peaked along the surface normal with a maximum translational energy of 1.35 eV, which is close to the exit channel's exoergicity (1.45 eV at low coverage). The dynamics is quite similar to that found under thermal reaction conditions, which suggests passage through similar transition state configurations after rapid quenching of the electronic excitation. Photochemistry of adsorbed O_2 molecules is believed to result from transient capture of an excited metal electron into the $3\sigma_u^*$ state of O_2, and the transient ion is then attracted to the surface while its bond is elongated. After decay to the electronic ground state, the translationally and vibrationally excited molecule may rapidly diffuse, desorb, or dissociate (*126–128*). Manifestation of the kinetic energies of the photogenerated hot adatoms was achieved by monitoring the collision-induced desorption of coadsorbed noble gas atoms, from which a lower bound of about 0.7 eV for the energy of the fastest O atoms was determined (*129*).

The photochemical effects discussed so far were characterized by a linear dependence between yield and light fluence, indicating single-photon excitation as the underlying mechanism. Before the next photogenerated electron arrives, the excited adsorbate complex has already relaxed into its (electronic and vibrational) ground state, unless reaction has occurred. The vibrational relaxation in the electronic ground state typically takes place within a few picoseconds. If the next electron arrives earlier, such as is the case with higher light intensities attained with femtosecond laser pulses, the adsorbate complex may be excited again before it is quenched down to the vibrational ground state (*130*). By a sequence of multiple excitation-quenching cycles, the system may climb up the ladder of vibrational levels and eventually undergo desorption, a process first identified with CO desorption from Cu(111) (*131*) and denoted as desorption induced by multiple electronic excitations (DIMET) (*132*). As a consequence, with increasing laser fluence the photoyield changes from linear (DIET) to superlinear (DIMET), as, for example, was verified experimentally with the system O_2/Pt(111) (*133*). In addition to the nonlinear increase of the yield with fluence, other differences from ordinary (single-excitation) photochemistry were found, e.g., increasing mean translational energy with increasing laser fluence and variation of the branching ratio between desorption and reaction (*133, 134*) as well as highly vibrationally excited desorption products (*135*). In addition to the previously discussed original DIMET mechanism (*131*), a conceptually alternative, nonadiabatic approach was proposed (*136*). Thereafter, the electronic system is coupled to the adsorbate vibrational degrees of freedom via electronic friction. Since with high laser fluences the high density of hot electrons rapidly equilibrates internally (in $\lesssim$100 fs), the electron gas can be represented by an electronic temperature T_{el} (different from the lattice temperature T_{ph}), and it is the coupling of T_{el} to the adsorbate temperature T_{ad} via a friction coefficient that drives the surface reaction. This thermal approach underlies the analysis of the CO oxidation on Ru(0001) initiated by infrared femtosecond pulses, as discussed in Section II.B.

The MGR model underlying the mechanism of DIET was originally developed for interpretation of surface reactions caused by direct impact of low-energy electrons (electron stimulated desorption) (*118, 119*). This effect provides perhaps the most direct evidence for the initiation of chemical transformations at metal surfaces by electronic excitations. A continuing series of DIET conferences, starting in 1982 (*137*), reflects the growing interest in this special field, a more detailed discussion of which would go far beyond the scope of this review.

Instead of an electron gun, the tip of a STM can be used as a source of electronic excitations of adsorbate complexes, and application of this principle opened up exciting new possibilities to probe the local dynamics of surface processes. The possibility of using the STM tip to manipulate atoms or molecules on surfaces had been previously demonstrated (*138, 139*), but its use to induce and view bond-selected chemistry on an atomic scale [“angstrochemistry” (*140*)] has only recently begun to be explored. Stipe *et al.* (*140*) placed an STM tip over an O_2 molecule adsorbed on a Pt(111) surface, dissociated the molecule with electrons tunneling from the tip, the action of which was confined to this molecule, and subsequently imaged the two chemisorbed O atoms that were formed. After dissociation, these atoms were found to be one to three lattice constants apart from each other, just as in the case of thermal dissociation, as discussed previously in connection with the effect of hot adatoms (*2*). The nonlinear dependence of the dissociation rate on the current was rationalized in the framework of a general model for dissociation induced by intramolecular vibrational excitation via inelastic electron tunneling.

In a next step, even the rotational motion of single adsorbed O_2 molecules was induced by the STM tip and imaged (*141*). It was found that a current pulse through an isolated adsorbed molecule caused its rotation by 60° into its second equivalent orientation, and yet another pulse led to another rotation into the third orientation. Above a certain threshold bias voltage, the rate of rotation varied linearly with the tunneling current, indicating a single-electron excitation mechanism, whereas for low voltages a ladder-climbing mechanism (such as with DIMET) was operative. The dissociation probabilities are generally lower than those for rotation, the energy barrier for which was estimated to be about 0.15 eV, whereas that for dissociation was found to be about 0.4 eV. These results imply that the potential energy surface includes multidimensional pathways with possible coupling between different modes, and they demonstrate how complex the full dynamics of an apparently rather simple surface process may be.

This coupling between different modes was demonstrated directly for acetylene adsorbed on a Cu(100) surface at 8 K (*142*). Excitation of the C–H (C–D) stretch mode of C_2H_2 (C_2D_2) at 358 meV (266 meV) by tunneling electrons with corresponding energies led to a 10-fold (60-fold) increase in rotation rate, and the effect was attributed to intramolecular energy transfer from the stretch vibration to the hindered rotation. Detailed analysis showed that about 1% of the tunneling electrons excite the C–H stretch, and for each C–H stretch excitation the probability for rotation is only on the order of 10^{-8}; thus, the results demonstrate the rather low efficiency for energy transfer into this mode of nuclear motion, which is obviously preferably damped into excitations of the substrate.

This latter aspect becomes clearer with the final example discussed in this context (*143*); electrons tunneling from an STM tip to individual CO molecules adsorbed on Cu(111) can cause their hopping from the surface to the tip if the bias exceeds a threshold value of 2.4 eV. There is again a linear dependence between hopping rate and tunneling current, indicating a single-electron excitation as the primary process. Probing the electronic density of states above the Fermi level by two-photon photoelectron spectroscopy showed that this excitation consists of transient population of the $CO2\pi^*$-derived level centered at 3.5 eV above E_F. From the measured isotope effect it was concluded that only about 0.1% of the tunneling electrons are captured by this level, the occupation of which comprises a transition to a repulsive potential from which desorption may start through the well-known DIET mechanism (Fig. 25). A modified theoretical model suggests that subsequent tunneling of the CO $2\pi^*$ electron into the metal may cause excitation of the Cu–CO stretch vibration, eventually leading

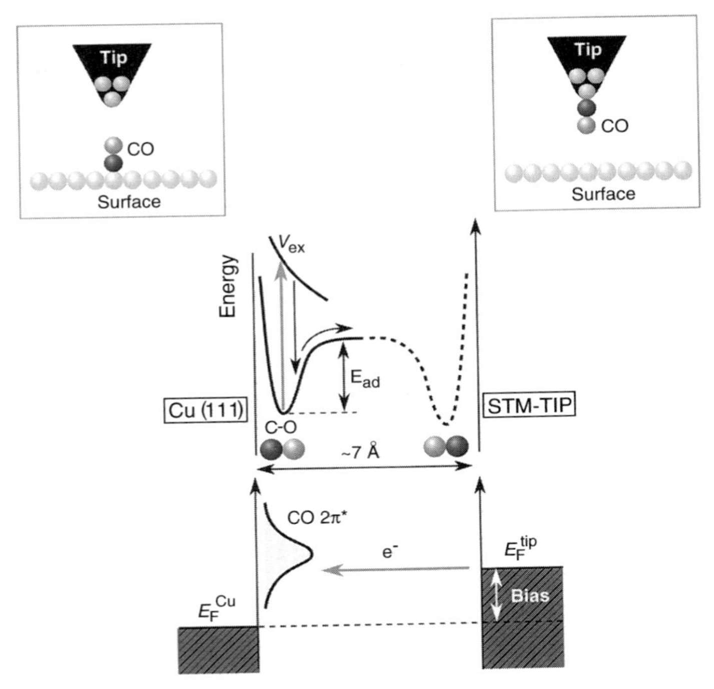

FIG. 25. The mechanism of desorption of a CO molecule adsorbed on Cu(111) induced by electrons from an opposite STM tip transiently occupying the CO $2\pi^*$-derived level (*143*).

to desorption (*144*). Since the desorption probability from this excited state is only about 5×10^{-9}, its lifetime is presumably very short. Time-resolved, two-photon experiments, together with the linewidth of the $2\pi^*$ level, indicate that this lifetime is indeed only 3 ± 2 fs, a result that is in complete accord with the findings regarding the lifetimes of hot electrons at metal surfaces in this energy range. This example of "atomic-scale femtochemistry" presumably marks the lower boundary of the time scale involved in chemical reactions.

III. The Atomic Level

A. The Local Geometry of the Adsorbate Complex

The original view of Langmuir (*145*), according to which the lattice of a single-crystal surface represents a periodic arrangement of equivalent adsorption sites, is correct only in exceptional cases. Because of the interactions between the adsorbed particles, almost never does saturation of the adsorbed layer correspond to a true monolayer in the sense that the density ratio of adsorbed species to surface atoms (θ) equals 1. Hence, it is appropriate to introduce, in addition to the *absolute* coverage (θ), a *relative* coverage ($\delta = \theta/\theta_{max}$). An example for which $\theta = 1$ was achieved is presented by the O/Ru(0001) system, the structure of which is reproduced in Fig. 26 (*146*). The O atoms are located in the threefold-coordinated hcp sites of the lattice of surface atoms. However, these are not randomly occupied with increasing coverage (as expected for noninteracting particles); rather, a whole series of ordered phases [2×2 O, 2×1 O, and $2 \times 2 - 3$O, with $\theta = 0.25$, 0.5, and 0.75, respectively (*147*)] is formed. When going from $\theta = 0.25$ (2×2 phase) to $\theta = 1$ (1×1), the energy for the process $\frac{1}{2}$ $O_2 \rightarrow O_{ad}$ decreases continuously from 2.6 to about 1.9 eV, and the distance between the two topmost Ru layers increases by 0.1 Å, reflecting the strong mutual interactions between the adatoms (*146, 147*). The kinetics of dissociative chemisorption does not follow a simple Langmuir expression, but, as a consequence of the decreasing bond strength, the sticking coefficient also decreases dramatically with increasing coverage so that very high O_2 exposures are needed to reach the 1×1 phase. This does not represent the termination of uptake because still higher doses lead to dissolution below the surface (“subsurface oxygen”) and eventually to oxide formation. The role of subsurface species in catalysis will be considered briefly in Section III.E.

At higher coverages, the formation of ordered phases is more the rule than the exception, and there is a wealth of information about the structural properties of such systems that for the most part was determined from analysis of low-energy electron diffraction (LEED) data and other diffrac-

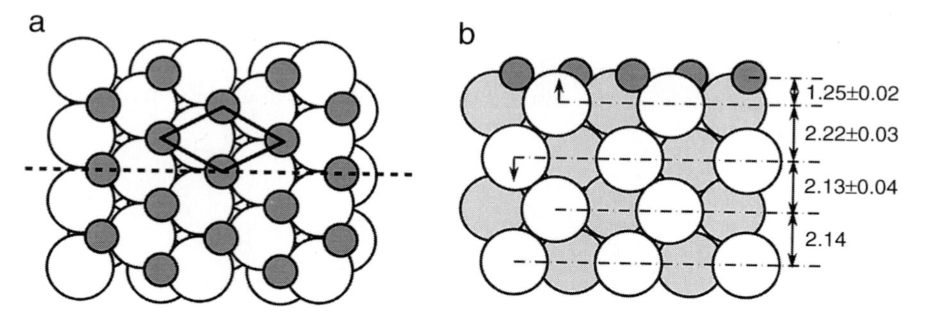

FIG. 26. Top view (a) and side view (b) of the 1×1 O/Ru(0001) phase, whereby all hcp hollow sites are occupied and the absolute coverage $\theta = 1$ is reached. The distances are in Å as derived by LEED (*146*).

tion data (*148–150*). At very low coverages, isolated adsorbed particles may be observed directly by STM. Diffusion is generally characterized by jumps to neighboring adsorption sites, and some values of the kinetic parameters for surface diffusion determined in this way are compiled in the contribution by J. Wintterlin in this volume. Detailed analysis of the migration of N atoms on Ru(0001), for example, indicated an activation energy of 0.95 eV and a preexponential factor of 10^{14} s^{-1} characterizing the exclusive operation of single isotropic hop (*151*). In contrast, essentially one-dimensional diffusion on the anisotropic Ag(110) surface was found to occur along the troughs in the $[1\bar{1}0]$ direction, with an activation energy of 0.22 eV (*152*).

The techniques for surface structural determination are now very advanced, so that even fairly complex and multicomponent systems may be successfully analyzed (*150*). For example, adsorption of CO on a Rh(111) surface causes the formation of a $\sqrt{3} \times \sqrt{3}R30°$ structure with $\theta_{CO} = 0.33$ (*153*). The CO molecules are located in on-top positions, and this phase inhibits completely the dissociative adsorption of oxygen so that the CO coverage has to decrease below this value in order to enable catalytic CO oxidation via the LH mechanism, $O_{ad} + CO_{ad} \rightarrow CO_2$. O atoms, in contrast, form 2×2 and 2×1 structures at $\theta = 0.25$ and $\theta = 0.5$, respectively [as in the O/Ru(0001) system], in which the O atoms are located in threefold coordinated sites and form an open mesh into which additional adsorption of CO is possible. Figure 27 shows the structure of the 2×2 O + CO phase, which contains equal densities of CO molecules on top sites and O atoms on threefold sites (*154*).

B. Interactions between Adsorbed Particles

Interactions between adsorbed particles are ubiquitous and may be either repulsive or attractive. Their physical orgin has been discussed in detail elsewhere (*155*), and their consequences regarding the variation of the adsorption energy and sticking coefficient with coverage were briefly discussed in the preceding section. With respect to the latter aspect, however, the effect of "extrinsic" precursor kinetics (*47, 48*) is reemphasized. In contrast to the original Langmuir concept, a particle colliding with an adsorbed species from the gas phase will not experience a hard wall but rather a weakly attractive (van der Waals-type) potential characteristic of the energetics of multilayer adsorption. Consequently, the incoming particle may be transiently trapped and may slide across the occupied sites until it either returns to the gas phase or finds an empty adsorption site. As a result, the sticking coefficient decreases less than linearly with coverage, as has frequently been observed for nondissociative adsorption (e.g., of

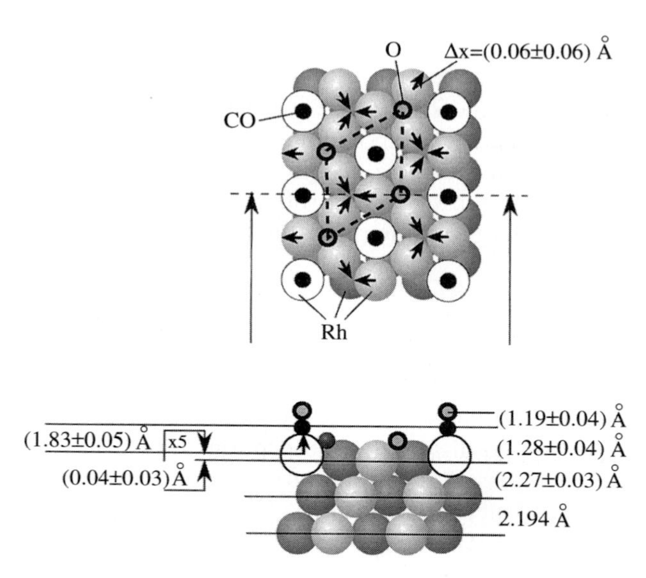

FIG. 27. The structure of the composite O + CO (2 × 2) phase on Rh(111) as analyzed by LEED (*154*).

CO). In the case of dissociative adsorption, the decrease of adsorption energy is paralleled by a rise of the activation barrier, and then the sticking coefficient may decrease much more strongly with coverage.

An example of the influence of only weak ($\varepsilon \lesssim 20$ meV) interactions on the lateral configuration of a two-component adsorbate phase is represented by the STM images of Fig. 28 (*156*). Equal amounts of chemisorbed N and O atoms on Ru(0001) (created by dissociative adsorption of NO) lead to the formation of a two-phase system consisting of a dense intermixed 2 × 2 phase (in which the O atoms appear darker than the N atoms so that the two species are distinguishable) and a diluted lattice gas in which only the N atoms are imaged since diffusion of the O atoms is too rapid to allow imaging. With increasing total concentration (Fig. 28a to 28c), the fraction of the surface covered by the dense phase grows at the expense of the dilute phase. Even without further details, this example clearly shows how the operation of interactions between adsorbed particles may cause pronounced deviations from the random uniform distribution underlying the mean-field approach of the Langmuir concept generally used for modeling the kinetics of bimolecular surface reactions. This discrepancy between the

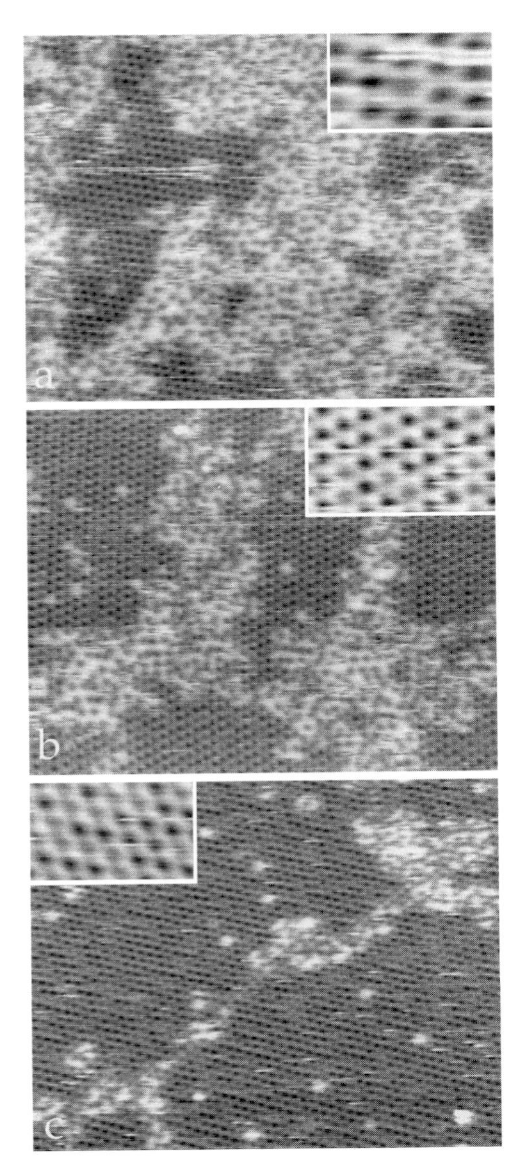

FIG. 28. A sequence of STM images with increasing concentration of N + O (with constant 1 : 1 ratio) adsorbed on a Ru(0001) surface at 300 K. The insets show details of the dense 2 × 2 phases in which the O atoms appear black and the N atoms gray (*156*).

observations and the Langmuir model becomes particularly relevant if attractive interactions cause the formation of adsorbate islands, the perimeters of which are the actual regions of reaction, as illustrated later by the oxidation of CO on Pt(111).

If the corrugation of the adsorption potential across the surface is only weak (as is frequently the case, for example, with adsorbed CO, for which there are only minor differences in the adsorption energies from site to site), then the interactions may cause a breakdown of the lattice gas concept. Either different high-symmetry sites (e.g., on-top, bridge, and threefold) become simultaneously occupied, or the registry with the substrate lattice is largely lost and incommensurate phases are formed. For example, Fig. 29 shows the phase diagram for the system CO/Pt(111) as a function of temperature and coverage (*157*). The c(4 × 2) phase around $\theta = 0.5$ contains CO molecules in on-top and bridge positions, and the strong repulsive

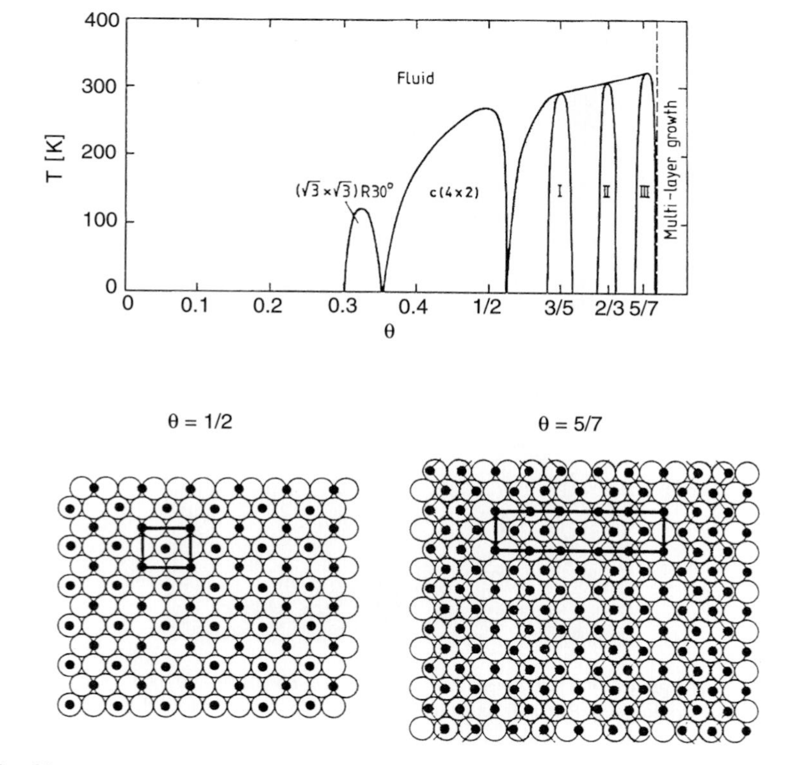

FIG. 29. Phase diagram for the CO/Pt(111) system together with structure models formed at low temperatures with $\theta = 1/2$ and $\theta = 5/7$ (*157*).

interactions at short distances force the molecules onto unsymmetric sites, also (as with phase III at $\theta = 5/7$) without reaching $\theta = 1$. Theoretical aspects of ordered structures and phase transitions in adsorbed layers are addressed by Persson (*158*) and Patrykiejew *et al.* (*159*).

Interactions between adsorbed particles also exert marked influences on their mobility. In the system O/Ru(0001), the mean residence time at 300 K varies from 60 ms for an isolated adatom to 220 ms when a second particle is located two lattice constants a_o apart from the first; this residence time is 13 ms when this separation distance is only $\sqrt{3}\, a_o$ (*1*). At not too small total coverages, attractive interactions cause the formation of a dense 2×2 phase in equilibrium with a dilute lattice gas, the atoms of which continuously condense at the islands, from which they evaporate again. The definition of a (microscopic) single-particle diffusion coefficient then becomes meaningless, all the more so if one considers that ubiquitous defects will act as intermediate traps for particles migrating over longer distances. It is believed that surface diffusion of adsorbates belongs to the most complex phenomena of dynamics in surface reactions, for which proper consideration of the relevant length scale is of decisive importance.

Another aspect concerns modification of the dissociation probability of an adsorbed molecule resulting from interaction with other adsorbates. The well-known promoter effects of adsorbed alkali metal atoms in catalysis fall into this category, as discussed in detail elsewhere (*160*). However, such an effect may be exerted even by the same kind of adsorbate. STM investigations showed that on Pt(111) the dissociation probability of an O_2 molecule becomes affected by the vicinity of already present O_{ad} species (*161*). Thus, a dynamic heterogeneity is introduced which leads to kinetically limited ordering of the adsorbate if the temperature is sufficiently low.

The LH process $O_{ad} + CO_{ad} \rightarrow CO_2$ on Pt(111) is the first example of a chemical reaction for which the rate was directly determined on the *atomic scale* (*162*). (The STM data are described in more detail in the chapter by J. Wintterlin, this volume.) Dissociative chemisorption of oxygen led to the formation of an O 2×2 phase, which was subsequently exposed to CO, and the progress of the reaction was monitored by recording a series of STM images. CO was adsorbed inside the 2×2 O unit cell to give a composite O + CO phase of the same type as shown in Fig. 27 for Rh(111). Interestingly, the reaction at 247 K did not occur uniformly across the whole $O_{ad} + CO_{ad}$ layer, but instead propagated from the boundaries of this phase with the adjacent pure c4×2 CO phase, which nucleated on empty patches and grew continuously with progress of the reaction. By this process, free sites were created and became occupied by CO that was continuously adsorbed from the gas phase. The atomic reaction rate was determined by counting the number of adsorbed O atoms, $r = -dN_O/dt$, and it was found to be proportional to the lengths of the domain boundaries, $r = kL$, rather than to the product of the densities of adsorbed O and adsorbed CO, $r' = k'N_ON_{CO}$, as would be expected for a LH reaction between randomly distributed adsorbates. Analysis of the data in terms of a LH process revealed, however, that over a limited range of intermediate coverages the ratio r'/N_ON_{CO} is approximately constant. Hence, LH kinetics is still a reasonably good approximation, and it becomes even better at higher temperatures, as thermal disorder assists in randomization of the adlayer. Measurements at different temperatures indicated for k an activation energy of 0.5 eV, which is identical with the *macroscopic* value at high coverages as determined in earlier molecular beam experiments (*163*). The latter study had already shown that the rate constant depends on coverage. At low coverages (i.e., when the interaction is between isolated O_{ad} and isolated CO_{ad}), the activation energy reaches a value of about 1 eV.

These findings are readily understood in view of the results of a recent theoretical treatment (*164*). Figure 30 shows a series of snapshots of the reaction pathway, starting from on-top CO and a nearby threefold-coordinated O atom as with the structure of the composite O + CO phase (Fig.

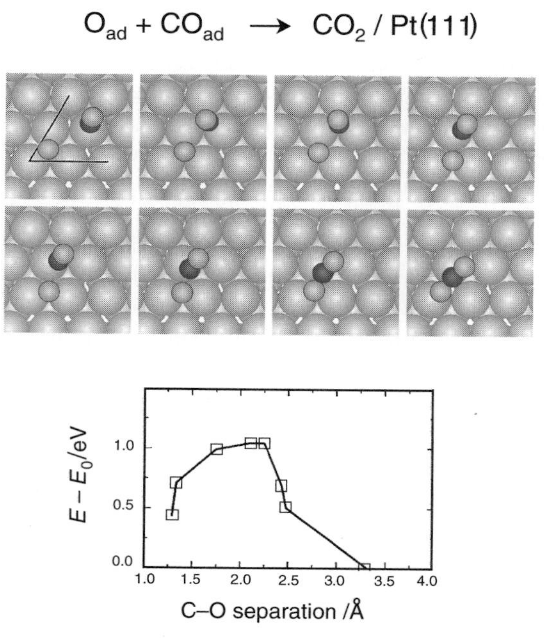

FIG. 30. Theoretical treatment of the $O_{ad} + CO_{ad}$ reaction on Pt(111) (*164*). Variation of the energy $E - E_o$ with separation between O_{ad} and C (from CO_{ad}) referred to large distance.

27) or when an individual CO molecule approaches an isolated O_{ad} atom. The calculated energies of these different configurations are plotted in the lower part of Fig. 27 as a function of the C–O distance and indicate a total activation barrier of 1 eV. However, if the reaction started from the second configuration in which the CO molecule is in a bridge position, a barrier of only 0.5 eV would have to be surmounted. This is the situation at the island boundaries, where some of the bridging CO molecules from the $c4 \times 2$ phase are close to O atoms. This explains why the reaction proceeds preferentially along the phase boundaries and why the on-top CO molecules inside the 2×2 O_{ad} mesh are less reactive; it also accounts for the fact that this low activation energy is only reached at coverages high enough to form the compressed CO phase.

This example demonstrates that even with a very simple reaction on a well-defined single-crystal surface, the assumptions underlying LH kinetics are not fulfilled. Nevertheless, satisfactory modeling within the LH framework can still be achieved if variations of the kinetics parameters with coverage are properly taken into account and if the analysis is restricted to certain ranges of parameters.

C. The Role of Defects: "Active Sites"

Even a clean and well-defined single-crystal surface is never absolutely perfect; rather, it contains atomic defects such as steps and dislocations which represent adsorption sites with altered coordination and hence energetics and kinetics different from those of a perfect crystal. For example, for adsorption of C_2H_2 on Pt(111), adsorption energies of 280 kJ/mol at steps and of 210 kJ/mol on terrace sites as well as sticking coefficients of 1 and 0.8 were determined (*165*). The concept of "active sites" was introduced by Taylor in 1925 (*166*) and reflects the experimental evidence that catalytic surfaces are frequently not uniformly active, which, led to the classification of structure-sensitive and structure-insensitive reactions (*167, 168*). In Schwab's "Adlineationstheorie" (*169*), it was speculated that one-dimensional defects such as those presented by atomic steps are of essential significance—a view which was later confirmed in single-crystal studies (*170*). An STM study of the dissociative chemisorption of NO on Ru(0001) enabled direct identification of the top metal atoms of atomic steps as the active sites for this reaction (*171*). After exposure to NO, the adsorbed atoms were found close to the steps, and the density of adatoms varied with the widths of the adjacent terraces. Hence, it was inferred that NO molecules trapped on the terraces rapidly diffuse across the surface and eventually become dissociated at steps. In addition, the hcp Ru(0001) surface offers two types of alternating steps. As illustrated in Fig. 31, a metal

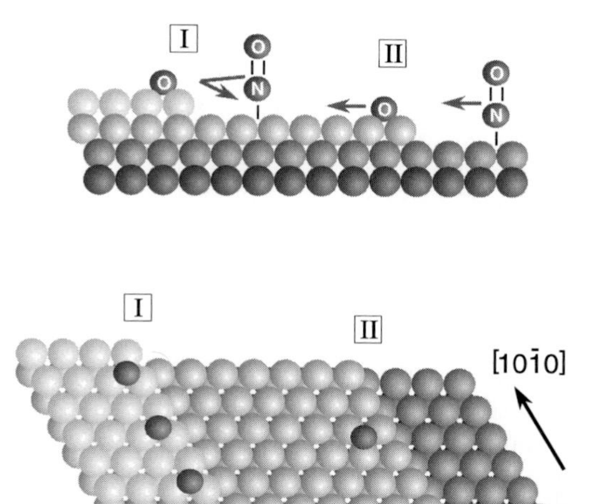

FIG. 31. The role of monoatomic steps on a Ru(0001) surface as active sites in dissociating NO. O atoms formed at step I are strongly held there and inhibit further reaction, whereas they diffuse away from step II (*171*).

atom at a type I step has one neighbor at the bottom of the step, whereas at a type II step it has two. O atoms are always located in threefold sites with another Ru atom underneath; at type I steps each of these sites includes two (coordinatively less saturated) front-row Ru atoms, but each of these sites at a type II step includes only one. Consequently, bonding at the former is stronger, and the site becomes blocked for further dissociation. The weaker bonding at type II steps, on the other hand, permits the O atoms to diffuse away so that this step retains its activity and is the actual active site.

A particularly striking example of the role of steps as active sites in a catalytic reaction was recently identified for the dissociative chemisorption of N_2 on Ru(0001) (*172*). This process is regarded as the decisive step in ammonia synthesis on Ru catalysts. Measurements with supported Ru catalysts indicated an activation energy of about 0.5 eV for this process, and extrapolation to 300 K led to a sticking coefficient of the order of 10^{-12}, as was determined from single-crystal experiments (*173–175*). In the cited work (*172*), these findings were confirmed, and from measurements at different temperatures an activation energy of 0.4 eV was determined. However, deposition of about 1% Au on the Ru surface led to a dramatic change. The activation energy increased to about 1.3 eV, and the sticking coefficient decreased to almost immeasurably small values. These latter data are considered to be representative of the terrace sites of the Ru(0001) surface on which the step sites are blocked by the small concentration of Au atoms. As confirmed by DFT calculations, in this case too the steps form the active sites, whereas the terrace sites are practically inactive. Most likely, similar effects also hold for N_2 chemisorption on the low-index planes of Fe, whereas the more open Fe(111) surface is indeed the active plane (*176*).

As a general conclusion, one should keep in mind that bond breaking is generally facilitated by defects, so that the overall kinetics of a catalytic reaction might be governed by the properties of these active sites rather than by those of the low-index crystal planes.

D. Surface Reconstruction

Since the bond between the adsorbate and the surface atoms is often of comparable strength to that between the surface atoms themselves, restructuring of the surface under the influence of adsorbates is a quite general phenomenon, the relevance of which for surface reactions has already been extensively emphasized (*177, 178*).

For example, consider CO adsorption on Pt(110); the influence of the adsorbed CO on the surface structure is also the key for the mechanism in kinetic oscillations during CO oxidation on this surface at low pressures. The clean Pt(110) surface is reconstructed into a missing-row 1×2 structure, as depicted in Fig. 32, and it transforms into the bulk-terminated 1×1 structure under the influence of adsorbed CO. [Similarly, the hexagonal configuration of the topmost atoms in the clean Pt(100) surface (hex) is transformed into the 1×1 phase if a critical CO coverage is reached, whereas the most densely packed (111) surface is not reconstructed and hence not affected by CO adsorption under ordinary conditions.] STM studies revealed that this structural transformation proceeds by local nucle-

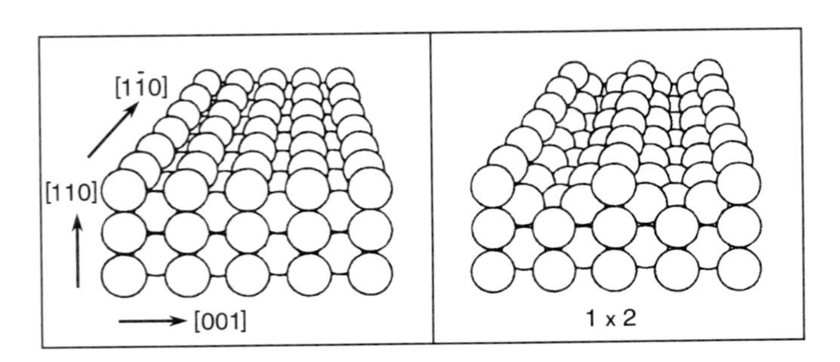

Fig. 32. Structure of the nonreconstructed and the reconstructed Pt(110) surfaces.

ation and displacements of the topmost Pt atoms over distances of only a few lattice constants without the need for mass transport over long distances whereby patches of the second layer with (bulk) 1 × 1 structure are exposed (*180*). Infrared measurements showed that CO occupies exclusively on-top sites on this surface without any change of the C–O stretch frequency associated with the 1 × 2 → 1 × 1 transition (*181*). At 90 K, the structural transformation is frozen so that CO molecules are also adsorbed in the troughs of the 1 × 2 structure (but with lower adsorption energy). Hence, as with CO/Pt(100) (*182*), surface restructuring occurs as soon as the gain in adsorption energy exceeds the energy difference between the two clean surface phases. Since the structural transformation proceeds through a nucleation and growth mechanism (as is common for first-order phase transitions), the kinetics of this process is generally not linear with the coverage of the adsorbate driving the transition. Molecular beam studies with CO/Pt(100) showed that the 1 × 1 island growth rate varies with an apparent reaction order of about 4.5 with respect to the local CO coverage on the hex phase, and consequently the net sticking coefficient becomes strongly dependent on the flux of impinging CO molecules (*183, 184*). Similar effects were observed with a series of other systems (*185–188*) so that it must be concluded that the concept of a sticking coefficient independent of the partial pressure (which underlies usual formulations of adsorption kinetics) becomes questionable whenever effects of adsorbate-induced surface structural transformations come into play.

E. Reactions Involving Subsurface Species

Adsorbate-induced restructuring affects not only the binding of the surface atoms but also the reactivity of the adsorbates. Oxygen chemisorption on Cu(110), for example, causes the formation of Cu–O–Cu chains in an added row structure (*189, 190*) in which the O atoms are much less reactive in CO oxidation than if present as a metastable overlayer at low temperatures on the unreconstructed surface (*191*).

Dissolution of particles in the bulk often represents a reservoir for adparticles through segregation back onto the surface. In addition, the altered coordination of the surface atoms may cause different binding properties for species located directly below the topmost layer, and such subsurface oxygen atoms have, for example, been found to play a significant role in the catalytic oxidation of CO on Pt(110) in the presence of excess O_2 (*112*).

A particularly interesting role in this context is played by subsurface hydrogen. Johnson *et al.* (*192*) found that on Ni(111) subsurface H atoms could hydrogenate adsorbed CH_3 to CH_4, whereas surface hydrogen could not. It was suggested that H atoms embedded into the Ni lattice directly

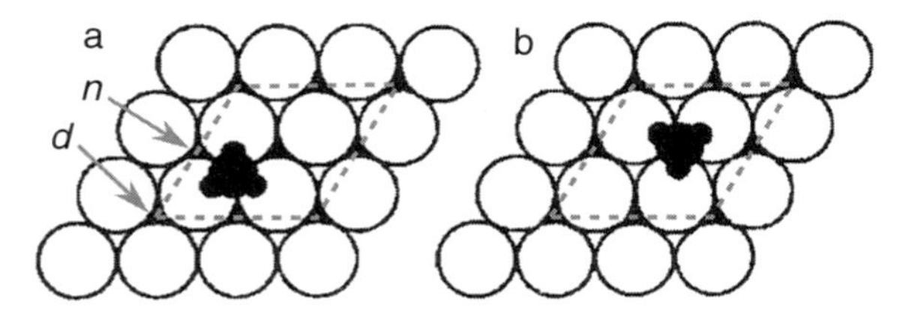

FIG. 33. Geometries for the theoretical treatment of the reaction between adsorbed methyl (a) and subsurface H (b) on Ni(111) (*196*).

below the methyl species may have a favorable reaction pathway not available to surface hydrogen. Similar effects were found with hydrogenation reactions of other hydrocarbons (*193–195*), and recent theoretical work (*196*) shed some light on the mechanism of this process. The treated geometries are shown in Fig. 33; in Fig. 33a the methyl is located on a hcp hollow site with H atoms below the surface, either in neighboring (*n*) or diametrically opposite (*d*) sites, whereas in Fig. 33b the CH_3 species sits on a fcc hollow site directly above a subsurface H atom. Reaction along the latter path [as suggested in the original work (*192*)] turned out to be energetically rather unfavorable, specifically, to exhibit an activation energy of 1.4 eV. Instead, the most favorable reaction pathway was identified as follows: An H atom in subsurface site *n* is energetically higher by about 0.5 eV relative to a particle adsorbed *on* the surface. To squeeze it out to the surface at the marked location, a barrier of 0.56 eV has to be overcome. The CH_3 species moves to the top site of one of the shared Ni atoms with only slight displacement of the H from where recombination requires only about 0.1 eV. This means that once the H atom in position *n* has moved to the surface, formation of CH_4 takes place practically instantaneously. It is indeed the activation barrier for bringing the metastable subsurface H atom on the *n* site back to the surface that determines the rate-limiting step. Since the transition state is similar to that for the reaction $H_{ad} + CH_{3,ad}$, the enhanced reactivity of H_{sub} has to be sought in its higher initial energy.

This is another manifestation of the fact that energetically higher, metastable surface species (such as the transient hot atoms discussed previously) also exhibit higher reactivity.

IV. The Mesoscopic Level

Knowledge of the kinetics and energetics of the elementary processes of surface reactions on the atomic scale forms the basis for modeling the macroscopic kinetics. Usually, it is assumed that the limited geometry of the surface area introduces no additional complications. Small particles exposing different crystal planes are then simply regarded as presenting a superposition of the contributions from different structural elements. Furthermore, it is commonly believed that under steady-state flow conditions the distribution of surface intermediates is constant, independent of time and location. Although these assumptions will generally form a reasonable approximation, closer inspection reveals that additional effects come into play that manifest themselves on the *mesoscopic* level (and are hence usually hidden) but may affect the *macroscopic* kinetics.

A conceptually simple situation is encountered if one considers a catalytic reaction proceeding with different kinetics on the facets of a small catalyst particle on which the adsorbed species may diffuse across the different crystal planes. Monte Carlo simulations of this effect were performed for the model reaction $2A + B_2 \rightarrow 2AB$ occurring on the two adjacent planes of a model particle as shown in Fig. 34 (*197*). It was assumed that B_{ad} was immobile, whereas particle A_{ad} diffused much faster than it reacted via $A_{ad} + B_{ad} \rightarrow AB$. Although the sticking coefficient of A was the same on both planes, it differed by a factor 10 for B_2. Figure 34a shows the variation of the reactivity with the relative partial pressure (p) of A for the two facets without diffusion over the facet boundary, and Fig. 34b demonstrates how the kinetics change completely if rapid diffusion across this boundary occurs. It is no longer a superposition of the properties of the two planes; instead, the activity is shifted into a new range of external parameters.

Since practical supported catalysts are relatively inaccessible to surface science techniques, several approaches to model studies were developed. In the current context, the use of electron beam lithography to produce controlled arrangements of mesoscopic catalyst particles appears to be an interesting technique (*197, 198*), but the results obtained are still scarce. Another possibility involves studying the progress of a reaction *in situ* on the tip of a field ion microscope (FIM). Figure 35 shows a sequence of FIM images from a Pt tip with a 180-nm radius under steady-state flow conditions for the catalytic oxidation of H_2 obtained by use of O_2 and H_2O as imaging gases (*199*). In the center, in the region of the (*100*) plane, a zone of high reactivity (which appears brighter due to imaging by H_2O^+ ions) nucleates and rapidly spreads across the whole surface (Fig. 35b–e). Subsequently, restructuring of the surface and build-up of a H_{ad} layer occur, whereby the surface becomes less reactive. However, the (*331*) zones adjacent to the central (*100*) region are still active and provide sources for concentration waves propagating toward the center, as becomes evident from the shrinking of the darker H_{ad}-covered central part, until after about 30 s the initial situation is restored and the periodic reaction cycle starts again. Parallel

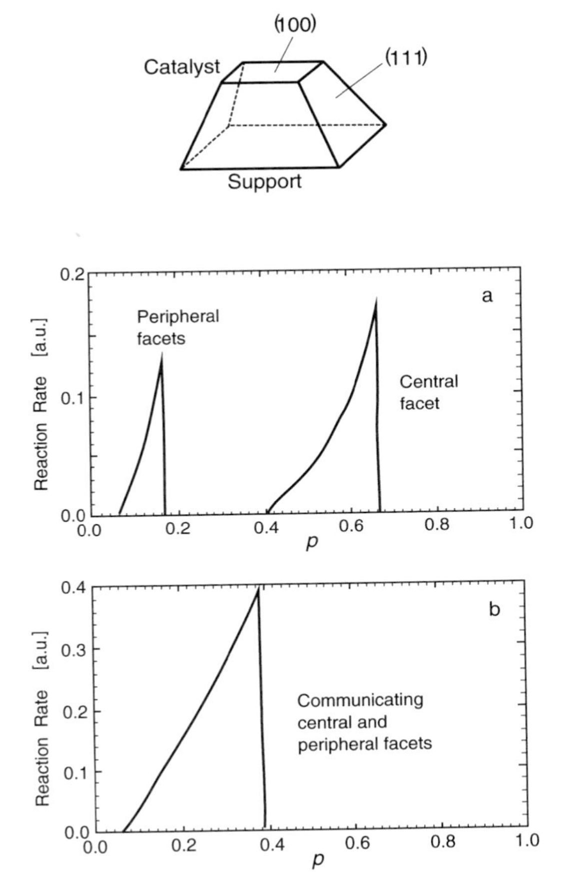

FIG. 34. Modeling the influence of adsorbate diffusion on the overall kinetics of reaction on a small catalyst particle exposing two different kinds of planes (*197*): (a) without diffusion; (b) with rapid diffusion across the plane boundaries.

experiments with an extended Pt(100) single-crystal surface exhibited constant steady-state reaction rates, whereas with the tip coupling of the (*100*) plane to adjacent facets caused the formation of propagating reaction waves and a profound modification of the overall kinetic behavior.

It is emphasized that this effect is fundamentally different from the concept of spillover. By contrast, it is the coupling between diffusion and reaction of adsorbed species that gives rise to the formation of reaction fronts which may propagate much faster than simple spreading of a concentration gradient by diffusion alone.

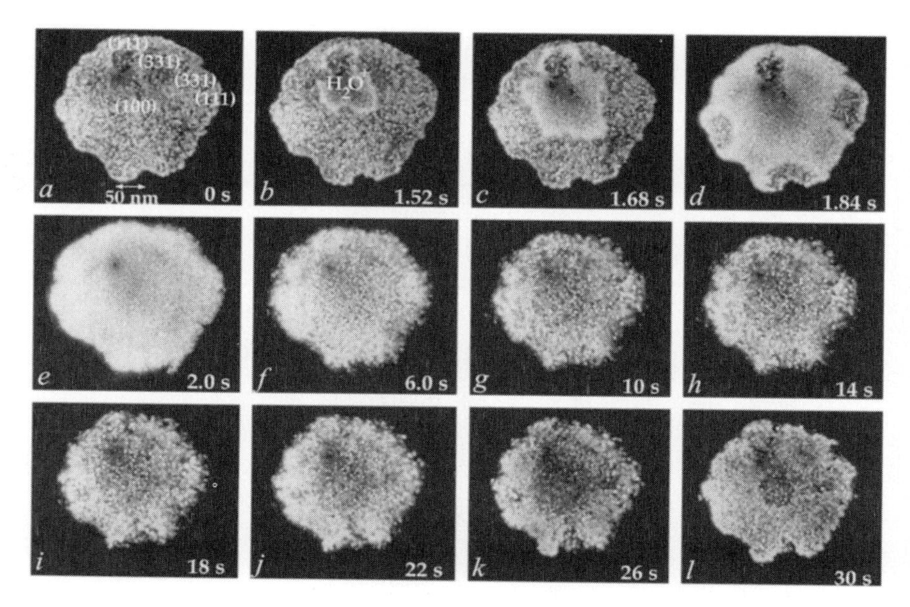

FIG. 35. A series of images from a Pt tip recorded *in situ* by field ion microscopy during catalytic oxidation of H_2 under steady-state conditions: $T = 300$ K, $pH_2 = 6 \times 10^{-4}$ mbar, $pO_2 = 5 \times 10^{-4}$ mbar (*199*).

The propagation of reaction waves again with H_2 oxidation was recently observed with atomic resolution by means of STM (*200*). Exposing an O-covered Pt(111) surface to H_2 caused first nucleation of OH_{ad} patches which grew with continuing hydrogen coverage until they covered almost the whole imaged area of about 20×20 nm^2. Within these islands, H_2O subsequently formed. At low temperatures ($\lesssim$150 K) water does not desorb, but instead it may react with neighboring O_{ad} according to $O_{ad} + H_2O_{ad} \rightarrow 2OH_{ad}$, followed by $OH_{ad} + H_{ad} \rightarrow H_2O_{ad}$, whereby an autocatalytic mechanism is created. Interestingly, on a mesoscopic scale (e.g., ~100 nm), the distribution of surface species is no longer uniform, but a reaction front with about 20-nm thickness consisting essentially of OH_{ad} propagates into the neighboring O_{ad}-rich area, leaving the reaction product H_2O_{ad}. This is manifested as shown in Fig. 36, in which the area with the small bright dots at the right of the image represents the largely unreacted O_{ad} phase with small bright OH_{ad} islands, and the large patches at the left arise from an H_2O_{ad} island (top) and mixed H_2O_{ad}/O_{ad} islands (bottom). The speckled region in between marks a zone with a high density of OH_{ad} which propagates as reaction front from left to right across the atomic steps. These

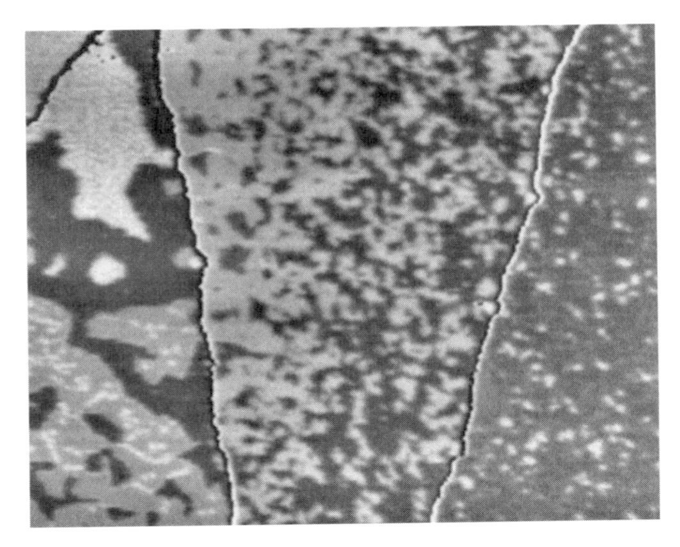

FIG. 36. STM snapshot from a Pt(111) surface (165 × 142 nm^2) during reaction of a preadsorbed O 2 × 2 structure with hydrogen at 131 K (*200*).

effects are clearly beyond the atomic scale and mark the transition to the mesoscopic level.

Effects of this type are always to be expected if the kinetic equations modeling the reaction mechanism are nonlinear, and hence these phenomena have their theoretical basis in the general framework of nonlinear dynamics. One important consequence is that the macroscopic kinetics under steady-state flow conditions is not necessarily constant but may become oscillatory or even chaotic, as has already been extensively reviewed (*201–206*). The most detailed studies were performed to characterize the catalytic oxidation of CO on Pt(110) and Pt(100). Both surfaces undergo reconstruction under the influence of CO adsorption, and the associated variation of the sticking coefficient gives rise to the mechanism of oscillatory kinetics at low pressures (*206*). This original model has been refined by taking into account the additional possibility for subsurface oxygen formation (*207*) and of the actual (nonlinear) kinetics of the CO-driven surface structural transformation (*208*).

Theoretical modeling of these phenomena has to take into account not only the temporal variation of the densities of surface species but also the fact that their lateral distribution might be nonuniform. Description in terms of a *continuum model* is usually performed in the framework of mean-field approximations. The surface concentrations (partial coverages,

θ_i) are then dependent on both time and spatial coordinates. The resulting partial differential equations consist of terms describing the kinetics as well as the transport processes. Under isothermal conditions, the latter consist of diffusion steps (driven by gradients in the partial coverages), and the characteristic length scales of the solutions of these reaction–diffusion equations are determined by a combination of rate constant and diffusion coefficient (*209*).

The dimensions of these patterns are frequently in the micrometer range and can most conveniently be investigated with photoemission electron microscopy, provided that the total pressure is lower than 10^{-4} mbar (*210*). For studies at higher pressures, appropriate optical imaging techniques have been developed (*211*). For example, Fig. 37 shows two images recorded

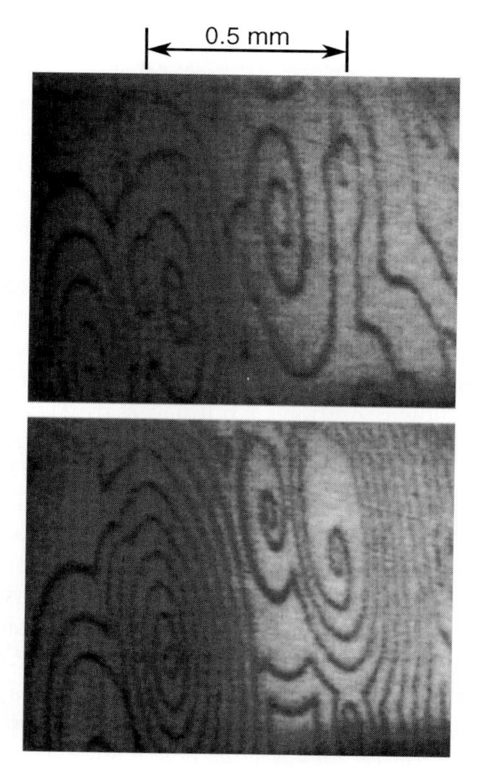

FIG. 37. The formation of propagating concentration patterns on a Pt(110) surface during the oxidation of CO under isothermal flow conditions as imaged by ellipsometric microscopy with an interval of 26 s. $T = 480$ K, $p_{O_2} = 1 \times 10^{-2}$ mbar, $p_{CO} = 1 \times 10^{-3}$ mbar (*212*).

in an interval of 26 s by means of ellipsometric microscopy from a Pt(110) surface during catalytic CO oxidation at 480 K with 1×10^{-2} mbar of O_2 and 1×10^{-3} mbar of CO as steady-state pressures, respectively; these images exhibit the propagation of spiral waves (*212*).

An example of the successful theoretical modeling within the framework of the cited continuum-level description in terms of appropriate coupled partial differential equations is the formation of standing wave patterns in the CO oxidation on Pt(110) associated with sustained kinetic oscillations (*213*). In addition to equations for the coverages of adsorbed O and CO, variables expressing the concentration of subsurface oxygen, and the degree of $1 \times 2 \rightarrow 1 \times 1$ phase transformation were taken into account, with all the parameters derived from experimental results. The results of the calculations reproduce the experimental observations of pattern formation fairly well. These include characteristic time scales on the order of 1 s and length scales on the order of 100 μm.

These values point to the main dilemma that is encountered if an alternative approach based on a microscopic view is adopted: Instead of working with continuum variables within a mean-field approximation, Monte Carlo simulations based on the properties of individual adsorbed particles can be performed. Several studies along these lines have been reported (*214–216*). Although the models are able to reproduce some of the observed qualitative features, quantitative modeling is still restricted by the problem of scaling up to realistic conditions.

The formation and propagation of concentration patterns on a mesoscopic scale are affected by the finite size of uniform crystal planes in polycrystalline samples. In practial catalysts, the supported particles are usually smaller than the characteristic dimensions of the (isothermal) patterns. However, nonuniform distributions of surface species may then be caused by coupling of reaction with heat conduction since local differences in reactivity will then give rise to temperature gradients (*217*).

The influence of the boundary conditions can be studied by creating microstructures of appropriate dimensions on a single-crystal surface by lithographic techniques (*218*). In addition to depositing inactive areas (e.g., Ti on Pt), domains with different reactivities [e.g., Pd on Pt (*219*) or submonolayers of Au on Pt (220)] have also been prepared. Investigations such as these serve not only to characterize the fundamental problems associated with the coupling of surface reaction and diffusion but also they allow exploration of novel possibilities for controlling activities and selectivities of catalytic systems.

Recently, the creation of localized structures on catalytic surfaces by coupling of the reaction with diffusion and adsorbate-induced structural transformation has been discovered theoretically (*221*). Such self-organized microreactors were found, under certain conditions, to develop spontaneously, exhibiting circular shapes, and having sizes ranging from nanometer to submicrometer scales. Although the reaction occurs everywhere on the surface, it would proceed predominantly inside the spots in which the adsorbed species are enriched. It will be interesting to determine whether experimental verification of this concept may be achieved.

V. Macroscopic Kinetics

The preceding sections demonstrated the complexity of the dynamics on various levels involved with reactions at solid surfaces. Nevertheless, attempts to model the kinetics of an actual catalytic reaction under practical conditions are usually based on concepts derived from the original Langmuir theory of adsorption. The resulting kinetics were first developed by Hinshelwood (*222*) (hence, the term Langmuir–Hinshelwood kinetics) and then adopted by Hougen and Watson (*223*) for technological process development. For the LH reaction $O_{ad} + CO_{ad} \rightarrow CO_2$ on Pt(111), for the first time a direct comparison of the macroscopic kinetics (as derived from molecular beam experiments) with the microscopic kinetics (resulting from analysis of STM data) was achieved, as summarized previously (*162*). It turned out that the LH concept is wrong insofar as the reaction does not occur between randomly distributed adparticles but instead at the boundaries of adsorbate islands. Nevertheless, LH kinetics was shown to be a fairly good approximation over a limited range of surface coverages. It is believed that similar situations are frequently encountered even with technological reactions, provided that the analysis is restricted to not too extended ranges of external parameters.

In addition to all the complications on the quantum level, the assumption of kinetic and energetic parameters independent of coverage is probably the most severe simplification. Even with reactions on well-defined single-crystal surfaces, defects and interactions between the adsorbed particles render such an approach questionable, as was clearly realized in experimental investigations of CO oxidation on Pd and Pt single-crystal surfaces (*59, 163*). In a recent detailed theoretical analysis (*224*) of this reaction on Pt, it was demonstrated that satisfying modeling of the reaction at pressures up to atmospheric can be achieved if, in addition to coverage dependence of the rate of CO desorption within a mean-field approximation, repulsive CO–CO interactions are explicitly taken into account. In this way, an earlier model that was successfully applicable to low pressures (*225*) was extended to the high-pressure regime. Similarly, the kinetics of this reaction on

Rh(111) were simulated successfully up to atmospheric pressures by proper adjustment of kinetic parameters determined from UHV experiments (*226*).

The situation becomes even more complex if practical catalysts with polycrystalline surface structures are considered. However, surprisingly, the LH concept (probably with certain modifications) may still be successfully employed. The classical example is that of ammonia synthesis, for which the famous Temkin–Pyzhev equation relating the rate to the external partial pressures was derived 60 years ago (*227*) on the basis of the assumption that adsorption of N_2 is rate determining and that nitrogen adsorption can be described by an isotherm according to which the heat of adsorption changes linearly with coverage. In the meantime, the elementary steps underlying this important industrial process have been elucidated in detail (*228, 229*), including the nature of the promoted Fe-based catalyst (*230*) and the particular role played by the Fe(111) surface (*19*). On the basis of this information, microkinetic modeling was performed by several groups.

Stoltze and Nørskov (*231, 232*) assumed that dissociative N_2 adsorption is rate-limiting and that all other steps are in equilibrium, whereby Langmuir isotherms served to correlate the coverages of the various surface species with the respective partial pressures. This approach revealed surprisingly good agreement between the theoretical predictions and experimental results recorded at pressures up to 304 bar, i.e., more than nine orders of magnitude higher than the surface science data used as input for the modeling.

Bowker *et al.* (*233, 234*) constructed a similar model without the *a priori* assumption of a rate-determining step (which, however, with the appropriate input parameters turned out to be dissociative N_2 adsorption). Although the agreement with experiment was again very good, it was concluded that there were serious discrepancies. This problem was satisfactorily resolved, however, in another microkinetics study (*235*). Further improvement and confirmations of the original Stoltze–Nørskov model were obtained later (*231, 236*). A careful analysis of the various kinetic models including the original Temkin–Pyzhev concept led to the conclusion that all of them provided fairly good agreement with experiment by proper adjustment of some of the input parameters (*237*). This result is attributed to the fact that under industrial conditions the surface is largely covered by atomic nitrogen so that the overall kinetics is essentially determined by a single rate constant.

This conclusion confirms the general experience that usually more than one set of kinetic parameters are able to simulate the experimental results and that a controversial mechanism can never be resolved on the basis of kinetics measurements alone.

In recent years, for ammonia synthesis the properties of Ru-based catalysts as alternatives for the classical Fe catalyst (*238*) have been explored in detail. With single-crystal surfaces, adsorption was investigated at low pressures (*175*), and the rate of ammonia synthesis was determined at various temperatures and at a total pressure of 2 bar (*239*). The kinetics of the reaction on MgO-supported, Cs-promoted Ru catalysts (*240*) was investigated in detail at higher pressures, and modeling was performed by using single-crystal information on the reaction mechanism (*241*). As shown in Fig. 38, good agreement between the model and experimental results at different gas compositions and total pressures was obtained (*241*). These results demonstrate the general power of microkinetics simulations in technological catalysis, notwithstanding, the numerous limitations and complexities involved with the elementary steps.

Other examples for which the transfer of kinetic data from single-crystals to high-surface-area catalysts were successfully performed concern the methanation reaction on Ni (*242*) and the synthesis of methanol on Cu (*243–245*).

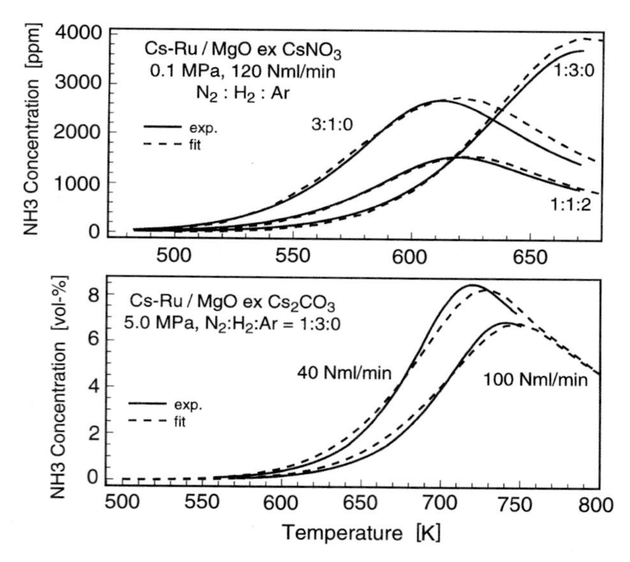

FIG. 38. Experimental results characterizing the rate of NH_3 formation on a Cs-promoted Ru catalyst supported on MgO as a function of temperature at different total pressures and gas compositions (solid lines) in comparison with the results of microkinetics modeling with parameters derived from surface science studies (dashed lines) (*241*).

VI. Conclusions

The philosophy underlying the surface science approach to heterogeneous catalysis was formulated by Langmuir (*145*) in 1922, as follows:

> Most finely divided catalysts must have structures of great complexity. In order to simplify our theoretical consideration of reactions at surfaces, let us confine our attention to reactions on plane surfaces. If the principles in this case are well understood, it should then be possible to extend the theory to the case of porous bodies. In general, we should look upon the surface as consisting of a checkerboard.

It may be concluded that substantial progress has been made, but the actual situation is far more complex and still offers demanding challenges for future research.

References

1. Renisch, S., Schuster, R., Wintterlin, J., and Ertl, G., *Phys. Rev. Lett.* **82,** 3839 (1999).
2. Wintterlin, J., Schuster, R., and Ertl, G., *Phys. Rev. Lett.* **77,** 123 (1996).
3. Stipe, B. C., Rezaei, M. A., and Ho, W., *J. Chem. Phys.* **107,** 6443 (1997).
4. Robota, H. J., Vielhaber, W., Lin, M. C., Segner, J., and Ertl, G., *Surf. Sci.* **155,** 101 (1985).
5. Bonn, M., Funk, S., Hess, C., Denzler, D., Stampfl, C., Scheffler, M., Wolf, M., and Ertl, G., *Science* **285,** 1042 (1999).
6. Michelsen, H. A., Rettner, C. T., and Auerbach, D. J., *in* "Surface Reactions" (R. J. Madix, Ed.), p. 123. Springer, New York, 1993.
7. Rettner, C. T., Michelsen, H. A., and Auerbach, D. J., *J. Chem. Phys.* **102,** 4625 (1995).
8. Rendulic, K. D., and Winkler, A., *Surf. Sci.* **299/300,** 261 (1994).
9. Wetzig, D., Rutkowski, M., David, R., and Zacharias, H., *Europhys. Lett.* **36,** 31 (1996).
10. Hou, H., Gulding, S. J., Rettner, C. T., Wodtke, A. M., and Auerbach, D. J., *Science* **277,** 80 (1997).
11. Gross, A., *Surf. Sci. Rep.* **32,** 291 (1998); Darling, G. R., and Holloway, S., *Rep. Prog. Phys.* **58,** 1595 (1995).
12. Ceyer, S. T., *Annu. Rev. Phys. Chem.* **39,** 479 (1988).
13. Lee, M. B., Yang, Q. Y., and Ceyer, S. T., *J. Chem. Phys.* **85,** 1693 (1986); **87,** 2724 (1987).
14. Yates, J. T., Zinck, J. J., Sheard, S., and Weinberg, W. H., *J. Chem. Phys.* **70,** 2266 (1979).
15. Hou, H., Huang, Y., Gulding, J. S., Rettner, C. T., Auerbach, D. J., and Wodtke, A. M., *Science* **284,** 1647 (1999).
16. Schlögl, R., *in* "Handbook of Heterogeneous Catalysis" (G. Ertl, H. Knözinger, and J. Weitkamp, Eds.), p. 1697. VCH, Weinheim, 1997.
17. Ertl, G., Lee, S. B., and Weiss, M., *Surf. Sci.* **114,** 515 (1982).
18. Rettner, C. T., and Stein, H., *Phys. Rev. Lett.* **59,** 2768 (1987).
19. Mortensen, J. J., Hansen, L. B., Hammer, B., and Nørskov, J. K., *J. Catal.* **182,** 479 (1999).
20. Puglia, C., Nilsson, A., Hernnas, B., Karis, O., Bennich, P., and Martensson, N., *Surf. Sci.* **342,** 119 (1995).
21. Nolan, P. D., Lutz, B. R., Tanaka, P. L., Davis, J. E., and Mullins, C. B., *Phys. Rev. Lett.* **81,** 3179 (1998).
22. Zangwill, A., *in* "Physics at Surfaces." Cambridge Univ. Press, Cambridge, UK (1988).
23. Barker, J. A., and Auerbach, D. J., *Surf. Sci. Rep.* **4,** 1 (1985).

24. Mödl, A., Gritsch, T., Budde, F., Chuang, T. J., and Ertl, G., *Phys. Rev. Lett.* **57,** 384 (1986).
25. Muhlhausen, C. W., Williams, L. R., and Tully, J. C., *J. Chem. Phys.* **83,** 2594 (1985).
26. Tully, J. C., *Surf. Sci.* **111,** 461 (1981).
27. Funk, S., Bonn, M., Denzler, D. N., Hess, Ch., Wolf, M., and Ertl, G., *J. Chem. Phys.,* in press (2000).
28. Segner, J., Robota, H., Vielhaber, W., Ertl, G., Frenkel, F., Häger, J., Krieger, W., and Walther, H., *Surf. Sci.* **131,** 273 (1983).
29. Comsa, G., and David, R., *Surf. Sci. Rep.* **5,** 145 (1985).
30. Kubiak, G. D., Sitz, G. O., and Zare, R. N., *J. Chem. Phys.* **81,** 6397 (1984).
31. Gadzuk, J. W., Landmann, U., Kuster, E. J., Cleveland, C. L., and Barnett, R. N., *Phys. Rev. Lett.* **49,** 426 (1982).
32. Rettner, C. T., Michelsen, H. A., and Auerbach, D. J., *J. Vac. Sci. Technol.* **111,** 1901 (1993).
33. Gross, A., Wilke, S., and Scheffler, M., *Phys. Rev. Lett.* **75,** 2718 (1995).
34. Diño, W. A., Kasai, H., and Okiji, A., *Surf. Sci.* **418,** L39 (1998).
35. Wetzig, D., Dopheide, R., Rudkowski, M., David, R., and Zacharias, H., *Phys. Rev. Lett.* **76,** 463 (1996).
36. Mödl, A., Robota, H., Segner, J., Vielhaber, W., Lin, M. C., and Ertl, G., *Surf. Sci.* **169,** L341 (1986).
37. Campbell, C. T., Ertl, G., Kuipers, H., and Segner, J., *J. Chem. Phys.* **73,** 5862 (1980).
38. Allers, K. H., Pfnür H., Feulner, P., and Menzel, D., *J. Chem. Phys.* **100,** 3985 (1994).
39. Mantell, D. A., Ryali, S. B., Halpern, B. L., Haller, G. L., and Fenn, J. B., *Chem. Phys. Lett.* **81,** 185 (1981).
40. Coulston, G. W., and Haller, G. L., *J. Chem. Phys.* **95,** 6932 (1991).
41. Wei, C., and Haller, G. L., *J. Chem. Phys.* **105,** 810 (1996).
42. Kunimori, K., Uetsuka, H., Iwade, T., Watanabe, T., and Ito, S., *Surf. Sci.* **283,** 58 (1993).
43. Uetsuka, H., Watanabe, K., and Kunimori, K., *Surf. Sci.* **363,** 73 (1996).
44. Matsushima, T., *Het. Chem. Rev.* **2,** 51 (1995).
45. Matsushima, T., *J. Chem. Phys.* **93,** 1464 (1990).
46. Taylor, J. B., and Langmuir, I., *Phys. Rev.* **44,** 423 (1933).
47. Ehrlich, G., *J. Phys. Chem.* **59,** 473 (1955).
48. Kisliuk, P. J., *J. Phys. Chem. Solids* **3,** 95 (1957); **5,** 78 (1958).
49. Weinberg, W. H., and Merrill, R. P., *J. Vac. Sci. Technol.* **10,** 411 (1972).
50. Harris, J., and Kasemo, B., *Surf. Sci.* **105,** L281 (1981).
51. Barth, J. V., Zambelli, T., Wintterlin, J., and Ertl, G., *Chem. Phys. Lett.* **270,** 152 (1997).
52. Brune, H., Wintterlin, J., Behm, R. J., and Ertl, G., *Phys. Rev. Lett.* **68,** 624 (1992).
53. Wahnström, G., Lee, A. B., and Strömqvist, J., *J. Chem. Phys.* **105,** 326 (1996).
54. Jacobsen, J., Hammer, B., Jacobsen, K. W., and Nörskov, J. K., *Phys. Rev.* **B52,** 14954 (1995).
55. Au, C. T., and Roberts, M. W., *Nature* **319,** 206 (1986).
56. Carley, A. F., Davies, P. R., Kulkarni, G. U., and Roberts, M. W., *Catal. Lett.* **58,** 93 (1999).
57. Matsushima, T., *Surf. Sci.* **127,** 403 (1983).
58. Wartnaby, C. E., Stuck, A., Yeo, Y. Y., and King, D. A., *J. Chem. Phys.* **102,** 1855 (1995).
59. Engel, T., and Ertl, G., *J. Chem. Phys.* **69,** 1267 (1978).
60. Mullins, C. B., Rettner, C. T., and Auerbach, D. J., *J. Chem. Phys.* **95,** 8649 (1991).
61. Campbell, C. T., Ertl, G., Kuipers, H., and Segner, J., *J. Chem. Phys.* **73,** 5862 (1980).
62. Kori, M., and Halpern, B. L., *Chem. Phys. Lett.* **110,** 223 (1984).
63. Ree, J., Kim, Y. H., and Shin, H. K., *J. Chem. Phys.* **104,** 742 (1996).
64. Rettner, C. T., *Phys. Rev. Lett.* **69,** 383 (1992).
65. Rettner, C. T., and Auerbach, D. J., *J. Chem. Phys.* **104,** 2732 (1996).

66. Eilmsteiner, G., and Winkler, A., *Surf. Sci.* **366,** 750 (1996).
67. Kratzer, P., and Brenig, W., *Surf. Sci.* **254,** 275 (1991).
68. Kratzer, P., *J. Chem. Phys.* **106,** 6752 (1997).
69. Persson, M., and Jackson, B., *J. Chem. Phys.* **102,** 1078 (1995).
70. Rettner, C. T., Auerbach, D. J., and Lee, J., *J. Chem. Phys.* **105,** 10115 (1996).
71. Eilmsteiner, G., Walkner, W., and Winkler, A., *Surf. Sci.* **352,** 263 (1996).
72. Kammler, Th., Lee, J., and Küppers, J., *J. Chem. Phys.* **106,** 7362 (1997).
73. Wehner, S., and Küppers, J., *J. Chem. Phys.* **108,** 3353 (1998).
74. Kammler, Th., Wehner, S., and Küppers, J., *J. Chem. Phys.* **109,** 4071 (1998).
75. Wehner, S., and Küppers, J., *J. Chem. Phys.* **109,** 294 (1998).
76. Kammler, Th., and Küppers, J., *Chem. Phys. Lett.* **267,** 391 (1997).
77. Dinger, A., Lutterloh, C., Biemer, J., and Küppers, J., *Surf. Sci.* **421,** 17 (1999).
78. Eley, D. D., and Rideal, E. K., *Nature* **146,** 401 (1940).
79. Beckerle, J. D., Yang, Q. Y., Johnson, A. D., and Ceyer, S. T., *J. Chem. Phys.* **86,** 7236 (1987).
80. Beckerle, J. D., Johnson, A. D., Yang, Q. Y., and Ceyer, S. T., *J. Chem. Phys.* **91,** 5756 (1989).
81. Szulczewski, G., and Lewis, R. J., *J. Chem. Phys.* **98,** 5974 (1993); **101,** 11070 (1994).
82. Åkerlund, C., Zorić, I., and Kasemo, B., *J. Chem. Phys.* **104,** 7359 (1996).
83. Åkerlund, C., Zorić, I., Kasemo, B., Cupolillo, A., Buatier de Mongeot, F., and Rocca, M., *Chem. Phys. Lett.* **270,** 157 (1997).
84. Åkerlund, C., Zorić, I., and Kasemo, B., *J. Chem. Phys.* **104,** 7359 (1996).
85. Yates, J. T., and Goodman, D. W., *J. Chem. Phys.* **73,** 5371 (1980).
86. Yamada, T., Onishi, T., and Tamaru, K., *Surf. Sci.* **133,** 533 (1983); **157,** L389 (1985).
87. Yamada, T., and Tamaru, K., *Surf. Sci.* **138,** L155 (1984); **146,** 341 (1984).
88. Lombardo, S. J., and Bell, A. T., *Surf. Sci.* **245,** 213 (1991).
89. Takagi, N., Yoshinobu, J., and Kawai, M., *Phys. Rev. Lett.* **73,** 292 (1994).
90. Hudgins, R. R., and Silveston, P. L., *Catal. Rev. Sci. Eng.* **11,** 167 (1975).
91. Langmuir, I., and Kingdon, K. H., *Phys. Rev.* **21,** 381 (1923).
92. Haber, F., and Just, G., *Ann. Phys.* **30,** 411 (1909); **36,** 308 (1911).
93. Kramer, J., *J. Phys.* **125,** 739 (1949); **129,** 34 (1951).
94. Greber, T., *Surf. Sci. Rep.* **28,** 1 (1997).
95. Hermann, K., Freihube, K., Greber, T., Böttcher, A., Grobecker, R., Fick, D., and Ertl, G., *Surf. Sci.* **313,** L806 (1994).
96. Kasemo, B., Törnqvist, E., Nørskov, J. K., and Lundqvist, B., *Surf. Sci.* **89,** 554 (1979).
97. Böttcher, A., Morgante, A., Gießel, T., Greber, T., and Ertl, G., *Chem. Phys. Lett.* **231,** 119 (1994).
98. Greber, T., Grobecker, R., Morgante, A., Böttcher, A., and Ertl, G., *Phys. Rev. Lett.* **70,** 1331 (1993).
99. Trowbridge, L. D., and Herschbach, D. R., *J. Vac. Sci. Technol.* **18,** 588 (1981).
100. Nørskov, J. K., and Lundqvist, B. I., *Phys. Rev.* **B19,** 5661 (1979).
101. Hellberg, L., Strömqvist, J., Kasemo, B., and Lundqvist, B. I., *Phys. Rev. Lett.* **74,** 4742 (1995).
102. Bergström, J., and Nordling, C., *in* "Alpha-, Beta-, and Gamma-Ray Spectroscopy" (K. Siegbahn, Ed.), Vol. 2. North Holland, Amsterdam, 1965.
103. Nienhaus, H., Bergh, H. S., Gergen, B., Majundar, A., Weinberg, W. H., and McFarland, E. W., *Phys. Rev. Lett.* **82,** 446 (1999).
104. Zhou, X. L., Zhu, X. Y., and White, J. M., *Surf. Sci. Rep.* **13,** 73 (1991).
105. H. L. Dai and W. Ho (Eds.), "Laser Spectroscopy and Photochemistry at Metal Surfaces," Vols. I and II. World Scientific, Singapore, 1995.

106. Petek, H., and Ogawa, S., *Progr. Surf. Sci.* **56,** 239 (1998).
107. Zimmermann, F. M., and Ho, W., *Surf. Sci. Rep.* **22,** 127 (1995).
108. Gadzuk, J. W., *in* "Femtochemistry" (J. Manz and L. Wöste, Eds.). Verlag Chemie, Weinheim, 1994.
109. Schoenlein, R. W., Fujimoto, J. G., Eesley, G. L., and Capehart, T. W., *Phys. Rev. Lett.* **61,** 2596 (1988).
110. Aeschlimann, M., Bauer, M., and Pawlik, S., *Chem. Phys.* **205,** 127 (1996).
111. Ogawa, S., and Petek, H., *Surf. Sci.* **357/358,** 585 (1996).
112. Hertel, T., Knoesel, E., Wolf, M., and Ertl, G., *Phys. Rev. Lett.* **76,** 535 (1996).
113. Knoesel, E., Hotzel, A., Hertel, T., Wolf, M., and Ertl, G., *Surf. Sci.* **368,** 76 (1996).
114. Schoene, W. D., Keyling, R., and Ekardt, W., personal communication.
115. Ho, W., *Surf. Sci.* **299/300,** 996 (1994).
116. Harris, S. M., Holloway, S., and Darling, G. R., *J. Chem. Phys.* **102,** 8235 (1995).
117. Hertel, T., Wolf, M., and Ertl, G., *J. Chem. Phys.* **102,** 3414 (1995).
118. Menzel, D., and Gomer, R., *J. Chem. Phys.* **41,** 3311 (1964).
119. Redhead, P. A., *Can. J. Phys.* **42,** 886 (1964).
120. Hasselbrink, E., Wolf, M., Holloway, S., and Saalfrank, P., *Surf. Sci.* **263,** 179 (1996).
121. Guo, H., and Seideman, T., *J. Chem. Phys.* **103,** 9062 (1995).
122. Bornscheuer, K. H., Nessler, W., Binetti, M., Hasselbrink, E., and Saalfrank, P., *Phys. Rev. Lett.* **78,** 1174 (1997).
123. Masson, D. P., Lanzendorf, E. J., and Kummel, A. C., *Phys. Rev. Lett.* **74,** 1799 (1995); *J. Chem. Phys.* **102,** 9096 (1995).
124. Mieher, W. D., and Ho, W., *J. Chem. Phys.* **91,** 4755 (1989).
125. Ukraintsev, V. A., and Harrison, I., *J. Chem. Phys.* **96,** 6307 (1992).
126. Zhu, X. Y., Hatch, S. R., Campion, A., and White, J. M., *J. Chem. Phys.* **91,** 5011 (1989).
127. Weik, F., de Meijere, A., and Hasselbrink, E., *J. Chem. Phys.* **99,** 682 (1993).
128. Harrison, I., *Acc. Chem. Res.* **31,** 631 (1998).
129. Artsyukhovich, A. N., and Harrison, I., *Surf. Sci.* **350,** L199 (1996).
130. Walkup, R. E., Newns, D. M., and Avouris, Ph., *Phys. Rev.* **B48,** 1858 (1993).
131. Prybyla, J. A., Tom, H. W. K., and Aumiller, G. D., *Phys. Rev. Lett.* **68,** 503 (1992).
132. Misewich, J. A., Nakabayashi, S., Weigund, P., Wolf, M., and Heinz, T. F., *Surf. Sci.* **363,** 204 (1996).
133. Busch, D. G., and Ho, W., *Phys. Rev. Lett.* **77,** 1338 (1996).
134. Misewich, J. A., Nakabayashi, S., Weigund, P., Wolf, M., and Heinz, T. F., *Surf. Sci.* **363,** 204 (1996).
135. Struck, L. M., Richter, L. J., Buntin, A., Cavanagh, R. R., and Stephenson, J. C., *Phys. Rev. Lett.* **77,** 4576 (1996).
136. Brandbyge, M., Hedegård, P., Heinz, T. F., Misewich, J. A., and Newns, D. M., *Phys. Rev.* **B52,** 6042 (1995).
137. Tolk, N. H., Traum, M. M., Tully, J. C., and Madey, T. E. (Eds.), "Diet I," Springer-Verlag, Berlin, 1983.
138. Stroscio, J. A., and Eigler, D., *Science* **254,** 1319 (1991).
139. Avouris, Ph., *Acc. Chem. Res.* **28,** 95 (1995).
140. Stipe, B. C., Rezaei, M. A., Ho, W., Gao, S., Persson, M., and Lundqvist, B. I., *Phys. Rev. Lett.* **78,** 4410 (1997).
141. Stipe, B. C., Rezaei, M. A., and Ho, W., *Science* **27a,** 1907 (1998).
142. Stipe, B. C., Rezaei, M. A., and Ho, W., *Phys. Rev. Lett.* **81,** 1263 (1998).
143. Bartels, L., Meyer, G., Rieder, K. H., Velic, D., Knoesel, E., Hotzel, A., Wolf., M., and Ertl, G., *Phys. Rev. Lett.* **80,** 2004 (1998).
144. Hasegawa, K., Kasai, H., Dino, W. A., and Okiji, A., *J. Phys. Soc. Jpn.* **67,** 4018 (1998).
145. Langmuir, I., *Trans. Faraday Soc.* **17,** 607 (1922).

146. Stampfl, C., Schwegmann, S., Over, H., Scheffler, M., and Ertl, G., *Phys. Rev. Lett.* **77,** 3371 (1996).
147. Kim, Y. D., Wendt, S., Schwegmann, S., Over, H., and Ertl, G., *Surf. Sci.* **418,** 267 (1998).
148. Van Hove, M. A., Hermann, K., and Watson, P. R., *Surf. Rev. Lett.* **4,** 1071 (1997).
149. Van Hove, M. A., and Somorjai, G. A., *J. Mol. Catal.* **131,** 243 (1998).
150. Gierer, M., and Over, H., *Z. Kristallogr.* **214,** 14 (1999).
151. Zambelli, T., Trost, J., Wintterlin, J., and Ertl, G., *Phys. Rev. Lett.* **76,** 795 (1996).
152. Barth, J. V., Zambelli, T., Wintterlin, J., Schuster R., and Ertl, G., *Phys. Rev.* **B55,** 12902 (1997).
153. Gierer, M., Barbieri, A., Van Hove, M. A., and Somorjai, G. A., *Surf. Sci.* **391,** 176 (1997).
154. Schwegmann, S., Over, H., De Renzi, V., and Ertl, G., *Surf. Sci.* **375,** 91 (1997).
155. Nørskov, J. K., *in* "The Chemical Physics of Solid Surfaces" (D. A. King and D. P. Woodruff, Eds.), Vol. 6, p. 1. Elsevier, New York, 1993.
156. Nagl, C., Schuster, R., Renisch, S., and Ertl, G., *Phys. Rev. Lett.* **81,** 3483 (1998).
157. Persson, B. N. J., Tushaus, M., and Bradshaw, A. M., *J. Chem. Phys.* **92,** 5034 (1990).
158. Persson, B. N. J., *Surf. Sci. Rep.* **15,** 1 (1992).
159. Patrykiejew, A., Sokolowski, S., and Binder, K., *Surf. Sci. Rep.,* in press (2000).
160. Bonzel, H. P., Bradshaw, A. M., and Ertl, G. (Eds.), "Physics and Chemistry of Alkali Metal Adsorption." Elsevier, New York, 1989.
161. Zambelli, T., Barth, J. V., Wintterlin, J., and Ertl, G., *Nature* **390,** 495 (1997).
162. Wintterlin, J., Völkening, S., Janssens, T. V. W., Zambelli, T., and Ertl, G., *Science* **278,** 1931 (1997).
163. Campbell, C. T., Ertl, G., Kuipers, H., and Segner, J., *J. Chem. Phys.* **73,** 5862 (1980).
164. Alavi, A., Hu, P., Deutsch, T., Silvestrelli, P. L., and Hutter, J., *Phys. Rev. Lett.* **80,** 3650 (1998).
165. Koe, R., Brown, W. A., and King, D. A., *J. Am. Chem. Soc.* **121,** 4845 (1999).
166. Taylor, H. S., *Proc. R. Soc.* **A108,** 105 (1925).
167. Boudart, M., *Adv. Catal.* **20,** 153 (1969).
168. Boudart, M., and Djéga-Mariadassou, G., "Kinetics of Heterogeneous Catalytic Reactions." Princeton Univ. Press, Princeton, NJ, 1984.
169. Schwab, G.-M., and Pietsch, E., *Z. Phys. Chem.* **131,** 385 (1929).
170. Davis, S. M., and Somorjai, G. A., *in* "The Chemical Physics of Solid Surfaces and Heterogeneous Catalysis" (D. A. King and D. P. Woodruff, Eds.), Vol. 4, p. 217. Elsevier, New York, 1982.
171. Zambelli, T., Wintterlin, J., Trost, J., and Ertl, G., *Science* **273,** 1688 (1996).
172. Dahl, S., Logadottier, A., Egeberg, R. C., Larsen, J. H., Chorkendorff, I., Törnqvist, E., and Nørskov, J. K., *Phys. Rev. Lett.* **83,** 1814 (1999).
173. Rosowski, F., Hinrichsen, O., Muhler, M., and Ertl, G., *Catal. Lett.* **36,** 229 (1996).
174. Hinrichsen, O., Rosowski, F., Hornung, A., Muhler, M., and Ertl, G., *J. Catal.* **165,** 33 (1997).
175. Dietrich, H., Geng, P., Jacobi, K., and Ertl, G., *J. Chem. Phys.* **104,** 375 (1996).
176. Nørskov, J. K., personal communication.
177. Somorjai, G. A., *Catal. Lett.* **12,** 17 (1992).
178. Titmuss, S., Wander, S., and King, D. A., *Chem. Rev.* **96,** 1291 (1996).
179. Ertl, G., *Science* **254,** 1756 (1991).
180. Gritsch, T., Coulman, D., Behm, R. J., and Ertl, G., *Phys. Rev. Lett.* **63,** 1086 (1989).
181. Sharma, R. K., Brown, W. A., and King, D. A., *Surf. Sci.* **414,** 68 (1998).
182. Behm, R. J., Thiel, P. A., Norton, P. R., and Ertl, G., *J. Chem. Phys.* **78,** 7438, 7446 (1983).
183. Hopkinson, A., Bradley, J. M., Guo, X. C., and King, D. A., *Phys. Rev. Lett.* **71,** 1597 (1993).

184. Hopkinson, A., Guo, X. C., Bradley, J. M., and King, D. A., *J. Chem. Phys.* **99,** 8262 (1993).
185. Pasteur, A. T., Dixonwarren, S. J., and King, D. A., *J. Chem. Phys.* **103,** 2251 (1995).
186. Walker, A. V., Klotzer, B., and King, D. A., *J. Chem. Phys.* **109,** 6879 (1998).
187. Ali, T., Klotzer, B., Walker, A. V., Ge, Q., and King, D. A., *J. Chem. Phys.* **109,** 6879 (1998).
188. Ali, T., Klotzer, B., Walker, A. V., and King, D. A., *J. Chem. Phys.* **109,** 10996 (1998).
189. Coulman, D. J., Wintterlin, J., Behm, R. J., and Ertl, G., *Phys. Rev. Lett.* **64,** 1761 (1990).
190. Jensen, F., Besenbacher F., Laesgaard, E., and Steensgaard, I., *Phys. Rev.* **B41,** 10233 (1990).
191. Sueyoshi, T., Sasaki, T., and Iwasawa, Y., *Surf. Sci.* **357,** 764 (1996).
192. Johnson, A. D., Daly, S. P., Utz, A. L., and Ceyer, S. T., *Science* **257,** 223 (1992).
193. Daley, S. P., Utz, A. L., Trautman, T. R., and Ceyer, S. T., *J. Am. Chem. Soc.* **116,** 6001 (1994).
194. Hang, K. L., Bürgi, T., Trautman, T. R., and Ceyer, S. T., *J. Am. Chem. Soc.* **120,** 8885 (1998).
195. Son, K. A., and Gland, J. L., *J. Am. Chem. Soc.* **117,** 5415 (1995).
196. Michaelides, A., Hu, P., and Alavi, A., *J. Chem. Phys.* **111,** 1343 (1999).
197. Johansson, S., Wong, K., Zhdanov, V. P., and Kasemo, B., *J. Vac. Sci. Technol.* **A17,** 297, 1721 (1999).
198. Yan, M. X., Gracias, D. H., Jacobs, P. W., and Somorjai, G. A., *Langmuir* **14,** 1458 (1998).
199. Gorodetskii, V., Lauterbach, J., Rotermund, H. H., Block, J., and Ertl, G., *Nature* **370,** 276 (1994).
200. Völkening, S., Bedürftig, K., Jacobi, K., Wintterlin, J., and Ertl, G., *Phys. Rev. Lett.* **83,** 2672 (1999).
201. Razon, L. F., and Schmitz, R. A., *Catal. Rev. Sci. Eng.* **28,** 89 (1986).
202. Ertl, G., *Adv. Catal.* **37,** 213 (1990).
203. Imbihl, R., *Prog. Surf. Sci.* **44,** 185 (1993).
204. Slinko, M. M., and Jaeger, N., "Oscillating Heterogeneous Catalytic Systems." Elsevier, Amsterdam, 1994.
205. Schüth, F., Henry, B. E., and Schmidt, L. D., *Adv. Catal.* **39,** 51 (1993).
206. Imbihl, R., and Ertl, G., *Chem. Rev.* **95,** 697 (1995).
207. Von Oertzen, A., Mikhailov, A., Rotermund, H. H., and Ertl, G., *Surf. Sci.* **350,** 259 (1996); *J. Phys. Chem.* **B102,** 4966 (1998).
208. Gruyters, M., Ali, T., and King, D. A., *Chem. Phys. Lett.* **232,** 1 (1995); *J. Phys. Chem.* **100,** 14417 (1996).
209. Mikhailov, A. S., "Foundations of Synergetics I," 2nd ed. Springer, New York, 1994.
210. Jakubith, S., Rotermund, H. H., Engel, W., von Oertzen, A., and Ertl, G., *Phys. Rev. Lett.* **65,** 3013 (1990).
211. Rotermund, H. H., Haas, G., Franz, R. U., Tromp, R. M., and Ertl, G., *Science* **270,** 608 (1995).
212. Ertl, G., and Rotermund, H. H., *Curr. Opin. Solid State Mat. Sci.* **1,** 617 (1996).
213. Von Oertzen, A., Rotermund, H. H., Mikhailov, A., and Ertl, G., *J. Phys. Chem. B,* in press.
214. Gelten, R. J., Jansen, T. P. J., van Santen, R. A., Lukkien, J. J., Segers, J. P. L., and Hilbers, P. A. J., *J. Chem. Phys.* **108,** 5921 (1998); *Isr. J. Chem.* **38,** 415 (1998).
215. Korthlücke, J., Kuzovkov, V. N., and von Niessen, W., *J. Chem. Phys.* **110,** 11523 (1999).
216. Zhdanov, V. P., *Phys. Rev.* **E59,** 6292 (1999).
217. Graham, M. D., Lane, S. L., and Luss, D., *J. Chem Phys.* **97,** 7564 (1993).
218. Graham, M. D., Kevrekidis, I. G., Asakura, K., Lauterbach, J., Rotermund, H. H., and Ertl, G., *Science* **264,** 80 (1994).

219. Lauterbach, J., Asakura, K., Rasmussen, P. B., Rotermund, H. H., Bär, M., Graham, M. D., Kevrekidis, I. G., and Ertl, G., *Physica* **D123,** 493 (1998).
220. Asakura, K., Lauterbach, J., Rotermund, H. H., and Ertl, G., *Phys. Rev.* **B50,** 8043 (1994); *J. Chem. Phys.* **102,** 8175 (1995).
221. Hildebrand, M., Kuperman, M., Wio, H., Mikhailov, A. S., and Ertl, G., *Phys. Rev. Lett.* **83,** 1475 (1999).
222. Hinshelwood, C., "Kinetics of Chemical Change." Oxford Univ. Press, Oxford, 1940.
223. Hougen, O. A., and Watson, K. M., "Chemical Process Principles. Vol. 3. Kinetics and Catalysis." Wiley, New York, 1947.
224. Zhdanov, V. P., and Kasemo, B., *Appl. Surf. Sci.* **74,** 147 (1994).
225. Bär, M., Zülicke, Ch., Eiswirth, M., and Ertl, G., *J. Chem. Phys.* **96,** 8595 (1992).
226. Oh, S. H., Fisher, G. B., Carpenter, J. E., and Goodman, D. W., *J. Catal.* **100,** 360 (1986).
227. Temkin, M., and Pyzhev, V., *Acta Physicochim. URSS* **12,** 327 (1940).
228. Ertl, G., *in* "Catalytic Ammonia Synthesis. Fundamentals and Practice" (J. R. Jennings, Ed.), p. 109. Plenum, New York, 1991.
229. Ertl, G., *in* "Encyclopedia of Catalysis." Wiley–VCH, New York, to be published.
230. Schlögl, R., *in* "Handbook of Heterogeneous Catalysis" (G. Ertl, H. Knözinger, and J. Weitkamp, Eds.), p. 1697. VCH, Weinheim, 1997.
231. Stoltze, P., and Nørskov, J. K., *Phys. Rev. Lett.* **55,** 2502 (1985).
232. Stoltze, P., and Nørskov, J. K., *J. Catal.* **110,** 1 (1988).
233. Bowker, M., Parker, I. B., and Waugh, K. C., *Appl. Catal.* **14,** 101 (1985).
234. Parker, I. B., Bowker, M., and Waugh, K. C., *J. Catal.* **114,** 457 (1988).
235. Dumesic, J. A., and Trevino, A. A., *J. Catal.* **16,** 119 (1989).
236. Stoltze, P., and Nørskov, J. K., *Topics Catal.* **1,** 253 (1994).
237. Aparicio, L. M., and Dumesic, J. A., *Topics Catal.* **1,** 233 (1994).
238. Tennison, S. R., *in* "Catalytic Ammonia Synthesis. Fundamentals and Practice" (J. R. Jennings, Ed.), p. 303. Plenum, New York, 1991.
239. Dahl, S., Taylor, P. A., Törnqvist, E., and Chorkendorff, I., *J. Catal.* **178,** 679 (1998).
240. Rosowski, F., Hornung, A., Hinrichsen, O., Herein, D., Muhler, M., and Ertl, G., *Appl. Catal.* **A151,** 443 (1997).
241. Hinrichsen, O., Rosowski, F., Muhler, M., and Ertl, G., *Chem. Eng. Sci.* **51,** 1683 (1996).
242. Goodman, D. W., Kelly, R. D., Madey, T. E., and J. T. Yates, *J. Catal.* **63,** 226 (1980).
243. Rasmussen, P. B., Holmblad, P. M., Askgaard, T., Ovesen, C. V., Stoltze, P., Nørskov, J. K., and Chorkendorff, I., *Catal. Lett.* **26,** 373 (1994).
244. Fujitani, T., Nakamura, I., Watanabe, T., Uchijima, T., and Nakamura, J., *Catal. Lett.* **35,** 297 (1995).
245. Yoshihara, J., and Campbell, C. T., *J. Catal.* **161,** 776 (1996).

ADVANCES IN CATALYSIS, VOLUME 45

Theoretical Surface Science and Catalysis—Calculations and Concepts

B. HAMMER

Institute of Physics
Aalborg University
DK-9220 Aalborg, Denmark

AND

J. K. NØRSKOV

Center for Atomic-Scale Materials Physics
Department of Physics
Technical University of Denmark
DK-2800 Lyngby, Denmark

The application of density functional theory to calculate adsorption properties, reaction pathways, and activation energies for surface chemical reactions is reviewed. Particular emphasis is placed on developing concepts that can be used to understand and predict variations in reactivity from one transition metal to the next or the effects of alloying, surface structure, and adsorbate–adsorbate interactions on the reactivity. Most examples discussed are concerned with the catalytic properties of transition metal surfaces, but it is shown that the calculational approach and the concepts developed to understand trends in reactivity for metals can also be used for sulfide and oxide catalysts. © 2000 Academic Press.

I. Introduction

When a solid catalyst is used to speed up a chemical process, the overall reaction usually consists of a series of elementary steps. These include

Abbreviations: B3LYP, Beckes hybrid description of exchange-correlation effects; bcc, body centred cubic; DFT, density functional theory; DOS, density of (one-electron) states; fcc, face centred cubic; GGA, generalized gradient approximation; hcp, hexagonal close packed; LDA, local density approximation; LEED, low energy electron diffraction; LMTO, linerized muffin tin orbital; PBE, Perdew, Burke, Ernzerhof exchange-correlation; PW91, Perdew, Wang 91 version of exchange-correlation; RPBE, Hammer, Hansen, Norskov modified PBE exchange-correlation; TS, transition state; E_{chem}, chemisorption energy; W, band width; ε_F, fermi energy; ε_d, center of d-band; ε_a, adsorbate energy level; V, coupling matrix element; V_{ad}, adsorbate-metal d coupling matrix element; V_{dd}, metal d–metal d coupling matrix element; r, chemical reaction rate; ν, frequency factor; E_a, activation energy; k, Boltzmann's constant; ΔE_{es}; electrostatic energy difference; $\Sigma_i \varepsilon_i$, sum of one-electron energies; ΔE_{dip}, dipole–dipole interaction energy; μ, dipole moment; E, electric field strength; E_{segr}, segregation energy.

0360-0564/00 $35.00

adsorption of the reactants on the surface of the solid, diffusion on the surface, breaking of some reactant bonds, and the creation of new ones to form the product molecules, which eventually desorb from the surface. The complexity of these processes and of the catalysts makes it a demanding task to establish a molecular-level understanding of heterogeneous catalysis. The appraoch taken in surface science has been to study elementary reactions on well-defined single-crystal surfaces in order to build an understanding of some of the basic processes involved in catalysis. Single-crystal surfaces are only crude models of the high-surface-area catalysts used industrially. Catalysts need to have a high surface area, and they often consist of mixtures of phases, some of which have the catalytically active surface, whereas others support the small particles of active phases or keep them from sintering. Recently, the methods of surface science have developed further and it has become possible to produce and study models of supported catalysts and to investigate them under realistic high-pressure and -temperature conditions (*1–8*). This rapidly closes the gap between the surface science studies and studies of real catalysts. Simultaneously, the methods used to characterize high-surface-area catalysts have been refined considerably, and it has become possible to characterize the structure and other properties of the working catalysts *in situ* (*9–12*). Together, these different types of experiments have provided us with a wealth of detailed information about surface structures, adsorption geometries, bond strengths, and elementary reaction steps.

The rapidly increasing amount of data emphasizes the need for a conceptual framework for understanding or rationalizing the results. It would be extremely helpful to have an understanding of the most important factors determining the ability of a surface to bind or react with a particular molecule.

An important step toward a fundamental description of surface chemical processes has been taken recently with the development of quantum theoretical methods allowing us to calculate equilibrium structures, adsorption energies, reaction paths, and activation energies for simple processes on metal surfaces. The accuracy is still not sufficient to calculate rates of chemical reactions, but it is sufficient for a semiquantitative description of adsorption and reaction processes, and in particular for comparing different systems. The latter is particularly important if one wants theoretical input into a search for surfaces with a desired activity or selectivity for a given chemical reaction.

In this review, we discuss how the calculational methods in close conjunction with experiments can be used to develop some useful concepts to describe and understand adsorption and reactions on surfaces. We concentrate mainly on metal surfaces because this is the area in which both the experiments and the calculations are most advanced. Many of the concepts developed for metal surfaces are more generally applicable also to semiconductors, oxides, and sulfides, and we illustrate this point at the end of this review. We show how we are beginning to understand which surface properties govern the variations in reactivity from one metal to the next and how adsorbate–adsorbate interactions, surface structure, strain, alloying, defects, and impurities may affect the reactivity.

The basis for our discussion of adsorbate–surface reactions is the density functional theory. We use density functional theory in accurate self-consistent calculations of adsorption properties and also as a basis for the development of models that discern the important physical quantities determining the reactivity of a given surface. We therefore start in Section II by giving a short description of the theoretical methods used and a discussion of their accuracy. We also discuss simple models of the electronic structure to be used in the following sections. In Section III, we discuss adsorption and surface reactions by treating in detail the simplest surface process, adsorption. We discuss in particular how the adsorption energy varies from one metal to the next. We also discuss the similar variation from one adsorbate to the next. We use the adsorption theory to introduce the parameters describing the surface which determine its ability to interact with an adsorbate. In Section IV we treat the more complex processes of molecules or atoms reacting on the surface. We focus in particular on trends in activation energies from one system to the next and show that the reactivity measures developed for simple atomic adsorption also work here. We use this as a starting point for a discussion in Section V of ways to change the reactivity of a given metal by changing the structure, by alloying, or by poisoning or promoting the surface by coadsorbed atoms and molecules. In Section VI, we summarize our results for metallic surfaces and establish a connection to the field of nonmetal catalysis.

II. Theory of Adsorbate–Surface Interactions

The theoretical description of adsorbate–surface interactions is divided into two levels: (i) the accurate, but computationally demanding, calculation of adsorption properties, and (ii) the model description which is approximate but computationally simple. The large-scale calculations can be viewed as computer experiments. They complement real experiments in several ways. Often, it is a good check of an experiment or an interpretation of an experiment to have a calculation for comparison. There are also cases for which the calculation is simpler than the experiment. For example, consider an important fundamental property of a surface such as the surface energy. It is very difficult to measure, and currently the best source of surface energies is calculations, which have been performed for all the metals in the periodic table (*13*). Finally, the calculation can sometimes be performed for situations that are not realizable experimentally. For instance, it is very simple to change the lattice constant of a metal in a calculation and bring out the effect of strain on the reactivity without concern about the problems associated with straining a crystal in real life (*14*). The ability to use the computer experiments to test models and theories of catalysis has recently breathed new life into the development of models and understanding of catalysis. The models are needed to bring out the concepts around which our description and understanding of catalysis revolves. What are the parameters of a surface determining its ability to adsorb a molecule or let one molecule react with another? Why are some adsorbates more reactive than others on a given surface? It is not possible, or even desirable, to have to perform a complete calculation or an experiment for each new system considered. We would like to be able to understand directly how a change in the catalyst composition or structure should change its reactivity. The models are essential in bringing out this understanding.

In the following sections, we first discuss briefly the calculational strategies and the strengths and limitations of the calculations. We then introduce some simple models and concepts to be used in the rest of the review to discuss and understand and the results of the calculations.

A. Density Functional Theory Calculations

There are two basically different approaches to the calculation of the electronic structure and total energies of molecules and solids: the wave function-based methods (*15*) and the density functional theory (DFT) methods (*16–18*). The former, which can be very accurate if a high level of configuration interactions is included, is currently limited to 10–100 electrons. If we consider transition metal surfaces, this limits the number of atoms that can be treated to about 10 since each transition metal has on the order of 10 valence electrons. This generally makes these methods unattractive for routine treatments of the complex systems needed to model catalysts. There are elegant ways of embedding a very accurately described region into a less accurately described surrounding region, which can increase the system sizes and make wave function-based methods very useful (*19*). The computational cost of such calculations and the limited size of systems that can be treated mean that, in a surface science and catalysis context, they are primarily useful as benchmarks, which can be used to gauge the accuracy of the less computationally demanding DFT methods. In the following, we therefore concentrate on the DFT methods.

Density functional theory describes how the ground state electron density and total energy can be obtained by solving a set of one-electron Schrödinger equations (the Kohn–Sham equations) instead of the complicated many-electron Schrödinger equation. This results in an enormous computational simplification, and systems with more than 1000 electrons can be treated.

There are several different methods used to solve the Kohn–Sham equations. In short, they can be characterized by the model used to describe the surface, by the kind of basis set used, and by the approximation used in the treatment of exchange-correlation effects:

1. *The Model Used to Describe the Surface*

The calculations cannot describe all the atoms in a solid or a catalyst particle, and a strategy must be chosen to limit the number of atoms treated explicitly. Two basic types of methods exist:

- *Cluster methods,* which describe only a limited cluster of the surface atoms in the hope that the surface atoms farther away from the adsorbates of interest are not important.
- *Slab methods,* whereby the surface is described as a slab with a periodic structure along the surface. The size of the surface unit cell determines the computational effort, and in principle the unit cell should be chosen to be large enough so that the adsorbates in neighboring unit cells do not interact.

The two types of models are illustrated in Fig. 1. The slab method can be viewed as a particular choice for the boundary conditions in the cluster approach, which, however, is generally found to describe the surface properties better than the cluster approach for a given number of atoms in the super cell or the cluster, respectively. In both methods, only a finite number of atomic layers can be included, and this will always represent an approximation. In addition, there are Greens function-based methods which can, in principle, treat a single adsorbate on a semi-infinite substrate (*20–22*), but they are not widely used.

2. *The Basis Set*

Two basic types of basis sets are most widely used:

- *Localized functions* [gaussians, atomic orbitals, linearized muffin tin orbitals (LMTOs), etc.]
- *Plane waves* [including augmented plane waves (LAPWs), etc.]

The choice depends on preference and the kind of problem studied. Localized functions are thus the usual choice in cluster-type models,

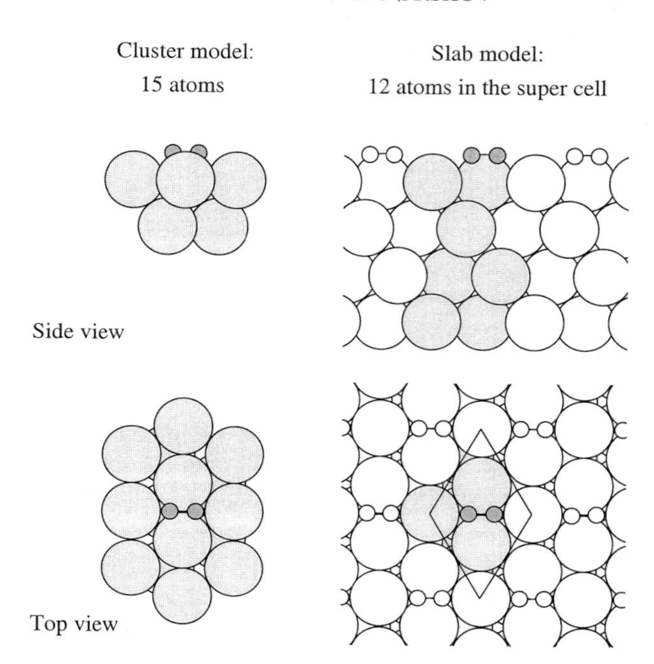

FIG. 1. Illustration of cluster and slab representations of an H_2 molecule (small circles) interacting with a semi-infinite surface. Replicas of the super cell atoms are shown with open circles.

whereas both localized (LMTOs) and plane waves are used in slab-type calculations. Usually, the core electrons are not treated explicitly, but instead are included as "frozen" or through a pseudopotential description of the ionic core (*23*). There are also elegant methods whereby the localized states are projected out and treated differently from the valence electrons (*24*).

3. *The Description of Exchange Correlation*

The error made by the choice of model used to describe the surface or the basis set can be controlled by increasing the size up to the point of convergence. This leaves the choice of exchange correlation functional as the main approximation in the DFT calculations. The most accurate method for treating exchange and correlation is probably Becke's hybrid description of exchange-correlation effects (B3LYP) (*25*). In this method, a Hartree–Fock calculation is performed to derive the exact exchange energy, which is mixed with the DFT-based exchange energy. This hybrid method is therefore computationally very demanding. Less computationally demanding are calculations which make use of one form or another of the generalized gradient approximation (GGA) (*26–28*). As illustrated in Table I, different GGA flavors can give quite different adsorption energies—they all, however, represent a great improvement over the local density approximation (LDA) description of the adsorption bond (*29–33*). The variations between different GGAs may be viewed as the intrinsic uncertainty of these methods. Adsorption energies or energy barriers are therefore not

TABLE I
Calculated Chemisorption Energies as a Function of the Exchange-Correlation Energy Functional, Compared with Measured Chemisorption Energies[a]

	E_{chem}				
	LDA	PW91	PBE	RPBE	E_{chem}^{exp}
O(fcc)/Ni(111)	−6.68	−5.38	−5.27	−4.77	−4.84
O(hol)/Ni(100)	−6.97	−5.66	−5.55	−5.03	−5.41
O(hol)/Rh(100)	−6.64	−5.34	−5.23	−4.71	−4.56
O(fcc)/Pd(111)	−5.34	−4.08	−3.98	−3.49	
O(hol)/Pd(100)	−5.39	−4.14	−4.04	−3.53	
σ_O	1.84	0.57	0.47	0.24	
CO(fcc)/Ni(111)	−2.85	−1.99	−1.88	−1.49	−1.35
CO(hol)/Ni(100)	−3.05	−2.11	−2.00	−1.58	−1.26
CO(brd)/Rh(100)	−3.02	−2.28	−2.16	−1.81	−1.19
CO(fcc)/Pd(111)	−2.95	−2.07	−1.96	−1.56	(−1.47)
CO(brd)/Pd(100)	−2.77	−1.98	−1.87	−1.50	−1.69
σ_{CO}	1.58	0.78	0.67	0.37	
	(1.49)	(0.64)	(0.54)	(0.23)	
NO(hol)[b]/Ni(100)	−6.31	−4.52	−4.41	−3.68	−3.99
NO(brd)/Rh(100)	−3.73	−2.76	−2.67	−2.28	
NO(fcc)/Pd(111)	−3.27	−2.20	−2.12	−1.67	(−1.86)
NO(hol)/Pd(100)	−3.19	−2.12	−2.04	−1.58	−1.61
σ_{NO}	1.98	0.52	0.43	0.22	
σ_{tot}	1.76	0.66	0.56	0.30	
	(1.76)	(0.58)	(0.48)	(0.23)	

[a] LDA is the local density approximation, and PW91 (Perdew, Wang 91 version of exchange correlation), PBE (Perdew, Burke, Ernzerhof exchange correlation), and RPBE (Hammer, Hansen, Nørshov modified PBE exchange correlation) are different GGAs. All values are in eV per adsorbate. The rms deviations for the calculated chemisorption energies for O, CO, and NO (σ_O, σ_{CO}, and σ_{NO}, respectively) and for all three adsorbates (σ_{tot}) have been compiled only against the highly accurate microcalorimetric experimental results from Brown *et al.* (34). Chemisorption energies derived from temperature-programmed desorption experiments are given in parentheses. The σ values in parentheses occur when the CO/Rh(100) data are neglected. The E_{chem} value is relative to atomic O in the gas phase. From Hammer *et al.* (28). [b] Dissociative adsorption.

determined with uncertainties less than about 0.25 eV or 25 kJ/mol. Fortunately, energy differences and variations among systems are usually more accurate than because the errors tend to cancel for like systems. This means that vibrational frequencies and structures are usually extremely good in any of the methods. It also means that trends are better, and this is what we mainly focus on.

B. Simple Models

We consider in this section the simplest one-electron description of the quantum mechanics of atoms and molecules interacting with a metal surface. The idea is not to give an accurate description (this was discussed previously) but instead to bring out the essential physics.

A one-electron state in an atom or molecule outside a metal surface will interact with all the valence states of the surface atoms. These states form a band or several bands of states. Figure 2 shows a typical density of states for a transition metal. The broad *s* band is half filled—all the transition metals have one *s* electron (in the metallic state)—and the *d* states are seen to form much narrower bands. The occupancy of the *d* bands varies along the transition metals as they shift through the Fermi level. The narrow *d* bands are a consequence of the small coupling matrix element V_{dd} between the localized *d* states; one of the important conclusions from tight binding theory is that the band width is proportional to V_{dd} *(35)*.

Since *d* bands are narrow, the interaction of an adsorbate state with the *d* electrons of a surface often gives rise to bonding and antibonding states just as in a simple two-state problem. This is illustrated in Fig. 3, which also illustrates what happens in the case of interaction with a broad band such as the *s* band of a metal. The adsorbate state only broadens. The broad band limit with a single resonance is often called "weak chemisorp-

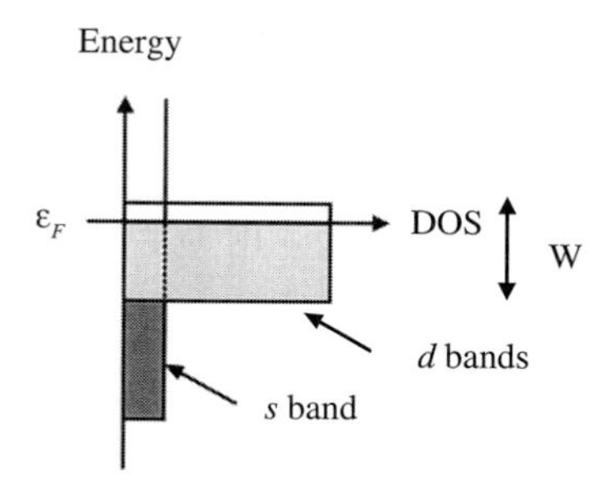

FIG. 2. Schematic illustration of the density of states of a transition metal, showing the broad *s* band and the narrow *d* bands (width *W*) around the Fermi level, ε_F.

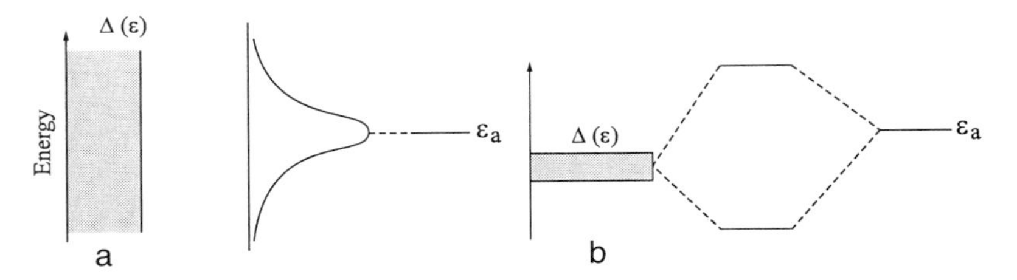

FIG. 3. The local density of states at an adsorbate in two limiting cases: (a) for a broad surface band; (b) for a narrow metal band. Case a corresponds to the interaction with a metal *s* band and case b is representative of the interaction with a transition metal *d* band.

tion," whereas when there are split-off bonding and antibonding states we refer to "strong chemisorption" (*36*).

Figure 4 shows the result of a model calculation whereby a single adsorbate level is coupled by a matrix element V to a band of states. In the calculation, V is kept fixed as the band is shifted up in energy. The filling of the band is kept fixed so that as the center of the band ε_d is shifted up toward the Fermi level, the band width W decreases. When the band is low and broad, only a single resonance can be seen at the bottom of the band, but as ε_d shifts up a distinctive antibonding state appears above the band. Since these antibonding states are above the Fermi level, they are empty, and the bond becomes increasingly stronger as the number of empty antibonding states increases. The model calculation illustrates the transition between "weak" and "strong" chemisorption. The model calculation also illustrates a general principle about bonding at a surface: Strong bonding occurs if antibonding states are shifted up through the Fermi level (and become empty). The same is true if bonding states are shifted down through the Fermi level (and become filled). We use this general principle extensively in the following section.

III. The Chemisorption Bond

In this section, we consider the bonding of an adsorbate to a surface. First, we discuss the simplest example of an atomic adsorbate and then increase the level of complexity by considering molecular adsorbates.

A. Atomic Adsorbates

To illustrate the general quality of the DFT calculations for the calculation of chemisorption properties, consider as a first example oxygen adsorp-

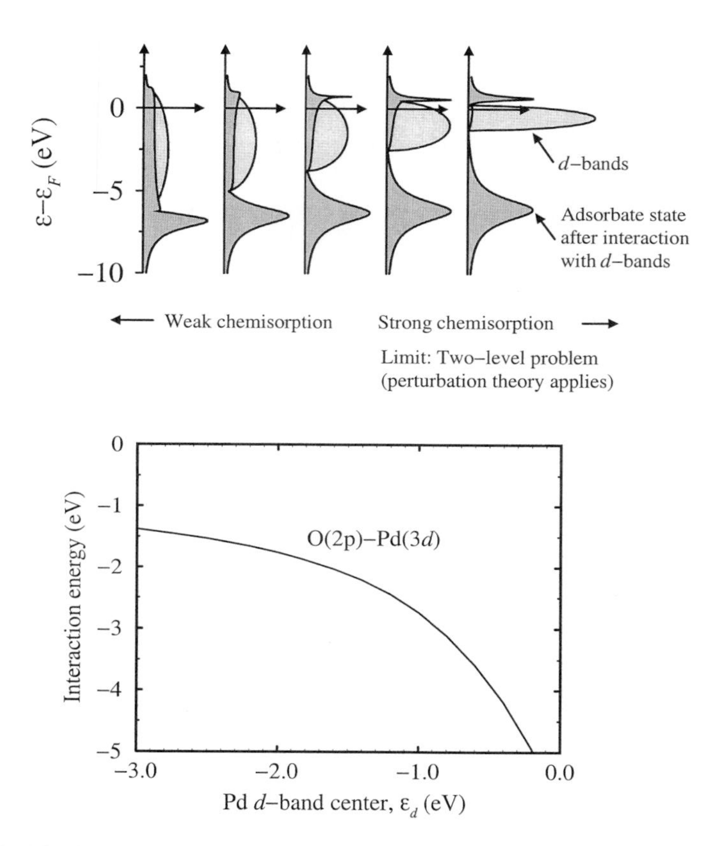

FIG. 4. The local density of states projected onto an adsorbate state interacting with the *d* bands at a surface. The strength of the adsorbate–surface coupling matrix element V is kept fixed as the center of the *d* bands ε_d is shifted up toward the Fermi energy ($\varepsilon_F = 0$) and the width W of the *d* bands is decreased to keep the number of electrons in the bands constant. As ε_d shifts up, the antibonding states are emptied above ε_F and the bond becomes stronger (bottom). The calculation was done by using the Newns–Anderson model (*37*). Adapted from Hammer (*38*).

tion on the Pt(111) surface. This system has been studied in detail experimentally (*39*) and in several theoretical investigations (*40, 41*). We concentrate on the ordered $p(2 \times 2)$ oxygen overlayer structure observed at a quarter of a monolayer of oxygen atoms. Figure 5 shows the geometry of this structure including relaxations of the platinum atoms at the surface and three layers down, as deduced from results of low-energy electron diffraction experiments. We also show the result of a DFT calculation (*40*). The equilibrium structure is determined by allowing all the coordinates of

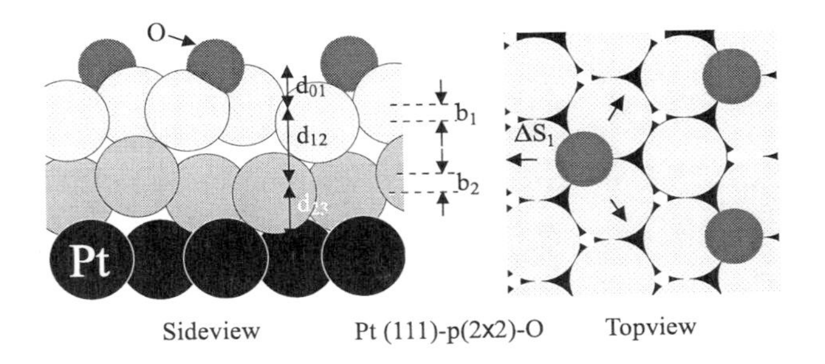

	Experiment	Theory
d_{01}	1.18 Å	1.21 Å
$(d_{12} - d)/d$	1.3 %	0.6 %
$(d_{23} - d)/d$	0.0 %	-0.1 %
b_1/d	3.1 %	3.5 %
b_2/d	4.0 %	3.9 %
$\Delta S_1/S_1$	1.6 %	2.9 %

FIG. 5. The experimental and theoretical (PW91) equilibrium structure of the Pt(111)–p(2 × 2)–O system. From Hammer and Nørskov (*40*). The experimental results are from Starke *et al.* [*39*].

the atoms in the oxygen adlayer and in the two first layers to vary until the lowest-energy geometry is found. Clearly, all bond lengths and the relaxation pattern are reproduced by the calculation down to the 1% level.

The chemisorption potential energy, including relaxations of the platinum substrate, is calculated to be $\Delta E_{\text{O}} = -4.29$ eV per oxygen atom relative to an atom in the gas phase [GGA and PW91 (Perdew, Wang 91 version of exchange correlation)]. The corresponding calculated bond energy per oxygen atom in O_2 is 2.95 eV, so the heat of adsorption is −2.68 eV per molecule [the value found by using the RPBE (Hammer, Hansen, Nørskov modified PBE exchange correlation) functional would be 0.2–0.3 eV higher (*28*)]. This result is in reasonable agreement with results from microcalorimetry, which gives an integral heat of adsorption for a quarter of a monolayer of oxygen of about -2.4 eV/O_2 (*34*).

We can also deduce the vibrational frequency for the oxygen vibration perpendicular to the surface. The experimental value is 58 meV (*42*), and the value calculated from the curvature of the total energy as a function of the height of the oxygen atoms above the surface is 59 meV.

To illustrate the variations in adsorption energies from one metal to the next, consider the adsorption of atomic oxygen on many transition metals

(Fig. 6). It is seen that copper, silver, and gold bind most weakly, and the unique nobleness of gold is clearly borne out. The O–Au surface bond is weaker than the O–O bond so that oxygen molecules will not readily dissociate on gold at all.

To understand the differences in chemisorption energy from one metal

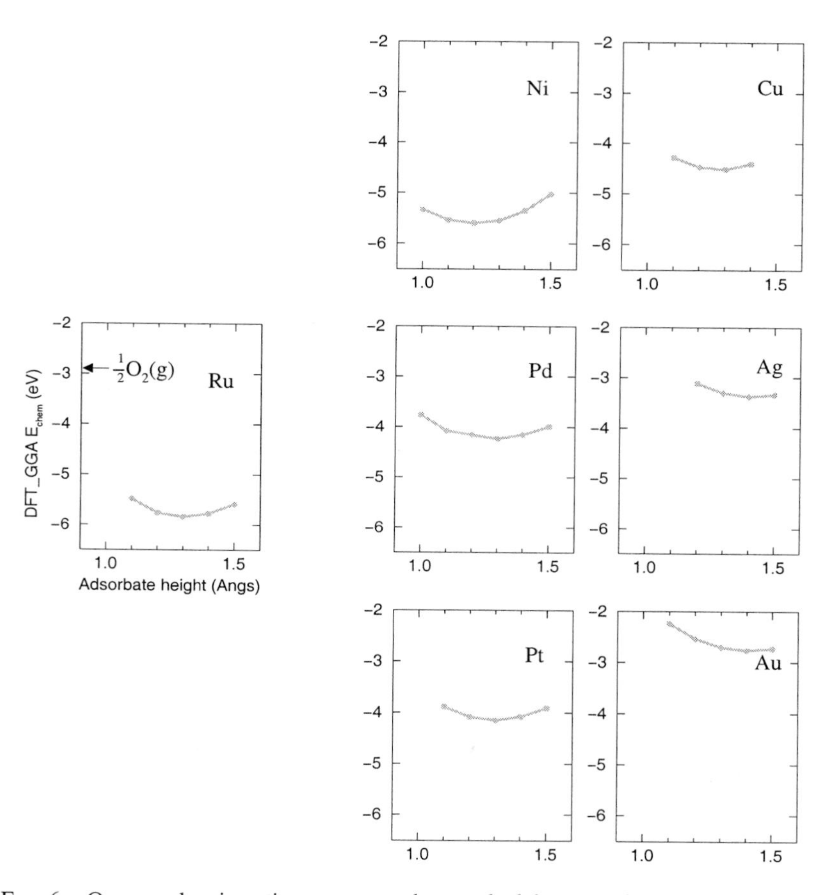

FIG. 6. Oxygen chemisorption at many close-packed late transition and noble metal surfaces {hexagonal close packed [hcp(0001)] for ruthenium, but face-centered cubic [fcc(111)] for the other metals}. A quarter monolayer of oxygen is adsorbed at the three-fold fcc site (the three-fold hcp for ruthenium) in a p (2 × 2) pattern. The adsorbate height is varied, and the metal ions are fixed at the truncated bulk positions. The chemisorption energy, E(O/surface) − E(O atom) − E (surface) is calculated by use of the PW91 functional. From Hammer and Nørskov (*40*).

to the next, consider in Fig. 7 the density of states projected onto one of the O 2*p* states for oxygen chemisorbed on copper, silver, gold, nickel, palladium, platinum, and ruthenium. In each case, the density of states has a similar structure, and in Fig. 8 we show how one can think about the origin of the main peaks in the density of states. We imagine that the bond formation takes place in two steps (*36, 40, 43*). First, we let the adsorbate valence 2*p* state that we consider interact with the metal *s* electrons. As discussed previously, this gives rise to a single resonance. All three oxygen 2*p*-derived resonances are well below the Fermi level and hence completely

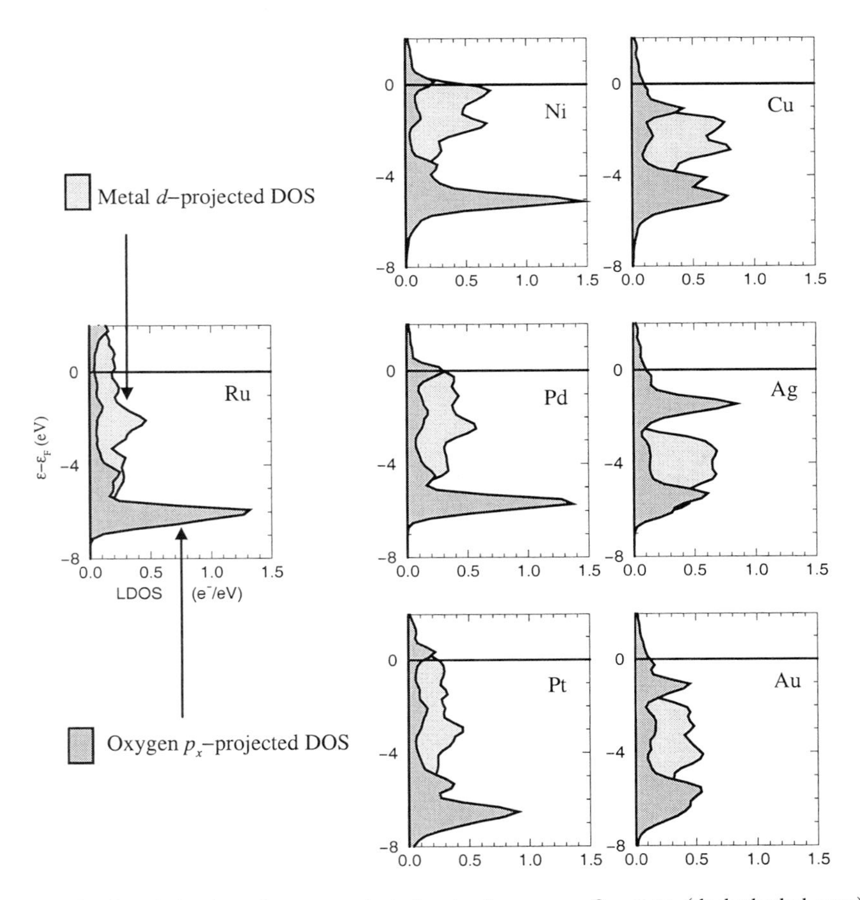

FIG. 7. Local density of states projected onto the oxygen $2p_x$ state (dark-shaded area) for atomic oxygen 1.3 Å above close-packed surfaces of late transition metals (cf. Fig. 6). The light-shaded areas give the metal *d*-projected DOS for the respective metal surfaces before the oxygen chemisorption. From Hammer and Nørskov (*40*).

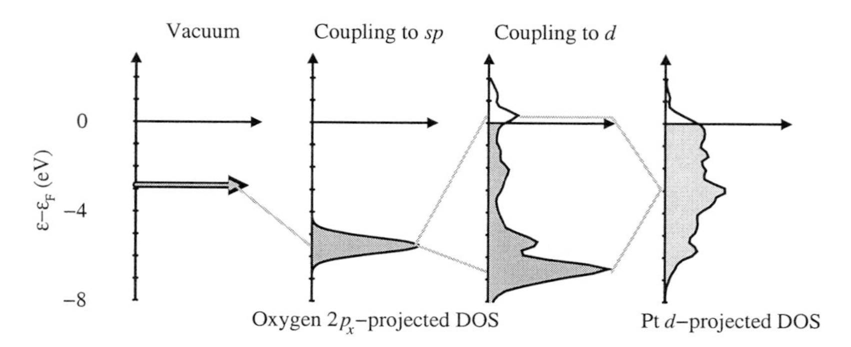

FIG. 8. Schematic illustration of the change in local electronic structure at an oxygen atom upon adsorption on simple and transition/noble metal surfaces. First, the sharp atomic states of the gas phase are broadened into resonances and shifted down due to the interaction with the metal *sp* states. Next, these renormalized states interact with the narrow *d* bands at the transition and noble metal surfaces, forming covalent bonding and antibonding states below and above the initial adsorbate and surface states. The coupling to the metal *d* electrons can roughly be viewed as a two-level coupling. The O/Pt(111) and Pt(111) DOS are from the self-consistent calculations in Fig. 7.

filled (one could say that the oxygen is in a 2^- state). We then turn on the coupling to the *d* electrons. Since the *d* bands are narrow, this gives rise to strong interaction; that is, it gives rise to a splitting of the oxygen resonance into two: one state which is bonding with respect to the adsorbate and metal *d* states and another above the *d* bands which is antibonding.

The interpretation of the changes in electronic structure discussed previously has some immediate consequences for our understanding of the trends in the binding energies. We can also think of the binding energy as having two components—one from the coupling to the metal *s* states and one due to the extra coupling to the *d* states (*36, 40, 43–45*). Judging from the calculated densities of states in Fig. 7, we arrive at two conclusions: (i) The coupling to the *d* states is essentially a two-level problem giving rise to a bonding and an antibonding state, and (ii) the *d* bands can to a large extent be characterized by the band center, ε_d, only.

There are two general trends in Fig. 6 that must be explained. First in general, the farther to the left in the periodic table, the stronger the bond. Second, the farther down the periodic table, the weaker the interaction; the 5*d* metals are more noble than the 4*d* and 3*d* metals.

To understand these two effects, we first note that since the contribution from the coupling to the metal *s* states is approximately the same for each of the metals considered, the main trends in the chemisorption energies should be given by the coupling to the *d* electrons (*45*).

The first effect is simple to understand in light of the previous discussion.

As we move to the left from copper, silver, or gold, the *d* bands move up in energy, and increasingly more antibonding adsorbate–metal *d* states become empty. This is clearly shown in Fig. 7. For copper, silver, and gold, the antibonding states are completely filled because the *d* bands are well below the Fermi level. As we move farther to the left in the 3*d*, 4*d*, or 5*d* series, the *d* bands shift up and the antibonding states become depopulated. The effect is illustrated more fully in Fig. 9, in which it is shown that the oxygen chemisorption energies become increasingly stronger as we move to the left in the 4*d* transition metal series. Figure 9 includes experimental values, and the effect is evident in both the calculated and the measured values. In Fig. 9, it is also shown that the adsorption energy varies with the position of the *d* band center relative to the Fermi level, just as in the model calculation in Fig. 4, showing that the *d* band center is one possible measure of the reactivity of the transition metals. We note that the band

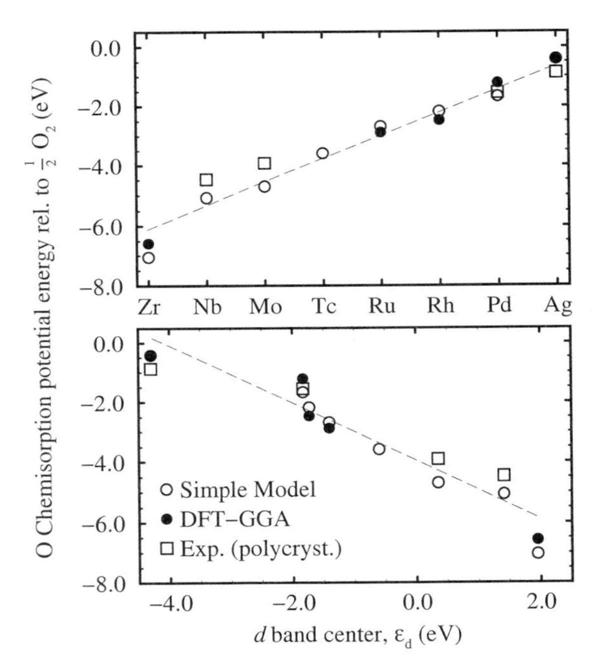

FIG. 9. Comparison of DFT-based oxygen chemisorption energies, $E(\text{O/surface}) - \frac{1}{2}E(\text{O}_2) - E\,(\text{surface})$ (PW91), experimental values, and model estimates of the bond strengths for the various close-packed transition and noble metal surfaces. Data represented by open circles were determined by using the Newns–Anderson model. The experimental values are from Toyoshima and Somorjai (*49*). (Bottom) The calculated adsorption energies correlate well with the *d* band center ε_d.

center, the filling, and the width of the d bands vary through the $4d$ series, i.e., the three quantities are strongly coupled. We choose to focus here on the band center, but we could also have chosen either of the other two parameters. It is shown below that this choice provides a very general picture. It is clear from the previous argument that the same trends should be expected of all simple atomic adsorbates with a filled valence level (after interaction with the metal s band), including hydrogen, carbon, nitrogen, fluorine, sulfur, and chlorine (*40, 46–48*).

To understand the second effect a slightly deeper analysis is required. We start by considering copper, silver, and gold, each of which has a filled d band and, according to Fig. 7, a negligible contribution to the bonding from the emptying of antibonding oxygen $2p$–metal d states. There is, however, still a contribution from the interaction between the metal d electrons and the oxygen $2p$ states. The Pauli principle states that no two electrons can be in the same state, which means that the oxygen $2p$ states have to become orthogonal to the metal d states when they come into contact (*40, 46*). This raises the kinetic energy by an amount that is approximately proportional to the square of the adsorbate–metal d coupling matrix element V_{ad}^2; the stronger the overlap, the larger the repulsion.

Idealized d- band filling

V_{ad}^2 [Relative to Cu]

0.1 20.8	0.2 7.90	0.3 4.65	0.4 3.15	0.5 2.35	0.6 1.94	0.7 1.59	0.8 1.34	0.9 1.16	1.0 1.0	1.0 0.46
Ca	Sc	Ti	V	Cr	Mn	Fe	Co	Ni	Cu	Zn
4.12	3.43	1.50 3.05	1.06 2.82	0.16 2.68	0.07 2.70	-0.92 2.66	-1.17 2.62	-1.29 2.60	-2.67 2.67	2.65
0.1 36.5	0.2 17.3	0.3 10.9	0.4 7.73	0.5 6.62	0.6 4.71	0.7 3.87	0.8 3.32	0.9 2.78	1.0 2.26	1.0 1.58
Sr	Y	Zr	Nb	Mo	Tc	Ru	Rh	Pd	Ag	Cd
4.49	3.76	1.95 3.35	1.41 3.07	0.35 2.99	-0.60 2.84	-1.41 2.79	-1.73 2.81	-1.83 2.87	-4.30 3.01	3.1
0.1 41.5	0.2 17.1	0.3 11.9	0.4 9.05	0.5 7.27	0.6 6.04	0.7 5.13	0.8 4.45	0.9 3.90	1.0 3.35	1.0 2.64
Ba	Lu	Hf	Ta	W	Re	Os	Ir	Pt	Au	Hg
4.65	3.62	2.47 3.30	2.00 3.07	0.77 2.95	-0.51 2.87	2.83	-2.11 2.84	-2.25 2.90	-3.56 3.00	3.1

Bulk Wigner-Seitz radius, s [au]

ε_d [eV]

FIG. 10. Section of the periodic table with the $3d$, $4d$, and $5d$ transition metals. Shown in the lower right corner for each element is bulk Wigner–Seitz radius, s. In the lower left corner, the center of the d band is calculated for the most close-packed surface for each of the metals [(111) for fcc, (001) for hcp, and (110) for bcc]. In the upper right corner is shown the behavior of the adsorbate (s or p)–metal d-coupling matrix element squared, V_{ad}^2. The V_{ad}^2's generally decrease for increasing nuclear charge within a row and increase down the groups. All the values, except for the properties of zinc, cadmium, and mercury were compiled from Andersen *et al.* (*50*). In the upper left corner, the idealized d band fillings are shown. These are similar to the actual, calculated bulk d band fillings considering the uncertainties in interpreting them (*50*).

The absolute magnitude of V_{ad}^2 depends on the metal, the adsorbate, and the position of the adsorbate relative to the metal, but for a given adsorbate and a fixed adsorbate geometry the variations of the matrix element depend only on the extent of the metal *d* states. The variation in V_{ad}^2 from one metal to the next for a fixed adsorbate and adsorbate geometry is therefore a property of the metal. For each element we know the *relative* strength of the coupling matrix element directly from tabulated values of the LMTO potential functions (*40, 50*). The values are shown in Fig. 10 for a large section of the periodic table.

Using the data in Fig. 10, we can test the hypothesis that the variation in adsorption strength from copper to gold is given mainly by the Pauli repulsion. In Fig. 11 we plot the calculated bond energies as a function of V_{ad}^2, and the proportionality is clearly seen. Gold has the most extended *d* states, the largest V_{ad}^2, and therefore the largest repulsion. This makes oxygen least stable on gold, and thus gold is the most noble metal (*46*). The same is true for hydrogen (*46*) and sulfur (*40*) and should, according to the model, hold for other electronegative adsorbates (such as the halogens) with deep-lying valence states after coupling with the metal *sp* states. For carbon and nitrogen, the picture is very similar, but here the adsorbate va-

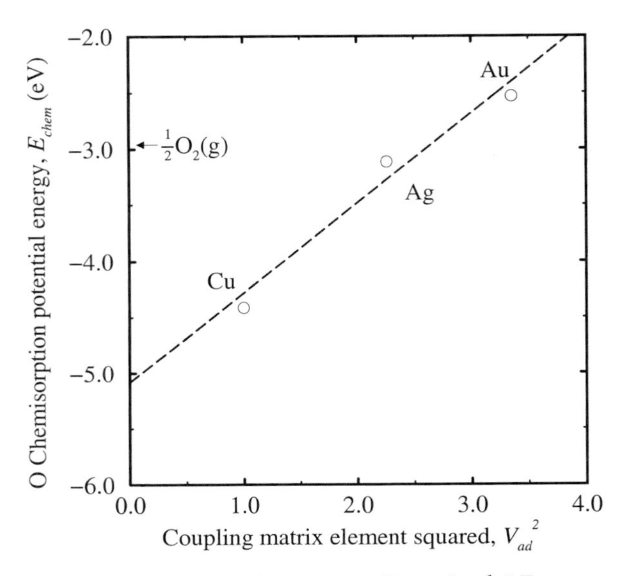

FIG. 11. The calculated chemisorption energy, E_{chem}, for $\frac{1}{4}$ ML oxygen on Cu(111), Ag(111), and Au(111) (cf. Fig. 6) plotted versus the coupling matrix elementes squared, V_{ad}^2, from Fig. 10. The proportionality confirms that the variations in bond strength are given by the strength of the Pauli repulsion between the oxygen 2*p* states and the metal *d* states.

lence states are sufficiently high-lying that the coupling to the *d* states is strong enough to push antibonding states above the Fermi level for gold. This is not possible for silver, which has a deeper-lying *d* band and a smaller coupling matrix element. For these adsorbates, silver is slightly more noble than gold.

Examination of the values of V_{ad}^2 for the metals to the right for copper, silver, and gold in the periodic table (Fig. 10) shows that the smallest value is for zinc. These metals also have low-energy *d* states and no bonding contribution due to the emptying of antibonding states, and the strength of the oxygen bonding should therefore also be given primarily by the size of V_{ad}^2. The previously discussed scenario would therefore suggest that oxygen and the other electronegative adsorbates should bind most strongly to zinc and that after gold, mercury should be the most noble.

The transition metals to the left of copper, silver, and gold will have the repulsive Pauli repulsion in addition to the attractive interaction due to the empty antibonding states disucssed previously. Both the Pauli repulsion and the attractive interaction due to bond formation become stronger as the matrix elements become larger. As long as the repulsion is stronger than the attraction, the 5*d* metals will be more noble than the 4*d* metals, which will be more noble than the 3*d* metals just above them in the periodic table. This reflects the fact that the 5*d* orbitals are always more extended than the 4*d* orbitals, which are in turn more extended than the 3*d* orbitals (Fig. 10). Since the bond strength increases as a result of moving to the left and up in the transition metal series, the platinum group metals in the lower right corner should be the most noble.

B. Molecular Adsorbates

The adsorption of molecules is only slightly more complicated to describe than atomic adsorption. The main complication arises from the fact that usually several adsorbate valence states are important for the interaction with a surface.

As the first example, consider the adsorption of CO. Several theoretical studies (*51–54*) suggest that the filled 5σ and the doubly degenerate, empty $2\pi^*$ electronic states are mainly responsible for the bonding to metal surfaces.

It is possible to understand the changes in electronic structure during CO chemisorption using exactly the same two-step process as for the atomic adsorbates. This is illustrated in Fig. 12. We consider here the change in the adsorbate density of states when a CO molecule interacts with a metal with only *s* electrons and the case in which there are also *d* electrons with which the CO states can interact. It is seen that the interaction with the metallic *s* electrons gives rise to a downshift and broadening of both the 2π and 5σ states, whereas the coupling to the metallic *d* states gives rise

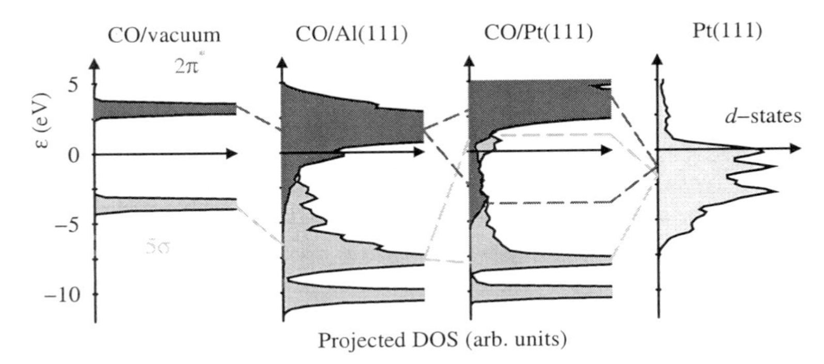

FIG. 12. The self-consistent electronic DOS projected onto the 5σ and $2\pi^*$ orbitals of CO: in vacuum and on Al(111) and Pt(111) surfaces. Also shown is the DOS from the *d* bands in the Pt(111) surface. The sharp states of CO in vacuum are seen to broaden into resonances and shift down in energy on the simple metal surface (mixing with the 4σ state causes additional structure in the 5σ resonance). On the transition metal surfaces the CO resonances further hybridize with the metal *d* states. This leads to shifts in the 5σ and $2\pi^*$ levels and to antibonding 5σ–*d* states at the top of the *d* bands and bonding $2\pi^*$–*d* states at the bottom. These states have low weight in the 5σ and $2\pi^*$ projections shown. Adapted from Hammer *et al.* (*55*).

to bonding and antibonding states below and above the two original states. As a consequence of the different symmetries, the 2π and 5σ states interact with different *d* orbitals, and the two interactions can be treated independently.

The current model of CO bonding is in complete agreement with the theoretical interpretations developed by Blyholder (*53*), Bagus and Pacchioni (*52*), and Van Santen and Neurock (*54*). Usually, the CO–metal bond is described in terms of electron donation from the CO 5σ to the metal and back-donation from the metal to the CO $2\pi^*$. With the current division of the donation and back-donation into separate metal *s* and *d* steps, which follows the reasoning of Bagus and Pacchioni (*52*), we obtain a simple picture that can even be developed into a quantitative model of the trends in the CO chemisorption energies on metal surfaces and overlayers (*55*).

Two important effects can be observed immediately from Fig. 12. First, the 5σ contribution to the bonding is quite small. There are few antibonding states shifted above the Fermi level, and if we include the Pauli replusion this interaction with the metal *d* electrons is repulsive (*54*). This is true for transition metals to the right in the periodic table and for copper, silver, and gold. The 2π interaction, on the other hand, is attractive and dominates the variations from one metal to the next to the right in the periodic table. The large difference from the 5σ case is that the 2π state is above the Fermi level before interaction with the *d* states, and in this case new bonding states are shifted below the Fermi level. Note that there is an attractive

2π–metal *d* interaction even for the noble metals. As we move to the left in the periodic table from the noble metals, the bond strength increases, again mainly because the *d* states move up in energy. The effect is considerably weaker than for the atomic adsorbates, however. This point is illustrated in Fig. 13, in which the calculated adsorption strength for adsorbed CO is compared with that of adsorbed carbon atoms and adsorbed oxygen atoms on some 4*d* transition metals. The molecular adsorption is strongest to the right, but farther to the left there is a crossover, and the dissociated state becomes the most stable. This type of behavior is quite general. In Fig. 13, a similar comparison for NO and adsorbed N + O is shown. The situation is the same, except that the crossover is farther to the right (*56*). For each of the simple molecular adsorbates there is such a crossover between atomic and molecular adsorption somewhere in the group of transition metals, as illustrated by the experimental results collected by Broden

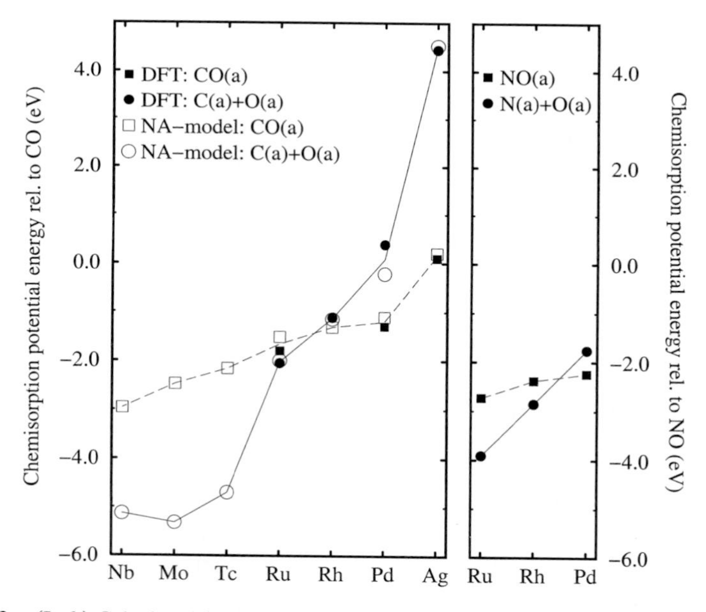

FIG. 13. (Left) Calculated (PW91) and model estimates of the variation in the adsorption energy of molecular CO compared to atomically adsorbed C and O for the most close-packed surface of the 4d transition metals. (Right) Calculated (PW91) molecular and dissociative chemisorption of NO. Solid symbols are DFT calculations; open symbols are Newns–Anderson model calculations. For CO, dissociative chemisorption appears to the left of rhodium. For NO, dissociative chemisorption appears farther to the right, i.e., also on rhodium.

et al. (*57*) (Fig. 14). The crossover point depends on the transition metal row—the 5*d*'s tend to cause molecular dissociation less willingly than the 4*d*'s and 3*d*'s. The origin of this effect is the increase in nobleness from the 3*d*'s to the 4*d*'s and 5*d*'s discussed previously. The crossover also depends on the molecule adsorbed, as reflected largely by the variations in atomic adsorption energies.

The above description of the bonding of molecules to surfaces is readily extended to even larger and more complicated molecules (*58, 59*). In Fig. 15, the analysis of Pallassana and Neurock (*58*) of the interaction between ethylene and a Pt(111) surface is shown.

C. Adsorbate–Adsorbate Interactions

So far, we have mostly considered trends in adsorption energies for a fixed surface coverage. The surface coverage of reactants, intermediates, and products on a catalyst can vary significantly depending on reaction

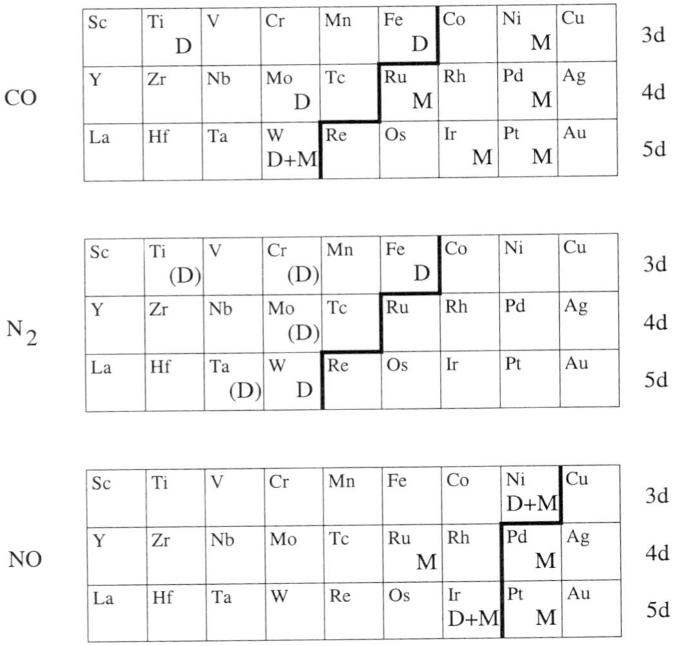

CO									
Sc	Ti D	V	Cr	Mn	Fe D	Co	Ni M	Cu	3d
Y	Zr	Nb	Mo D	Tc	Ru M	Rh	Pd M	Ag	4d
La	Hf	Ta	W D+M	Re	Os	Ir M	Pt M	Au	5d

N_2									
Sc	Ti (D)	V	Cr (D)	Mn	Fe D	Co	Ni	Cu	3d
Y	Zr	Nb	Mo (D)	Tc	Ru	Rh	Pd	Ag	4d
La	Hf	Ta (D)	W D	Re	Os	Ir	Pt	Au	5d

NO									
Sc	Ti	V	Cr	Mn	Fe	Co	Ni D+M	Cu	3d
Y	Zr	Nb	Mo	Tc	Ru M	Rh	Pd M	Ag	4d
La	Hf	Ta	W	Re	Os	Ir D+M	Pt M	Au	5d

Fig. 14. Compilation of experimental data for the ability of transition metals to adsorb and dissociate CO, N_2, and NO molecules. M, molecular adsorption; D, dissociative adsorption. Adapted from Broden *et al.* (*57*).

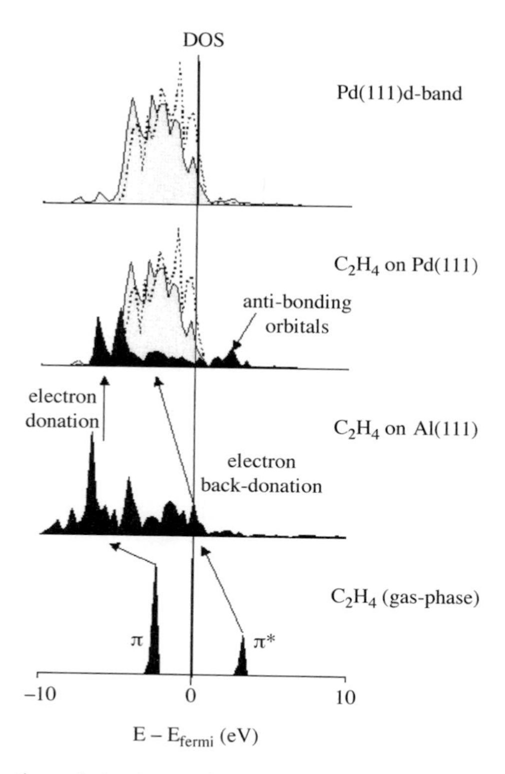

FIG. 15. Illustration of the interaction between C_2H_2 and a Pd(111) surface. From Pallassana and Neurock (*58*).

conditions, and often the coverages are so large that adsorbate–adsorbate interactions are significant and should be included in the description. In fact, these interactions are often so large that a variation in the coverage can change the state of the surface more than the variation from one metal to the next in the periodic table.

As an example of a calculated coverage dependence, consider in Fig. 16 the nitrogen adsorption energy as a function of nitrogen atom coverage on various iron surfaces (*60*). The results illustrate that interactions can be both attractive and repulsive. The former leads to island formation at low coverages (and temperatures); even at relatively low coverages, adsorbates prefer to cluster together to take advantage of the attractive interaction. Such effects have been observed for the N/Fe(100) system (*61*), as suggested by the results shown in Fig. 16. Repulsive interactions do not lead to island formation but instead to dispersed overlayers and a strongly coverage-

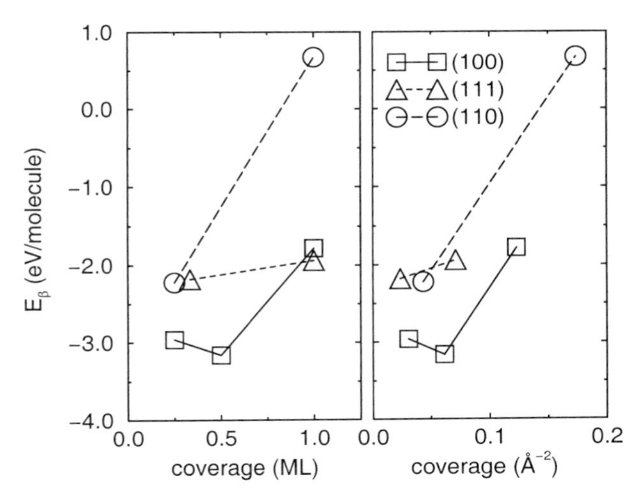

FIG. 16. Adsorption energies (PW91) for dissociated N_2 on bcc-iron as a function of coverage. The squares correspond to (1 × 1), c(2 × 2), and (2 × 2)–N/Fe(100) structures; the triangles represent (1 × 1) and ($\sqrt{3} \times \sqrt{3}$)R30°–N/Fe(111) structures; and the circles represent (1 × 1) and (2 × 2)–N/(110) structures. From Mortensen *et al.* (*60*).

dependent heat of adsorption. Although attractive interactions are often quite weak and depend on the details of the system, repulsive interactions are very common, particularly at high coverages (*62–64*). They are often observed experimentally as a strong decrease in the heat of adsorption with coverage, as for example in the calorimetric results reported by Brown *et al.* (*34*) (Fig. 17).

The attractive interactions are usually dependent on details in the electronic structure. Attraction may also be a result of reconstructions of the surface. Consider for example the N/Fe system of Fig. 16. It is evident that the nitrogen atoms strongly prefer the local geometry provided by the Fe(100) surface relative to the two other low-index surfaces. The nitrogen atoms may therefore prefer to completely restructure the (110) or (111) surfaces to obtain a similar local geometry. For such a major restructuring to occur, an island of a certain size is needed in order for the reconstruction to be stable, and the result may be an effective attractive interaction between the nitrogen atoms mediated by the reconstruction on these surfaces, even though the intrinsic N–N interaction is strongly repulsive.

There are four common causes for interactions between like adsorbates:

1. Direct interactions due to overlap of wavefunctions: Direct overlap between adsorbate states may lead to attraction if there are states close

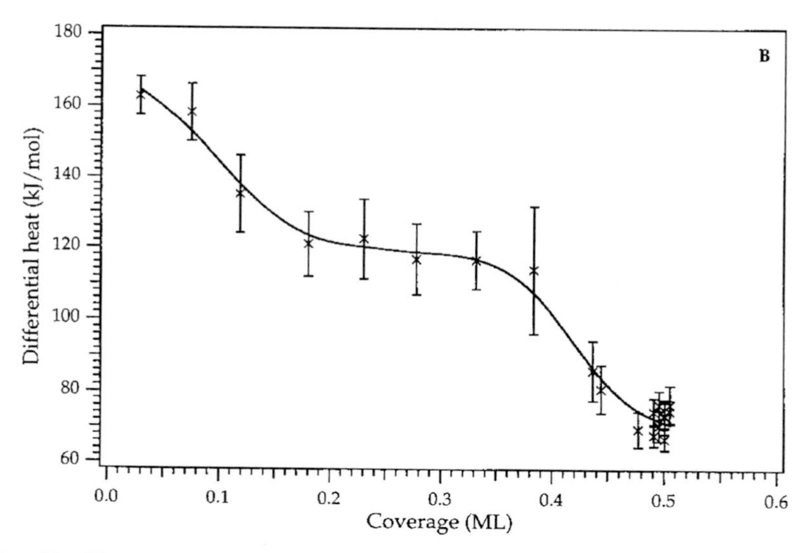

FIG. 17. The measured differential heat of adsorption of NO on Pd(100) as a function of coverage. Reprinted with permission from Brown *et al.* (*34*) © 1998 American Chemical Society.

enough to the Fermi level that the interaction can shift states through the Fermi level. This is often not the case, and the interaction is then dominated by the Pauli repulsion. The detailed analysis of O–O interactions on a Pt(100) surface by Ge and King (*65*) illustrates this effect in detail (Fig. 18).

2. Indirect interactions: One adsorbate may change the electronic structure of the surface in such a way that the adsorption energy of a second adsorbate is changed. One commonly observed effect is that adsorption leads to a downshift of the *d* states of the neighboring transition metal atoms (*47, 66*). As shown previously, a downshift of the *d* states usually leads to a weaker interaction with an adsorbate. This means that a second adsorbate trying to bond to the same transition metal atoms as the first may be bound less strongly. This effect is discussed later.

3. Elastic interactions: Adsorption usually leads to local distortions of the surface lattice. This distortion is experienced by other adsorbates as a repulsion (*67, 68*).

4. Nonlocal electrostatic effects: These can, to lowest order, be described as dipole–dipole interactions.

In addition to interactions with other adsorbates of the same kind, interactions with other kinds of adsorbates may also be important. We return to

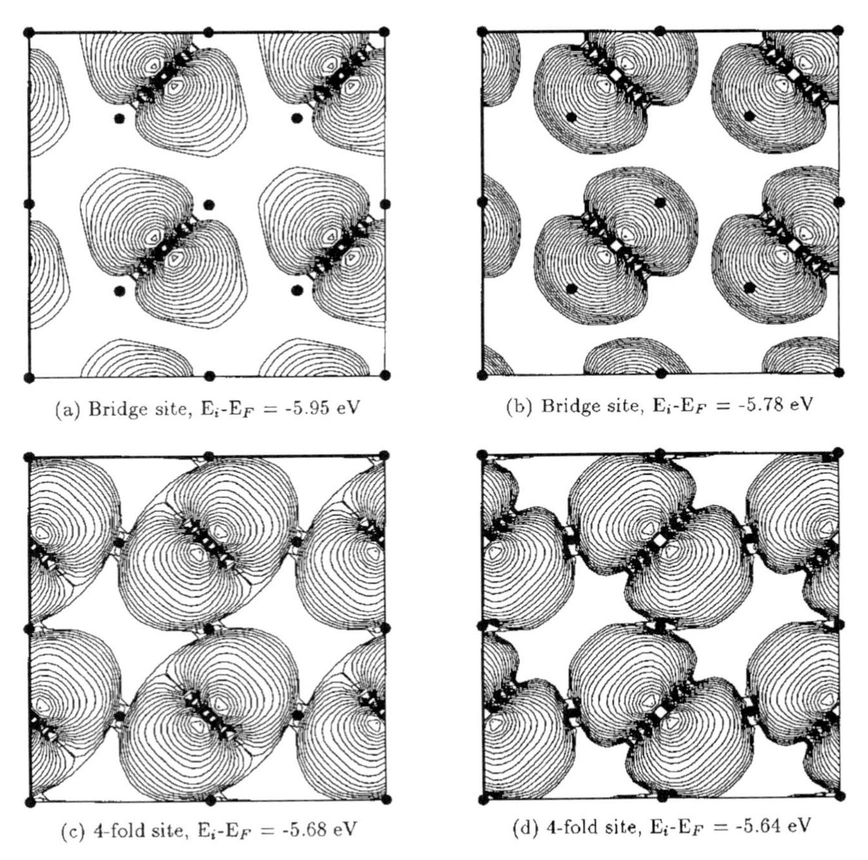

FIG. 18. Charge density contours for oxygen-induced states 5 or 6 eV below the Fermi level on Pt(100). For adsorption in the fourfold sites, in which the O–O repulsion is large, the oxygen p-like states are seen to overlap substantially more than for oxygen atoms in the twofold sites, in which the repulsion is weaker. From Ge *et al.* (*65*).

this topic in connection with our discussion of poisoning and promotion in Section V.C.

The strong coverage dependence of adsorption energies profoundly affects the reactivity of a surface. A more weakly bound adsorbate may be more reactive, and changing the reaction conditions (temperature and/or partial pressures) can therefore change the reactivity considerably. This has recently been exploited by Bradlay *et al.* (*69*) to suggest new reaction conditions for the ammonia oxidation on platinum catalysts.

IV. Bond-Making and -Breaking at a Surface

We now discuss the question of how to describe reactions at surfaces. Any chemical reaction can be described as a transition between two local minima on the potential energy surface of the system as a function of the coordinates of all the involved atoms. The reaction path we define as the minimum energy path leading from the reactant minimum to the product minimum. The saddle point on this path defines the transition state, and the energy difference between the saddle point and the reactant minimum is the activation energy (E_a) of an elementary process. Figure 19 illustrates a potential energy surface in two important degrees of freedom for a dissociative adsorption process (or, equivalently, a recombinative desorption process). Figure 19 also shows the minimum energy path as well as the usual one-dimensional representation of such a reaction, in which the energy is shown as a function of the distance along the reaction path (the reaction coordinate). We stress that in general there are not just a few degrees of freedom taking part in the reaction. Usually, several of the adsorbate degrees of freedom are important, and there are cases for which deformation of the surface is also an important part of the reaction (*47, 70*).

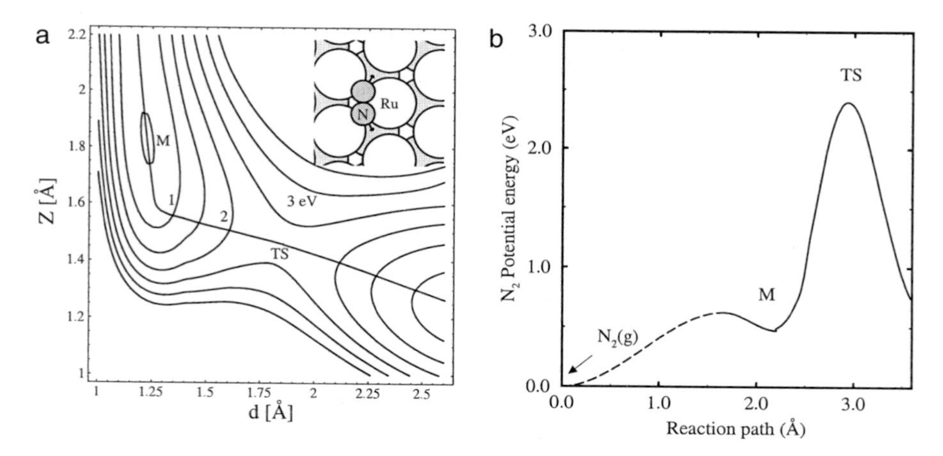

FIG. 19. (a) The potential energy surface (RPBE) for N_2 dissociating on a Ru(0001) surface. The energy zero is a molecule far from the surface. The adsorption geometry is shown in the inset. The distance of the center of mass of the molecule above the surface, Z, and the N–N bond length, d, are varied. The minimum energy path is indicated, and in (b) the energy along the path is shown. Note that here only two degrees of freedom have been included. When the rest are included, the minimum energy path has a lower energy barrier (Fig. 34). Adapted from Murphy *et al.* (*71*).

The rate of an elementary step is given in general by an Arrhenius expression such as

$$r = \nu e^{-E_a/kT}, \tag{1}$$

where E_a is the activation energy, k is Boltzmann's constant, and T is the temperature. The prefactor ν is also given by the properties of the potential energy surface. We do not discuss this aspect but refer, for example, to the review by Van Santen and Neurock (*54*). We also note that the modeling of particular experiments, e.g., molecular beam and state-resolved desorption studies, may require that the *dynamics* of a surface reaction be considered. For a further discussion of this topic, see recent reviews by Darling and Holloway (*72*), Groß (*73*), and Kroes (*74*). In the following sections, we concentrate on the factors controlling the activation energy. We show that trends in activation energies can be understood by using the same concepts developed previously to understand atomic and molecular adsorption energies.

A. The Reaction Path and the Nature of the Transition State

In Fig. 20, the transition states of some elementary reaction steps on close-packed surface of transition metals are shown. Two types of reaction

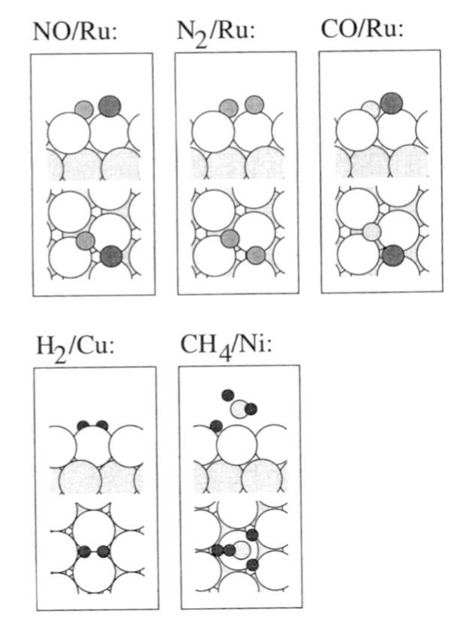

Fig. 20. Transition states for some dissociation reactions of small molecules on transition metal surfaces. The geometries are taken from Hammer (*84*) (NO/Ru), Dahl *et al.* (*83*) (N_2/Ru), Mavrikakis *et al.* (*14*) (CO/Ru), Hammer *et al.* (*32*) (H_2/Cu), and Kratzer *et al.* (*86*) (CH_4/Ni).

paths can be observed. For the dissociation of diatomics such as CO, NO, and N_2, the transition state is a highly stretched molecule (*75–78*). One of the atoms is already close to the final, threefold-coordinated site, whereas the other is on its way to the nearest similar (fcc or hcp) site. At the transition state it is more or less in a twofold coordination site corresponding to the top of the barrier for diffusion of one of the atoms. The same is true for the transition state of, e.g., the CO + O reaction (or, equivalently, the CO_2 dissociation) (*79–81*) and for C–C bond breaking or formation (*82*).

The other type of path shown in Fig. 20 is one in which the transition state is much less stretched. This type of reaction is characterized by an initial state with a very short bond (typically involving hydrogen), and the stretched transition state observed for the "longer bonds" is well beyond the point at which the bond is broken.

The electronic structure at the transition state reflects the two different kinds of reaction paths. For "short" transition states, such as for H_2 dissociation, the transition state has bonding and antibonding states, as in the molecule (Fig. 21). Clearly, the interaction of these states with the surface electrons can be thought of in the same way as for the atomic and molecular adsorbates. The coupling to the metal *s* electrons gives rise to a broadening, and the coupling to the transition metal *d* electrons gives rise to bonding and antibonding states. Just as for CO adsorption, the adsorbate states of different symmetry (the bonding σ_g and the antibonding σ_u H_2 states) can be treated independently.

For the "stretched" transition states, the electronic structure of the transition state is much more like that of the adsorbed atoms than that of the

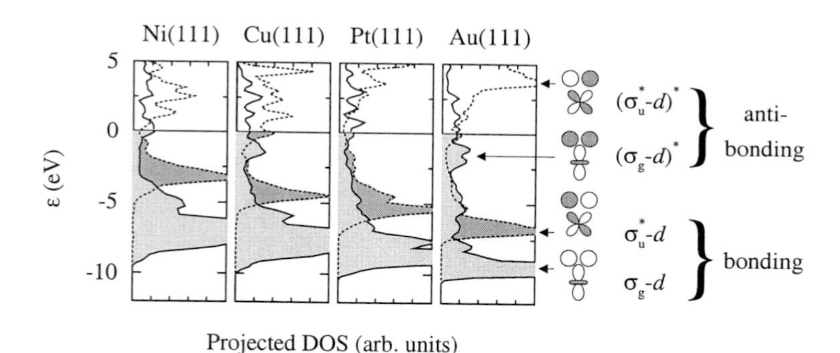

FIG. 21. The DOS projected onto σ_g and σ_u^* for H_2 in the dissociation transition state on Cu(111), Ni(111), and Au(111), and Pt(111) surfaces. From Hammer and Nørskov (*40*).

adsorbed molecule. This is illustrated in Fig. 22, in which the density of states projected onto the two nitrogen atoms in the transition state for N_2 dissociation on Ru(0001) is shown. Becuase of the highly stretched N–N bond length in the transition state (1.9 Å), the electronic structure of the molecule is very final state-like. The distinct splitting of the nitrogen $2p$ states into $2\sigma_u$, $1\pi_v$, $3\sigma_g$, and $1\pi_g$ molecular-like states for adsorbed N_2 (not shown) is thus disappeared in the transition state.

B. Variations in Reactivity from One Metal to the Next

Since the electronic structure is understandable in the same language as that developed to describe atomic and molecular adsorption, the variation in the energy of the transition states—or, equivalently, the activation barriers—follows along the same lines.

Consider as the first example the H_2 dissociation on copper, nickel, gold, and platinum. The energy along the reaction path is shown in Fig. 23. In agreement with experimental knowledge, nickel and platinum each has a low barrier and dissociates H_2 readily (*87, 88*), whereas copper has a sizable barrier (*89*) and H_2 dissociation is impossible on gold (*90*). Just as for the adsorption cases discussed previously, there are two effects involved. Nickel and platinum are more reactive than copper and gold, mostly because more bonding states between the initially empty antibonding H_2 state and the metal *d* states are shifted below the Fermi level. Again, the general rule is that the higher the *d* band center, the more reactive the metal (that is, the lower the transition state energy) (*46, 45, 91*).

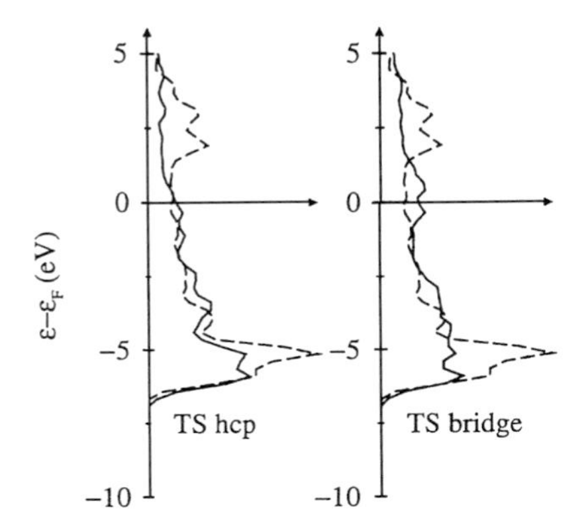

Fig. 22. DOS projected onto $2p$ orbitals of the two nitrogen atoms in the transition states for N_2 dissociation on Ru(0001). One atom is close to an hcp site and one close to a bridge site. Both DOS are compared with the DOS projected onto the $2p$ orbitals of a nitrogen atom absorbed at an hcp site (dashed lines). From Mortenson *et al.* (*47*).

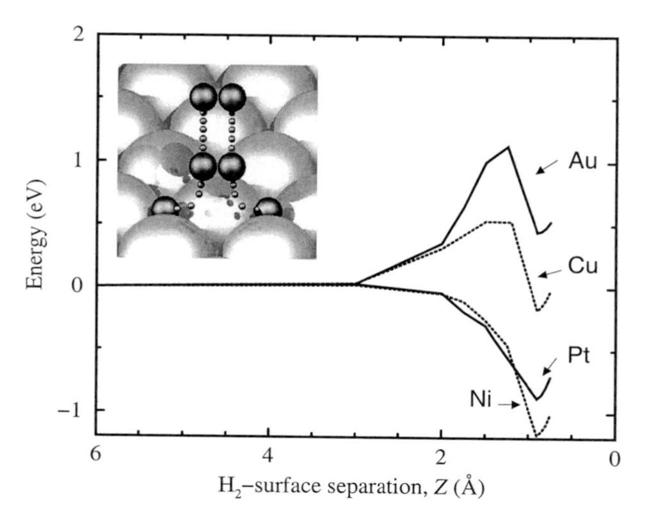

FIG. 23. The potential energy (PW91) along the atop reaction path for H_2 dissociating on copper, nickel, gold, and platinum. Z is the H_2 height above the plane of the surface atomic position. From Hammer and Nørskov (*46*). Reprinted by permission from *Nature* (*46*) © 1995 Macmillan Magazines Ltd.

Gold (and to a lesser extend platinum) is less reactive than copper (or nickel) because the Pauli repulsion is larger for the 5*d* metals than for the 3*d* metals, just as discussed previously for atomic adsorbates. The dissociation of H_2 on rhodium, palladium, silver, and tungsten surfaces has been investigated with DFT methods, with the results being consistent with the current picture (*91–95*).

The picture is even simpler for the streched transition states. Since they are very final state-like, the variations in the transition state energies closely follow the variations in the adsorption energies of the products, as illustrated in Fig. 24.

The strong coupling between transition state energies and final state energies has as an important consequence, namely, that the search for the perfect transition metal to catalyze a certain reaction is always a compromise. Moving to the left in the transition metal series gives a lower activation energy but also stronger bonding of the reactants and thus less free surface area. The "Sabatier principle"-type behavior usually leads to "volcano curves" describing the relation between catalytic activity and position in the periodic table (*2*).

For the "late" barrier processes, the transition state energy is considerably higher than the final state energy for two reasons: (i) In the transition state, one of the atoms is in the twofold site, which is higher in energy than

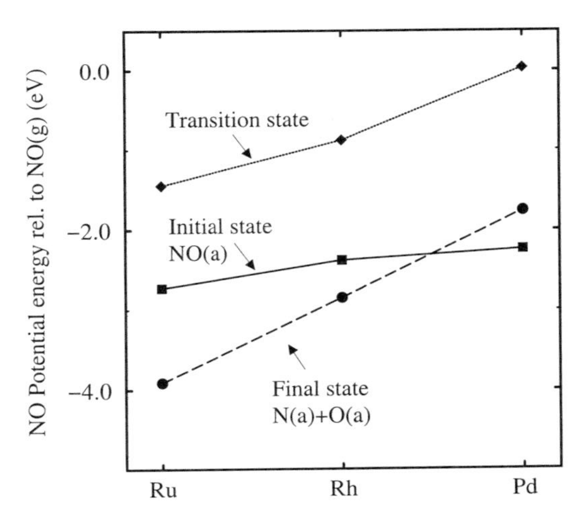

FIG. 24. The variation in energy (PW91) of the initial, transition, and final state for NO adsorption on the tcc(111) surfaces of the 4d transition metals Ru, Rh, and Pd. The correlation between the transition and final state energies is clearly evident. Data from Mavrikakis *et al.* (*75*) and Hammer (*84, 110*).

the equilibrium threefold site, and (ii) the two dissociating atoms share two surface metal atoms in the transition state, giving rise to a large repulsion.

V. Changing the Reactivity

In the preceding sections, we discussed variations in the stability of adsorbed atoms and molecules or of the transition state complex for surface chemical reactions from one metal to the next. Here, we focus on even more subtle variations in reactivity. Often, the reactivity of a given metal can be changed substantially by changing the surface structure, by alloying, or by introducing additional adsorbates onto the surface. In the following sections, we discuss each of these cases separately and show that the underlying mechanisms responsible for the variations in reactivity are the same.

A. Structure Sensitivity

The chemisorption and reaction properties of a metal surface depend on the electronic as well as the geometric structures of the surface. Although the electronic and geometric structures of a surface usually cannot be varied independently, it is very useful to consider the two as causing independent effects on the reactivity of a surface. The two effects are usually termed "electronic" and "geometric" effects. The electronic effects originate from the local electronic structure of the surface and are given by the one-electron spectrum of the metal states that interact with the adsorbates. In most cases, in continuation of the previous discussion, the local average of the *d* electron energies, ε_d, suffices to describe the electronic effect.

The geometric effect is "the rest," so to speak. If adsorbates or reaction complexes interact in different geometric arrangements with surface atoms with identical local electronic properties, the differences in chemisorption bonds and energy barrier heights are ascribed to the differences in the bonding geometry. The simplest measure of the geometric effect is thus the coordination number of the adsorbate with respect to the surface atoms.

1. *Strain*

We start by considering a transition metal surface subject to strain. The geometric arrangement of the surface atoms changes only slightly when the interatomic distances parallel to the surface are modified. Threefold adsorption sites, for example, are still threefold adsorption sites (only the bond lengths have changed), and an adsorbate may compensate for this change by relaxing in the direction perpendicular to the surface. Studying surface processes as a function of strain therefore offers a means of evaluating the magnitude of the electronic effect.

When a surface undergoes compressive or tensile strain, the overlap of metal *d* states at neighboring sites will either increase or decrease and so will the *d* bandwidths. It is a general finding of our density functional calculations that no charging or decharging of the *d* states of late transition metal sites follows from a change in *d* bandwidth. Rather, the *d* bands move in energy to maintain a constant filling. Compressive or tensile strain therefore leads to downshifts and upshifts of the *d* band centers, respectively. The effect is illustrated in Fig. 25.

Figure 26 illustrates two ways in which strained surfaces can be realized experimentally in connection with single-crystal surfaces. Gsell *et al.* (*96*) modified a ruthenium surface through ion implantation by a noble gas. This treatment causes regions of the surface to buckle out. The middle of these regions is subject to tensile strain, whereas the periphery of the regions is subject to compressive strain. Another experimental possibility involves evaporation of one metal onto another (*97–102*). By choosing combinations of metals that give pseudomorphic growth of a monolayer of one metal on the other, highly strained surfaces of the one metal can be realized. Both of these strained metals are easily modeled by DFT calculations—either as slabs under uniform lateral strain or as "sandwich slabs" of one metal on another with the surface unit cell area determined by the substrate

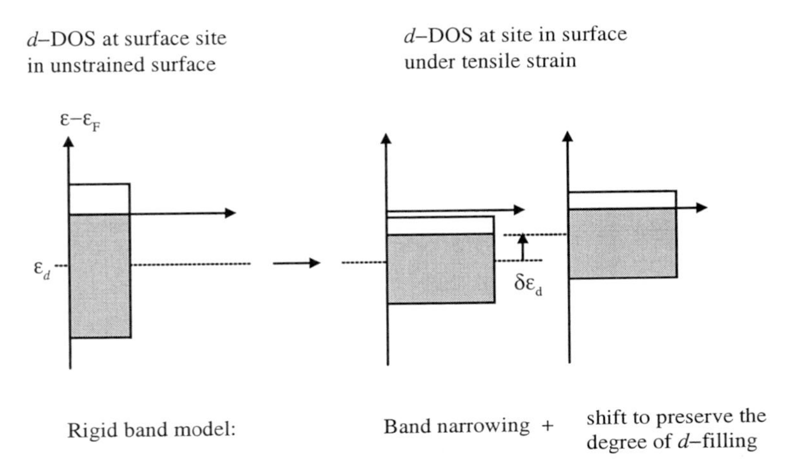

Fig. 25. Illustration of the effect of tensile strain on the d band center. Increasing the lattice constant shrinks the band width, and, to keep the number of d electrons fixed, the d states have to move up in energy.

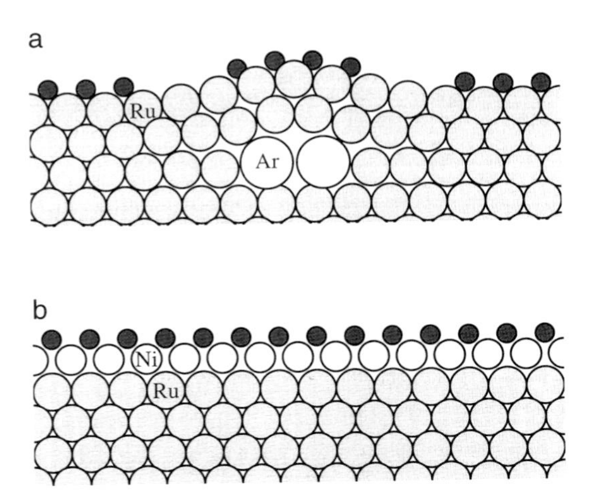

Fig. 26. Two ways of inducing strain on a single-crystal surface: (a) by Ar bubbles; (b) by the growth of strained pseudomorphic overlayers. Dark circles indicate adsorbates bound in regions of tensile strain.

metal. A third possibility involves characterization of dislocations ending at the surface (*103*).

Figure 27 shows the DFT results for adsorption and dissociation of CO on strained Ru(0001). The chemisorption potential energies and reaction energy barriers are plotted versus the *d* band center. A clear correlation between the chemisorption energies or energy barriers and the *d* band center is established. This correlation can be understood by comparison with the model calculation in Fig. 4. Here it was found that the higher the energy of the *d* bands, the more likely antibonding metal *d* adsorbate states are to be pushed above the Fermi level and the more likely the metal *d*–adsorbate interaction is to become net attractive.

The strain effect is not limited to CO adsorption and dissociation on ruthenium. Figure 28 shows the activation energy for N_2 dissociation on hexagonally close-packed iron surfaces with different lattice constants. Again, large lattice constants with high-lying *d* states are the most reactive.

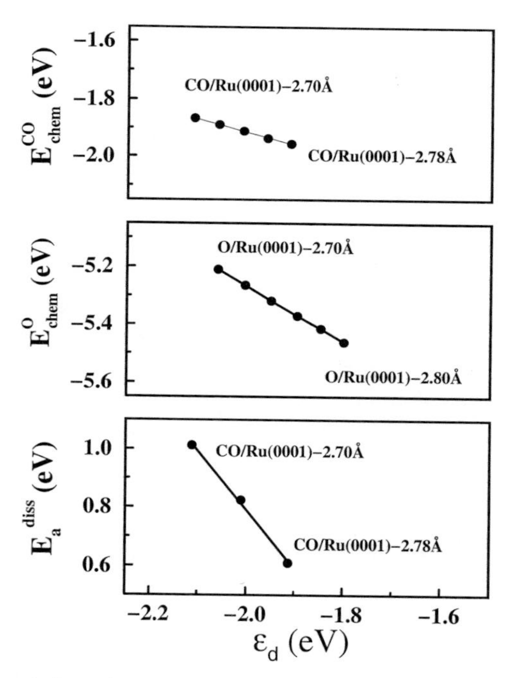

FIG. 27. Calculated (PW91) variation in the adsorption energy of CO, the adsorption energy of atomic oxygen, and the energy barriers for CO dissociation on a Ru(0001) surface for different lattice constants. The variations in energy are shown to be correlated with the variations in the *d* band center ε_d. Adapted from Mavrikakis *et al.* (*14*).

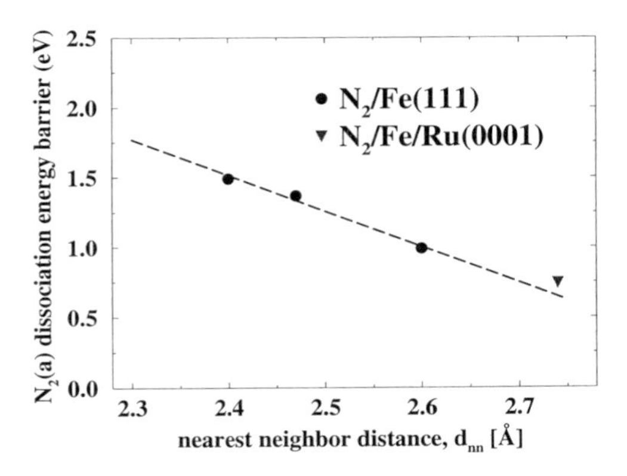

FIG. 28. Variation in the energy barrier for N_2 dissociation on a close-packed fcc–Fe(111) surface as a function of the nearest-neighbor separation in the surface. Also included is the result for a single layer of iron on Ru(0001). This point is placed at the equivalent ruthenium nearest-neighbor separation. The fact that it falls in line with the other points suggests that the main effect of the ruthenium substrate is to stretch the iron overlayer. From Logadottir (*104*).

Furthermore, one point corresponding to a monolayer of iron on Ru(0001) is included. It can be seen that the effect is the same and that the Fe/Ru(0001) point falls in line with the general trends. This result shows that in this case the main effect of the ruthenium is to increase the iron lattice constant. One would get the same result even without the ruthenium underneath.

Additional examples are shown in Fig. 29. Here, results from Pallassana and Neurock (*58*) for the activation energy for dehydrogenation of ethylene and of ethyl on different palladium overlayers are shown. Palladium on gold is more reactive than pure palladium because the gold lattice constant is larger than the palladium lattice constant and vice versa for ruthenium. The palladium/rhenium result represents an exception to the general rule that the main effect of the substrate is to change the lattice constant of the overlayer. The two metals have essentially the same lattice constant, but the palladium overlayer is less reactive than pure palladium. We note, however, that the shift in ε_d does capture the correct trend. In this case, the shift must be caused primarily by the interaction between the electrons in the palladium overlayer and the rhenium atoms underneath. We therefore conclude this section by noting that the reactivity scales very well with shifts in ε_d for both strained crystals and overlayers.

For nanosized supported catalysts, not just overlayers of one metal on another can have a changed reactivity due to strain. Small metal particles on an oxide support may also have a lattice constant (and thus a reactivity)

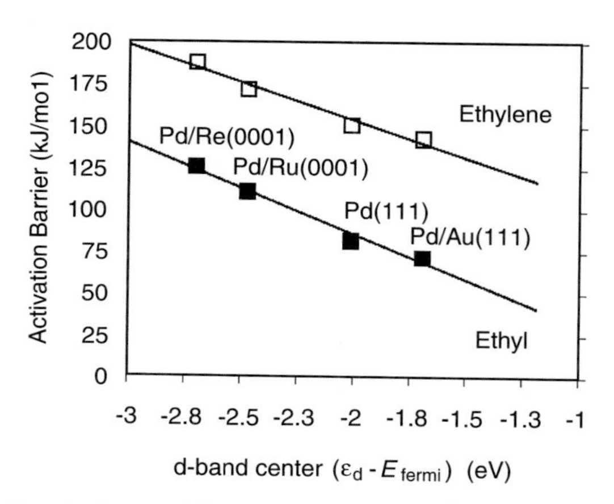

FIG. 29. Variations in the transition state energy for dehydrogenation of ethylene and of ethyl on Pd(111) and on palladium overlayers on various other metals. The variations in activation energy correlate well with the variations in the *d* band centers for the surface palladium atoms. From Pallassana and Neurock (*58*).

different from that of a large facet. Two factors play a role: (i) the surface tension and (ii) the interaction with the support. The former usually tends to decrease the lattice constant, whereas the latter may both increase and decrease the lattice constant. There are many examples in which lattice constants different from the bulk value have been observed for supported metal nanoparticles (*105–107*), and we expect this to be a very general phenomenon. Both the surface tension effect and the strain induced by the oxide support will be most important for very small particles, and the effect will decrease with increasing particle size.

2. *Different Facets*

When comparing the reactivities of different surface facets [e.g., fcc(111) and fcc(100)], both the electronic and the ion structure change. The change of the electronic structure is a consequence of the change of the ionic structure. We maintain, however, the strict division into electronic and geometric effects.

If an adsorption or reaction configuration is sufficiently simple, it is sometimes possible to disentangle the electronic and geometric effects, even when comparing different facets. Figure 30 shows examples of this: an atop bonding of CO on a range of platinum surfaces (*108*) and an H_2 transition state complex on two different copper surfaces (*109*). Because the local chemisorption configuration of CO is the same on all platinum

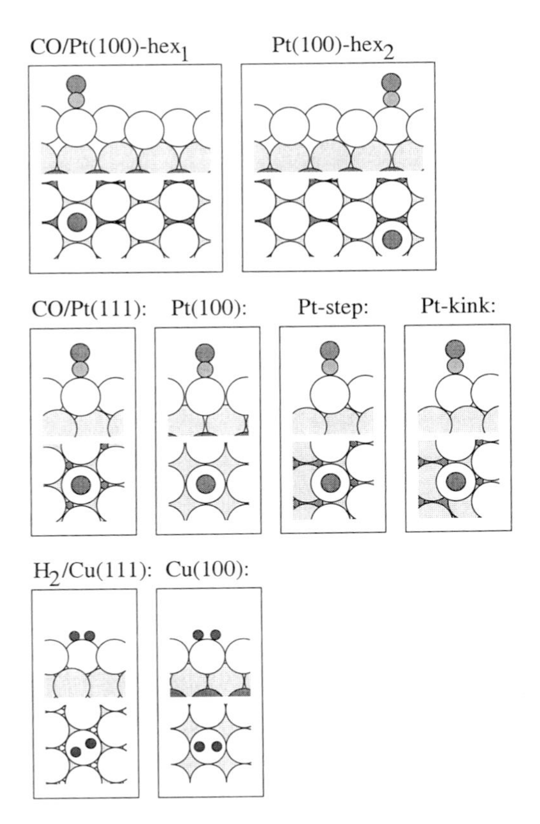

FIG. 30. Different ways of changing the adsorption or reaction geometry while not changing the local geometry. The chemisorbed CO and the H_2 transition state complex are probing the local electronic structure effects.

surfaces in Fig. 30, the variations in the CO chemisorption potential energy must be ascribed primarily to the electronic effects (although second-neighbor effects may also be of some importance). Likewise, because the H_2 reaction geometry (over the atop site into bridge sites) is the same on the two copper surfaces considered, the variations in the H_2 dissociation energy barriers must be ascribed to the changes in the electronic structure between the facets. The correlation of the CO chemisorption energy and H_2 dissociation energy barrier with our parameter for the electronic effect, the local center of the metal d states, is presented in Fig. 31, which shows that when the adsorption or reaction geometry is kept fixed, the effect of the facet is well described by the d band center.

The magnitude of the geometric effect can be estimated by allowing the

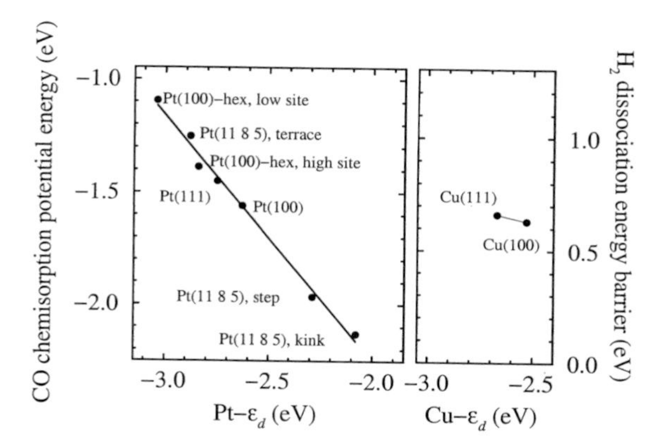

FIG. 31. (Left) Calculated variations in the chemisorption energy (PW91) for CO adsorbed atop platinum atoms in different surroundings. (Right) DFT energy barriers (PW91) for dissociative adsorption of H_2 atop copper atoms in two different copper surfaces. The correlation between chemisorption energy and the *d* band center for the relevant platinum atoms is evident. Adapted from Hammer *et al.* (*108*) and Kratzer *et al.* (*109*).

adsorbate or transition state complex to exploit all available ionic degrees of freedom in the bond formation rather than restricting the geometry as was done previously. For the H_2 dissociation, new optimum transition state complexes result on the different copper surfaces. These are shown in Fig. 32. On Cu(100), a tilted configuration of the H_2 transition state complex is found in which a hydrogen atom is almost in a fourfold hollow site (*33*). On Cu(111), the H_2 axis is parallel to the surface and both hydrogen atoms are in threefold-like positions. The transition state potential energies are reduced by 0.14 and 0.03 eV for Cu(111) and Cu(100), respectively, com-

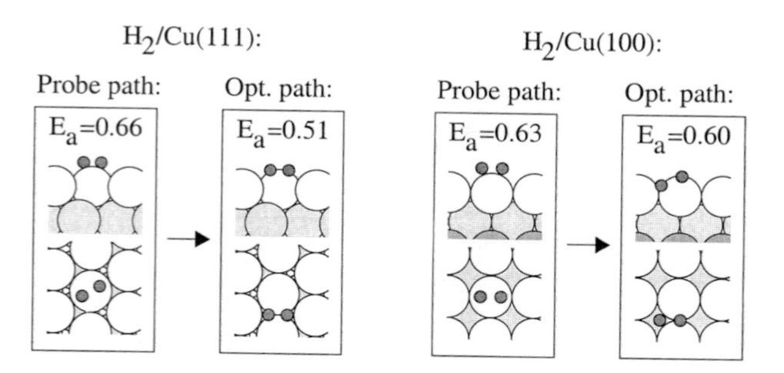

FIG. 32. The calculated transition states and (PW91) energy barriers (in eV) for H_2 dissociation on Cu(100) and Cu(111). Both the transition state for the atop path used to probe the electronic effect and the optimum energy path are shown. Adapted from Kratzer *et al.* (*109*).

pared with the values obtained by using the restricted transition state configurations of Fig. 30. The difference, 0.11 eV, is thus a measure of the geometric effect for H_2 dissociation on a low-index copper surface.

3. *Steps and Defects*

When considering high-Miller-index surfaces with steps and kinks, or surfaces with other defects such as adatoms or vacancies, the variations in the electronic structure become larger (i.e., ε_d varies more). Again, the smaller the metal coordination number of the surface atom, the smaller the *d* band width and hence the higher energy of the *d* band center (if the *d* band is more than half filled). The electronic effect is completely analogous to what we have previously demonstrated. This may be seen from Fig. 31, in which the CO bonding to step and kink platinum atoms is also included.

The electronic effect may become so large at the step edges of metal surfaces that it can change the site preference of atomic adsorbates. For example, on a flat Ru(0001) surface, an oxygen atom prefers threefold bonding at an hcp site by 0.54 eV over twofold bonding at a bridge site. On Ru(0001), all ruthenium atoms have the same *d* band center, and therefore the site preference is a purely geometrical effect, originating from the change in coupling matrix element, adsorbate coordination, and nonlocal electrostatic effects caused by the change in adsorption site. On a monoatomic step on Ru(0001), however, the oxygen is as stable (within the calculational accuracy) in a hang-off-the-edge, twofold bonding configuration at the step edge as at an hcp threefold site behind the step edge, and it is preferred in the hang-off-the-edge configuration over the fcc threefold configuration by 0.27 eV. The configurations are depicted in Fig. 33 (*110*). The reason for the change in site preference from threefold at Ru(0001) to twofold at a monoatomic step can (with our division) be ascribed to the electronic effect (i.e., the finding that at the step edge the lower coordination of the surface atoms leads to energetically higher *d* bands and hence to higher reactivity). Similar effects have been observed, for example, by Feibelman *et al.* (*111*) for the O/Pt(111) system.

The electronic effect at steps is quite general. In addition to the examples stated previously, it has been shown in calculations that steps increase the stability of NO, N, and O on palladium (*38, 66*) and of N_2 on platinum (*112*) and that they lower the barrier for CH_4 dissociation on nickel (*113*), increase the stability of various C_2H_x species on platinum, and lower the barriers for C–H and C–C bond breaking on the same surfaces (*82*).

The geometric effect, however, may also become more prominent at a step edge. This has been illustrated in two recent studies of activation of NO and N_2 at monoatomic steps on Ru(0001) (*83, 84*). It was found that

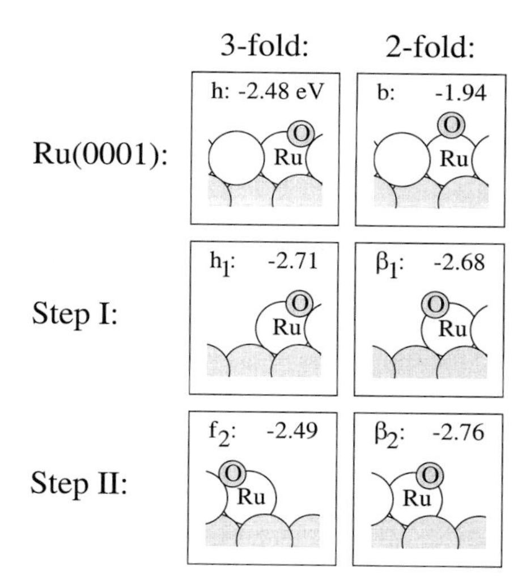

FIG. 33. Chemisorption of oxygen on a flat and stepped Ru(0001) surface. The chemisorption potential energy is given in eV. On the flat Ru(0001) surface threefold hcp chemisorption (h) is highly preferred over twofold bridge chemisorption (b). At the step edges, hcp chemisorption (h_1) is comparable to bridge chemisorption (β_1) for one-step geometry (step I), and bridge chemisorption (β_2) even becomes preferred over fcc chemisorption (f_1) at the other step geometry (step II). Adapted from Hammer (*110*).

the N–O and N–N bond activation barriers almost vanish for a reaction geometry in which the molecules initially are positioned at the base of the step. The effect is too large to be ascribed solely to the electronic effect at the step edge, the size of which is known from the NO and N_2 dissociation in configurations at the top of the step edges. Rather, a stabilization of the transition state complex on the order of 0.3 eV must be explained as originating from the geometric effect. In Fig. 34, the NO and N_2 transition state complexes are shown for reaction on flat and stepped Ru(0001), and the potential energy diagram is given for N_2. The NO and N_2 are seen to coordinate to five ruthenium atoms in the highly reactive configurations at the base of the step edges, whereas they coordinate to only four ruthenium atoms on flat Ru(0001). The coordination to more ruthenium atoms at the step increases the total coupling matrix element between metal d states and adsorbate valence states. Simultaneously, the indirect repulsion between the reaction products (the chemisorbed nitrogen and oxygen atoms) is reduced (see Section III.C).

There is extensive experimental evidence that bonding is stronger at steps than at facets (*2, 114, 115*). The generality of the effect suggests that it may even be more important than has been realized. In the case of NO

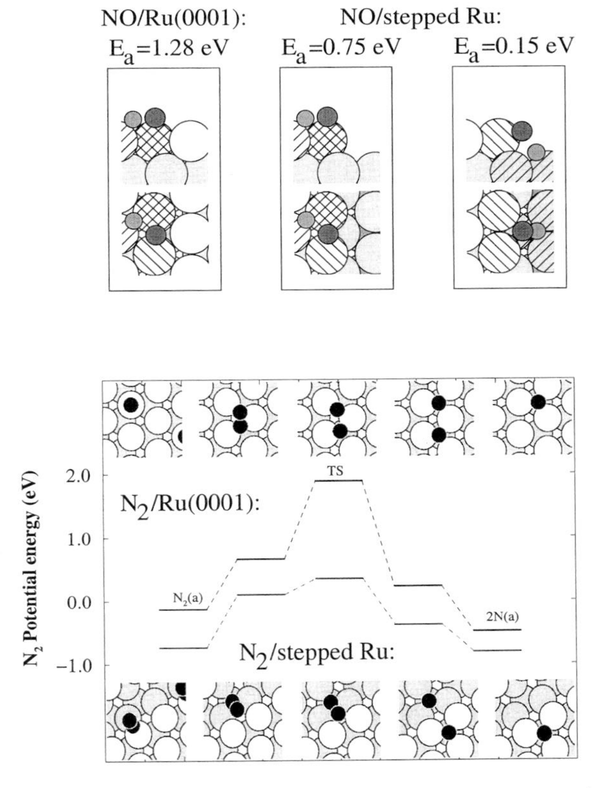

FIG. 34. (Top) Transition state geometries and corresponding energies (PW91) for NO dissociation on Ru(0001) terraces and steps. (Bottom) Reaction path and energies (RPBE) for N_2 dissociation on the same surface. Again, both a terrace and a step configuration have been considered. In both cases the energy zero is the molecule far from the surface. Adapted from Dahl *et al.* (*83*) and Hammer (*84*).

dissociation on Ru(0001) steps, it has been shown by direct observation of the dissociation products by scanning tunneling microscopy that the dissociation occurs primarily at steps (*116*). In the case of N_2 dissociation on Ru(0001), the reactivity of the steps is so much greater than that of the terraces that even the few steps present on any single-crystal surface (on the order of 1%) are enough to completely dominate the experiments. In this case, it was the large barrier found in the calculations that led to the discovery that the experiments were actually not measuring the reactivity of the terrace atoms (*83, 117*).

Traditionally, reactions in heterogeneous catalysis have been divided into structure-sensitive and structure-insensitive reactions, depending on whether the reaction rate depends on particle size (*118–120*). A particle-size-dependent activity is usually a simple consequence of the structure dependence discussed previously. For small catalyst particles, the ratio of the different facets and the density of steps and other defects will be strongly dependent on particle size. Such structure sensitivity, however, may also be a consequence, for example, of the strain effect discussed in Section V.A.1.

B. Alloying

Another way of changing the reactivity is through alloying. The addition of one or more chemical elements to a metallic surface increases the possible bonding geometries of adsorbates and reaction complexes and simultaneously changes the electronic structure of the alloy surface from that of the pure metallic surface alone. The effect of alloying may also be more indirect, however, if, for example, one of the alloy elements segregates to the surface. In this case, effects such as strain in the surface due to a difference in bulk lattice parameters of the alloy and its elemental components can be used to understand the effect of alloying (see Section V.A.1).

The reactivities of many alloy surfaces have been studied by density functional theory. H_2 dissociation, which is nonactivated on Ni(111), was found to experience large energy barriers—even on nickel sites—on NiAl(110), and the increased repulsion can be assigned the electronic effect and traced to a downshift of the nickel d bands in the alloy (*45, 121*). Similarly, the platinum sites in a $Cu_3Pt(111)$ surface were found to be more reactive with respect to H_2 bond activation than platinum sites in the Pt(111) surface, and again it is an electronic effect originating from an upshift of the platinum d bands on alloying (*45*).

An interesting type of alloy, Au/Ni, was investigated by Kratzer *et al.* (*86*) for its ability to dehydrogenate methane. Gold and nickel are immiscible in the bulk, but a two-dimensional (2-D) surface alloy exists in which gold alloys into the outermost layers of a Ni(111) crystal (Fig. 35). The alloy is interesting from a synthesis point of view because the gold does not dissolve into the nickel bulk, which implies that only small traces of gold are required to create the alloy (*122, 123*). The gold is inert for methane decomposition, and because the gold causes a downshift of the nickel d bands at neighboring sites, the ability of the alloy to catalyze the C–H bond activation is slightly inferior to that of the clean Ni(111) (Fig. 36). A nickel-based catalyst, however, suffers from carbon deposition (growth of carbon whiskers) during operation (*124*). Due to the electronic effect, carbon is found to bond less favorably to the Au/Ni(111) surface than to the Ni(111) surface. The inhibition of the carbon formation process is often more important than a

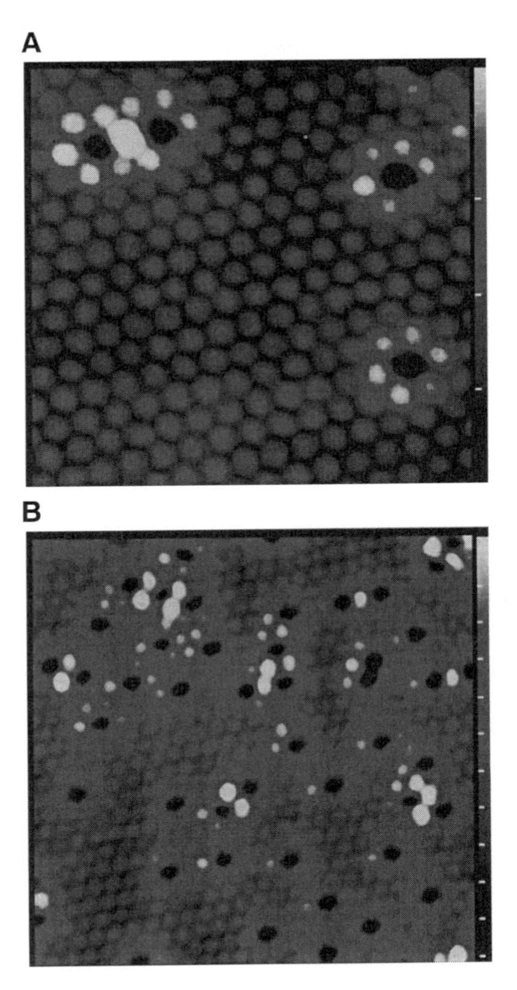

FIG. 35. Two STM images of a Ni(111) surface with 2% (A) and 7% (B) of a monolayer of gold. The gold atoms appear black in the images. The nickel atoms next to the gold atoms appear brighter due to a change in geometry and electronic structure, indicating that the chemical reactivity of the nickel atoms may be modified by nearest-neighbor gold atoms. Reprinted with permission from Besenbacher *et al. Science* **279,** 1913 (1998). Copyright 1998 American Association for the Advancement of Science.

high reactivity with respect to the primary reaction channel, the methane decomposition. On the basis of DFT studies it was therefore suggested that the 2-D alloy of Au/Ni would be a better catalyst for methane dehydrogenation. Subsequent synthesis of an Au/Ni catalyst has shown this to be true (Fig. 37) (*122*).

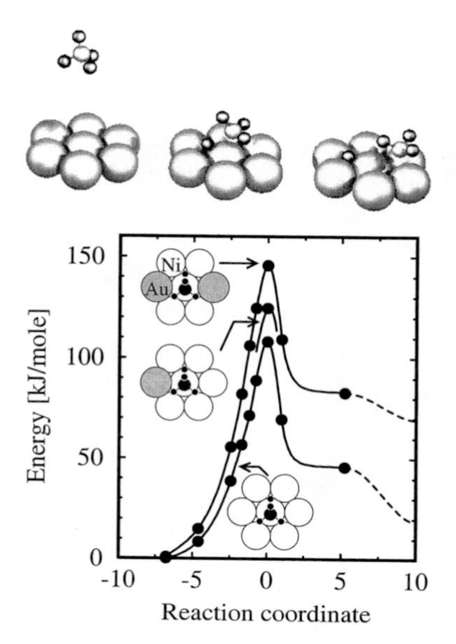

FIG. 36. Illustration of the calculated minimum energy path (reaction coordinate, in Å) for the CH_4 dissociation on a Ni(111) surface. The energy (PW91) among the path is shown below. The shift in the barrier for dissociation when one or two gold atoms are nearest neighbors to the nickel atom under the dissociating CH_4 molecule is also included. Adapted from Holmblad *et al.* (*123*).

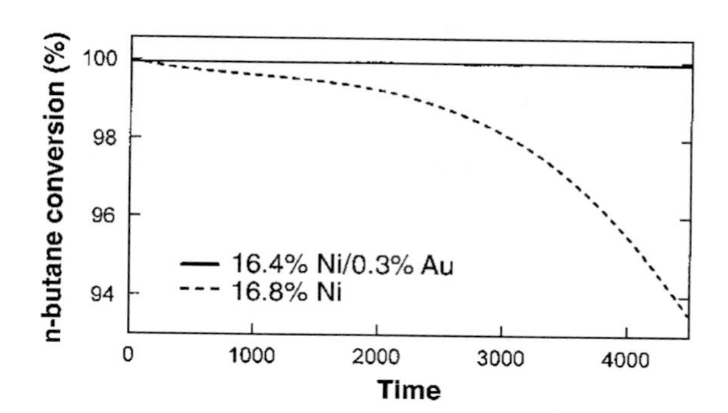

FIG. 37. Conversion of *n*-butane as a function of time (in seconds) during steam reforming in a 3% *n*-butane/7% hydrogen/3% water in helium mixture at a space velocity of 1.2 h^{-1}. The dashed curve shows the *n*-butane conversion for the nickel and the solid curve is for the gold/nickel supported catalyst. Reprinted with permission from Besenbacher *et al. Science* **279,** 1913 (1998). Copyright 1998 American Association for the Advancement of Science.

Several DFT calculations have been reported that show the effect of alloying on adsorption and reactions. Delbecq and Sautet (*125, 126*) investigated the bonding of CO and of NO on surfaces of Pd_3Mn alloys. Pallassana *et al.* (*127*) also investigated the effect of alloying on hydrogen chemisorption on Pd–Re alloys and found that the adsorption energy varies with the average *d* band center, as has been previously determined.

The strong correlation between ε_d and bond energies and activation barriers makes it interesting to have an overview of the way the *d* band centers change when one metal forms a monolayer on top of (or is alloyed into) the first layer of another metal. These shifts can be calculated by using the DFT, and Table II summarizes results for many combinations of catalytically interesting metals (*128*). We stress that the results are for

TABLE II

Shifts in d Band Centers of Surface Impurities and Overlayers Relative to the Clean Metal Values (Bold)[a]

	Fe	Co	Ni	Cu	Ru	Rh	Pd	Ag	Ir	Pt	Au
Fe	**−0.92**	0.05	−0.20	−0.13	−0.29	−0.54	−1.24	−0.83	−0.36	−1.09	−1.42
		0.14	−0.04	−0.05	−0.73	−0.72	−1.32	−1.25	−0.95	−1.48	−2.19
Co	0.01	**−1.17**	−0.28	−0.16	−0.24	−0.58	−1.37	−0.91	−0.36	−1.19	−1.56
	−0.01		−0.20	−0.06	−0.70	−0.95	−1.65	−1.36	−1.09	−1.89	−2.39
Ni	0.09	0.19	**−1.29**	0.19	−0.14	−0.31	−0.97	−0.53	−0.14	−0.80	−1.13
	0.96	0.11		0.12	−0.63	−0.74	−1.32	−1.14	−0.86	−1.53	−2.10
Cu	0.56	0.60	0.27	**−2.67**	0.58	0.32	−0.64	−0.70	0.58	−0.33	−1.09
	0.25	0.38	0.18		−0.22	−0.27	−1.04	−1.21	−0.32	−1.15	−1.96
Ru	0.21	0.26	0.01	0.12	**−1.41**	−0.17	−0.82	−0.27	0.02	−0.62	−0.84
	0.30	0.37	0.29	0.30		−0.12	−0.47	−0.40	−0.13	−0.61	−0.86
Rh	0.24	0.34	0.16	0.44	0.04	**−1.73**	−0.54	0.07	0.17	−0.35	−0.49
	0.31	0.41	0.34	0.22	0.03		−0.39	−0.08	0.03	−0.45	−0.57
Pd	0.37	0.54	0.50	0.94	0.24	0.36	**−1.83**	0.59	0.53	0.19	0.17
	0.36	0.54	0.54	0.80	−0.11	0.25		0.15	0.31	0.04	−0.14
Ag	0.72	0.84	0.67	0.47	0.84	0.86	0.14	**−4.30**	1.14	0.50	−0.15
	0.55	0.74	0.68	0.62	0.50	0.67	0.27		0.80	0.37	−0.21
Ir	0.21	0.27	0.05	0.21	0.09	−0.15	−0.73	−0.13	**−2.11**	−0.56	−0.74
	0.33	0.40	0.33	0.56	−0.01	−0.03	−0.42	−0.09		−0.49	−0.59
Pt	0.33	0.48	0.40	0.72	0.14	0.23	−0.17	0.44	0.38	**−2.25**	−0.05
	0.35	0.53	0.54	0.78	0.12	0.24	0.02	0.19	0.29		−0.08
Au	0.63	0.77	0.63	0.55	0.70	0.75	0.17	0.21	0.98	0.46	**−3.56**
	0.53	0.74	0.71	0.70	0.47	0.67	0.35	0.12	0.79	0.43	

[a] The impurity/overlayer atoms are listed horizontally, and the host entries are listed vertically. For each combination, the first of the two numbers listed represents the isolated surface impurity and the second the overlayer. The surfaces considered are the most close packed, and the overlayer structures are pseudomorphic. No relaxations from the host lattice positions are included. All values are in eV, and the elemental *d* band centers are relative to the Fermi level. From Ruban *et al.* (*128*).

overlayers and surface impurities occupying lattice positions of the substrate, and we have not taken into account here whether these structures are actually stable under experimental conditions. However, Table II can provide an idea about possible ways of modifying the electronic structure of a surface.

Consider for example the case of platinum on Au(111). According to the results of Table II, both platinum overlayers and platinum impurities in a Au(111) surface have higher-lying *d* states than platinum atoms in a Pt(111) surface and should therefore bind adsorbates more strongly and have lower transition state energies. We note that this may still lead to higher activation barriers for certain elementary steps if the initial state for this step is stabilized more than the transition state. The somewhat counterintuitive prediction that gold should *increase* the reactivity of platinum has been tested experimentally (*102*). In Fig. 38, it is clearly shown that CO is bonded more strongly to a monolayer of platinum on Au(111) than to Pt(111), in accordance with the prediction from Table II. The

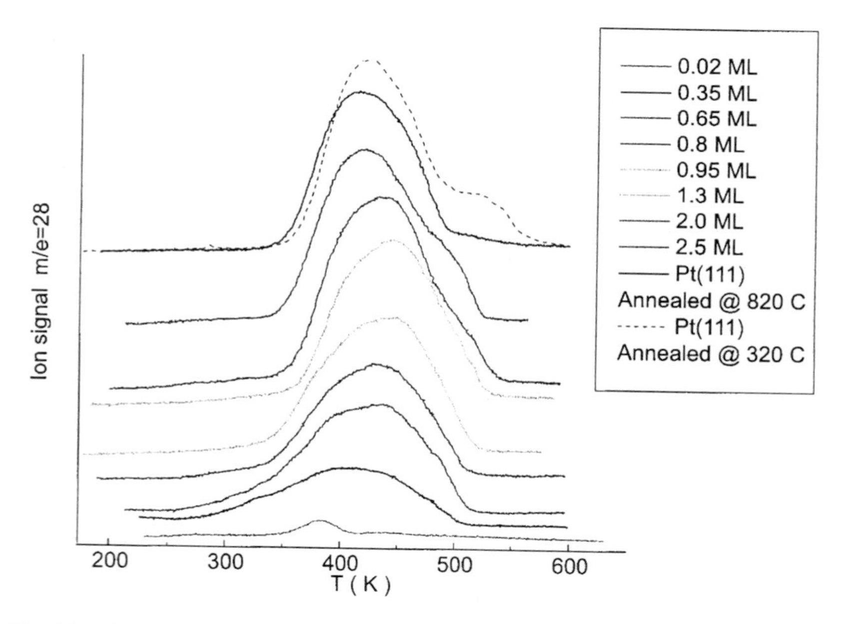

FIG. 38. CO thermal desorption spectra for different coverages of platinum Au(111). The spectrum for the clean Pt(111) surface is shown for comparison. Approximately one monolayer of platinum on the Au(111) surface is shown to bind CO more strongly (with a higher desorption temperature) than the clean Pt(111) surface. Reprinted from Pedersen *et al.* (*102*). Copyright 1999 with permission from Elsevier Science.

experimental results also show that platinum impurities do not bind CO as strongly as Pt(111). This would not be expected from Table II. This discrepancy can be traced to contributions to the bonding from second-nearest neighbors in the surface (*102*), and the example shows that one must be careful in using Table II when the second-nearest neighbors significantly change the reactivity. In the current case, platinum and gold are so different that even though only about 10% of the bonding comes from the second neighbors, it is enough to be important. This problem does not occur when pseudomorphic overlayers are considered.

It is known from several theoretical investigations that the variation in the center of the *d* bands for metal overlayers is accompanied by a similar variation in the surface core level shifts—at least toward the right in the transition metal series (*129–131*). It has also been shown that the *trends* in variations in the surface core level shifts for different overlayers are often given by the initial state shift, i.e., by the changes in the electronic structure of the unperturbed surface. Therefore, we conclude that we can view the variation in the surface core level shifts as a measure of the variation in the *d* band center. This is fortuitous because it means that in the surface core level shifts there is an *in situ* experimental measure of the valence *d* band center shifts that are important for the chemical properties of metal surfaces, as discussed previously. As an outstanding example of the information carried by the surface core level shifts, Rodriguez and Goodman (*97*), in their detailed studies of metallic overlayers, observed a strong correlation between the surface core level shift of such overlayers and the CO chemisorption energy at these overlayers.

When using Table II, for example, to decide which surface composition will be best for a given reaction, it must be borne in mind that the actual surface composition of a real catalyst particle cannot be varied at random. The tendency of one metal to segregate to the surface of another largely controls the surface composition. In the design of a catalyst, it therefore is of interest to have an overview of the tendency of different metals to segregate to the surface. Again, DFT calculations are a useful source of such information. Figure 39 shows a compilation of surface segregation energies of all transition metals (*132*). It can be seen from Fig. 39, for example, that even though the Pt/Au(111) system has interesting chemical properties is not a potential catalyst because the platinum is driven to the bulk by the energetics [the segregation energy for the Pt/Au(111) system is positive].

In many cases, alloying effects have been shown experimentally to change the reactivities of single-crystal surfaces (*99, 133–135*), and alloying effects have been explored extensively in catalyst development (*136*). Testing of such catalysts may benefit especially from new fast screening techniques

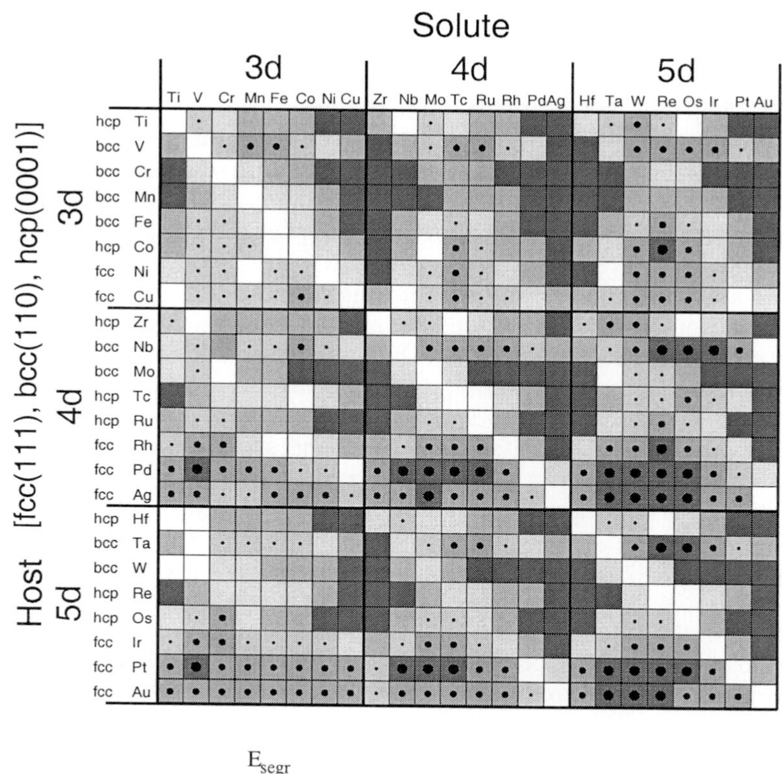

E_{segr}

(dark)	< - 0.7 eV	very strong segregation
	-0.7 ... -0.3 eV	strong segregation
	-0.3 ... -0.05 eV	moderate segregation
	-0.05 ... 0.05 eV	no segregation
•	0.05 ... 0.3 eV	moderate antisegregation
•	0.3 ... 0.7 eV	strong antisegregation
●	> 0.7 eV	very strong antisegregation

FIG. 39. Compilation of calculated segregation energies on the most close-packed surface of all binary combinations of transition metals. Adapted from Christensen *et al.* (*132*).

(*137*). The concepts developed previously, in conjunction with databases such as those in Fig. 39 and Table II, may be useful in guiding the choice of candidate combinations of metals to test experimentally.

C. Promotion and Poisoning

The reactivity of a metal surface is modified by the presence of coadsorbates. Alkali metal adsorbates, for instance, are added to the iron- and ruthenium-based industrial ammonia synthesis catalysts, causing an enhanced reactivity with respect to N_2 bond activation (*138–140*). Another example of a coadsorbate-modified catalyst is the nickel-based catalyst used in the SPARG process for steam reforming (*124*). In this process, H_2S is added to the feed gas flow, resulting in adsorbed sulfur atoms that inhibit coking of the surface in much the same way that gold inhibits coking in the Au/Ni catalyst discussed previously.

The interaction of identical coadsorbates was considered in Section III.C, and four different interaction types were described. The direct overlap of orbitals is operative for small adsorbate–adsorbate distances. It always occurs and is usually repulsive. The same is true of the indirect elastic interaction through the substrate. The other two interaction types are more interesting:

- Indirect interaction through the metal *d* bands
- Direct electrostatic interaction

The indirect interaction through the metal *d* bands is illustrated in Fig. 40 (*141*). Here, the Ru 4*d* bands at a ruthenium atom next to a threefold site are shown before and after the adsorption of sulfur in the threefold site. Figure 40 shows how the ruthenium 4*d* band center is shifted down upon adsorption of sulfur. The sulfur atom effectively provides some of the coordination that a ruthenium atom in the surface lacks. Once the ruthenium atom is more highly coordinated, the *d* band width of the surface atom increases and the *d* band shifts down. This electronic effect weakens the bonding of other adsorbates in sites involving ruthenium atoms that have

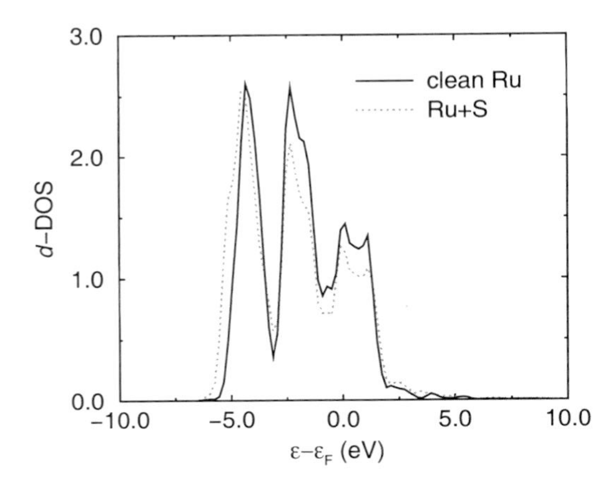

FIG. 40. The Ru(0001) *d* DOS including the effect of sulfur. Adapted from Mortensen *et al.* (*141*).

already reacted. For N_2 dissociation on a Ru(0001) surface with coadsorbed sulfur, the energy barrier is increased substantially by the presence of the surface sulfur atoms (Fig. 41). Farther away, the effect is weaker, however, because it involves only the shift of *d* bands at immediate surface neighbor atoms.

So far, we have implicitly assumed that the changes in adsorbate bond strengths and energy barriers resulting from changes in the electronic structure of a metal surface can be traced to the Kohn–Sham eigenvalue spectra of the DFT calculations, i.e., to the change in the sum of Kohn–Sham eigen energies resulting from adsorption, $\Delta \sum_i \varepsilon_i$. A formal derivation, however, reveals that electronic structure changes may also be manifest in changes of the (nonlocal) electrostatic contributions, ΔE_{es}, to the adsorbate bond strengths and energy barriers (*40, 44, 142*). The potential energy change from the electronic effect must therefore be written as follows:

$$\Delta E_a = \Delta \sum_i \varepsilon_i + \Delta E_{es}. \tag{2}$$

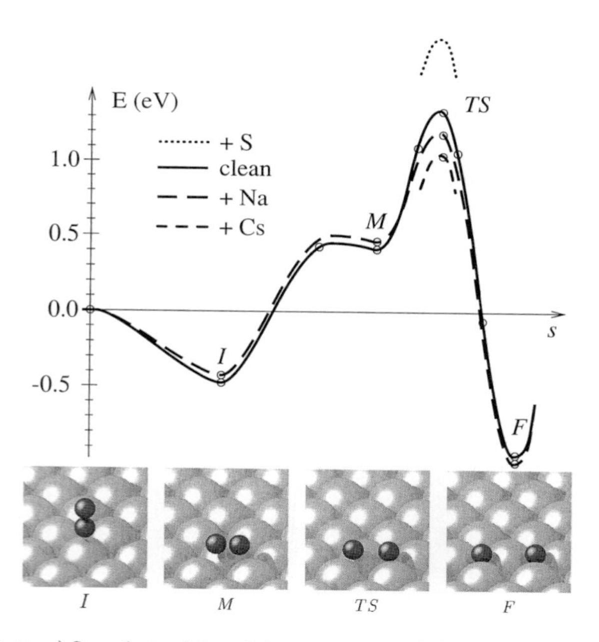

FIG. 41. (Bottom) Snapshots of the minimum energy path for the N_2 dissociation reaction: Initial state in which the molecule is standing perpendicular to the surface (*I*), metastable state (*M*), transition state (*TS*), and final state (*F*). (Top) The energy along the path. The effect on the transition state of adding one-eighth of a monolayer of sulfur or sodium or one-sixth of a monolayer of cesium is also shown. Adapted from Mortensen *et al.* (*141*).

When sodium or cesium is adsorbed on Ru(0001), there is no *d* band shift (relative to the Fermi level) at ruthenium sites farther than one lattice constant away. Consequently, the first term in Eq. (*2*) is approximately zero. Because of charge transfer from the alkali metal adsorbates to the ruthenium surface, however, there are electrostatic effects. This is shown in Fig. 42, in which the *d* bands are shown to be essentially unchanged, whereas an electrostatic field, E, has been induced outside the ruthenium surface. In Fig. 41 the result of a full DFT description of the N_2 dissociation on a Ru(0001) surface with coadsorbed sodium or cesium is also shown. The energy barrier is reduced by some amount $|\Delta E_{TS}|$ as a consequence of the presence of either sodium or cesium. These atoms thus act as promoters for the N_2 bond activation. The effect can be traced to the electrostatic interaction between the dipole, μ, of the N_2 transition state complex interacting with the induced electrostatic field, E, due the adsorbed alkali atoms.

In Fig. 43, ΔE_{TS} is plotted as a function of the quantity

$$\Delta E_{\text{dip}} = -\text{E}\mu. \tag{3}$$

A clear correlation is evident, confirming that the electrostatic interaction is dominating (*143*).

In the examples discussed previously, special care was taken to consider low coverages of poisons or promoters. This is usually the situation encoun-

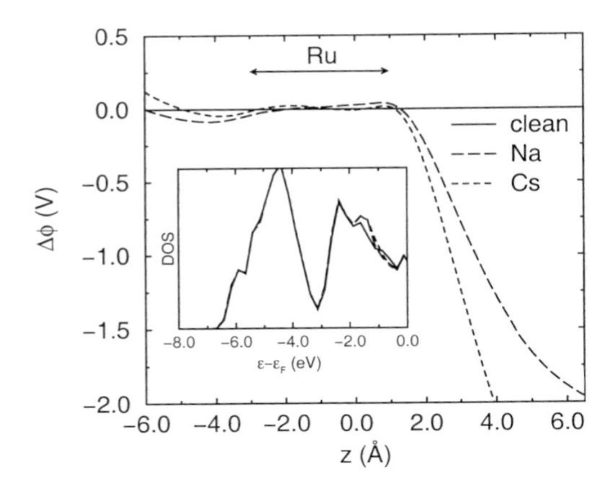

FIG. 42. Calculated change in the electrostatic potential outside a Ru(0001) surface as a result of the adsorption of a sodium or a cesium atom. The potential is shown along a line perpendicular to the ruthenium surface at a site 5.48 Å from the alkali metal atom. In the inset, the *d* DOS is shown with (dashed line) and without (solid line) the sodium illustrating that the *d* DOS is almost unaffected by the presence of the alkali metal atom. From Mortensen *et al.* (*141*).

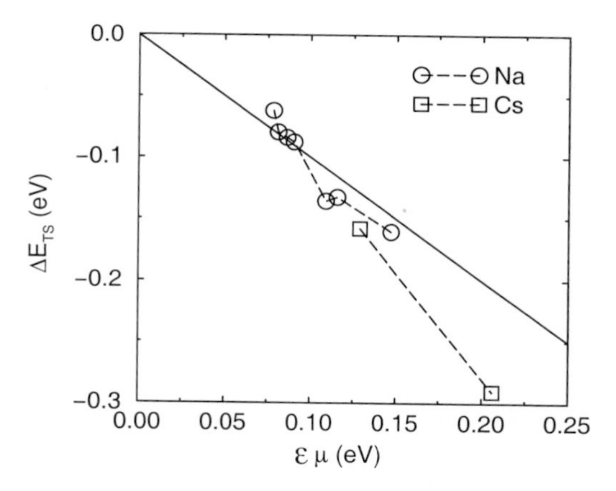

FIG. 43. Interaction energy between alkali metal atoms and N_2 in the transition state for dissociation as a function of $E\mu$. The alkali metal atoms are sodium (circles) and cesium (squares). Many different geometries have been considered corresponding to different distances between the transition state complex and the alkali metal atoms. Adapted from Mortensen *et al.* (*141*).

tered in practical catalysis, and it singles out the long- and short-range effects as well. If one instead considers high coverages, one finds that all the interaction types work simultaneously, and the picture becomes much more complicated (*144, 145*).

VI. Summary and Extension of Concepts to Metal Sulfides

Density functional calculations of adsorption and reactions at metal surfaces have reached a point at which the complexity of real catalytic surfaces and adsorbates can be treated at a level of accuracy that is sufficient for understanding bonding mechanisms, for determining reaction pathways, and for comparing different systems. The calculations therefore provide a new and powerful source of input to the description of heterogeneous catalysis. The calculations will never replace experiments, but they may in many cases be the simplest or even the only possible way of obtaining some information.

The calculations also provide new ways of testing whether the concepts we use to describe surface chemical processes are correct. For instance, it has been shown in this way that earlier reactivity measures such as the density of states at the Fermi level or the number of *d* holes cannot in general describe variations in reactivity from one transition metal to the next (*45*).

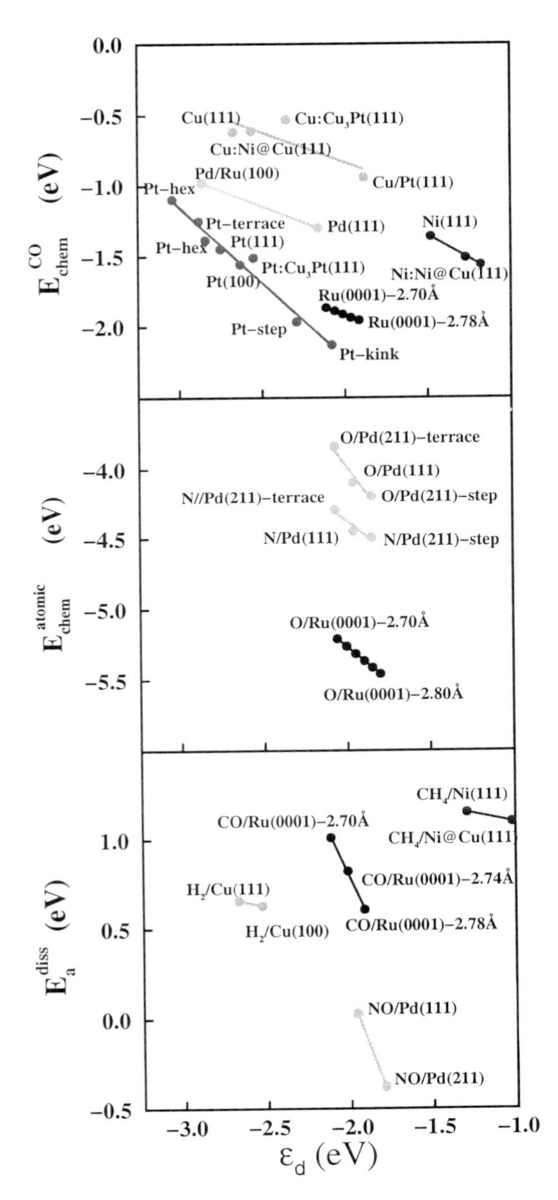

FIG. 44. Molecular (E^{CO}_{chem}) and atomic (E^{atomic}_{chem}) binding energy as a function of the d band center (ε_d) of the metal surface (top and middle, respectively). The barrier for dissociation of small molecules, referred to gas-phase zero, as a function of ε_d is shown in the bottom. Common shadings are used for data corresponding to the same metal in each of the three panels. Lines drawn represent best linear fits. *X*:*XY* reflects chemisorption or dissociation at atom *X* in an *XY*–alloy surface. *X*@*Y* means an *X* atom impurity in a *Y* surface. From Mavrikakis *et al.* (*14*).

In this review, we have focused on some simple concepts that can be used to classify the reactivities of transition metal surfaces. The goal is to gain an understanding of the properties of the clean surface that determine its reactivity. If we can single out these factors, we have powerful concepts that can be used to develop new and more effective catalysts.

We have shown that electronic and geometrical factors can be separated and that both are important in determining the reactivity. The main conclusion of our work is that when considering variations in the reactivity of a particular metal or group of metals, a single parameter, the center of the *d* band (ε_d), is strongly related to both the stabilities of atoms and molecules on the surface and the energies of transition states for surface processes. A summary of many calculations illustrating this point is shown in Fig. 44. We have shown in detail why this is the case, and we have shown what determines variations in ε_d; low-coordinated or "expanded" metal atoms are more reactive than highly coordinated or "compressed" metal atoms.

These concepts have been developed specifically for transition metal surfaces, but there are good reasons to believe that they are more general. To exemplify this, we show here as the final example how very similar concepts can be used to understand the reactivities of transition metal sulfides. MoS_2 (on supports) is used, extensively as a catalyst for hydrotreating processes. It is known that cobalt and nickel promote the catalysts, whereas iron (next to these elements in the periodic table) does not. It is also known that the promoter atoms are situated at the perimeter of the supported MoS_2 nanoparticles. In this connection, it is interesting to know

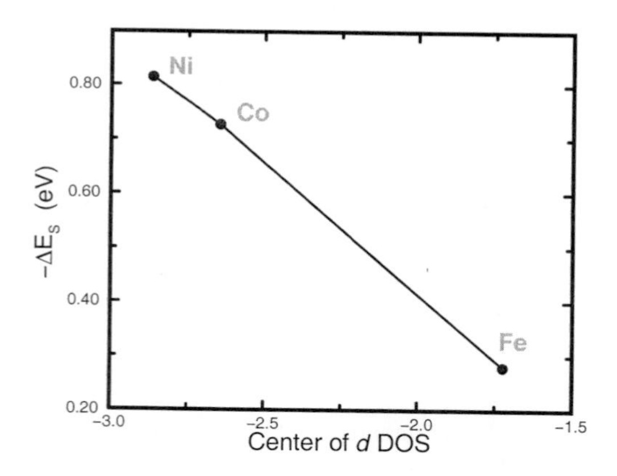

FIG. 45. Plot of the change in the S-binding energy to the edge of a MoS_2 slab, with Fe, Co, and Ni substituted for one of the Mo atoms next to the S atom in question. The result is shown as a function of the center of the projected *d* density of states on the Fe, Co, or Ni atoms in questions. From Byskov *et al.* (*146*).

how iron, cobalt, and nickel affect the stability of edge sulfur atoms. If, for example, the sulfur atoms are more easily removed in the presence of cobalt, this might create more active sites with coordinatively unsaturated metal atoms and thus explain the promoting effect. In Fig. 45, the variations in the calculated sulfur binding energy (the reaction energy for H_2S formation from H_2) for different promoter atoms are shown. Again, the variations correlate with the position of the *d* band center.

Acknowledgments

We thank many colleagues and collaborators for contributing to the work described in this review, including H. Bengaard, F. Besenbacher, L. Byskov, I. Chorkendorff, B. S. Clausen, J. A. Dumesic, M. V. Ganduglia-Pirovano, A. Groß, L. B. Hansen, K. W. Jacobsen, P. Kratzer, J. H. Larsen, A. Logadottir, B. I. Lundqvist, E. Lægsgaard, M. Mavrikakis, Y. Morikawa, J. J. Mortensen, M. Neurock, O. H. Nielsen, V. Pallassana, J. Rostrup-Nielsen, A. Ruban, M. Scheffler, H. L. Skriver, Ž. Šljivančanin, I. Stensgaard, K. Stokbro, P. Stoltze, E. Törnqvist, H. Topsøe, R. M. Watwe, and J. T. Yates. JKN gratefully acknowledges the hospitality of H. Metiu and the support from the Center for Quantized Electronic Structures, University of California, Santa Barbara, during the writing of the manuscript. This work was in part financed by the Danish Research Councils through The Center for Surface Reactivity and Grants 9501775, 9800425, and 9803041. The Center for Atomic-Scale Materials Physics is sponsored by the Danish National Research Foundation.

References

1. Barteau, A. M., and Vohs, J. M., *in* "Handbook of Heterogeneous Catalysis" (G. Ertl, H. Knözinger, and J. Weitkamp, Eds.), p. 873. VCH, Weinheim, 1997.
2. Somorjai, G. A., *in* "Introduction of Surface Chemistry and Catalysis." Wiley, New York, 1994.
3. Gunter, P. L. J., Niemantsverdriet, J. W., Ribeiro, F., and Somorjai, G. A., *Catal. Rev.—Sci. Eng.* **39,** 77 (1997).
4. Cambell, C. T., *Surf. Sci. Rep.* **27,** 1 (1997).
5. Valden, M., Lai, X., and Goodman, D. W., *Science* **281,** 1647 (1998).
6. Henry, C. R., *Surf. Sci. Rep.* **31,** 235 (1998).
7. Sandell, A., Libuda, J., Brühwiler, P. A., Andersson, S., Maxwell, A. J., Bäumer, M., Mårtensson, N., Freund, H.-J., *J. Vac. Sci. Technol. A* **14,** 1546 (1996).
8. Højrup Hansen, K., Worren, T., Stempel, S., Lægsgaard, E., Bäumer, M., Freund, H.-J., Besenbacher, F., and Stensgaard, I., *Phys. Rev. Lett.* **83,** 4120 (1999).
9. Clausen, B. S., Topsøe, H., and Frahm, R., *Adv. Catal.* **42,** 315 (1998).
10. Thomas, J. M., and Lambert, R. M. (Eds.), "Characterization of Catalysts." Wiley, New York, 1980.
11. Delgass, W. N., Haller, G. L., Kellerman, R., and Lunsford, J. H. (Eds.), "Spectroscopy in Heterogeneous Catalysis." Academic Press, New York, 1979.
12. Niemantsverdriet, J. W., "Spectroscopy in Catalysis: An Introduction." VCH, Weinheim, 1993.
13. Vitos, L., Ruban, A. V., Skriver, H., and Kollar, J., *Surf. Sci.* **411,** 186 (1998).
14. Mavrikakis, M., Hammer, B., and Nørskov, J. K., *Phys. Rev. Lett.* **81,** 2819 (1998).
15. Here, W. J., Radom, L., Schleyer, P. v. R., and Pople, J. A., "Ab Initio Molecular Orbital Theory." Wiley, New York, 1987.

16. Hohenberg, P., and Kohn, W., *Phys. Rev.* **136,** B864 (1964); Kohn, W., and Sham, L., *Phys. Rev.* **140,** A1133 (1965).
17. Payne, M. C., Teter, M. P., Allan, D. C., Arias, T. A., and Joannopoulos, J. D., *Rev. Mod. Phys.* **64,** 1045 (1992).
18. Kresse, G., and Forthmüller, J., *Comput. Mat. Sci.* **6,** 15 (1996).
19. Yang, H., and Whitten, J. L., *J. Chem. Phys.* **96,** 5529 (1992).
20. Inglesfield, J. E., and Benesh, G. A., *Phys. Rev. B* **37,** 6682 (1988).
21. MacLaren, J. M., Crampin, S., Vvedensky, D. D., and Pendry, J. B., *Phys. Rev. B* **40,** 12164 (1989).
22. Skriver, H. L., and Rosengaard, N. M., *Phys. Rev. B* **43,** 9538 (1991).
23. Vanderbilt, D., *Phys. Rev. B* **41,** 7892 (1990).
24. Blöchl, P. E., *Phys. Rev. B* **50,** 17953 (1994).
25. Becke, A. D., *J. Chem. Phys.* **98,** 5648 (1993).
26. Perdew, J. P., Chevary, J. A., Vosko, S. H., Jackson, K. A., Pederson, M. R., Singh, D. J., and Fiolhais, C., *Phys. Rev. B* **46,** 6671 (1992).
27. Perdew, J. P., Burke, K., and Ernzerhof, M., *Phys. Rev. Lett.* **77,** 3865 (1996).
28. Hammer, B., Hansen, L. B., and Nørskov, J. K., *Phys. Rev. B* **59,** 7413 (1999).
29. Hammer, B., Jacobsen, K. W., and Nørskov, J. K., *Phys. Rev. Lett.* **70,** 3971 (1993).
30. Philipsen, P. H. T., te Velde, G., and Baerends, E. J., *Chem. Phys. Lett.* **226,** 583 (1994).
31. Hu, P., King, D. A., Crampin, S., Lee, M.-H., and Payne, M. C., *Chem. Phys. Lett.* **230,** 501 (1994).
32. Hammer, B., Scheffler, M., Jacobsen, K. W., and Nørskov, J. K., *Phys. Rev. Lett.* **73,** 1400 (1994).
33. White, J. A., Bird, D. M., Payne, M. C., and Stich, I., *Phys. Rev. Lett.* **73,** 1404 (1994).
34. Brown, W. A., Kose, R., and King, D. A., *Chem. Rev.* **98,** 797 (1998).
35. Pettifor, D., "Bonding and Structure of Molecules and Solids." Clarendon, Oxford, 1995.
36. Lundqvist, B. I., Gunnarsson, O., Hjelmberg, H., and Nørskov, J. K., *Surf. Sci.* **89,** 196 (1979).
37. Newns, D. M., *Phys. Rev.* **178,** 1123 (1969); Grimley, T. B., *J. Vac. Sci. Techno.* **8,** 31 (1971); **9,** 561 (1971).
38. Hammer, B., *Faraday Discuss.* **110,** 323 (1998).
39. Starke, U., Van Hove, M. A., and Somorjai, G. A., *Prog. Surf. Sci.* **46,** 305 (1994).
40. Hammer, B., and Nørskov, J. K., *in* "Chemisorption and Reactivity on Supported Clusters and Thin Films" (R. M. Lambert and G. Pacchioni, Eds.), pp. 285–351. Kluwer Academic, Dordrecht, 1997.
41. Bogicevic, A., Stromquist, J., and Lundqvist, B. I., *Phys. Rev. B* **57,** R4289 (1998).
42. Steininger, H., Lehwald, S., and Ibach, H., *Surf. Sci.* **123,** 1 (1982).
43. Nørskov, J. K., *Rep. Prog. Phys.* **53,** 1253 (1990).
44. Holloway, S., Lundqvist, B. I., and Nørskov, J. K., *in* "Proceedings of the International Congress on Catalysis," Vol. 4, p. 85, Verlag Chemie Weinheim Berlin, 1984.
45. Hammer, B., and Nørskov, J. K., *Surf. Sci.* **343,** 211 (1995); **359,** 306 (1996).
46. Hammer, B., and Nørskov, J. K., *Nature* **376,** 238 (1995).
47. Mortensen, J. J., Morikawa, Y., Hammer, B., and Nørskov, J. K., *J. Catal.* **169,** 85 (1997).
48. Stokbro, K., and Baroni, S., *Surf. Sci.* **370,** 166 (1997).
49. Toyoshima, I., and Somorjai, G. A., *Catal. Rev. Sci. Eng.* **19,** 105 (1979).
50. Andersen, O. K., Jepsen, O., and Glötzel, D., *in* "Highlights of Condensed Matter Theory," LXXXIX, p. 59. Corso Soc. Italiana di Fisica, Bologna, 1985.
51. Hoffmann, R., *Rev. Mod. Phys.* **60,** 601 (1988).
52. Bagus, P. S., and Pacchioni, G., *Surf. Sci.* **278,** 427 (1992).
53. Blyholder, G., *J. Phys. Chem.* **68,** 2772 (1964).

54. Van Santen, R. A., and Neurock, M., *Catal. Rev.—Sci. Eng.* **37,** 557 (1995).

55. Hammer, B., Morikawa, Y., and Nørskov, J. K., *Phys. Rev. Lett.* **76,** 2141 (1996).

56. Loffreda, D., Simon, D., and Sautet, P., *J. Chem. Phys.* **108,** 6447 (1998).

57. Broden, G., Rhodin, T. N., Brukner, C., Benbow R., and Hurysh, Z., *Surf. Sci.* **59,** 593 (1976).

58. Pallassana, V., and Neurock, M., *J. Catal.* in press (2000).

59. Ge, Q., and King, D. A., *J. Chem. Phys.* **110,** 4699 (1999).

60. Mortensen, J. J., Ganduglia-Pirovano, M. V., Hansen, L. B., Hammer, B., Stoltze, P., and Nørskov, J. K., *Surf. Sci.* **422,** 8 (1999).

61. Bozso, F., Ertl, G., Grunze, M., and Weiss, M., *J. Catal.* **49,** 18 (1977); Imbihl, R., Behm, R. J., Ertl, G., and Moritz, W., *Surf. Sci.* **123,** 129 (1982).

62. Stampfl, C., Schwegmann, S., Over, H., Scheffler, M., and Ertl, G., *Phys. Rev. Lett.* **77,** 3371 (1996).

63. Stampfl, C., Kreuzer, H. J., Payne, S. H., Pfnür, H., and Scheffler, M., *Phys. Rev. Lett.* **83,** 2993 (1999).

64. Diekhöner, L., Baurichter, A., Mortensen, H., and Luntz, A. C., submitted for publication.

65. Ge, Q., King, D. A., Lee, M.-H., White, J. A., and Payne, M. C., *J. Chem. Phys.* **106,** 1210 (1997).

66. Hammer, B., and Nørskov, J. K., *Phys. Rev. Lett.* **79,** 4441 (1997).

67. Lau, K. H., and Kohn, W., *Surf. Sci.* **65,** 607 (1977).

68. Österlund, L., Pedersen, M. Ö., Stensgaard, I., Lægsgaard, E., and Besenbacher, F., *Phys. Rev. Lett.* **83,** 4812 (1999).

69. Bradley, J. M., Hopkinson, A., and King, D. A., *J. Phys. Chem.* **99,** 17032 (1995).

70. Somorjai, G. A., and Borodko, Y., *Catal. Lett.* **59,** 89 (1999).

71. Murphy, M. J., Skelly, J. F., Hodgson, A., and Hammer, B., *J. Chem. Phys.* **110,** 6954 (1999).

72. Darling, G. R., and Holloway, S., *Rep. Prog. Phys.* **58,** 1595 (1995).

73. Groß, A., *Surf. Sci. Rep.* **32,** 291 (1998).

74. Kroes, G. J., *Prog. Surf. Sci.* **60,** 1 (1999).

75. Mavrikakis, M., Hansen, L. B., Mortensen, J. J., Hammer, B., and Nørskov, J. K., *in* "Transition State Modelling for Catalysis" (D. G. Truhlar and K. Morokuma, Eds.), ACS Symposium Series, 721, Oxford University Press, 1999.

76. van Daelen, M. A., Li, Y. S., Newsam, J. M., and van Santen, R. A., *Chem. Phys. Lett.* **266,** 100 (1994).

77. Eichler, A., and Hafner, J., *Phys. Rev. Lett.* **79,** 4481 (1997).

78. Gravil, P. A., Bird, D. M., and White, J. A., *Phys. Rev. Lett.* **77,** 3933 (1996).

79. Alavi, A., Hu, P., Deutsch, T., Silvestrelli, P. L., and Hutter, J., *Phys. Rev. Lett.* **80,** 3650 (1998).

80. Eichler, A., and Hafner, J., *Phys. Rev. B* **59,** 5960 (1999).

81. Stampfl, C., and Scheffler, M., *Phys. Rev. Lett.* **78,** 1500 (1997).

82. Watwe, R. M., Cortright, R. D., Nørskov, J. K., and Dumesic, J. A., to be published.

83. Dahl, S., Logadottir, A., Egeberg, R. C., Larsen, J. H., Chrokendorff, I., Törnqvist, E., and Nørskov, J. K., *Phys. Rev. Lett.* **83,** 1814 (1999).

84. Hammer, B., *Phys. Rev. Lett.* **83,** 3681 (1999).

85. Morikawa, Y., Mortensen, J. J., Hammer, B., and Nørskov, J. K., *Surf. Sci.* **386,** 67 (1997).

86. Kratzer, P., Hammer, B., and Nørskov, J. K., *J. Chem. Phys.* **105,** 5595 (1996).

87. Berger, H. F., Leisch, M., Winkler, A., and Rendulic, K., *Chem. Phys. Lett.* **175,** 425 (1985); Robota, H. J., Vielhaber, W., Lin, M. C., Segner, J., and Ertl, G., *Surf. Sci.* **155,** 101 (1985).

88. Brown, J. K., Luntz, A., and Schultz, P. A., *J. Chem. Phys.* **95,** 3767 (1991).

89. Hayden, B. E., and Lamont, C. L., *Phys. Rev. Lett.* **63,** 1823 (1989); Rasmussen, P. B., Holmblad, P. M., Christoffersen, H., Taylor, P. A., and Chorkendorff, I., *Surf. Sci.* **287/288,** 79 (1993).
90. Sault, A. G., Madix, R. J., and Cambell, C. T., *Surf. Sci.* **169,** 347 (1986).
91. Eichler, A., Kresse, G., and Hafner, J., *Surf. Sci.* **397,** 116 (1998).
92. Wilke, S., and Scheffler, M., *Phys. Rev. B* **53,** 4926 (1996).
93. Groß, A., Wilke, S., and Scheffler, M., *Phys. Rev. Lett.* **75,** 2718 (1995).
94. White, J. A., Bird, D. M., and Payne, M. C., *Phys. Rev. B* **53,** 1667 (1996).
95. Kay, M., Darling, G. R., Holloway, S., White, J. A., and Bird, D. M., *Chem. Phys. Lett.* **245,** 311 (1995).
96. Gsell, M., Jakob, P., and Menzel, D., *Science* **280,** 717 (1998).
97. Rodriguez, J. A., and Goodman, D. W., *Science* **257,** 897 (1992).
98. Madey, T. E., Nien, C.-H., Pelhos, K., Kolodziej, J. J., Abdelrehim, I. M., and Tao, H.-S., *Surf. Sci.* **438,** 191 (1999).
99. Bautier de Mongeot, F., Scherer, M., Gleich, B., Kopatzki, E., and Behm, R. J., *Surf. Sci.* **411,** 249 (1998).
100. Kampshoff, E., Hahn, E., and Kern, K., *Phys. Rev. Lett.* **73,** 704 (1994).
101. Larsen, J. H., and Chorkendorff, I., *Surf. Sci.* **405,** 62 (1998).
102. Pedersen, M. Ø., Helveg, S., Ruban, A., Stensgaard, I., Lægsgaard, E., Nørskov, J. K., and Besenbacher, F., *Surf. Sci.* **426,** 395 (1999).
103. Zambelli, T., Wintterlin, J., and Ertl, G., to be published.
104. Logadottir, A., to be published.
105. Clausen, B. S., and Nørskov, J. K., *Topics Catal.,* in press (1999).
106. Klimenkov, M., Nepijko, S., Kuhlenbeck, H., Baumer, M., Schlögl, R., and Freund, H.-J., *Surf. Sci.* **391,** 27 (1997); Nepijko, S., Klimenkov, M., Kuhlenbeck, H., Zemlyanov, D., Herein, D., Schlögl, R., and Freund, H.-J., *Surf. Sci.* **412/413,** 192 (1998).
107. Giorgio, S., Chapon, C., Henry, C. R., Nihoul, G., and Penisson, J. M., *Phil. Mag. A* **64,** 87 (1991).
108. Hammer, B., Nielsen, O. H., and Nørskov, J. K., *Catal. Lett.* **46,** 31 (1997).
109. Kratzer, P., Hammer, B., and Nørskov, J. K., *Surf. Sci.* **359,** 45 (1996).
110. Hammer, B., *Surf. Sci.,* in press (2000).
111. Feibelman, P. J., Esch, S., and Michely, T., *Phys. Rev. Lett.* **77,** 2257 (1996).
112. Tripa, C. E., Zubkov, T. S., Yates, J. T., Mavrikakis, M., and Nørskov, J. K., *J. Chem. Phys.* **111,** 8651 (1999).
113. Bengaard, H. S., Rostrup-Nielsen, J. R., and Nørskov, J. K., to be published.
114. Ramsier, R. D., Gao, Q., Neergaard Waltenburg, H., and Yates, J. T., Jr., *J. Chem. Phys.* **100,** 6837 (1994).
115. Kose, R., Brown, W. A., and King, D. A., *J. Am. Chem. Soc.* **121,** 4845 (1999).
116. Zambelli, T., Wintterlin, J., Trost, J., and Ertl, G., *Science* **273,** 1688 (1996).
117. Dietrich, H., Geng, P., Jacobi, K., and Ertl, G., *J. Chem. Phys.* **104,** 375 (1996).
118. Taylor, H. S., *Proc. R. Soc. London A* **108,** 105 (1925).
119. Boudart, M., *Adv. Catal. Relat. Subj.* **20,** 153 (1969).
120. Shekhar, R., and Barteau, M. A., *Catal. Lett.* **31,** 221 (1995).
121. Hammer, B., and Scheffler, M., *Phys. Rev. Lett.* **74,** 3487 (1995).
122. Besenbacher, F., Chorkendorff, I., Clausen, B. S., Hammer, B., Molenbroek, A., Nørskov, J. K., and Stensgaard, I., *Science* **279,** 1913 (1998).
123. Holmblad, P. M., Hvolbæk Larsen, J., Chorkendorff, I., Nielsen, L. P., Besenbacher, F., Stensgaard, I., Lægsgaard, E., Kratzer, P., Hammer, B., and Nørskov, J. K., *Catal. Lett.* **40,** 131 (1996).

124. Rostrup-Nielsen, J., *J. Catal.* **85,** 31 (1994); Andersen, N. T., Topsøe, F., Alstrup, I., and Rostrup-Nielsen, J., *J. Catal.* **104,** 454 (1987).
125. Delbecq, F., and Sautet, P., *Chem. Phys. Lett.* **302,** 91 (1999).
126. Delbecq, F., and Sautet, P., *Phys. Rev. B* **59,** 5142 (1999).
127. Pallassana, V., Neurock, M., Hansen, L. B., Hammer, B., and Nørskov, J. K., *Phys. Rev. B* **60,** 6146 (1999); Pallassana, V., Neurock, M., Hansen, L. B., and Nørskov, J. K., to be published.
128. Ruban, A., Hammer, B., Stoltze, P., Skriver, H. L., and Nørskov, J. K., *J. Mol. Catal. A* **115,** 421 (1997).
129. Weinert, M., and Watson, R. E. *Phys. Rev. B* **51,** 17168 (1995).
130. Hennig, D., Ganduglia-Pirovano, M. V., and Scheffler, M., *Phys. Rev. B* **53,** 10344 (1996).
131. Ganduglia-Pirovano, M. V., Natoli, V., Cohen, M. H., Kudrnovsky, J., and Turek, I., *Phys. Rev. B* **54,** 8892 (1996).
132. Christensen, A., Ruban, A. V., Stoltze, P., Jacobsen, K. W., Skriver, H. L., Nørskov, J. K., and Besenbacher, F., *Phys. Rev. B* **56,** 5822 (1997); Ruban, A. V., Skriver, H. L., and Nørskov, J. K., *Phys. Rev. B,* **59,** 15900 (1999).
133. Bertolini, J. C., Tardy, B., Abon, M., Billy, J., Delichere, P., and Massardier J. *Surf. Sci.* **135,** 117 (1983).
134. Nieuwenhuys, B. E., *Surf. Rev. Lett.* **3,** 1869 (1996).
135. Linke, R., Schneider, U., Busse, H., Becker, C., Schröder, U., Castro, G. R., and Wandelt, K., *Surf. Sci.* **307–309,** 4074 (1994).
136. Sinfelt, J. H., "Bimetallic Catalysts: Discoveries, Concepts, and Applications." Wiley, New York, 1983; Ponec, V., *Adv. Catal.* **32,** 149 (1983).
137. Cong, P., Doolen, R. D., Fan, Q., Giaquinta, D. M., Guan, S., McFarland, E. W., Poojary, D. M., Selt, K., Turner, H. W., and Weinberg, W. H., *Angew. Chem., Int. Ed. Engl.,* **38,** 484 (1999).
138. Jennings, J. R. (Ed.), "Catalytic Ammonia Synthesis." Plenum, New York, 1991; H. Topsøe, H. Boudart, and J. K. Nørskov (Eds.), *Topics Catal.* **1,** 185–415 (1994).
139. Ertl, G., Lee, S. B., and Weiss, M., *Surf. Sci.* **111,** L711 (1981); Ertl, G., *Catal. Rev.—Sci. Eng.* **21,** 201 (1980).
140. Spencer, N. D., Schoonmaker, R. C., and Somorjai, G. A., *J. Catal.* **74,** 129 (1982).
141. Mortensen, J. J., Hammer, B., and Nørskov, J. K., *Phys. Rev. Lett.* **80,** 4333 (1998); Mortensen, J. J., Hammer, B., and Nørskov, J. K., *Surf. Sci.* **414,** 315 (1998).
142. Jacobsen, K. W., Nørskov, J. K., and Puska, M. J., *Phys. Rev. B* **35,** 7423 (1987).
143. Nørskov, J. K., Holloway, S., and Lang, N. D., *Surf. Sci.* **137,** 65 (1984); Nørskov, J. K., *in* "Physics and Chemistry of Alkali Metal Adsorption" (H. P. Bonzel, A. M. Bradshaw, and G. Ertl, Eds.), p. 253. Elsevier, Amsterdam, 1989.
144. Wilke, S., and Scheffler, M., *Phys. Rev. Lett.* **76,** 3380 (1996).
145. Wilke, S., and Cohen, M. H., *Surf. Sci.* **380,** 1441 (1997).
146. Byskov, L., Clausen, B. S., Topsøe, H., and Nørskov, J. K., *J. Catal.* **187,** 109 (1999).

Scanning Tunneling Microscopy Studies of Catalytic Reactions

JOOST WINTTERLIN
Fritz-Haber-Institut der Max-Planck-Gesellschaft
D-14195 Berlin, Germany

In the past 10 years, scanning tunneling microscopy (STM) has developed into a technique that can monitor catalytic surface reactions with atomic resolution. This review summarizes this work, most of which was performed with single-crystal metal surfaces under ultrahigh vacuum conditions. Emphasis is placed on the mechanisms of the processes underlying catalytic reactions. The instrumental developments that made such studies possible are addressed, as are the possibilities for chemical identification of the adsorbate features in the STM topographs. STM investigations have revealed information about each of the elementary processes involved in heterogeneous catalytic reactions. The primary adsorption and dissociation steps and surface diffusion of adsorbed particles are discussed briefly. The main part of the article concerns individual surface reactions. In almost all cases, qualitative—and even the first quantitative—deviations from standard Langmuir descriptions of surface reactions have been observed, with pronounced local effects on the reactivities. © 2000 Academic Press.

I. Introduction

Since Binnig *et al.* published their first papers in 1982 (*1, 2*), scanning tunneling microscopy (STM) has spread into almost all fields of surface science and has had an enormous impact on our understanding of the structures and electronic properties of surfaces. Because STM was also found to provide atomic resolution of adsorbed particles, it was inevitable that attempts would be made to develop STM into a tool for monitoring chemical reactions on surfaces. It was hoped that in this way a direct view would be obtained of the atomic processes that form the basis of

Abbreviations: D, diffusion coefficient; E^*, activation energy; E_F, Fermi energy; HOMO, highest occupied molecular orbital; HREELS, high resolution electron energy loss spectroscopy; IETS, inelastic electron tunneling spectroscopy; L, Langmuir; 1 L = 1.33×10^{-6} mbar s; LDOS, local density of states; LEED, low energy electron diffraction; LUMO, lowest unoccupied molecular orbital; MC, Monte Carlo; STM, scanning tunneling microscopy; STS, scanning tunneling spectroscopy; $\Theta(i)$, coverage of surface by species i; UHV, ultrahigh vacuum.

0360-0564/00 $35.00

heterogeneous catalysis and other surface reactions, such as those of chemical vapor deposition, electrochemical processes, and oxidation of metal and semiconductor surfaces. This article is a review of what has been achieved by STM in the field of heterogeneous catalysis, namely, of reactions on metal surfaces, from 1990 (when results of the first successful experiments were published) to 1999. The goal is to show what kind of insight can be obtained with this technique and also to discuss the current limitations. [Shorter reviews of reaction studies were published in 1996 and 1997 (*3, 4*).]

It is obvious what we can expect to learn about catalytic reactions by employing a technique with atomic resolution. It can provide microscopic information about the locations of the dissociation and reaction steps, the atomic processes at active sites, the role of defects and contaminants, etc.—issues about which we have had only indirect information.

However, STM can have a more general impact on catalysis research. Our knowledge of catalytic reactions is almost exclusively derived from macroscopic measurements, mostly kinetics and spectroscopy. The quantities obtained from these measurements represent averages over huge particle ensembles so that information about the individual particles is necessarily lost. In a strict sense, the models developed from these observations cannot provide unique reaction mechanisms because we cannot rule out the possibility that some species, some reaction steps, or some unforeseen phenomena have been overlooked.

For example, in kinetics models, the usual formulation of the surface recombination step of a reaction as a second-order process [rate $\sim \Theta(A)\,\Theta(B)$] implies that the adsorbate layer forms a Langmuir gas (an ideal lattice gas with site exclusion). On the other hand, surface science investigations have often indicated the formation of islands of adsorbates, in contrast to the Langmuir assumption. However, successful application of a Langmuirian kinetics model to a surface reaction does not allow one to exclude the possible role of islands. In contrast, STM provides an atomic view of such phenomena. Thus, STM has the potential of revealing reaction mechanisms that are not subject to the nonuniqueness problem associated with ensemble averaging. Furthermore, one can build up ensemble averages from the statistics of the atomic events observed by STM and arrive at quasi-macroscopic quantities (such as reaction rates or diffusion constants).

This review does not include the more structural aspects of STM studies of adsorption on metals, such as the determination of adsorption sites of particles and the structure of ordered phases and of reconstructions. These topics are addressed in several review articles (*5–11*). Nor are STM studies of the structures of model catalysts included here (*12–15*).

The topic of this review is *processes* on surfaces. Although not all of the investigations discussed actually achieved dynamic resolution, the focus is on the mechanisms of the basic surface steps of catalysis, namely, adsorption of reactants, dissociation of molecules, surface diffusion of particles, and reaction between adsorbed reactants. STM investigations of this subject have been largely restricted to single-crystal metal surfaces under ultrahigh vacuum (UHV) conditions; investigations under conditions similar to those of practical catalysis are rare. After a short description of instrumental developments important for this kind of experiment (Section II), the possibilities for identification of adsorbates on metal surfaces by STM (a prerequisite for investigations of surface reactions) are presented (Section III). Section IV is a review of STM results characterizing the first steps of catalytic reactions, the adsorption of the reactants from the gas phase and their dissociation on the surface. The following step, the surface diffusion of the adsorbates, is discussed in Section V, which is also a review of experiments characterizing interactions between adsorbates, which affect their mobility and distribution on the surface. Work on the central step, the reaction between adsorbed reactants to give new chemical compounds, is discussed in Section VI. This main part is organized according to the various new mechanistic effects discovered, whereas the individual reactions investigated are treated in subsections serving as examples for the more general aspects.

II. Instrumentation

There are special instrumental requirements for STM investigations of chemical processes on metal surfaces. Adsorbed particles on metal surfaces are often highly mobile so that the STM should be able to work at low temperatures to resolve these adsorbates. Low temperatures are mandatory for experiments characterizing reaction intermediates and precursors. To determine kinetic parameters of surface processes, the instrument should be able to work at variable temperatures. Elevated temperatures are required for mechanistic investigations of activated processes. Because this requirement is in conflict with the need to reduce the surface mobility of adsorbates, it is desirable to increase the recording speed of the instrument to allow one to monitor the particles even under such conditions.

When the work reviewed here was started, most experimental setups were restricted to room-temperature operation as a consequence of the difficulties of cooling the system without coupling in mechanical vibrations and of managing thermal drift. STM imaging rates were usually on the order of one frame per minute, determined by the maximum frequency at which the tip sample distance could be controlled. Time resolution was therefore a major problem; in many cases the surface particles moved too fast across the surface to be resolved by STM. Nevertheless, there have been many successful studies of surface reactions carried out with room-temperature scanning tunneling microscopes. The investigations were focused on systems in which at least one of the adsorbed components was sufficiently immobile at 300 K, as was often the case when this adsorbate was involved in a surface reconstruction.

In approximately 1990, helium-cooled UHV STMs appeared, working at constant low temperatures (at ~4 K) (*16–21*). These instruments, which are often extremely stable and almost free of drift, allowed investigation of metastable precursors, study of the imaging of molecules, manipulation of adsorbed particles, and performance of spectroscopic measurements. At about the same time, variable-temperature STMs became available (*22–30*). By adjustment of the temperature, the rates of chemical processes could then, in principle, be shifted into the time window of STM operability. In this way much more general reactions have become accessible, and temperature coefficients of surface processes could be determined. There has also been progress toward higher recording speeds in STM, widening the time window for dynamic experiments (*11, 31–33*); imaging rates of up to 20 frames per second have been achieved recently (*33*).

Most STM investigations of catalytic reactions were performed under typical surface science conditions, with single-crystal samples and under UHV conditions, because of the advantages of working with well-defined systems. In principle, STM does not require a vacuum environment, in contrast to most other surface science techniques. The first high-pressure instruments were recently developed (*34–36*), providing the possibility for investigation of reactions under conditions more similar to those of practical catalysis.

A possible problem with STM reaction studies is that the surface processes may be affected by the presence of the tunneling contact. Adsorbates are expected to be influenced by the forces from the orbital overlap between the tip and the surface, by the tunneling current (leading to excitations of the adsorbate complex), and by the electric field between the tip and the sample. The tip could also perturb the flow of particles to and from the gas phase by geometric shielding of the space above the area under investigation. All of these effects can indeed occur, but in the majority of cases the experimental parameters can be chosen so that these perturbations are reduced below detection limits. In turn, such effects can be induced intentionally by variation of the tunneling parameters beyond critical thresholds for short periods. In this way chemical processes can be stimulated nonthermally and with spatial control, providing new ways for characterization of chemical reactions.

III. Chemical Identification of Adsorbates with STM

In general, STM does not show the positions of adsorbed atoms, nor does it directly offer chemical sensitivity. Without additional input from other experimental methods or from calculations, the various features in STM images cannot be identified with certainty with specific atoms or molecules. This limitation is a major obstacle for studies of chemical reactions and occurs because STM probes the electronic structure near the Fermi energy rather than the total charge density, which would reflect the geometry in the sense of positions of ion cores and which is usually not related straightforwardly to adsorbate orbitals that would give an indication of the chemical nature and bonding of the adsorbate.

A. Imaging of Adsorbates by STM

STM images are usually interpreted in the framework of the Tersoff–Hamann model (*37, 38*), according to which an STM image represents a contour map of constant local density of states (LDOS) of the sample at the Fermi energy (E_F) at the position of the tip. Two major approximations underlie this description. The tunneling current is calculated by Bardeen's time-dependent first-order perturbation treatment (*39*), according to which the wavefunctions of sample and tip that carry the current do not affect each other, and the tip is modeled by a spherically symmetric (s-like) wavefunction placed in front of the surface. The main advantage of this model is that the STM image becomes an exclusive property of the sample—the sample LDOS at E_F—whereas the tip does not enter (except by its radius and distance).

That this model can provide qualitatively correct predictions of adsorbates on metal surfaces followed first from calculations by Lang (*40*). These made use of the perturbation approach for the current but treated the tip more realistically, as an Na atom on a jellium surface. Figure 1 (*40*) shows results for the imaging of Na_{ad} and S_{ad} atoms, also adsorbed on jellium. The calculated state density of Na (Fig. 1, left) shows an enhancement at E_F, caused by the Na 3s resonance just above E_F. The resulting contour of constant current (Fig. 1, right) displays a large protrusion at the position of the Na_{ad} atom. In contrast, the 3p resonance maximum of S_{ad} is considerably below E_F, the enhancement at E_F is less, and the protrusion above the atom is smaller. The difference results in part because S_{ad} resides closer to the surface than Na_{ad}. Helium, with its highest occupied state being very low on the energy scale, only polarizes the metal electrons away from E_F and appears as a hole. The investigation of Lang (*40*) also showed that for Na_{ad} the STM contour is quite similar to the total charge density

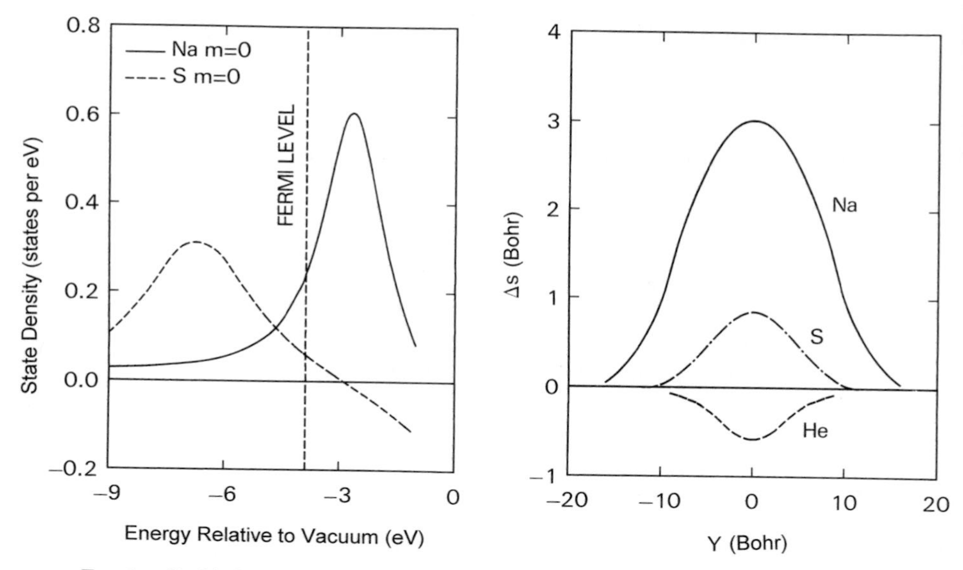

FIG. 1. (Left) Calculated state densities of Na and of S adsorbed on jellium (actually state density differences between the adsorbate/metal system and bare metal). Only $m = 0$ components are shown; the $m \neq 0$ components contribute much less to the tunneling current. (Right) Calculated lines of constant tunneling current over an Na atom, an S atom, and an He atom adsorbed on jellium [reprinted with permission from Lang (*40*). Copyright © 1986 The American Physical Society].

contour, but that for S_{ad} the STM profile is significantly lower than that of the total charge density. However, in both cases the lines of constant current follow the E_F LDOS contours very well, in agreement with the Tersoff–Hamann model. The LDOS for adsorbed oxygen atoms at E_F was found to be reduced with respect to the value for the bare metal so that a depression in the STM contour was predicted. Experiments with adsorbed sulfur (*41*) and oxygen atoms (*42*) confirmed these predictions qualitatively.

On the other hand, it is a common experimental experience that the appearance of adsorbates in STM images depends markedly on the state of the tip. During an experiment the tip can pick up foreign atoms, leading to changes at the tip apex that are not treated by the Tersoff–Hamann model. Published examples are contrast inversions in various adsorbate systems resulting from tip state changes (*41, 43, 44*). Moreover, in several theoretical investigations (*45–50*) it was pointed out that the approximations of the Tersoff–Hamann model are not well fulfilled in the STM configuration. In later theoretical work, one or both of the Tersoff–Hamann approximations were thus dropped. The tunneling current was more appropriately represented as a scattering process that includes interactions between the tip and the sample, and for the tips realistic atomic structures were assumed. For detailed discussions of the theoretical description of imaging in STM, see the articles in Wiesendanger and Güntherodt (*51*), and for a recent review of the imaging of adsorbates, see Sautet (*52*).

For example, Fig. 2 (*53–55*) shows experimental STM images of benzene molecules on a Pt(111) surface [obtained by Weiss and Eigler (*53*)] and calculated contours of constant tunneling current reported by Sautet and Bocquet (*54*). The experimental STM data show three different shapes of benzene molecules—one with three lobes, one with a volcano shape, and one with a simple cone shape. It was proposed that these correspond to different adsorption sites of the flat-lying benzene molecules. The calculations, based on the scattering formalism and an extended Hückel treatment for the tip–sample system, reproduce these experimental shapes quite well and confirm that the different shapes are caused by different adsorption sites. It was clear from the beginning that the contours would not show the carbon or hydrogen atoms, but the results also show that there is no obvious correlation with the benzene highest occupied molecular orbital or lowest unoccupied molecular orbital formed by the π/π^* orbitals. These are the orbitals closest to E_F, and they might have been expected to dominate the appearance of the molecule. Although the π/π^* orbitals are important, the authors found that lower-lying orbitals contribute significantly to the tunneling current.

This example demonstrates, on the one hand, that the current theoretical models allow a good description of STM images of molecular adsorbates. On the other hand, it shows that it will not generally be possible to interpret STM images of adsorbed particles by simple visual inspection or on the basis of qualitative arguments about molecular orbitals. For adsorbed atoms there appears to be a correlation between the height at which the atom is imaged and the electronegativity and the polarizability of the atom. Electropositive polarizable atoms appear as bumps and small electronegative atoms as holes (*56*).

It is emphasized that the data of Fig. 2 were derived from experiments performed under extremely stable conditions with a 4-K STM—clearly an ideal situation. Under conditions of chemical reactions, at higher temperatures and during dosing with reactants, the noise level is much higher, particles are mobile, and adsorbates stick not only to the sample but also to the tip, leading to contrast changes. In experimentation, researchers have therefore mostly chosen a pragmatic approach to the problem of adsorbate identification. Most of the reactions investigated have involved only a few species that had previously been well characterized by spectroscopic and other macroscopic methods, usually leaving only a few possibilities for

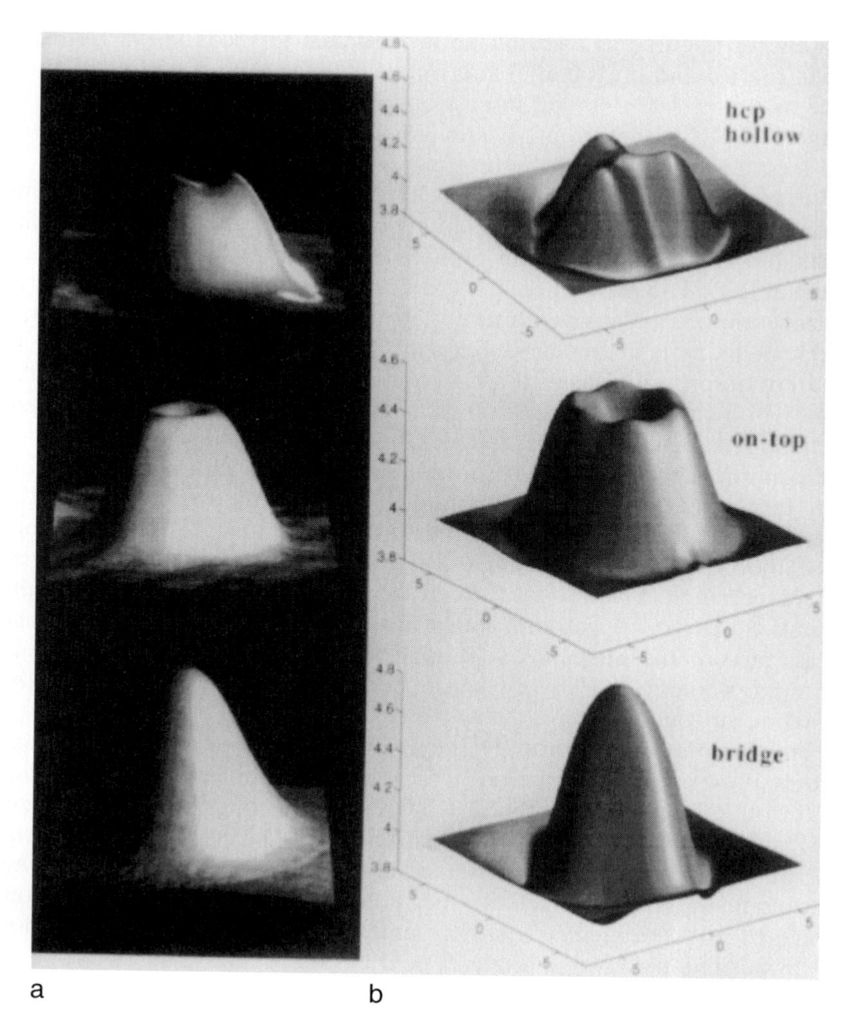

FIG. 2. (a) STM images of three different types of adsorbed benzene molecules on a Pt(111) surface (recorded at 4 K, 15 × 15 Å) [reprinted with permission from Weiss and Eigler (*53*). Copyright © 1993 The American Physical Society]. (b) Calculated constant current contours of benzene molecules adsorbed on a hollow site, an on-top site, and a bridge site (12 × 12 Å) [reprinted with permission from Sautet and Bocquet (*54*). Copyright © 1996 Laser Pages Publishing].

interpretation of the observed features. Even when the imaging contrast changed, the experimental conditions and the behavior of the particles (e.g., formation of ordered structures and mobility) usually allowed a unique interpretation. This situation may be different in the future, when more complicated systems are investigated.

B. Spectroscopy of Adsorbates by STM

In addition to recording of the topography of surfaces, STM also provides the opportunity for characterization of surfaces by their electronic structures by the use of scanning tunneling spectroscopy (STS). In STS, the STM measures, in addition to the topographic image, the tunneling conductance dI/dV at selected locations of the image or over extended 2-D arrays of points. Plots of the normalized conductance $(dI/dV)/(I/V)$ vs V provide spectra of the electronic state density (occupied and empty) at these locations, with atomic resolution (*57, 58*). The method has been very useful for characterization of semiconductors (*59*); for metals, the surface states occurring on several bare surfaces were investigated (*60, 61*), and it was possible to detect electronic states of adsorbed transition metal atoms (*62*). However, STS has played almost no role in the characterization of systems more relevant to catalytic reactions. Spectra of adsorbed oxygen on copper and on tungsten surfaces showed intensity variations between oxygen-covered and empty areas without displaying clear oxygen-induced features (*63, 64*). An investigation of CO adsorbed on Cu(110) (*65, 66*) showed an increasing empty state density above the CO molecules that was attributed to the CO $2\pi^*$ resonance, although the peak maximum, expected at 3.5 eV above E_F, was outside the accessible energy range. Because the sample and the tip are not stable at high electric fields (the tip–sample distance is kept constant as the spectra are recorded), the tunneling voltages in STS are typically restricted to ± 2 V, limiting the spectra to an energy range of this magnitude near E_F. Because electronic resonances of adsorbates on metals are rather wide and mostly outside this energy range, the spectra are usually almost featureless and of little analytic value for adsorbate–metal systems. Another restriction is that the wavefunctions have to extend sufficiently into the vacuum to be detectable. This difficulty has prevented detection of the metallic d states in the spectra, which otherwise would be of interest for the bonding of adsorbates to transition metals.

A straightforward method to characterize adsorbates would be provided by vibrational spectroscopy. It had been expected that this could be achieved by STM in the same way as with traditional inelastic electron tunneling spectroscopy (IETS), which is performed with planar solid tunneling junctions (67). In IETS, inelastic excitations of vibrations appear as peaks in the second derivative of the tunneling current with respect to the voltage. Recording d^2I/dV^2 vs V during scanning of the surface with the STM could therefore provide atomically resolved vibrational spectra of adsorbed molecules. However, after an early report of vibrational spectra of sorbic acid molecules on graphite (*68*), the physical mechanism of which was not quite clear, further attempts were unsuccessful.

A breakthrough has recently been achieved by Stipe *et al.* (*69, 70*), who reported STM inelastic tunneling spectra of adsorbed acetylene molecules on Cu(100) and Ni(100). In the topography [Fig. 3a (*70*)] the acetylene molecules appear as dumbbell-shaped dark dots; the $C_2H_{2,ad}$ and the singly and doubly deuterated molecules are imaged identically, as expected. It had previously been determined that the molecules reside in the fourfold hollow positions, with the C–C bond parallel to and the C–H bonds tilted away from the surface, i.e., the molecules rehybridize upon adsorption. The molecules are oriented perpendicularly to the dark dumbbells. The d^2I/dV^2 spectrum [Fig. 3b, top (*70*)] recorded above the $C_2H_{2,ad}$ molecules shows a peak at 358 meV that, according to previous studies by high-resolution electron energy loss spectroscopy (HREELS), indicates the C–H stretch mode. Evidence that the peak is actually caused by a C–H vibration was derived from the expected isotope shift (to 266 meV) observed for the $C_2D_{2,ad}$ isotope. The singly deuterated isotope accordingly shows both modes. [Equivalent observations were also made for the Ni(100) surface (Fig. 3c).] Moreover, the authors were able to map spatial intensity distributions of the inelastic tunneling signals. For the singly deuterated acetylene the inelastic tunneling signal at 269 meV was found to be centered at a region

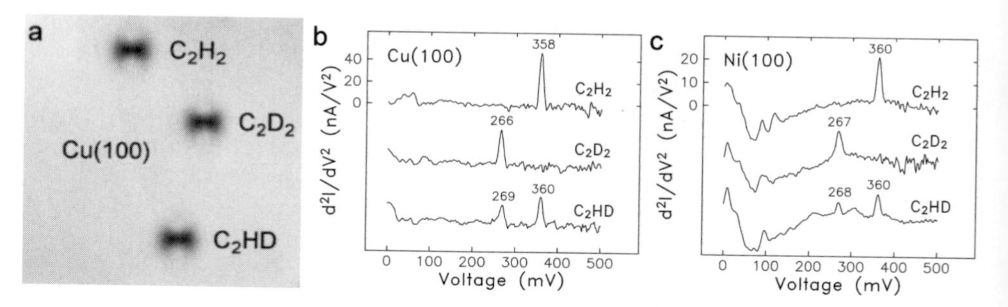

FIG. 3. (a) STM image of adsorbed acetylene molecules (dumbbell-shaped features) on Cu(100) at 8 K (56 × 56 Å). (b) Inelastic tunneling spectra recorded while holding the tip at fixed positions above individual acetylene molecules on Cu(100). (c) Inelastic tunneling spectra of acetylene molecules on Ni(100). All spectra were averaged over several runs, and bare surface spectra were subtracted [reprinted with permission from Stipe *et al.* (*70*). Copyright © 1999 The American Physical Society].

on one side of the dumbbell, in agreement with the assumed orientation of the molecules. These results show that the method can in fact determine vibrational modes associated with the locations of individual chemical bonds.

It must be recognized that inelastic tunneling effects are small (the conductance changes by only a few percent), which demands an extremely stable tunneling gap. To allow averaging of spectra, adsorbate mobility and thermal drift must be minimal. Such stringent stability requirements are usually fulfilled only in low-temperature instruments. (The aforementioned investigation was performed at 8 K.) Low temperatures are also required because the peak widths are determined by the width of the Fermi function. Inelastic tunneling spectroscopy will therefore probably not be applicable as an on-line method during chemical reactions, but instead it will require freezing out of the particles under investigation.

IV. Initial Steps of Catalytic Reactions

The results of several investigations have demonstrated that STM can provide information about the initial steps of catalytic reactions, including the adsorption of the reactant molecules and the dissociation steps that often follow. Although it is clear that the picosecond timescale of the fundamental processes is beyond the time resolution of STM, information about mechanisms and energetics can be extracted from the spatial distribution of the particles after adsorption and possible dissociation.

A. Adsorption from the Gas Phase

When particles become attached to a surface, the adsorption energy is released. It was found in STM experiments that this energy may be converted into kinetic energy of a "hot" adsorption state. For example, after dosing of a Pt(111) surface at 4 K with small amounts of Xe, nearly all of the Xe_{ad} atoms were found at the atomic steps (*71*). Because the adsorbed Xe atoms were not thermally mobile at this temperature, they could not have reached the steps from their impact points on the terraces by diffusion. It was concluded that, after landing on the surface, the atoms must have "skated" several hundred angstroms across the terraces until they collided with steps at which they eventually became accommodated. Such effects are not restricted to physisorbed particles. After the (molecular) adsorption of oxygen molecules on Ag(110), one-dimensional clusters of $O_{2,ad}$ molecules were observed (*72*), which at 60–70 K could not have been formed as a result of simple thermally induced diffusion. A similar skating mechanism was proposed by which the molecules traveled along the close-packed rows of the (110) surface until they were stopped by collision with other molecules. [On Cu(110), clusters of $O_{2,ad}$ were observed even after adsorption at 4 K (*73*).]

B. Dissociation of Molecules on Surfaces

Extensive information about the dissociation step that precedes most surface catalytic reactions has also been obtained by STM. It can be assumed that dissociation usually involves an adsorbed, molecular precursor. Precursor dissociation was investigated in detail for oxygen on Pt(111). The thermodynamically stable state is atomic oxygen, but there is also a chemisorbed, peroxo-like molecular state. The $O_{2,ad}$ state is metastable and dissociates at temperatures higher than ~95 K. At higher temperatures, the molecular state acts as a precursor to O_{ad}. The adsorbed atoms become mobile at ~200 K so that dissociation experiments could be performed over a range of ~100 K without subsequent thermal mobility of the atoms (which would cause the loss of the information of the dissociation). After exposure of O_2 at 160 K (Fig. 4a), Wintterlin *et al.* (*42*) found that the resulting O_{ad} atoms occurred in pairs, as expected. The O_{ad} atoms are imaged dark, in agreement with theoretical predictions, and were found to occupy exclusively the fcc sites that are the thermodynamically stable sites for O_{ad}. In contrast to the usual Langmuir model, according to which two adjacent empty sites are required for dissociative adsorption, Fig. 4a shows that the atoms in the individual pairs were more than one lattice constant apart. Larger data sets revealed distances of up to three lattice sites, with two lattice sites being the most frequent separation distance (the distances observed are indicated in Fig. 4b). [A similar distance distribution was also reported for Cu(110)/oxygen (*73*).] Because the atoms were thermally immobile, it was concluded that this distribution resulted from the release of the atomic binding energy of 1.1 eV (per oxygen atom) upon dissociation, which caused the atoms to scatter at the metallic potential and to perform several jumps until they were accommodated on the surface.

The distances covered by the atoms in this way are small for Pt(111), but the kinetic energy stored in this motion is large enough to induce chemical reactions with neighboring particles. In previous temperature-programmed desorption experiments (*74*), a low-temperature channel for the reaction of $O_{2,ad}$ with coadsorbed CO_{ad} was reported which was apparently caused by this process. In the case of the Al(111) surface, an extreme case of hot oxygen atoms resulting from O_2 dissociation was reported that was attributed to the high binding energy of the oxygen atoms to the aluminum surface that was released in this process (*75*).

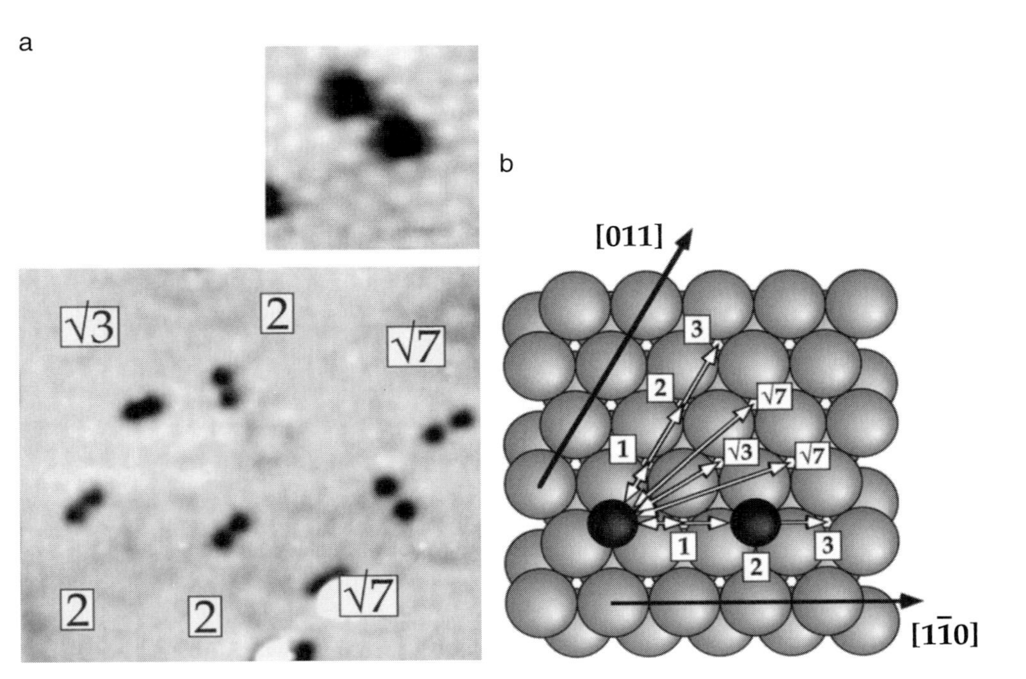

FIG. 4. (a) STM images of oxygen atoms (black dots) on Pt(111) recorded after dissociative adsorption of O_2 at approximately 160 K. The small inset (19 × 19 Å) also shows the metal lattice. In the larger image (110 × 92 Å) the separations between the O atoms are indicated in units of lattice constants. (b) Model of the Pt(111) surface with a pair of oxygen atoms from the dissociation of O_2. All separations that were observed in the experiments are indicated; the most frequent separation was at two lattice constants [reprinted with permission from Wintterlin *et al.* (*42*). Copyright © 1996 The American Physical Society].

Experiments performed over a range of temperatures by Zambelli *et al.* (*76*) revealed additional complexities of the O_2 dissociation on Pt(111). At 160 K (Fig. 4a), the atom pairs appeared more or less randomly distributed, indicating a constant dissociation probability on the empty surface. This was no longer the case when the temperature was reduced. At 105 K [Fig. 5a (*76*)], the atom pairs were seen to be aligned in chains, which, on a mesoscopic scale, led to net-like patterns. Moreover, the sticking coefficient increased substantially (the exposures in the experiments represented in Figs. 4a and 5a were approximately equal).

This result can only mean that the dissociation probability of the adsorbed molecular precursor is enhanced near already present oxygen atoms. As indicated in Fig. 5b, a molecular precursor approaches a chain of oxygen atoms and dissociates with a higher probability at the end than at the side

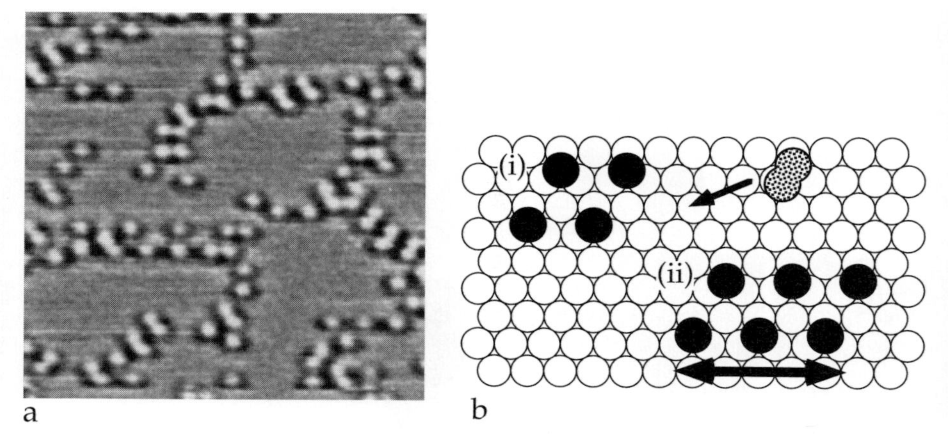

FIG. 5. (a) Chains of adsorbed oxygen atoms formed by dissociative adsorption of O_2 on Pt(111) at 105 K. O_{ad} atoms appear bright with dark borders (70 × 70 Å). (b) Model illustrating the growth of O_{ad} chains by the action of O_2 precursors that diffuse to the chains and dissociate preferentially at the terminal O_{ad} atoms [reprinted with permission from Zambelli *et al.* (*76*). Copyright © 1997 Macmillan Magazines Limited].

of the chain (or on the empty surface). This mechanism was found to occur at temperatures lower than ~110 K when the lifetime of the molecular precursor (with respect to desorption) was apparently long enough to allow it to reach a chain end by diffusion. Because the average distance the precursor has to travel decreases with the number of O_{ad} atoms already present, the effect also explains the unusual earlier finding (*77*) that at 100 K the sticking coefficient increased with increasing oxygen coverage on the surface. The STM thus provides information about accommodation and dissociation mechanisms of adsorbing particles, which actually remain on the surface; the information complements that obtained from traditional molecular beam experiments.

Furthermore, STM has the capacity to induce adsorption–desorption and dissociation processes nonthermally and with spatial control. By application of tunneling voltage pulses, particles can be transferred from the surface to the tip and back; this has also been achieved for several metal–adsorbate systems (*65, 78–81*). Although such experiments have usually been performed with the intention of developing methods to move atoms and molecules in a controlled way, such effects can also provide chemical information, as was shown by Stipe *et al.* (*82, 83*) for the cleavage of the O–O bond of oxygen molecules on Pt(111). Figure 6 (*82*) shows the results of an experiment in which the (metastable) chemisorbed, molecular adsorption state of O_2 was prepared at 50 K. After the recording of frame a in Fig. 6,

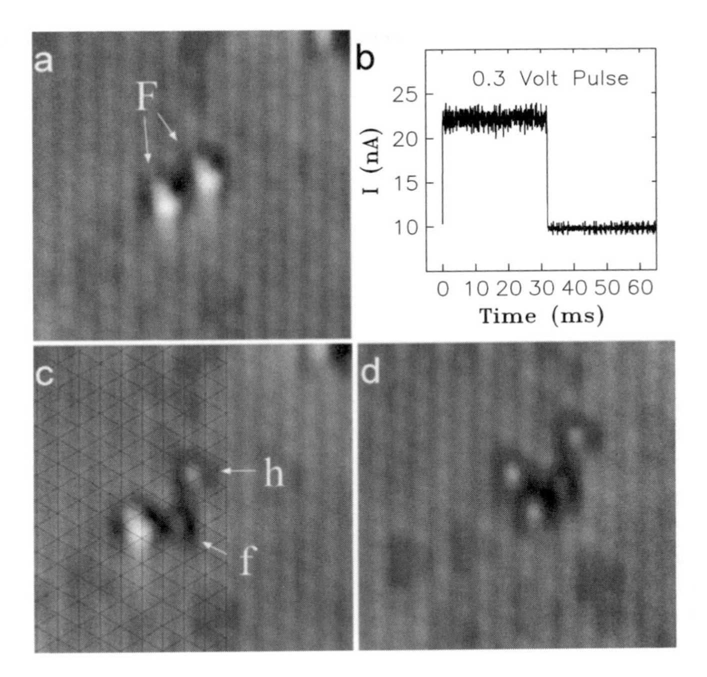

FIG. 6. Results of an experiment showing the controlled dissociation of adsorbed O_2 by the tunneling current. (a) STM image of two chemisorbed O_2 molecules at 50 K. (b) Tunneling current recorded during a 0.3-V pulse while the tip was positioned above the right molecule. The step at 30 ms indicates the moment of dissociation. (c) The same area after the pulse, with one O_{ad} atom on an hcp site and one on an fcc site (see overlaid grid). (d) The same area after a pulse above the second molecule [reprinted with permission from Stipe *et al.* (*82*). Copyright © 1997 The American Physical Society].

showing two molecules, the tip was placed above the right-hand molecule and a voltage pulse was applied (Fig. 6b). The current, which increased at first because of the higher voltage, suddenly decreased after 30 ms, indicating that the tunneling barrier had changed. The subsequent image (Fig. 6c) shows that the right-hand molecule had dissociated into two atoms (under these conditions O_{ad} appeared as bright dots with dark borders). As shown by the overlaid grid, only one atom was on an fcc site, whereas the other occupied an hcp site. After positioning of the tip over the second molecule and application of a pulse, that molecule was dissociated as well (Fig. 6d). In this case both atoms were found on hcp sites. By analyzing the parameter dependence of the dissociation, the authors showed that the bonds were broken by vibrational excitations by inelastic electron tunneling. Whereas the separations between the two atoms were in the same range (between one and three lattice constants) as for the thermal dissociation, the tunneling-induced dissociation also caused the occupation of hcp sites. Hence, the atoms underwent the same scattering process as after thermal dissociation ("hot atoms"), but the atoms that landed in the hcp sites did not jump to the energetically lower fcc sites at the lower temperature of this experiment. That the thermal dissociation led only to atoms on fcc sites was apparently a consequence of a secondary process that could not be suppressed at the temperatures necessary to overcome the activation energy for dissociation. By inducing chemical processes with the tunneling current, one can thus freeze out metastable intermediate states.

An important aspect of dissociation is its structure sensitivity. This has traditionally been investigated with stepped surfaces, with the steps acting as models for active sites. With STM the influence of steps can also be investigated with low-index surfaces by choosing areas with occasional steps. Whereas O_2 dissociation on Pt(111) did not show any correlation with atomic steps [in contrast to findings from previous macroscopic experiments (*84*)], a clear step effect was found by Zambelli *et al.* (*85*) for NO dissociation on a Ru(0001) surface. The work thus showed an active site in action. Figure 7 (*85*) is a large-scale STM image of the ruthenium surface with an atomic step (black vertical stripe) that was taken 30 min after exposure of 0.3 L of NO at 315 K. The many dark dots centered around the step are N_{ad} atoms, and the small islands of black dots at greater distances from the step are islands of O_{ad}. This identification of N_{ad} and O_{ad} atoms was based on results of experiments with the two individual adsorbates. The distribution of the atoms indicates that the NO_{ad} molecules that had adsorbed randomly on the terraces had diffused to the steps, where they dissociated into N_{ad} and O_{ad}. Because of its high (thermal) mobility, the NO_{ad} precursor was not seen. The diffusion of the O_{ad} atoms at room temperature was followed by high-speed STM (see Section V), but the mobility was so high that at the time the image was recorded the oxygen had already moved away from the step and clustered into islands. However, the diffusion profile of the less mobile N_{ad} atoms can still be recognized. There is an N_{ad} concentration gradient away from the step, indicating that the dissociation had taken place there. An unexpected finding was that both sides of the step were covered with N_{ad} atoms. This observation is incompatible with a reaction at the lower edge of the step (the N_{ad} atoms do not cross the step by diffusion), in which case N_{ad} atoms would have been found only on the lower adjoining terrace. It was thus concluded that the dissociation must have occurred at the upper step edge, from which the atoms migrated to both adjacent terraces. This finding is important for a theoretical understanding of the influences of steps on surface reactions.

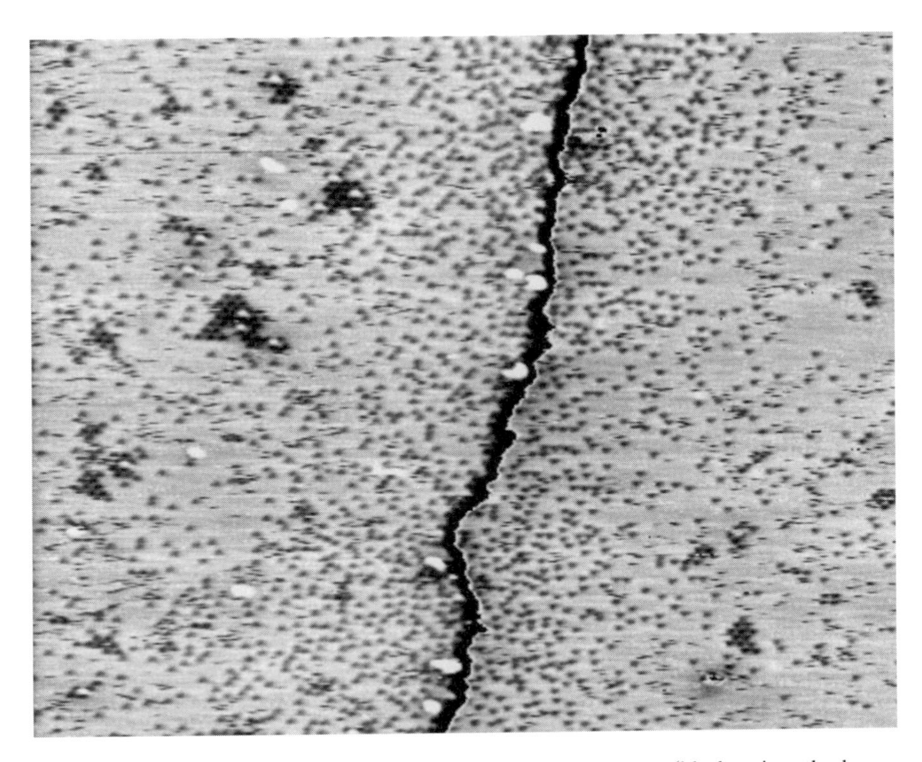

FIG. 7. STM image of a Ru(0001) surface with an atomic step (black stripe; the lower terrace is to the right) recorded after dissociative adsorption of a small amount of NO at 315 K. Disordered dots are N_{ad} atoms; islands contain the O_{ad} atoms (380 × 330 Å) [reprinted with permission from Zambelli *et al.* (*85*). Copyright © 1996 American Association for the Advancement of Science].

It was suggested that the altered d band width at the upper edge of the step is responsible for its relatively high reactivity (*85*).

V. Dynamic Behavior of Adsorbates and Their Distributions on Surfaces

The behavior of adsorbed reactants, intermediates, and products before and after reaction on the surface is determined by their mutual interactions and surface mobilities, which influence the distribution of the adsorbates on the surface and the rate by which reactants collide or reach active sites. In the traditional kinetics models of surface reactions, these effects are neglected. Surface diffusion is assumed to be fast and not rate determining, and the particles are assumed to be interaction free and randomly distributed (except for site exclusion), i.e., the reacting adsorbate layer is treated as a Langmuir gas in thermodynamic equilibrium.

A. Adsorbate–Adsorbate Interactions

In contrast to the assumptions underlying a Langmuir gas, many STM studies of metal–adsorbate systems have demonstrated island formation of the adsorbate particles, indicating attractive interactions. However, there has been only relatively little work that could quantify the interactions, which would be important for predictions of their relevance under real catalytic conditions. For a quantitative analysis, the system must be in equilibrium, which makes it difficult to monitor the particles because of their mobility. Figure 8 (*33*) shows an example for which a fast-recording technique allowed investigation of equilibrium fluctuations in an O_{ad} layer on a Ru(0001) surface. The two STM images (Figs. 8a and 8b; two successive frames from a six frames/s "movie") show oxygen atoms forming islands of a (2×2) structure in addition to many individual atoms. (In contrast to the usual finding, O_{ad} appears bright here because of the negative voltage and the constant height mode applied.) The atomic positions (Figs. 8c and 8d) show that all the single atoms and several of the low-coordinated atoms at the island perimeters had jumped in the period between the two frames. The whole series shows that atoms permanently detached from the islands or became reattached to the perimeters. Whereas these processes led to a complete rearrangement of the atom distribution over longer times, the average order did not change. It was therefore concluded that the observed changes represent equilibrium fluctuations, which could be followed because of the fast-recording technqiue.

O_{ad}–O_{ad} interactions were determined by Monte Carlo (MC) simulations by comparison of the experimental with calculated configurations. Qualitative agreement was obtained with an attractive interaction of -25 meV between two atoms at a distance of two lattice sites [this is operative in the (2×2) structure], in agreement with an upper limit of -0.05 eV from previous low-energy electron diffraction (LEED) studies (*86*). The energies at distances of one and $\sqrt{3}$ lattice constants were assumed to be repulsive and zero, respectively; the absolute values were not significant for the ordering behavior at low coverages. The experimental and simulated distributions show differences in details, indicating that the MC simulations were still too simple, but for a first estimate the attraction was a good approximation. Similar comparisons between STM and MC configurations were made for Cu(110)/O_{ad} (*87*), Cu(110)/alkali$_{ad}$ (*88*), and Re(0001)/S_{ad} (*89*).

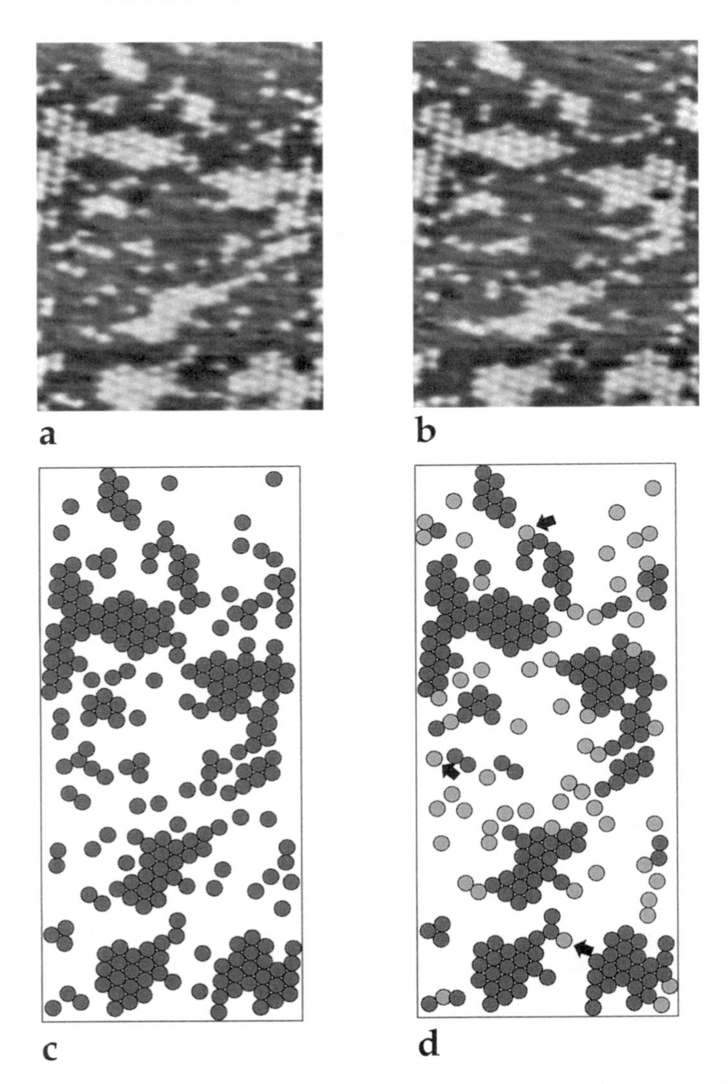

FIG. 8. (a and b) Constant-height STM images of O_{ad} atoms (bright dots) on Ru(0001) at 300 K taken with a "fast" STM at six frames/s. Time difference is 0.17 s (80 × 190 Å). (c and d) The atomic positions in the STM images are indicated. Atoms shown brighter in d have moved with respect to positions in c, and the arrows mark atoms that moved along island edges or became attached to or detached from islands [reprinted with permission from Wintterlin *et al.* (*33*). Copyright © 1997 Elsevier Science].

Another basis for elucidation of lateral interactions between adsorbates was provided by the pair correlation obtained by statistical analysis of images of dilute adsorbate layers. The pair correlation function represents the occupation probabilities (normalized to the average occupation probability) of the sites around a reference atom. The interaction energies follow from static particle distributions, i.e., no time resolution is required. The method had been applied much earlier in field ion microscopy, which is largely restricted to adsorbed metal atoms (*90, 91*). An example from STM for N_{ad} atoms on Ru(0001) is shown in Fig. 9 (*92*). The N_{ad} atoms appear dark; their triangular shape reflects the threefold adsorption site. All atoms occupy equivalent hcp sites (Fig. 9a). Figure 9b shows the pair correlation function g vs distance j obtained from a similar (but much larger) image (the black dots in Fig. 9b are experimental results). The term j numbers the sites [here $j = 1$ denotes the site at a distance of one lattice constant (a), $j = 2$ at $\sqrt{3}a$, $j = 3$ at $2a$, etc.]. That g is almost zero at $j = 1$ and small at $j = 2$ reflects the qualitative observation that these sites were either not occupied or occupied with a lower than average probability, respectively, whereas the $j = 3$ site was occupied with a higher than average probability.

The pair correlation function g can be translated into an effective interaction potential V_{eff} (potential of mean force; Fig. 9c), which accordingly displays a repulsion at $j = 1$ and 2 and a small attractive interaction (−18

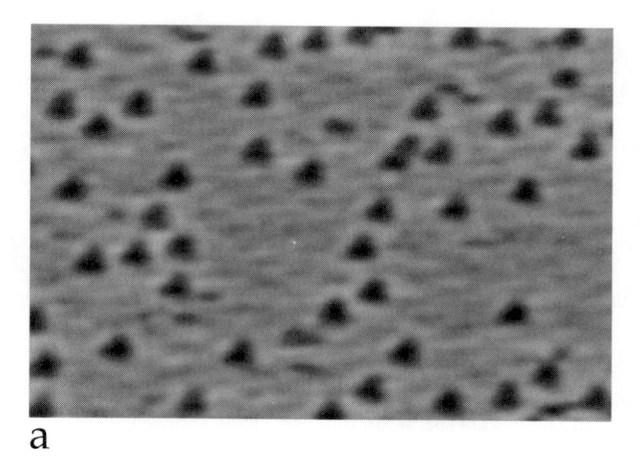

FIG. 9. (a) STM topograph of N_{ad} atoms (dark dots) on Ru(0001) at 300 K (86 × 54 Å). (b) Pair correlation function g vs site j from an image with 1344 nitrogen atoms (●). Open circles are from an MC calculation for hard spheres, blocking the sites at $j = 1$ and 2. (c) Potential of mean force V_{eff} obtained from the pair correlation function in b by $g(j) = \exp[-V_{eff}(j)/kT]$ [reprinted with permission from Trost *et al.* (*92*). Copyright © 1996 The American Physical Society].

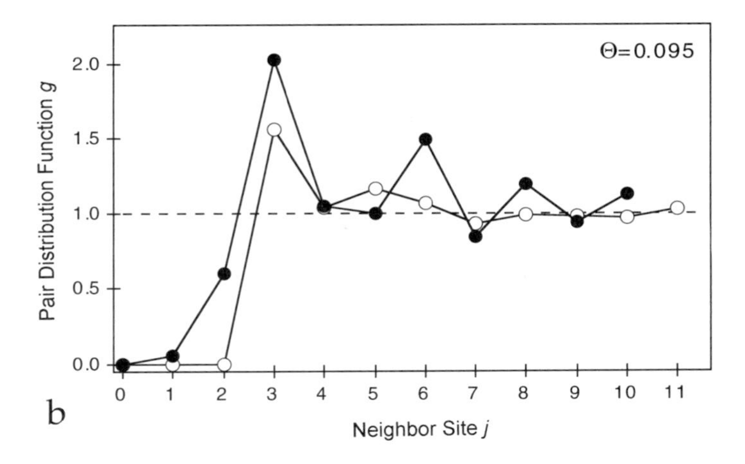

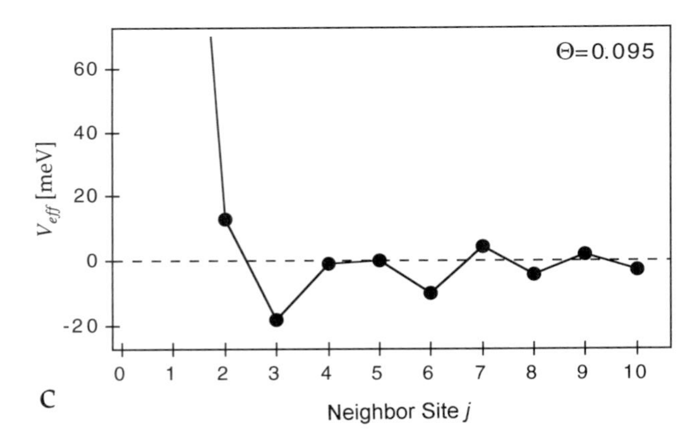

FIG. 9. *(continued)*

meV) at j = 3. Hence, the N_{ad} atoms experience a potential that drives them into a (2 × 2) structure, which is too small to lead to islands of such a structure at 300 K. Accordingly, in the STM images the N_{ad} atoms appear largely disordered (see Fig. 7). The actual attractive pair potential at j = 3 was found to be even weaker than suggested by Fig. 9c because V_{eff} contains contributions by "entropic forces" that occur for statistical reasons at finite coverages. Their operation became apparent from MC simulations with hard spheres (g = 0 at j = 1 and 2; open circles in Fig. 9b), which led to a g maximum at j = 3 notwithstanding the absence of attractive interactions. Part of the experimental maximum at $j = 3$ is a consequence of such an effect. The potential at greater distances does not deviate significantly from zero; the minima at $j = 6$ and 8 arise from entropic forces.

Similar analyses of STM data were performed for Ni(100)/O_{ad} (*93*), Cu(110)/O_{ad} (*94*), Re(0001)/S_{ad} (*95, 96*), Cu(111)/$C_6H_{6,ad}$ (*97, 98*), and Fe(100)/N_{ad} (*99*) with similar values and ranges of the interactions. However, the latter study demonstrated that a pair correlation analysis failed to account for the observed N_{ad} cluster sizes and shapes and that three-body interactions have to be included in MC simulations.

A more complicated (but nonetheless relevant) situation for surface catalytic reactions involves coadsorption of different adsorbates. Both cases known from LEED investigations (*100*) were observed—a phase separation of the two components (competitive adsorption) and the formation of a mixed phase (cooperative adsorption). No attempts were made to obtain values for the underlying repulsive or attractive interaction energies, except in an investigation of coadsorbed layers of O_{ad} and N_{ad} on Ru(0001) in which an energy of mixing of N_{ad} atoms in the $(2 \times 2)O_{ad}$ phase was determined (*101*).

The investigations of the interactions between adsorbates revealed only small attractive interactions (with magnitudes of a few kT) of short range (several lattice constants) that are not expected to affect reaction energies significantly, unless the coverages become so high that the stronger (repulsive) interactions at very short distances come into play (mostly at the nearest or next nearest sites; e.g., Fig. 9c). [These investigations cannot rule out additional smooth, long-range, possibly elastic interactions that were proposed in one case to explain mesoscopic superstructures of adsorbate islands (*102*).] On the other hand, the interactions are often large enough so that the distribution of the particles can be expected to deviate from that of a Langmuir gas even at the higher temperatures of real catalytic reactions.

B. Surface Diffusion of Adsorbates

Surface diffusion of adsorbates on metals has also been characterized on the atomic scale by STM. There was an early attempt to extract diffusion coefficients from the flicker noise of the tunneling current, attributed to atoms rapidly diffusing underneath the static tunneling tip (*103*), and diffusion coefficients were also determined from growth rates of islands (*104*). However, the most straightforward access was achieved by monitoring the motion of individual particles in successive frames, which after averaging yielded the hopping rates (ν) and, by the Einstein–Smoluchowski equation

$D = (\langle r^2 \rangle/4)\nu$ (for quadratic and hexagonal lattices), the diffusion coefficients. Table I provides a list of studies performed with nonmetal adsorbates on metal surfaces. (There are many additional investigations with metal-on-metal systems and with semiconductors.) The poor time resolution of STM is probably the reason why only a few of these investigations led to determinations of D over a range of temperatures (with the corresponding limitation of the reliability of the E^*_{diff} values).

The data listed in Table I are from experiments in which the particles were (or were assumed to be) sufficiently far apart so as not to affect each other. Under catalytic reaction conditions, the coverages may be appreciable so that such independent motion can no longer be expected. Even for the usually assumed Langmuir gas, one would have to include corrections at nonzero coverages because of the site blocking (hard sphere potential).

The complex situation with real interactions was studied for O_{ad} atoms on Ru(0001) by Renisch *et al.* (Fig. 10) (*111*). For this purpose the behavior of O_{ad} pairs was investigated, which were far away from all other oxygen atoms. The measurements were performed with a "fast" STM at room temperature (*33*), with series of more than 1000 successive STM images of the type shown in Fig. 8 but with lower coverages. The two atoms performed jumps between subsequent frames, leading to changes in the distance which were analyzed statistically. Figure 10a (*111*) shows the paths along which the hopping rate of one of the atoms was found to be affected by

TABLE I

Results of STM Studies of Diffusion of Nonmetal Adsorbates on Metal Surfaces

Metal/ Adsorbate	Temperature range (K)	Preexp. factor (s^{-1})	E^*_{diff} (eV)	Reference
Ni(100)/O_{ad}	353	10^{12}–10^{13} (assumed)	0.57–0.64	Binnig *et al.* (*103*)
Ni(100)/O_{ad}	300	10^{13} (assumed)	0.99	Kopatzki and Behm (*93*)
Re(0001)/S_{ad}	300	10^{14} (assumed)	0.79 ± 0.1	Dunphy *et al.* (*95, 96*)
Ru(0001)/N_{ad}	300–350	$10^{14\pm1.5}$	0.94 ± 0.15	Zambelli *et al.* (*105*)
Pd(110)/$C_6H_{6,ad}$	212–228	10^{12} (assumed)	0.57 along [001]	Yoshinobu *et al.* (*106*)
Pt(111)/O_{ad}	191–205	$10^{9.4\pm1}$	0.43 ± 0.04	Wintterlin *et al.* (*42*)
Ag(110)/$O_{2,ad}$	65–100	$10^{13\pm3}$	0.22 ± 0.05 along $[1\bar{1}0]$	Barth *et al.* (*107*)
Cu(110)/CO_{ad}	42–53	$10^{7.9\pm0.37}$	0.097 ± 0.004 along $[1\bar{1}0]$	Briner *et al.* (*108*)
Al(111)/O_{ad}	440	10^{13} (assumed)	1.0–1.1	Trost *et al.* (*104*)
Pd(111)/$C_2H_{2,ad}$	65–70	10^{13} (assumed)	0.18	Dunphy *et al.* (*109*)
Pd(110)/$C_2H_{2,ad}$	250–260	10^{12} (assumed)	0.56–0.58	Ichihara *et al.* (*110*)
Ru(0001)/O_{ad}	300	10^{10}–10^{13} (assumed)	0.55–0.7	Renisch *et al.* (*111*)
Pd(110)/PVBA	330–370	$10^{10.3\pm0.4}$	0.83 ± 0.03 along $[1\bar{1}0]$	Weckesser *et al.* (*112*)

the second. The result is that for positions on the indicated sites (Fig. 10a) the residence time differed significantly from that of an isolated atom (the nearest-neighbor site is strongly repulsive and was never occupied). The mutual influence extends up to a distance of 3*a* so that, together

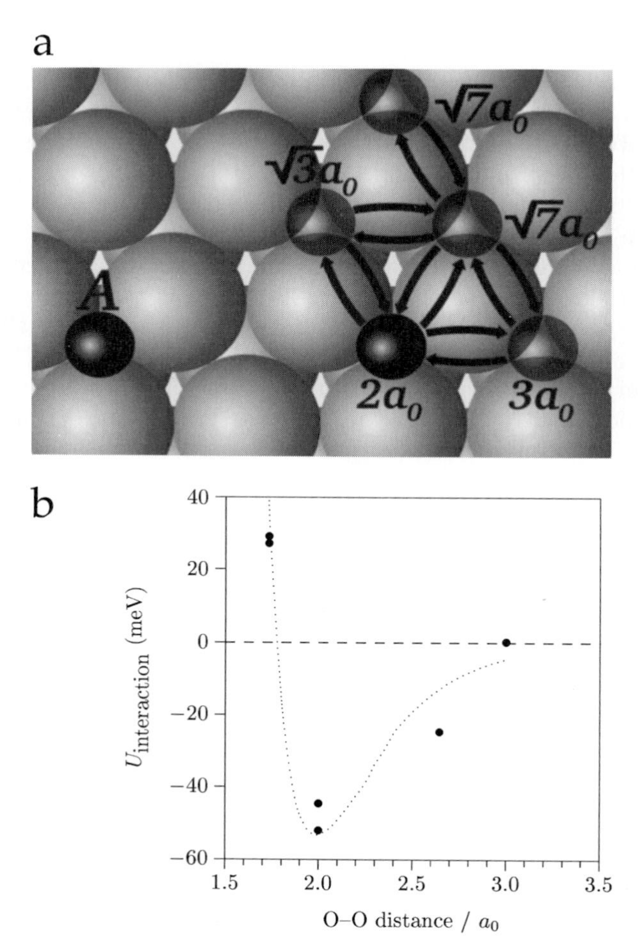

FIG. 10. (a) Model illustrating the mutual effects of two O_{ad} atoms on Ru(0001) on their motions. The arrows mark the directions in which the jumps of one atom are affected by the second atom localized at A. The nearest-neighbor site is not occupied because of strong repulsion. (b) Interaction potential between two O_{ad} atoms resulting from the analysis of the processes indicated in (a) [reprinted with permission from Renisch *et al.* (*111*). Copyright © 1999 The American Physical Society].

with the symmetrically equivalent sites, each atom affects 36 sites around it, on which the residence times vary by more than one order of magnitude. This result emphasizes the importance of including interactions to describe the dynamic behavior of adsorbates, even at moderate surface coverages.

The data could be translated further into an interaction potential between two O_{ad} atoms (Fig. 10b) that extends up to three lattice constants and has the expected shape, with an attractive and a repulsive branch. The minimum at $2a$ is in accord with the (2×2) structure formed by O_{ad} on this surface (Fig. 8). Such analyses of mutual effects on the dynamic behavior thus provide another method for obtaining adsorbate–adsorbate interactions, as had previously been demonstrated for the simpler cases of one-dimensional diffusion of atoms in the atomic troughs of the fcc(110) surface [platinum atoms on Pt(110) (*113*) and O_2 molecules on Ag(110) (*107*)].

Mutual influences between adsorbates on their mobilities can go beyond the deformation of the surface potential experienced by the particles. At low temperatures, CO_{ad} molecules on Cu(110) were found to arrange partially into short chains in the [001] direction (*108*), evidencing attractive interactions. In contrast to expectation and to observations of O_{ad} pairs on Ru(0001), it was found that the chain molecules jumped with a higher rate than single atoms. It was determined that this effect was not caused by a lower diffusion barrier of the chain molecules but instead by a higher preexponential factor. A reduced entropy of the molecules in the chains was suggested as an explanation.

Dunphy *et al.* (*109*) succeeded in mapping the adsorbate surface potential for the more complicated motions of larger molecules, including rotations. Figure 11a shows an adsorbed acetylene molecule on a Pd(111) surface at 44 K; the molecule is imaged as a bright dot with an adjacent depression. This asymmetric appearance was caused by the adsorption geometry shown in Fig. 11b, which was inferred from calculations (*109*); the spectroscopically determined geometry (*114*) is approximately the same. The molecules rehybridize to an ethylene-like geometry, with a tilt of the H-C-C-H plane with respect to the surface normal. The lowest energy site was found to be the hcp site, but the energy characterizing the fcc site is almost the same. Calculations of the STM contour in the same investigation showed that the geometric position of the molecules is between the bright and the dark feature seen in the images, at about the position of the cross in Fig. 11a; the bright dot corresponds approximately to the bridge site adjacent to the adsorption site. (The cross marks the center of rotation.) Because of the adsorption geometry, the bright/dark features in the STM images display three rotational orientations (at 120° to each other for the molecules on the hcp site) and three additional orientations for the molecules on the fcc

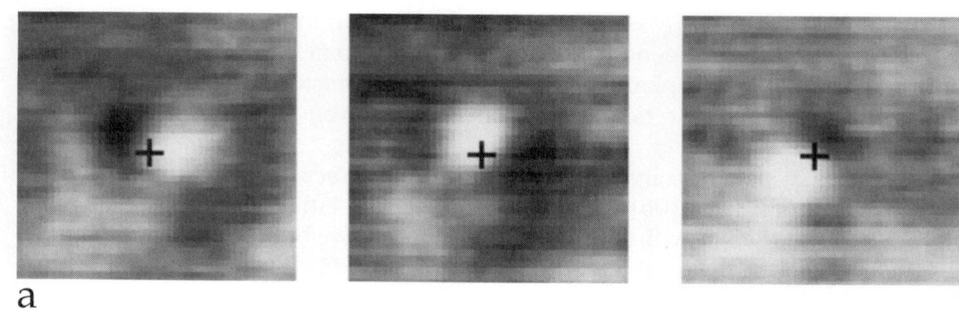

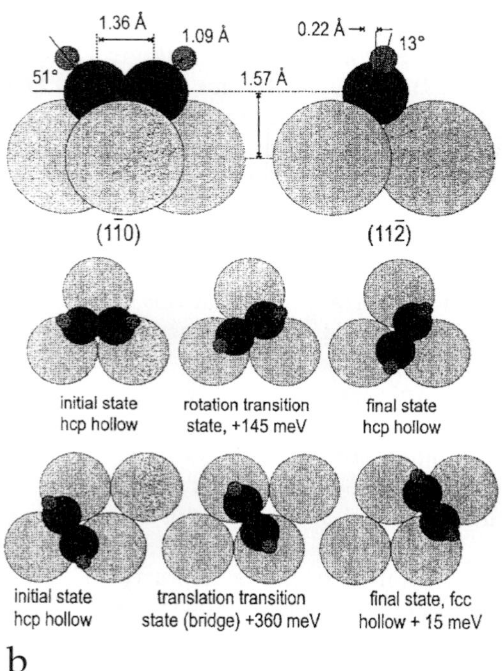

FIG. 11. STM images of an acetylene molecule on Pd(111) in three different rotational orientations; 44 K (12 × 13 Å). Crosses mark a fixed point on the surface near the center of rotation. It is also approximately the geometric position of the molecule according to calculations of the imaging of acetylene by STM. (b) Top, adsorption geometry of acetylene from total energy calculations; center, molecular rotation about the hcp adsorption site; bottom, transition from hcp to fcc site [reprinted with permission from Dunphy *et al.* (*109*). Copyright © 1998 The American Physical Society].

sites (three orientations are shown in Fig. 11a). The authors observed that at ~44 K the bright/dark features started to switch between the three equivalent rotational orientations, indicating the onset of rotation of the molecules about their adsorption sites at this temperature (Fig. 11b, center). With an assumed preexponential factor of 10^{13} s^{-1}, a rotational energy barrier of 0.11 ± 0.01 eV was estimated, which is in reasonable agreement with the calculated value for the energy of the transition state. The molecules could also exchange between hcp and fcc sites, with an estimated activation energy of approximately half of the calculated value (Fig. 11b, bottom).

Many authors reported tip-induced motions of atoms and molecules, which reflect the low values of the diffusion barriers. When performed in a controlled way, induced motion is a major method for the creation of artificial ensembles of atoms and molecules in adsorbate–metal systems (*16, 73, 115–118*). In most cases the mechanism is probably a simple pushing or dragging of the particles with the tip, which can be controlled by the tip–surface distance. Inelastic tunneling mechanisms have also been reported for tunneling-induced rotations of molecules (*70, 119, 120*). It is clear that in diffusion measurements care must be taken so that such effects can be ruled out (*121*).

VI. Surface Reactions

The following sections comprise a review of investigations of catalytic reactions giving evidence of the formation of new chemical compounds from adsorbed reactants.

A. Mass Transport and Mesoscopic Restructuring on Reconstructed Surfaces

Several metal surfaces, particularly the (110) surfaces of fcc metals, are structurally unstable and are reconstructed in the presence of some adsorbates. In turn, there are surfaces that are reconstructed in the uncovered state but transform into the nonreconstructed state upon adsorption. Such transformations and reversals of reconstructions have been a major subject of STM adsorption studies (*5, 7*). An essential contribution of STM was the solution of the mesoscopic restructuring problem associated with these processes. In most reconstructed phases the density of substrate atoms differs from that of the unreconstructed surface. Adsorbate-induced reconstruction therefore leads to transport of substrate material and a mesoscopic restructuring of the surface, which can be monitored by STM. It is therefore not surprising that reconstructed surfaces were among the first with which STM investigations of chemical reactions were performed. An important advantage of such systems is that the mobility of adsorbates is reduced when they are involved in reconstructed structures, allowing the performance of STM experiments at room temperature.

The best-studied systems are the so-called added-row reconstructions of the (110) surfaces of nickel, copper, and silver that are induced by atomic oxygen. The general structure model was first solved by STM for Cu(110) (Fig. 12) (*122, 123*). The reconstruction consists of long chains of copper and oxygen atoms, running in the [001] direction, on top of the first unchanged metal layer. Because copper atoms are incorporated into the chains, the formation of the phase upon adsorption of oxygen is connected with copper mass transport. The copper atoms originate from step edges and, to a minor degree, from pits on the terraces. The structures are quite similar on nickel, copper, and silver, differing, however, in the periodicity of the chains in the [1$\bar{1}$0] direction. On the copper surface there is a weak attractive interaction between the chains, leading to a

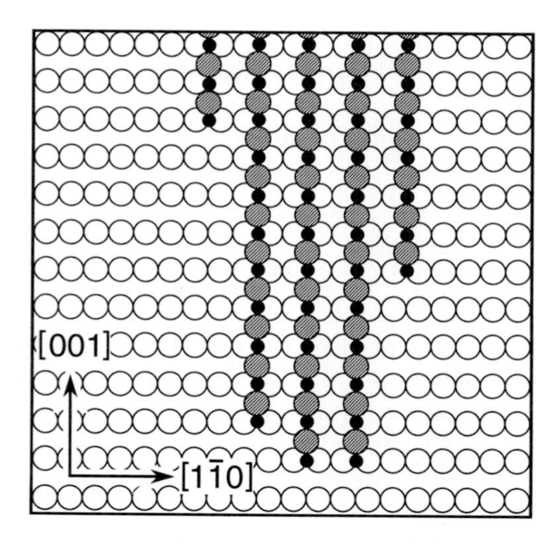

FIG. 12. Structure model of an island of the $(2 \times 1)O_{ad}$ added-row reconstruction of Cu(110). The large open spheres are the Cu atoms of the first substrate layer, and large hatched spheres and small black spheres are the Cu and oxygen atoms of the added-row chains, respectively.

(2 × 1) structure with a chain periodicity $2a$ in the $[1\bar{1}0]$ direction. At submonolayer coverages, long, narrow islands of this phase form; Fig. 12 shows a model of the end of such an island. Upon annealing, these islands order into regular stripes, which was explained by an additional long-range repulsion mediated by elastic interactions (*102*). On Ag(110) (*124*), the chains repel each other, causing a series of (n × 1) structures, with n decreasing from 7 to 2 as $\Theta(O_{ad})$ increases. On Ni(110) (*125*), there is, with increasing $\Theta(O_{ad})$, a (3 × 1) phase with a $3a$ periodicity of the chains, a (2 × 1) phase with a $2a$ periodicity, and a second (3 × 1) phase with a $3a$ periodicity of pairs of chains. In the STM experiments the atomic structure along the rows was often resolved. However, it was shown that whether the metal or the oxygen atoms appeared bright depended on the state of the tip (*44*).

1. *Reaction of H_2S with Oxygen on Ni(110)*

The first reaction study with STM, involving an added-row reconstructed surface, was performed by Ruan *et al.* (*126*), who investigated the reaction of H_2S with oxygen on Ni(110). The $(2 \times 1)O_{ad}$-reconstructed Ni(110) surface was exposed to H_2S at 300 K, leading to S_{ad} and H_2O:

$$H_2S \rightarrow H_2S_{ad} \tag{1a}$$

$$H_2S_{ad} + O_{ad} \rightarrow S_{ad} + H_2O\uparrow \tag{1b}$$

Figure 13 (*126*) shows a series of STM images recorded for the same surface area during adsorption of H_2S on the oxygen-precovered surface $[\Theta(O_{ad}) = 0.5]$. Figure 13a displays the typical striped structure of the Ni–O added rows of the fully $(2 \times 1)O_{ad}$-reconstructed surface. Upon H_2S adsorption (Fig. 13b), bright islands with a characteristic quasi-hexagonal pattern formed, along with darker areas covered with the same pattern. From other experiments it was known that this pattern indicates the $c(2 \times 2)S_{ad}$ structure, which indicates that part of the O_{ad} had reacted with H_2S_{ad} and was replaced by S_{ad}. (The water desorbs.) The formation of the bright islands and dark areas was explained by the lifting of the added-row reconstruction when the oxygen was removed by the reaction. The added-row nickel atoms released in this way condensed into one-atomic-layer-high islands which became covered by S_{ad} atoms, giving rise to the bright $c(2 \times 2)S_{ad}$ islands. The areas from which the added rows had been removed also became covered by S_{ad} atoms, but they were on a one-atomic-layer deeper level than the islands and therefore appear dark. [The $c(2 \times 2)S_{ad}$ structure is not reconstructed.] The mesoscopic roughening (Fig. 13b) is

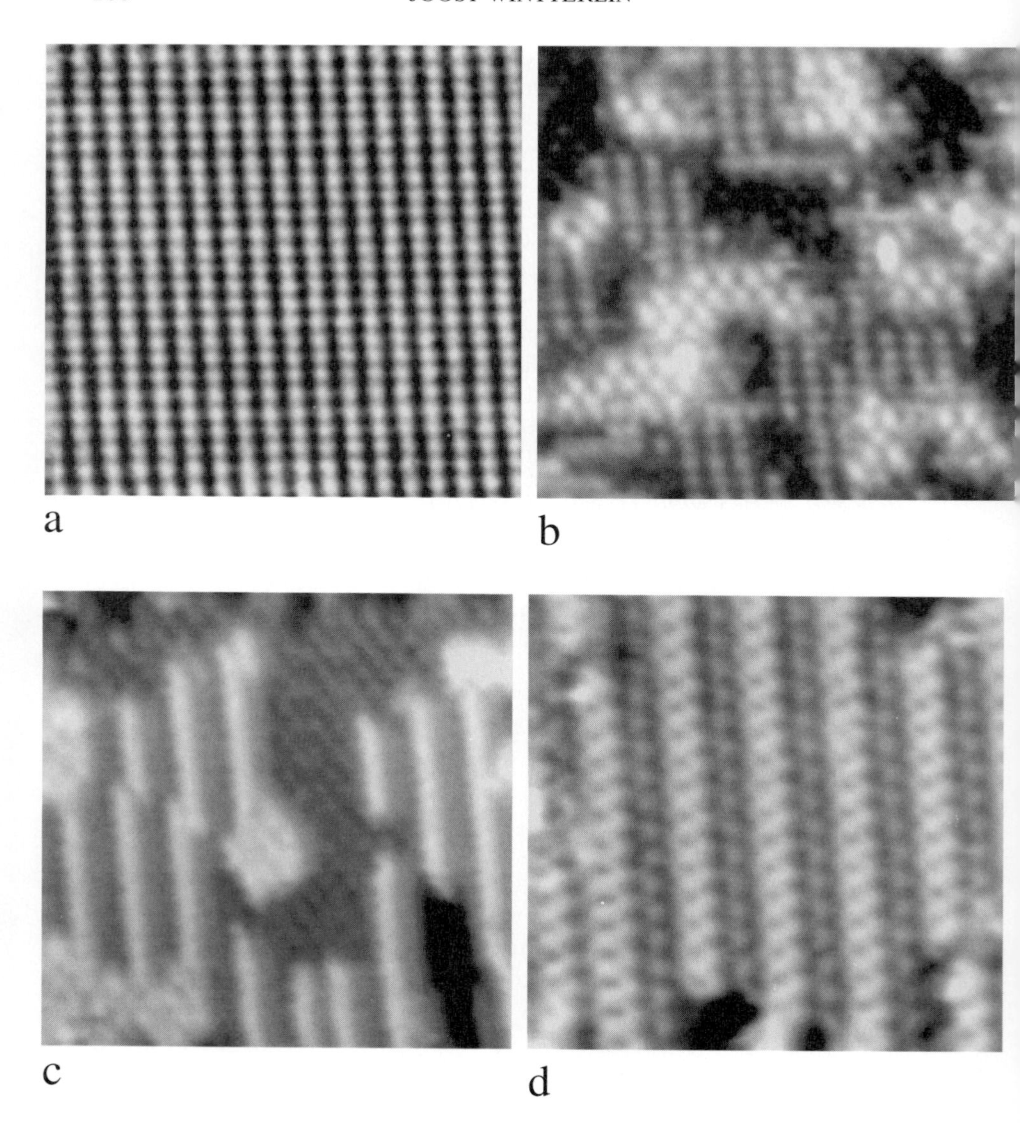

FIG. 13. Series of STM images recorded during H_2S adsorption on the oxygen-covered Ni(110) surface at 300 K. (a) $(2 \times 1)O_{ad}$ reconstruction before H_2S adsorption. (b) Partial transformation into the $c(2 \times 2)S_{ad}$ structure. (c and d) $(4 \times 1)S_{ad}$ structure. All from the same area; 85 × 91 Å (a–c) and 59 × 63 Å (d) [reprinted with permission from Ruan *et al.* (*126*). Copyright © 1992 The American Physical Society].

therefore a consequence of (i) the lifting of the added-row reconstruction that has a nickel density of one-half relative to that of the nonreconstructed surface and (ii) the relatively low mobility of the released nickel atoms under these conditions, which cannot form larger islands or reach steps. Upon further dosing with H_2S (Figs. 13c and 13d), this structure was finally transformed into the $(4 \times 1)S_{ad}$ phase. The $(4 \times 1)S_{ad}$ structure is reconstructed and has the same nickel density as the original $(2 \times 1)O_{ad}$ phase and was therefore flat again (Fig. 13d). When the experiment was performed starting with an annealed, flat $c(2 \times 2)S_{ad}$ structure, the $(4 \times 1)S_{ad}$ phase did not form at room temperature. The authors concluded that the high surface energy of the roughened intermediate state led to a higher reactivity than that of the flat surface.

The data show that a catalytic surface is not necessarily a static substrate to which the adsorbates bind. When adsorbed reactants or intermediates are involved in reconstructions, the mesoscopic topography of the surface may change, which in turn may influence structure-sensitive chemical reactions.

Later, the authors (*127*) also investigated the same reaction on Cu(110). As on the nickel surface, the added-row reconstruction was lifted and replaced by the $c(2 \times 2)S_{ad}$ structure. However, the reaction on Cu(110) also showed marked differences from that on Ni(110). In particular, the reconstructed rows on copper reacted off less isotropically than on nickel, but instead in a pronounced row-by-row fashion. This is actually the more typical behavior of these structures, as discussed in Section VI.B

2. *Formic Acid Oxidation on Cu(110)*

The topography resulting from the transport processes during the formation of a reconstruction or the lifting of a reconstruction depends not only on the mobility of the metal atoms involved but also on other properties that determine the restructuring processes during surface reactions on such surfaces. Several such influences were discovered in investigations of the oxidation of formic acid to give formate on the Cu(110) surface (*128–134*):

$$\mathrm{HCOOH} \rightarrow \mathrm{HCOOH_{ad}} \tag{2a}$$

$$2\,\mathrm{HCOOH_{ad}} + \mathrm{O_{ad}} \rightarrow 2\,\mathrm{HCOO_{ad}} + \mathrm{H_2O} \uparrow \tag{2b}$$

The investigations were motivated by the importance of methanol synthesis from syngas, whereby formate is the most abundant surface intermediate under steady-state conditions (*135*).

Figure 14, taken from work by Bowker *et al.* (*131*), shows a series of STM images recorded during formate adsorption on $Cu(110)/O_{ad}$ at 300 K. The reaction was started with a partially (2×1) reconstructed surface,

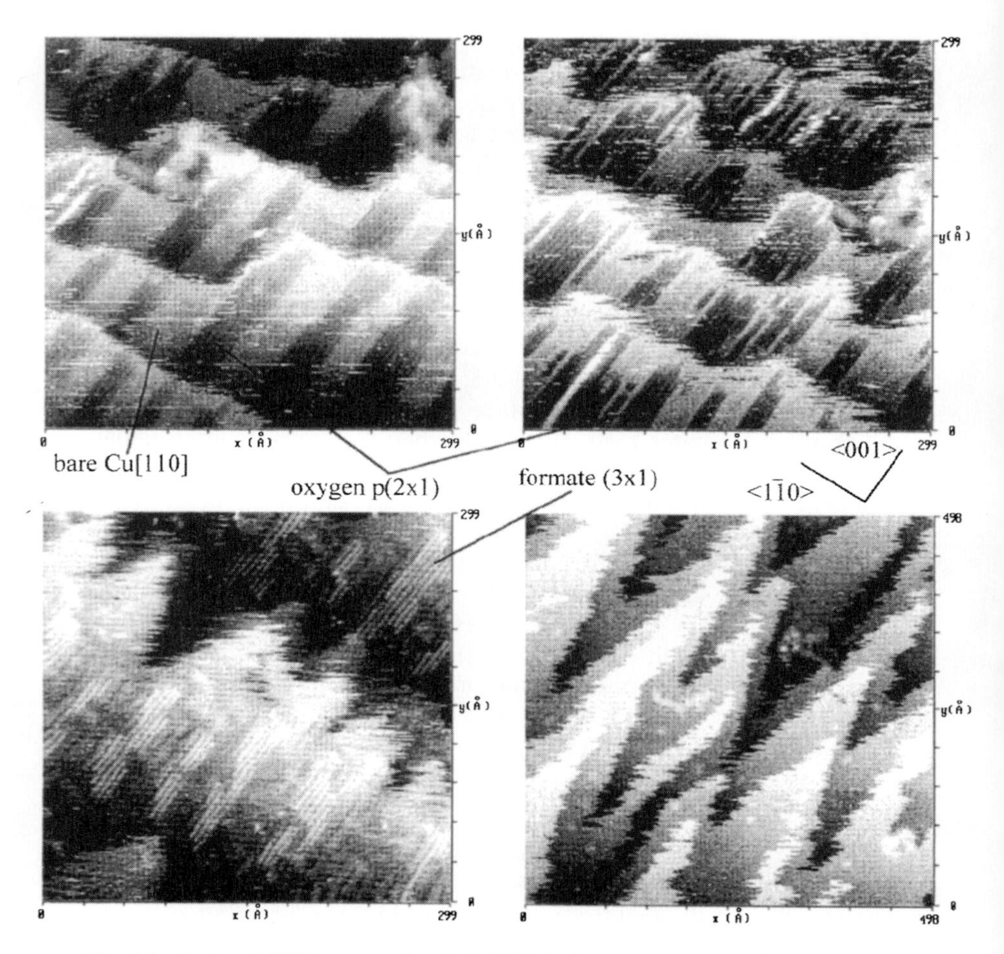

FIG. 14. Series of STM topographs of Cu(110) during exposure of a partially $(2 \times 1)O_{ad}$ added-row reconstructed surface to formic acid at 300 K. Surface phases and image sizes as indicated [reprinted with permission from Bowker *et al.* (*131*). Copyright © 1998 Baltzer Science Publishers].

and the O_{ad} coverage was approximately $\Theta(O_{ad}) = 0.25$. The dark bands in the first frame are islands of the (2×1) reconstruction, with individual rows being weakly visible. The relatively uniform widths of the islands and their even spacings are typical of the annealed (2×1) phase and had been explained by long-range elastic repulsions (*102*). At an intermediate stage (frames 2 and 3), a (3×1) phase occurred (bright stripes) that is formed by $HCOO_{ad}$. Finally, the terraces appeared flat (frame 4), as a consequence of another formate phase, the c(2 × 2)$HCOO_{ad}$ structure, which covered the surface completely (but was not resolved on this scale). The structure model is shown in Fig. 15a (*130*). During the formation of the c(2 × 2) structure the terraces developed a characteristic sawtooth pattern, indicating transport processes of copper atoms. Because the c(2 × 2)$HCOO_{ad}$ structure does not involve a reconstruction, copper atoms were released

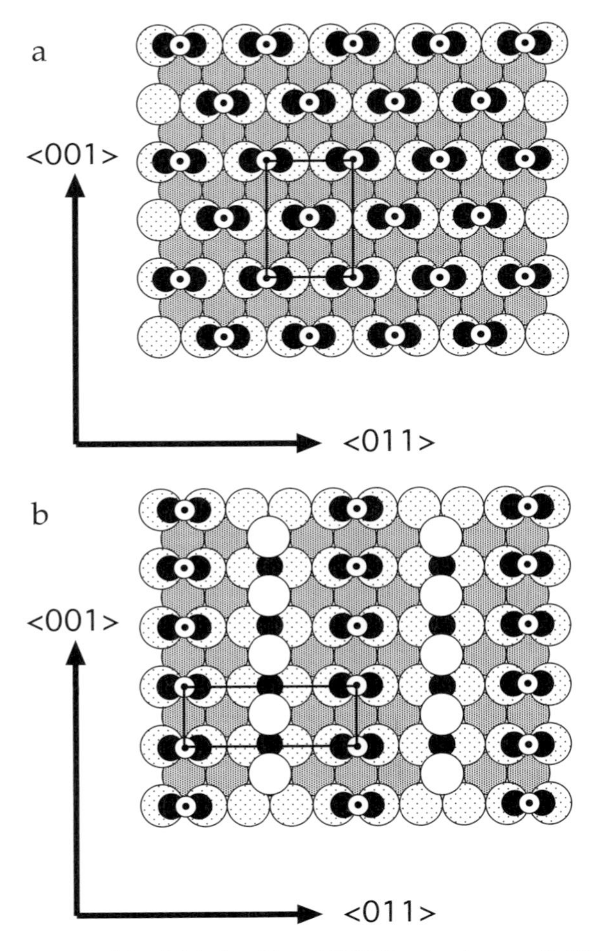

FIG. 15. Structure models of (a) the c(2 × 2)$HCOO_{ad}$ structure on Cu(110) and (b) the (4 × 1) mixed O_{ad}/$HCOO_{ad}$ structure. Large circles are Cu atoms, solid circles oxygen atoms, small open circles carbon atoms, and small dots hydrogen atoms [reprinted with permission from Haq and Leibsle (*130*). Copyright © 1997 Elsevier Science].

as the oxygen atoms reacted with formic acid, and the reconstruction was lifted. The copper atoms apparently had become attached to the step edges. This led to a sawtooth-like growth of the terraces that was rationalized by the fact that the copper atoms had moved to the steps only along the stripes of empty surface between the (2 × 1) islands. Hence, the terraces were growing in stripes that reproduced the periodicity of the initial (2 × 1) islands. The unusual triangular terrace edges might have been additionally stabilized in some way by the c(2 × 2)$HCOO_{ad}$ structure. The rate of the reaction also played a role. Sawtooth patterns occurred only for fast adsorption of formic acid, whereas slow adsorption led to less drastic topography changes, and the reaction appeared to be more concentrated to the areas inside the oxygen islands. Moreover, the availability of empty surface was important. When the experiment was performed with $\Theta(O_{ad}) = 0.5$ [i.e., with a fully (2 × 1) reconstructed surface], the c(2 × 2)$HCOO_{ad}$ structure could no longer form because for each O_{ad} atom two $HCOO_{ad}$ groups are formed [Reaction (2b)], and the c(2 × 2)$HCOO_{ad}$ structure has a formate coverage of 0.5. Part of the oxygen did not react in this case, and the surface roughened microscopically, forming a patchwork of different surface phases, mostly of the (3 × 1)$HCOO_{ad}$ and the c(6 × 2)O_{ad} structures (*136, 137*).

B. ONE-DIMENSIONAL REACTIVITIES ON RECONSTRUCTED fcc(110) SURFACES

An important property of fcc(110) surfaces is their only twofold rotational symmetry, which can be expected to cause anisotropies in chemical reactions. Anisotropies are also displayed by other processes on these surfaces, such as adsorbate diffusion, which is often characterized by pronounced differences in the diffusion coefficients parallel and perpendicular to the close-packed atomic rows (*107, 108, 121, 138*). Some of the reconstructed phases, such as the oxygen-induced added-row structures of the Ni, Cu, and Ag(110) surfaces, have extreme aspect ratios of the rows as basic units and of the islands. The oxygen atoms are on almost linear positions between the metal atoms of the rows, corresponding to anisotropic chemical bonds so that particularly strong effects can be expected.

1. *CO Oxidation on Rh(110)*

The first system for which a pronounced reaction anisotropy was observed by STM was the reconstructed Rh(110) surface in an investigation by Leibsle *et al.* (*139*). Oxygen on Rh(110) forms several reconstructed phases that are responsible for the striped pattern in the STM image of Fig. 16a (*139*). Most of the oxygen-covered parts have the c(2 × 6)O_{ad} structure, which is of the missing-row type [model shown in Fig. 16b (*139*)]. The segmented stripes in the STM images result from another oxygen-induced reconstruction—a (1 × 2) structure.

After preparation of this state, O_{ad} was reacted off by dosing of carbon monoxide at 300 K:

$$CO \rightarrow CO_{ad} \tag{3a}$$

$$CO_{ad} + O_{ad} \rightarrow CO_2 \uparrow \tag{3b}$$

CO_2 desorbs under these conditions, and the changes were monitored by STM. Figure 16a represents an intermediate stage. The medium-gray stripes correspond to areas covered by unreacted oxygen, and the brighter stripes are empty areas or areas covered by CO_{ad}. The long extension of the reacted (brighter) areas shows that the oxygen was removed by the reaction

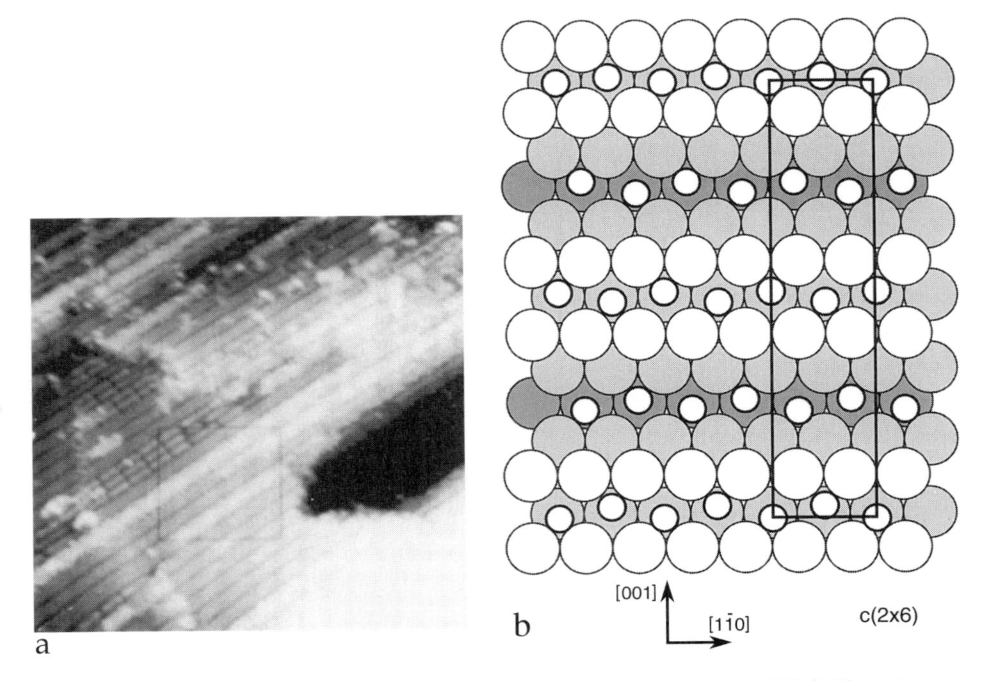

FIG. 16. (a) STM image taken after CO exposure to an oxygen-covered Rh(110) surface at 300 K. Reacted areas appear as bright blurred lines (400 × 400 Å). (b) Model of the c(2 × 6)O_{ad} structure covering most of the unreacted parts in the STM image (a) [reprinted with permission from Leibsle *et al.* (*139*). Copyright © 1993 Macmillan Magazines Limited].

in an extremely one-dimensional fashion in long narrow stripes in the $[1\bar{1}0]$ direction. The width of these stripes corresponds to the area between two missing rows, i.e., to half the $c(2 \times 6)$ unit cell width in the [001] direction (Fig. 16b). Hence, the direction into which the reaction progressed was fully governed by the reconstruction; there was no reaction perpendicular to the row structure. (From the data it is not clear whether the reconstruction was lifted after removal of the oxygen and whether there were transport processes of rhodium atoms.) According to the structure model (*140*), the oxygen atoms form rows in the $[1\bar{1}0]$ direction, with adsorption sites between close-packed rows of metal atoms so that the O_{ad} atoms are in a highly anisotropic surrounding. The potential energies experienced by CO molecules approaching the rows from the sides or from the ends must therefore be quite different, making the one-dimensional reactivity in this system understandable.

2. *Oxidation of Methanol and of Formic Acid on Cu(110)*

An extensively investigated reaction with STM is the oxidation of methanol to adsorbed methoxy on Cu(110):

$$CH_3OH \rightarrow CH_3OH_{ad} \tag{4a}$$

$$2\,CH_3OH_{ad} + O_{ad} \rightarrow 2\,CH_3O_{ad} + H_2O\uparrow \tag{4b}$$

The experiments, performed by Leibsle *et al.* (*141–144*), were motivated by the methanol synthesis from syngas on copper catalysts, in which the reverse process (the reduction of methoxy to methanol) represents the final step (*135*). At 300 K, Reaction (4b) is followed by a slow reaction of the methoxy group to give formaldehyde, which desorbs:

$$2\,CH_3O_{ad} \rightarrow 2\,HCHO\uparrow\ + H_2\uparrow \tag{5}$$

In the methanol synthesis, formaldehyde formation also occurs as a side reaction.

In the first experiments, submonolayers of oxygen were preadsorbed, leading to a partially $(2 \times 1)O_{ad}$-reconstructed surface. Then the sample was exposed to methanol to induce Reaction (4b). At temperatures >300 K, no ordered methoxy structures were seen, presumably because Reaction (5) was too fast or because the CH_3O_{ad} molecules were too mobile. When the experiments were performed during warm-up of the sample from ~270 K to room temperature, a $c(2 \times 2)$ structure was observed which transformed into a (5×2) structure. Both phases were attributed to methoxy.

Figure 17a (*142*), recorded after partial reaction, shows a domain boundary between unreacted added-row reconstructed areas with the typical row structure and the $(5 \times 2)CH_3O_{ad}$ structure. The bright dots of this phase

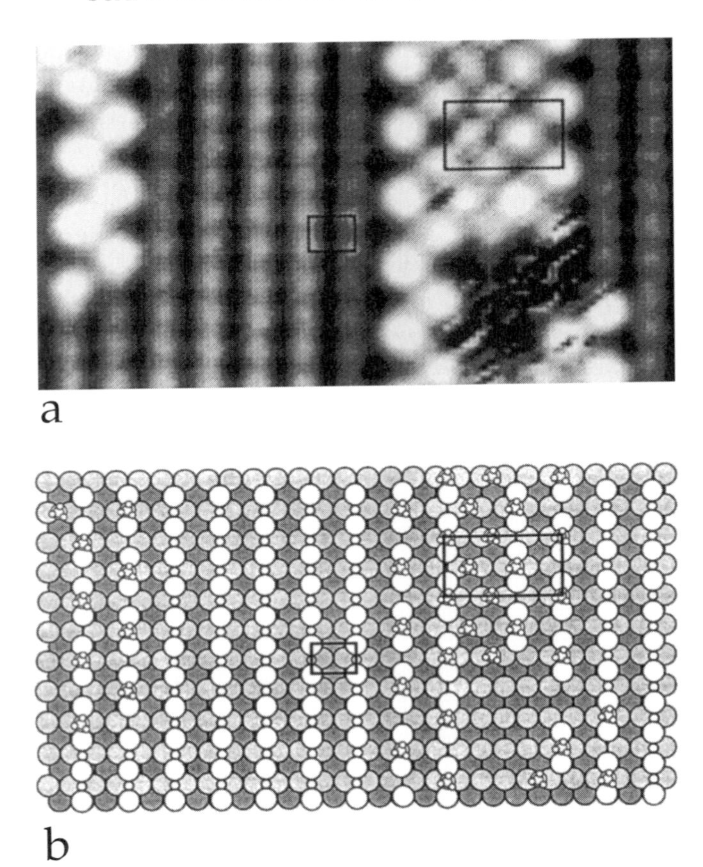

FIG. 17. (a) STM topograph of the $(2 \times 1)O_{ad}$ added-row reconstructed Cu(110) surface after partial reaction with methanol at approximately 270 K. Vertical stripes are unreacted added rows, and areas with bright dots are the (5×2) methoxy structure $(70 \times 40$ Å). (b) Ball model of the STM topograph. The model for the $(5 \times 2)CH_3O_{ad}$ structure is only one of several possibilities [reprinted with permission from Leibsle *et al.* (*142*). Copyright © 1994 Elsevier Science].

were interpreted as the CH_3O_{ad} groups. The structure model (Fig.17b) accounts for the observation that the brightness of the methoxy features in the data varied, suggesting that several different sites must have been occupied. Since no condensation of copper atoms at steps or on terraces was observed, the structure must have incorporated the copper atoms from the added rows (i.e., it must also have involved a reconstruction). In the $(5 \times 2)CH_3O_{ad}$ model of Fig. 17b, $\Theta(Cu)$ is 0.4, close to the value of one-half for the added-row reconstruction [other models with $\Theta(Cu) = 0.6$ were also considered]. The exact arrangement of the methoxy groups and the copper atoms in the unit cell, however, could not be determined on the basis of these data. The initially formed c(2 × 2) structure is probably quite similar because the (5 × 2) structure model contains c(2 × 2) fragments.

The time series of Fig. 18 (*141*), recorded during exposure of the partially O_{ad}-covered surface to methanol at ~270 K, shows the transformation of (2 × 1) added-row islands into the (5 × 2) methoxy structure. The (2 × 1)O_{ad} islands appear as long, narrow islands in which the individ-

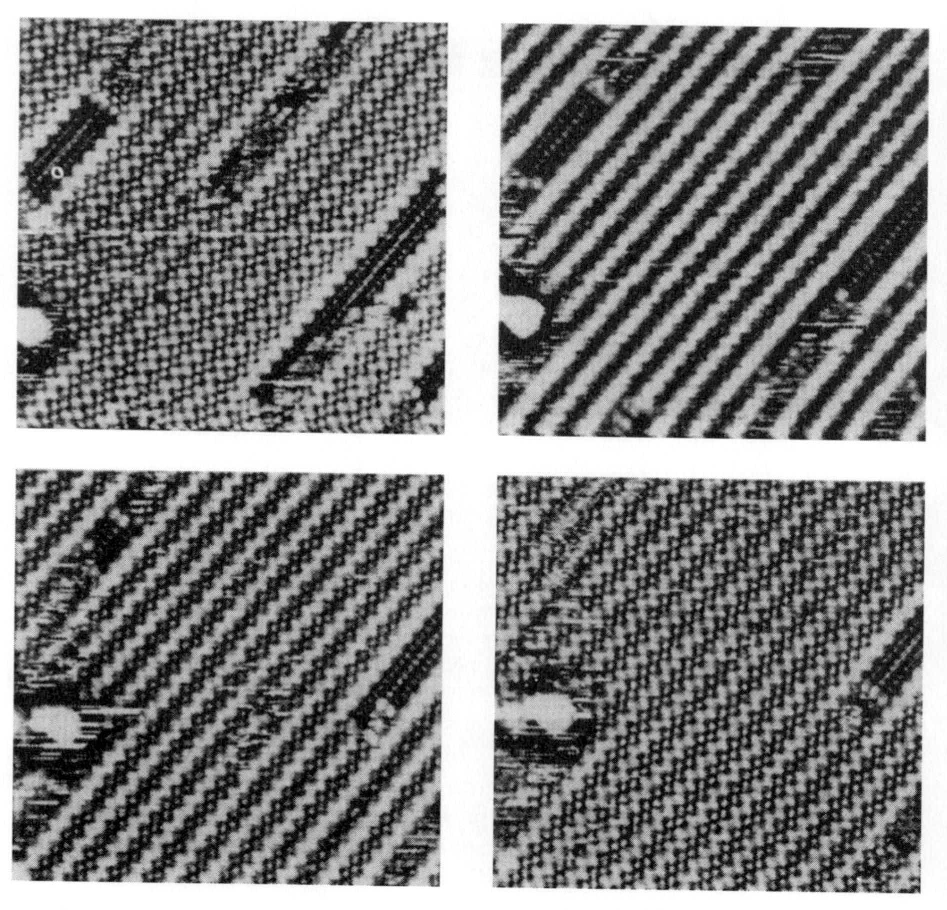

FIG. 18. Sequence of STM images of a partially O_{ad}-covered Cu(110) surface during dosing with methanol. T = ~270 K (200 × 200 Å). Narrow, rectangular islands are the (2 × 1)O_{ad} structure, and most of the surface is covered by the (5 × 2)CH_3O_{ad} phase [reprinted with permission from Leibsle *et al.* (*141*). Copyright © 1994 The American Physical Society].

ual added rows are resolved; the rest of the surface was covered by the $(5 \times 2)CH_3O_{ad}$ structure. The $(2 \times 1)O_{ad}$ islands shrank with time, but only along the long axis, whereas the widths remained unchanged. It was concluded that the methanol molecules reacted only with the oxygen atoms at the ends of the chains, with the sides remaining intact.

According to this mechanism, the rate-limiting factor is the availability of open ends of the added rows. This fact can explain the behavior of the reactive sticking coefficient of methanol on the oxygen-covered surface that was determined by molecular beam measurements [Fig. 19 (*143*)]. For a fully (2 × 1) covered surface $[\Theta(O_{ad}) = 0.5]$, an extended induction period was observed that was caused by the good order of the $(2 \times 1)O_{ad}$ structure and the resulting absence of open ends. Only at atomic steps, where open ends exist, could the reaction start. Because methoxy is not stable at the experimental temperature of 353 K [it decays into formaldehyde, which desorbs according to Reaction (5)], it was concluded that O_{ad} vacancies were created at the steps that diffused into the ordered areas, creating further open ends. In this way the sticking coefficient increased. In agreement with this view, the initial sticking coefficient at lower oxygen precoverages $[\Theta(O_{ad}) = 0.25]$ was much higher.

Another reaction studied in the context of the methanol synthesis was

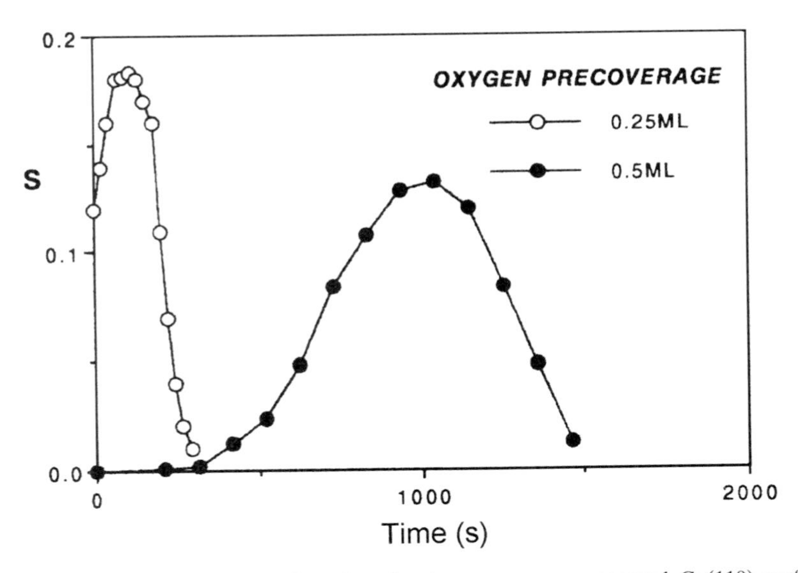

FIG. 19. Sticking coefficient of methanol onto an oxygen-precovered Cu(110) surface obtained by molecular beam measurements. Two different coverages; $T = 353$ K [reprinted with permission from Francis *et al.* (*143*). Copyright © 1994 Elsevier Science].

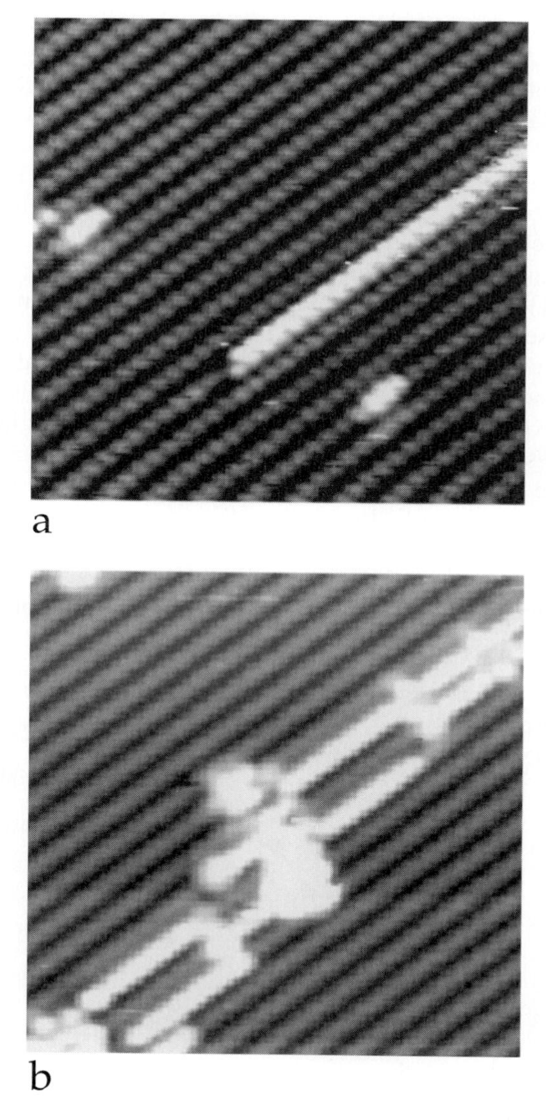

FIG. 20. STM topographs taken after exposing the oxygen-reconstructed Cu(110) surface to small amounts of formic acid at 300 K (different areas). The stripe structure is the $(2 \times 1)O_{ad}$ added-row reconstruction; single, brighter rows are due to rows of formate (100 × 100 Å) [reprinted with permission from Haq and Leibsle (*130*). Copyright © 1997 Elsevier Science].

the oxidation of formic acid to formate on Cu(110) according to Reaction (2) (*128–134*). Figure 20 shows STM images from experiments by Haq and Leibsle (*130*) in which a fully $(2 \times 1)O_{ad}$-covered surface had been exposed to small amounts of formic acid at 300 K. In Fig. 20a a bright stripe is evident on the $(2 \times 1)O_{ad}$-reconstructed surface, and in Fig. 20b pairs of stripes are evident with one remaining Cu–O row in between. These data mark the beginning of the reaction to give formate. The double rows in Fig. 20b represent a local (4×1) structure, which had previously been detected by LEED as one of the formate structures. The tentative structure model of this phase [Fig. 15b (*130*)] has rows of formate on bridge sites [as was shown by spectroscopic work (*145*)] between unreacted added rows. Figure 20 displays a similar one-dimensional reaction, as was found for the methanol-to-methoxy reaction, again indicating that the O_{ad} atoms were reacted off from the ends of the Cu–O rows. Qualitatively, the higher reactivity of the terminal atoms is of course easy to understand since only one Cu–O bond has to be broken instead of two bonds that have to be broken for an atom in the interior of the chains. However, in contrast to the methanol oxidation (Fig. 19), the initial reactive sticking coefficient of formic acid was high on the fully O_{ad}-covered surface, which was explained by the greater proton-donor acidity of formic acid relative to methanol. The resulting higher reactivity must have also allowed attack on the O_{ad} atoms in the interior of the Cu–O chains.

Poulston *et al.* (*146*) and Jones *et al.* (*147*) discovered that at 300 K formate could be formed not only by oxidation of formic acid [Reaction (2)] but also by oxidation of methanol. In the experiments described previously, in which the oxygen-covered surface was exposed to methanol, methanol oxidation either stopped at the level of methoxy [Reaction (4)] or went to formaldehyde [Reaction (5)]. When oxygen was not preadsorbed, but instead codosed with methanol, structures indicating the presence of formate were observed. Figure 21a (*146*) shows an STM image recorded during dosing of the Cu(110) surface with a methanol–oxygen mixture. It includes an area with a $c(2 \times 2)$ structure that was associated with formate (see Section VI.A.2) and an area with the (5×2) methoxy structure. Continued dosing (Fig. 21b) led to the striped (3×1) formate phase and the $(2 \times 1)O_{ad}$ reconstruction. The interpretation was that under codosing conditions the O_{ad} atoms formed by O_2 dissociation were not immediately incorporated into the added-row structure. It is quite likely that such an O_{ad} species, which is not stabilized by the strong bonding to the copper atoms in the added-row reconstruction, is more reactive and can oxidize the intermediately formed methoxy to formate before it decays to formaldehyde and desorbs. This interpretation is supported by the observation that the $(2 \times 1)O_{ad}$ phase occurred only in a later stage of the reaction (Fig. 21b).

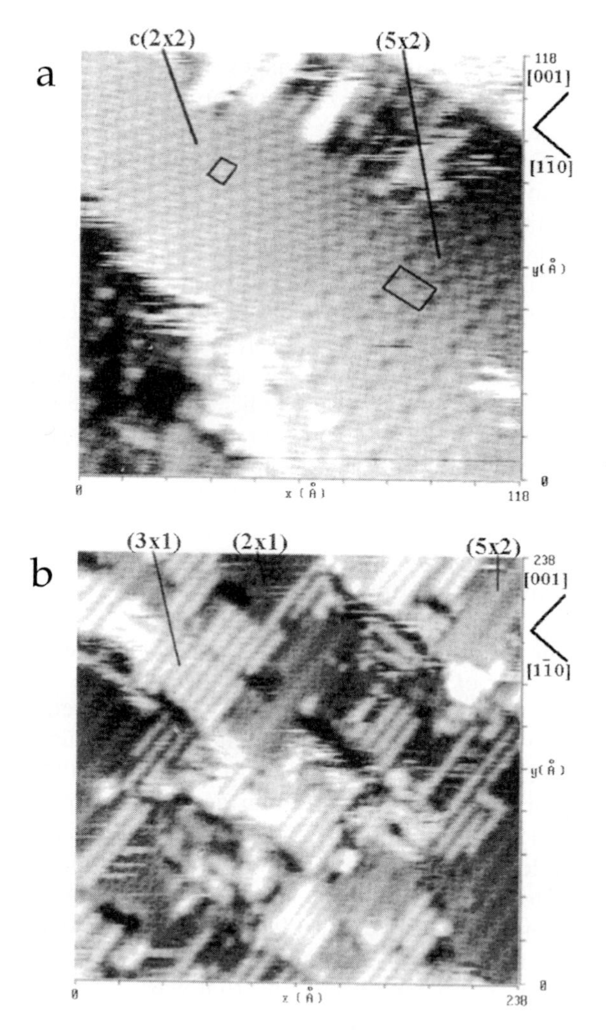

FIG. 21. (a) STM image taken while exposing a Cu(110) surface to a methanol : O_2 mixture (mixing ratio >5 : 1) at 300 K (118 × 118 Å). (b) After continued dosing (238 × 238 Å). Areas with the (5 × 2) methoxy structure and the c(2 × 2) and (3 × 1) formate phases are marked [reprinted with permission from Poulston *et al.* (*146*). Copyright © 1996 Institute of Physics Publishing Limited].

There has been dispute regarding this interpretation (*148–152*), which illustrates the various problems that may be encountered in STM investigations of catalytic reactions. The methanol oxidation provides a rare example illustrating that a surface phase may not be uniquely attributed to a certain adsorbate species. Leibsle (*148*) indicated that the c(2 × 2) structure observed in the codosing experiments (Fig. 21a) and associated with formate could just as well be ascribed to methoxy. It was previously mentioned that methoxy also formed a c(2 × 2) structure before the (5 × 2) phase appeared. [On the other hand, the (3 × 1) structure shown in Fig. 21b is quite clearly due to formate.] Moreover, it was suggested that in the initial experiments, in which oxygen and methanol had been dosed sequentially, formate may also have been present, possibly escaping detection because of its high mobility at the low coverages. Finally, it was pointed out that, for the codosing experiments, tip effects might have played a role because the c(2 × 2) and (5 × 2) areas that had indicated the reaction to formate (Fig. 21a) were not seen when, after the dosing, the tip was moved to other locations.

3. *Oxidation of Ammonia on Ni(110), Cu(110), and Rh(110)*

Ammonia reacts with preadsorbed oxygen atoms on a variety of metals. With increasing surface temperature the hydrogen atoms are successively abstracted from the molecules, leading to amide ($NH_{2,ad}$), imide (NH_{ad}), and finally nitrogen atoms:

$$NH_3 \rightarrow NH_{3,ad} \tag{6a}$$

$$NH_{3,ad} + O_{ad} \rightarrow NH_{2,ad} + OH_{ad} \tag{6b}$$

$$2\,OH_{ad} \rightarrow H_2O\uparrow + O_{ad} \tag{7}$$

$$2\,NH_{2,ad} + O_{ad} \rightarrow 2\,NH_{ad} + H_2O\uparrow \tag{8}$$

$$2\,NH_{ad} + O_{ad} \rightarrow 2\,N_{ad} + H_2O\uparrow \tag{9}$$

These reactions on the (110) surfaces of nickel (*153*), copper (*154–157*), and rhodium (*158*) have been studied by STM. The first investigation was performed with Ni(110) by Ruan *et al.* (*153*) and serves as an example. After predosing of the nickel surface with oxygen, ammonia was adsorbed at 300 K. Under these conditions ammonia decomposes to give $NH_{2,ad,}$ according to Reaction (6b), and the intermediate OH_{ad} decomposes to give water [Reaction (7)] or reacts with additional NH_3 to give water. Figure 22 shows a starting configuration after oxygen adsorption, with $\Theta(O_{ad}) = 0.14$; the rows correspond to the added-row structure of O_{ad} on this surface. The two consecutive frames show that the Ni–O rows were not fixed but

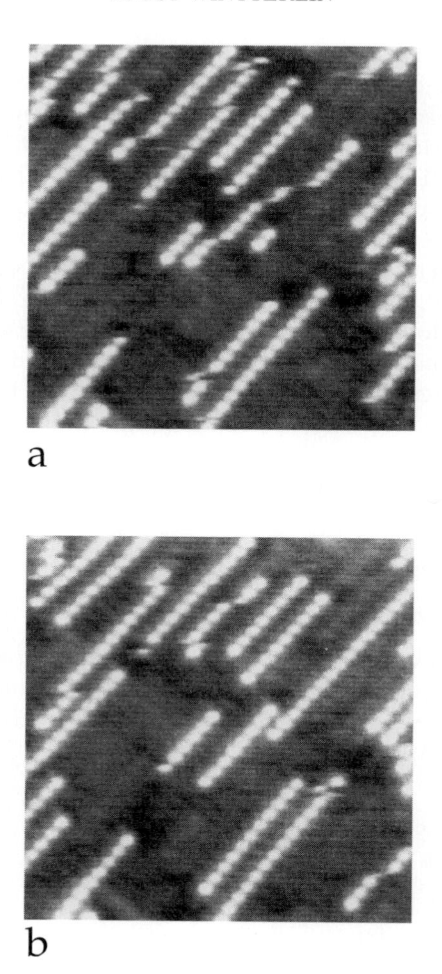

FIG. 22. STM images of a Ni(110) surface at 300 K with short Ni–O added rows (chains of bright dots). The two images were taken at a time interval of 10 s, indicating the mobility of the added rows (100 × 100 Å) [reprinted with permission from Ruan *et al.* (*153*). Copyright © 1994 Elsevier Science].

changed their positions; sometimes this happened while the tip scanned over a row, resulting in the segmented appearance of several rows. The sample was then dosed with ammonia (Fig. 23), leading to additional row structures running in the same direction as the residual Ni–O added rows, but these were imaged less brightly (arrow in Fig. 23a). The authors interpreted these structures as rows of OH_{ad} species that were formed by transfer of hydrogen from NH_3 to O_{ad} at the ends of the Ni–O rows. The $NH_{2,ad}$ groups were not resolved at this stage, probably because of their mobility.

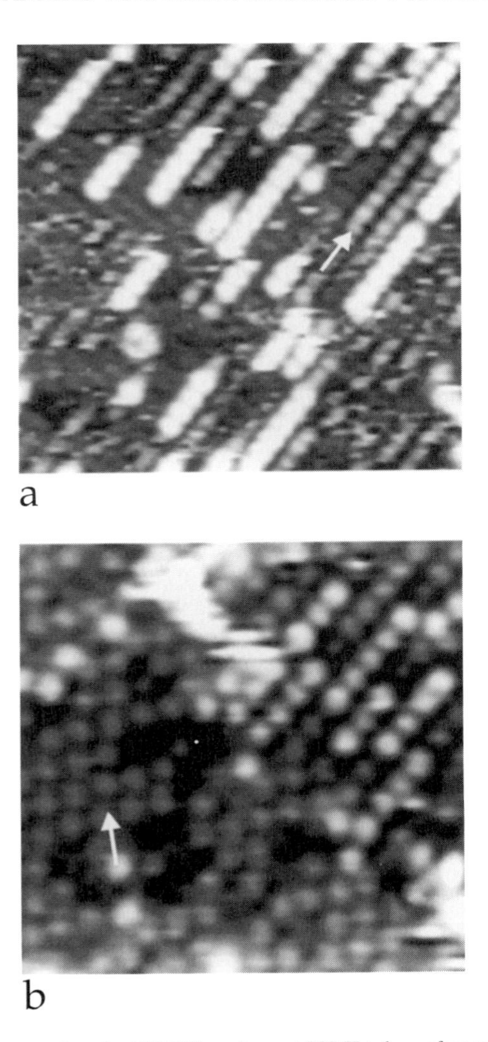

FIG. 23. STM topographs of a Ni(110) surface at 300 K taken after preadsorption of 0.17 monolayers of oxygen and subsequent exposure to NH_3. (a) After a small dose of ammonia: Bright structures are unreacted added rows, less bright structures are rows of OH_{ad} (arrow). $NH_{2,ad}$ is not visible because of rapid motion (90 × 90 Å). (b) After larger amounts of NH_3: The $NH_{2,ad}$ molecules have formed a dense, c(2 × 2) structure (arrow) (55 × 55 Å) [reprinted with permission from Ruan *et al.* (*153*). Copyright © 1994 Elsevier Science].

Continued dosing with NH_3 removed the OH_{ad} by reaction to give H_2O by the combined Reactions (6b) and (7). The $NH_{2,ad}$ groups finally became immobilized by forming a dense c(2 × 2) structure and became visible in the images (Fig. 23b, arrow). The released nickel atoms were found to condense into islands visible in larger area scans. The central step of the reaction, the transfer of hydrogen to oxygen, was again concluded to occur at the terminal atoms of the added rows. The fact that the added rows did not react off completely (Fig. 23b) was explained by their reduced mobility at higher densities of $NH_{2,ad}$. The rows then could no longer segment, which apparently is an essential mechanism for the creation of terminal oxygen atoms. This result demonstrates that the reactivities of the added-row reconstructions are not only determined by their static defect structures but that structural fluctuations have to be considered as well. Thus, previous observations (*159*) of a maximum in the reactivity of the surface at an O_{ad} coverage of ~0.16 were explained. This maximum can be understood as a maximum in the number of open ends of the reconstruction, both static and dynamic, at that coverage. The data in Fig. 22 also illustrate an important property of these reconstructions: Even single chains moved slowly enough to be followed by STM, making them particularly attractive for studies at room temperature.

The ammonia oxidation on the Cu(110) surface was also investigated with greater variations in the experimental conditions (*154–157*). On Cu(110), the reaction goes to NH_{ad} at 300 K, according to Reaction (8), and to N_{ad} at 400 K [Reaction (9)]. In both cases the terminal atoms of the Cu–O added rows were again more reactive than the others. Furthermore, it was found that the reaction ceased when the row ends became blocked by the reaction products NH_{ad} or N_{ad}, both of which form chain structures perpendicular to the Cu–O added rows.

4. *Oxidation of CO on Cu(110)*

Crew and Madix (*160–162*) investigated the reaction of CO with oxygen atoms on the Cu(110) surface:

$$CO \rightarrow CO_{ad} \tag{10a}$$

$$CO_{ad} + O_{ad} \rightarrow CO_2 \uparrow \tag{10b}$$

This reaction can also be viewed in the context of the methanol synthesis on copper surfaces, in which the CO of the syngas is converted into CO_2, which then becomes reduced to methanol. The results again showed the strong influence of the reconstruction, but the work contributed additional insights as well.

Whereas most other studies were restricted to room temperature, part of the work of Crew and Madix was performed at elevated temperatures. At 400 K, the partially oxygen-covered surface still displayed the typical one-dimensional islands of the $(2 \times 1)O_{ad}$ reconstruction. When CO was dosed, these were seen to shorten, which again pointed to preferential reaction at the row ends, but there was also some reaction from the sides. Simple MC simulations, in which the reaction probabilities of terminal vs chain atoms of the added rows were varied, reproduced the island shapes quite well when it was assumed that the row ends were 500–1000 times more reactive than the sides. From this ratio, an interaction energy along the chains of 0.22–0.32 eV was estimated, which is in reasonable agreement with the kink formation energy at the long sides of islands of ~0.18 eV obtained from equilibrium measurements. These values demonstrate the considerable lateral interactions connected with the added-row reconstructions so that island formation and the resulting inhomogeneities can play a role at quite high temperatures.

This conclusion was supported by the results of experiments performed by Crew and Madix under steady-state conditions. Such experiments are unique because all other experiments have been either titration experiments, in which preadsorbed oxygen was reacted off, or codosing experiments, such as the methanol oxidation experiments represented by Fig. 21 in which the reaction was also terminated after the surface became covered by reaction products. Figure 24 shows sequential STM topographs, recorded at 400 K, under constant pressures of O_2 and CO. The conditions were isosteric, i.e., the oxygen adsorption rate equaled the rate of oxygen consumption by reaction with CO. The dark bands are added-row islands (there is no atomic resolution of the added rows). The islands fluctuated in size and shape, and the number of islands also varied. However, their general shape was similar to that in the titration experiments, indicating that the oxygen removed from the islands by reaction was permanently replaced. The mechanism under steady-state conditions was therefore quite similar to that in the titration case.

In other work on the added-row structures it had been implicitly assumed that the respective reactions occur directly at the locations of the terminal O_{ad} atoms of the chains. Another possibility, particularly at these higher temperatures, is that an equilibrium is established between O_{ad} atoms in added-row islands and isolated O_{ad} atoms on the areas in between. Because the isolated O_{ad} atoms are not stabilized by the reconstruction, they should be particularly reactive (this was later also suggested for the methanol–oxygen codosing experiments; see Section VI.B.2). Unfortunately, these atoms could not be resolved because of their rapid motion. However, assuming that the islands were in equilibrium with isolated oxygen atoms

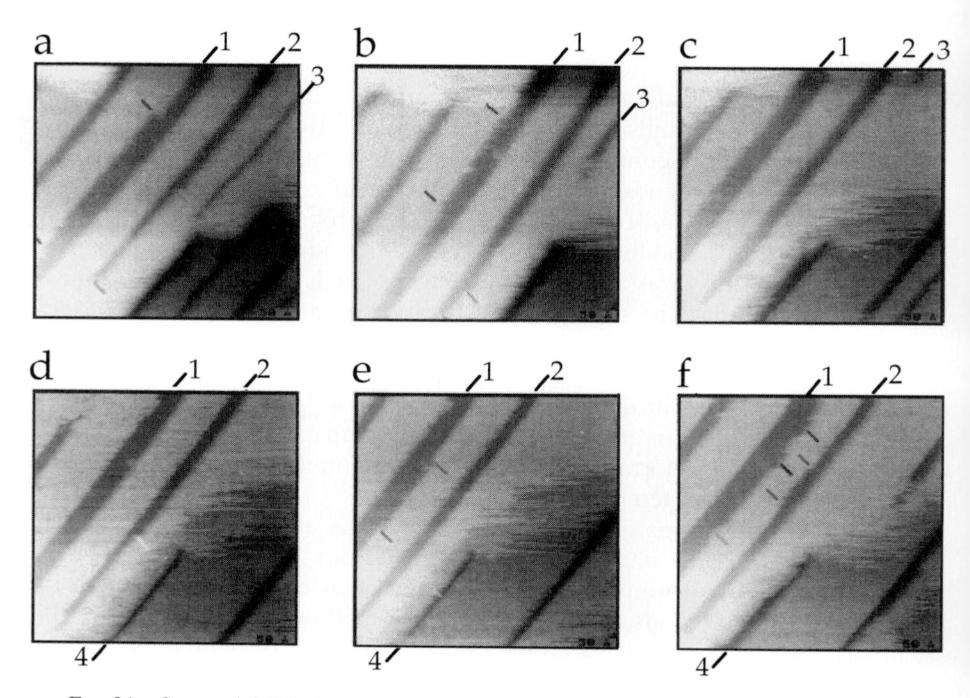

FIG. 24. Sequential STM images of Cu(110) recorded under isosteric conditions with CO and O_2 in the gas phase at 400 K. Dark bands are $(2 \times 1)O_{ad}$ islands in which the individual Cu–O rows are not resolved; identical islands are denoted by numbers (356 × 359 Å) [reprinted with permission from Crew and Madix (*161*). Copyright © 1996 Elsevier Science].

and that the rate of CO_2 formation by reaction with the isolated O_{ad} atoms was proportional to $\Theta(O_{ad})\Theta(CO_{ad})$, the experimentally observed rate could be reproduced. The conclusion was supported by experiments carried out at 150 K, at which temperature the added-row structure was completely unreactive. However, reaction was induced when O_2 was dosed *after* the CO. The O_{ad} atoms formed were not directly incorporated into the reconstruction and reacted with CO even at these low temperatures.

C. Reactions at Atomic Steps: Hydrogen Oxidation on Ni(110)

Influences of atomic steps were detected not only for dissociative adsorption (*85, 105*) (see Section IV.B) but also found in reaction studies (*143, 154, 155, 163*). Step effects were quite dominating in an investigation reported by Sprunger *et al.* (*163*) of the hydrogen reaction with preadsorbed O_{ad} on Ni(110):

$$H_2 \rightarrow 2\,H_{ad} \tag{11a}$$

$$2\,H_{ad} + O_{ad} \rightarrow H_2O\uparrow \tag{11b}$$

Figure 25 shows images recorded after experiments that were started with the high-coverage $(3 \times 1)O_{ad}$ structure [$\Theta(O_{ad}) = 2/3$; as described

a

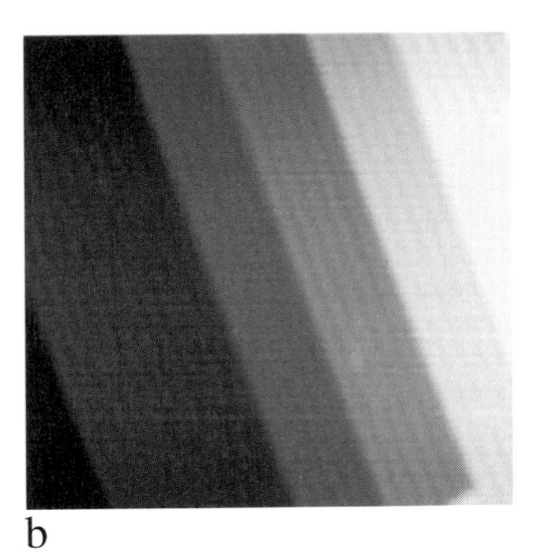

b

FIG. 25. STM images of Ni(110) after preparation of the $(3 \times 1)O_{ad}$ phase [with $\Theta(O_{ad}) = 2/3$] and subsequent reaction with H_2 at 475 K. Recorded at 300 K. (a) Almost fully reached region (400 × 400 Å). (b) Region with intact $(3 \times 1)O_{ad}$ phase (streaks) and steps running in the [001] direction (220 × 220 Å) [reprinted with permission from Sprunger *et al.* (*163*). Copyright © 1995 Elsevier Science].

in Section VI.A, the structure consists of added-row pairs]. After exposure of 800 L of H_2 at 475 K (water desorbs under these conditions), the sample was cooled to 300 K and investigated with the STM. Figure 25a is derived from an area with a high step density, with steps running in both [001] and [1$\bar{1}$0] directions. The terraces appear flat, without the O_{ad} row structure, i.e., oxygen was almost completely reacted off from this area except for individual added-row strings decorating the [001]-oriented steps (bright lines). Another area on the same sample (Fig. 25b) displayed the striped structure characteristic of the intact double-row (3×1) phase, indicating that no reaction had occurred. An obvious difference from the topography of Fig. 25a is the direction of the steps, which run in the [001] direction in this case (i.e., parallel to the added rows). The authors concluded that hydrogen had reacted with O_{ad} only at [1$\bar{1}$0]-oriented steps, whereas flat terraces and [001] steps were unreactive. [1$\bar{1}$0]-oriented steps are perpendicular to the added rows, which suggests that the reaction started at the open ends of the added rows. The origin of the step effect is hence similar to that of the one-dimensional reactivities discussed in the preceding section. The specific mechanism suggested by the authors in this case was that at first terminal O_{ad} atoms at the steps were removed by the reaction to water. This created empty nickel sites on which additional H_2 could adsorb and react with the adjacent oxygen atoms. In this way the reaction progressed to the terraces. This model also explains the induction period observed in previous macroscopic experiments (*164*).

D. Stoichiometries of Reactions: Reaction of CO_2 with Oxygen to Give Carbonate on Ag(110)

A particular advantage of STM is that it can provide absolute coverages simply by determination of the areas covered by the respective adsorbates. Macroscopic experiments usually require calibration to yield coverages, which is a significant source of uncertainty. An example in which the coverage is important to determine the stoichiometry of a chemical reaction is the reaction of CO_2 with oxygen atoms to give carbonate on Ag(110):

$$CO_2 \rightarrow CO_{2,ad} \tag{12a}$$

$$CO_{2,ad} + O_{ad} \rightarrow CO_{3,ad} \tag{12b}$$

The reaction plays a role in the ethylene oxidation on silver catalysts, in which CO_2 is a by-product (*165*). Carbonate formation from CO_2 has been shown to be important for the selectivity of the ethylene oxidation because an unwanted O_{ad} species is masked in this way. Stensgaard *et al.* (*166*) and Okawa and Tanaka (*167*) studied the reaction on Ag(110) by STM. A major unresolved question had been whether the reaction followed the 1:1 stoichiometry according to Reaction (12b) or instead required two oxygen atoms for each CO_2 molecule, implying greater complexity.

Figure 26a (*166*) shows an STM topograph recorded after CO_2 had been adsorbed on the O_{ad}-covered surface at 300 K. As mentioned previously, oxygen on Ag(110) forms an added-row reconstruction with a $(n \times 1)$ lattice. The straight lines are residual added rows; the pairing was caused by the compression by the carbonate species. The areas in between were covered with carbonate adsorption complexes (white dots). It was found that the white dots numbered only half of the oxygen atoms removed, i.e., the reaction in fact consumed two O_{ad} atoms per CO_2 molecule. This means that the carbonate adsorption complex incorporates an additional O_{ad} atom per CO_3 particle. Another result was that no silver atoms were released during the reaction with CO_2, which means that the added-row silver atoms must have remained in the carbonate structure (two silver atoms per carbonate molecule). Hence, this phase must be reconstructed as well. Together with the atomic structure resolved in the carbonate areas (short rows of atomic features, mostly triplets in the topograph), the structure model shown in Fig. 26b (*166*) was proposed. The large open circles represent the planar carbonate molecules, with their molecular planes parallel to the surface, which were the white features seen by STM. The triplet structures in the topographs may be due to the one O_{ad} and two silver atoms connected with the individual $CO_{3,ad}$ groups, but a definite atomic structure model is not possible using STM.

E. Reactions at Domain Boundaries and Other Defects in the Adsorbate Layer

Most metal surfaces are not reconstructed and do not reconstruct upon adsorption of foreign particles, which is therefore also most likely the more general case under conditions of catalytic reactions. Nonetheless, there are considerably fewer reaction studies by STM characterizing these systems than reconstructed surfaces. As previously mentioned, the major reason for the paucity of STM results is the generally higher mobility of adsorbates on nonreconstructed surfaces, which usually requires cooling to resolve the particles by STM, at least for ordinary low recording speeds.

In principle, the observations should be easier to interpret because the complicated restructuring processes are absent so that the processes are more characteristic of the "free" adsorbates. Deviations from the simple Langmuir description, which can be expected for unreconstructed surfaces, are effects of islands and domains, which were in fact observed in the systems investigated to date.

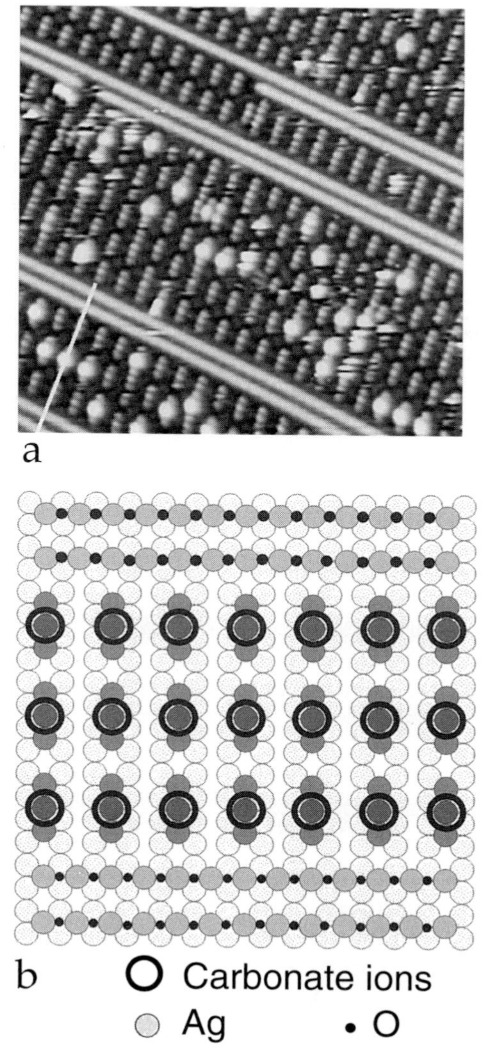

FIG. 26. STM image of Ag(110) taken after adsorption of oxygen and subsequent dosing with CO_2 at 300 K. Long parallel lines are unreacted Ag–O added rows. The other features are carbonate (bright dots) and a reconstruction (triplet features) underneath the carbonate (160 × 160 Å). (In the area shown, many triplets do not have carbonate attached to them; however, this is not representative of the whole surface.) (b) Structure model [reprinted with permission from Stensgaard *et al.* (*166*). Copyright © 1995 American Institute of Physics].

1. *Decomposition of Ethylene on Pt(111)*

The first STM reaction study of an unreconstructed system was performed to investigate the isomerization and decomposition of ethylene on Pt(111). Reactions of adsorbed ethylene, in particular on the Pt(111) surface, had been a frequently used model system for hydrogenation–dehydrogenation reactions in catalysis. By means of a variable-temperature STM, Land *et al.* (*168, 169*) studied the reactions indicated in Fig. 27a (*169*) that occurred on a Pt(111) surface. The first step, the isomerization and partial decomposi-

"CARBIDIC" CARBON C_xH_y → GRAPHITE

T < 230K — 230K – 450K — 450K – 770K — T > 770K

a

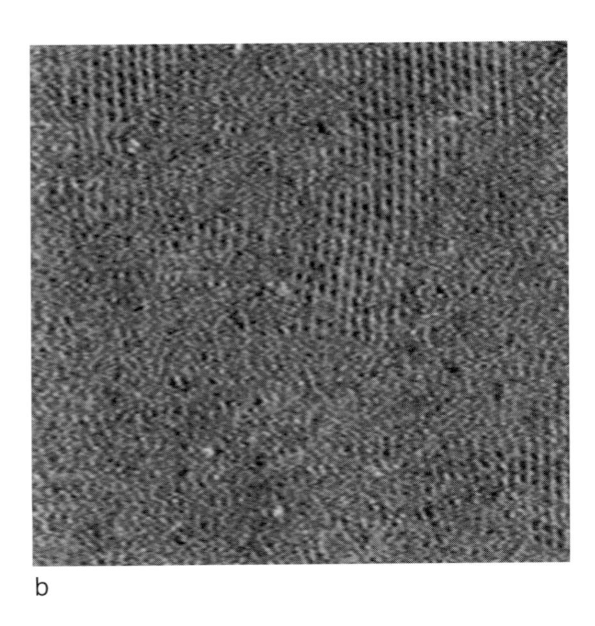

b

FIG. 27. (a) Processes by which ethylene decomposes on Pt(111). (b) STM image recorded during decomposition of ethylene to ethylidyne at 230 K. Ordered areas are due to ethylene; disordered areas are due to ethylidyne (400 × 400 Å) [reprinted with permission from Land *et al.* (*169*). Copyright © 1992 American Institute of Physics].

tion of adsorbed ethylene to give ethylidyne, was monitored at 230 K while the process was occurring. The STM image in Fig. 27b (*169*), recorded after the sample had been kept for several minutes at 230 K, shows areas with an ordered structure and areas with a disordered structure. From data at lower temperature the ordered structure was identified with adsorbed ethylene. The disordered structures were associated with the reaction product ethylidyne. By repeatedly scanning the area at this temperature, the authors observed that within several minutes the area with the ordered structure shrank, and the ordered structure was replaced by the disordered structure, indicating that indeed the reaction of ethylene to give ethylidyne had been monitored. That the reaction progressed by the shrinking of islands suggested that only the ethylene molecules at the island perimeters were reacting. In contrast, previous kinetics measurements revealed a first-order macroscopic reaction, suggesting a random reaction probability. The same study also treated the succeeding reaction steps at higher temperatures to give clusters of carbidic carbon and finally graphite, which could not be investigated during the process but only after cooling down from the reaction temperature.

2. *Trimerization of Acetylene on Pd(111)*

Another hydrocarbon reaction, the trimerization of acetylene to give benzene on Pd(111), was investigated by Janssens *et al.* (*170*):

$$C_2H_2 \rightarrow C_2H_{2,ad} \tag{13a}$$

$$3\,C_2H_{2,ad} \rightarrow C_6H_{6,ad} \tag{13b}$$

The reaction had been known to occur on Pd(111) at temperatures as low as 140 K, whereas on most other surfaces acetylene decomposes to give carbon and hydrogen atoms. The reaction is unique because it is one of the few reactions of hydrocarbons that can be studied under UHV conditions whereby new C–C bonds are formed. Figure 28 (*170*) shows an STM image recorded after exposure of the surface to a saturation coverage of acetylene at 140 K. Most of the surface displayed a hexagonal overlayer formed by the $(\sqrt{3} \times \sqrt{3})R30°$ acetylene structure. The structure model of this phase, which had evolved from previous spectroscopic investigations (*114*), is shown in Fig. 29. The acetylene bonding configuration is the same as that of the individual molecules referred to previously [Fig. 11b (*109*)], i.e., the molecules are rehybridized to an ethylene-like configuration and the C–H plane is tilted from the surface normal. That the STM image of the $(\sqrt{3} \times \sqrt{3})R30°$ structure (Fig. 28) showed a round dot for each molecule, in contrast to the asymmetric shape found in the work of Dunphy *et al.* (Fig. 11a), might be a consequence of the limited resolution in close-packed

FIG. 28. STM topograph of a saturated layer of acetylene on Pd(111) (140 K, 190 × 190 Å). The hexagonal structure is the $(\sqrt{3} \times \sqrt{3})R30°$ acetylene structure; bright dots are benzene molecules formed by acetylene trimerization at 140 K [reprinted with permission from Janssens *et al.* (*170*). Copyright © 1998 American Chemical Society].

layers, but it could also be explained by rapid rotation of the molecules at 140 K (discussed in Section V).

The finding in previous work that benzene formed only when the acetylene coverage exceeded values of approximately 0.3 suggested that the close packing in the $(\sqrt{3} \times \sqrt{3})R30°$ structure $[(\Theta(C_2H_{2,ad}) = 1/3$; Fig. 29)] was responsible for the bond formation between acetylene molecules. However, the view that a reaction of three molecules to give benzene in one step suggested by the arrangement of the molecules in the $(\sqrt{3} \times \sqrt{3})R30°$ structure (Fig. 29) had already been ruled out by previous experiments. The reaction most likely proceeds in two steps via an intermediate C_4H_4 species (*171*). The STM data also showed that the correlation between the threshold coverage for reaction and the formation of the $(\sqrt{3} \times \sqrt{3})R30°$ structure was misleading. In the STM image (Fig. 28), benzene molecules formed by the trimerization appear as large bright dots. They all reside at defects in the adlayer, at the dark areas (probably vacancies), or along domain boundaries of the $(\sqrt{3} \times \sqrt{3})R30°$ structure. There are no benzene molecules in the interior of ordered $(\sqrt{3} \times \sqrt{3})R30°$ domains. Moreover, continued dosing with acetylene did not lead to further benzene formation. These observations mean that molecules in the interior of the $(\sqrt{3} \times \sqrt{3})R30°$ phase are not reactive and that the reaction occurs during

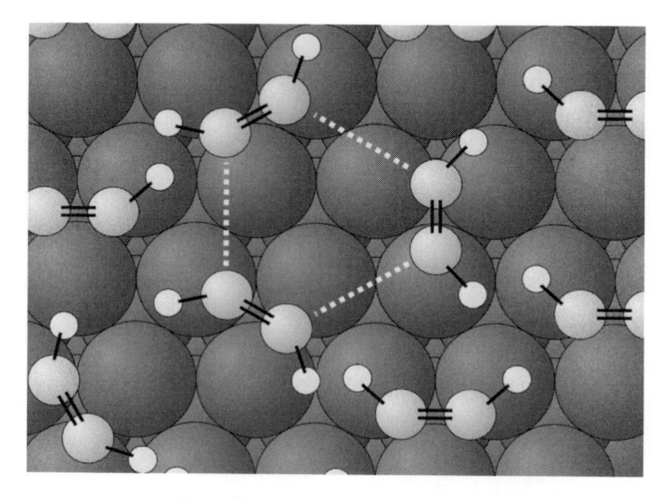

FIG. 29. Model of the $(\sqrt{3} \times \sqrt{3})R30^\circ$ acetylene structure on Pd(111), with rehybridized molecules rotating about their threefold hollow adsorption sites. (Medium-sized and small bright balls are C and H atoms, respectively; large dark balls are Pd atoms.) The intuitive mechanism of benzene formation from three molecules in the $(\sqrt{3} \times \sqrt{3})R30^\circ$ structure (bright dashed lines) was ruled out by the STM experiments.

the formation of the $(\sqrt{3} \times \sqrt{3})R30^\circ$ structure, which was actually observed in time series recorded during acetylene adsorption. Most likely, the reaction involves a transient acetylene species that occurs only during the ordering process of the $(\sqrt{3} \times \sqrt{3})R30^\circ$ structure, whereas the reaction stops when all acetylene molecules have adopted their stable configurations in the $(\sqrt{3} \times \sqrt{3})R30^\circ$ structure. Such a transient species could be a π-bonded molecule that reacts, at domain boundaries and other defects, with the more stable rehybridized σ-bonded molecules of the $(\sqrt{3} \times \sqrt{3})R30^\circ$ phase. The presence of the unreactive $(\sqrt{3} \times \sqrt{3})R30^\circ$ molecules also explains previous findings (*114*) that the majority of the acetylene molecules did not react to give benzene, an observation that seemed to be in contrast to the fact that all acetylene molecules in the $(\sqrt{3} \times \sqrt{3})R30^\circ$ structure are equivalent. Hence, as in the ethylene decomposition on Pt(111), not all molecules are equally reactive, although this is not correlated with any structure defects of the metal.

3. *Oxidation of CO on Pt(111)*

The oxidation of CO on Pt(111) is one of the best studied surface reactions, and it is regarded as an important model reaction. It was this reaction for which the validity of the Langmuir–Hinshelwood mechanism of surface reactions was first demonstrated (*172*), i.e., that reactants generally first adsorb and then react. Wintterlin *et al.* (*173*) investigated the reaction of preadsorbed oxygen with CO by STM:

$$CO \rightarrow CO_{ad} \tag{14a}$$

$$CO_{ad} + O_{ad} \rightarrow CO_2 \uparrow \tag{14b}$$

The system appeared particularly interesting because oxygen atoms were known to form islands on Pt(111) (*174*) so that this was a promising candidate for investigating nonrandom reaction probabilities. Previous measurements by temperature-programmed desorption (*175*) indeed indicated that the reaction was restricted to island perimeters. In contrast, isotope mixing studies (*176*) showed that the reaction occurred randomly, consistent with a Langmuir gas description.

Because O_{ad} atoms on Pt(111) were found to be too mobile to allow monitoring of the reaction at room temperature, the temperature had to be reduced to ~250 K. The STM experiments were performed by preparing layers of O_{ad}, cooling to selected temperatures, and monitoring the removal of the oxygen atoms while CO was continuously dosed (the CO_2 desorbed immediately upon formation). Figure 30 (*173*) shows an example in which an O_{ad} submonolayer had been prepared, which was then exposed to a constant CO pressure at 247 K. The hexagonal pattern is the $(2 \times 2)O_{ad}$ structure. Up to about 140 s (the first frame in Fig. 30), no reaction was observed, but the many small $(2 \times 2)O_{ad}$ islands that had existed before had ordered into the large (2×2) domains visible at that stage. This observation was explained by the presence of carbon monoxide on the surface, although not resolved up to this stage, which by its known repulsive interaction with O_{ad} led to a compression of the oxygen layer. In the following period the O_{ad} areas were seen to shrink, i.e., oxygen was being removed by reaction. Simultaneously a stripe pattern became visible in the areas between the (2×2) domains which was caused by the $c(4 \times 2)CO_{ad}$ structure. The STM thus resolved both reactants—O_{ad} and CO_{ad}. Figure 31 shows structure models of the two phases, with a domain boundary. In the $c(4 \times 2)$ phase (Fig. 31, right) the CO_{ad} molecules are relatively closely packed $[\Theta(CO_{ad}) = 1/2]$ and immobilized so that they are visible in the STM images. CO_{ad} molecules also occupy sites between the O_{ad} atoms of the (2×2) phase (Fig. 31, left), but since no vacancy formation was observed in the (2×2) domains, these "interstitial" CO_{ad} molecules apparently did not take part in the reaction. The $c(4 \times 2)$ phase grew with time (Fig. 30), until it covered the surface completely at the end of the reaction (~2000 s; not shown). Empty sites created by

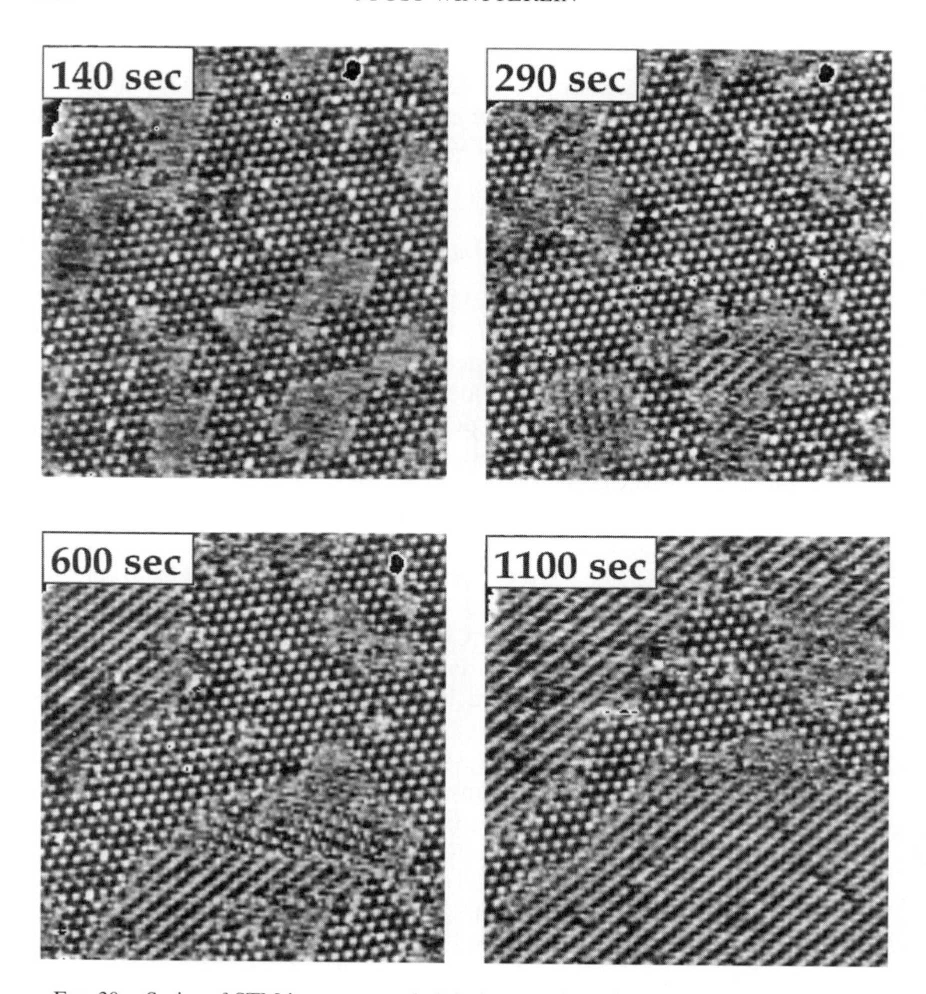

FIG. 30. Series of STM images recorded during reaction of oxygen atoms with adsorbed CO at 247 K, all from the same area of a Pt(111) surface. Oxygen had been preadsorbed; CO was continuously supplied from the gas phase. Hexagonal structure, $(2 \times 2)O_{ad}$ phase; streaky structure; $c(4 \times 2)CO_{ad}$ phase (180 × 170 Å) [reprinted with permission from Wintterlin *et al.* (*173*). Copyright © 1997 American Association for the Advancement of Science].

the reaction were immediately filled by CO from the gas phase. From these data it was obvious that O_{ad} and CO_{ad} were not randomly distributed during the reaction but were separated in domains. Hence, the reaction was not taking place randomly but instead was restricted to the (2×2)/ $c(4 \times 2)$ domain boundaries, a clear deviation from Langmuirian descrip-

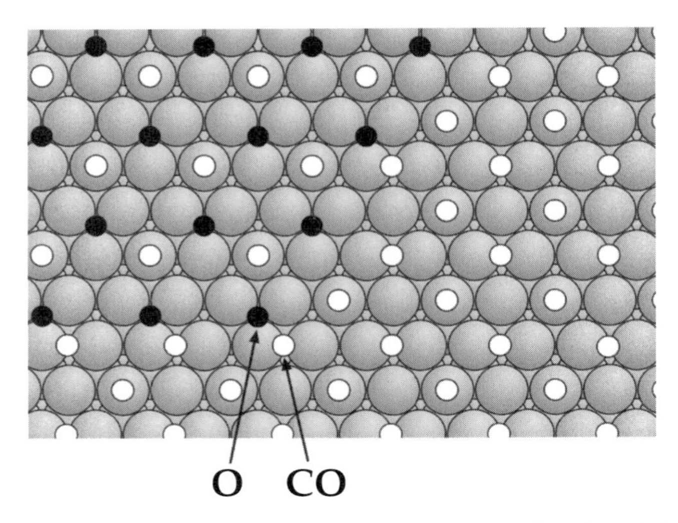

FIG. 31. Structure model of a domain boundary between the (2×2) and the $c(4 \times 2)$ phases from Fig. 30. Oxygen and CO sites as indicated.

tions of surface reactions. The conclusion about a random reactivity from the isotope experiments (*176*) was based on the assumption of an immobile oxygen layer, whereas O_{ad} was in fact quite mobile, even at these low temperatures.

F. Macroscopic Reaction Rates from Atomic Processes: Oxidation of CO on Pt(111)

Almost all the information regarding the reactions discussed previously is qualitative. A quantitative analysis to obtain reaction rates was performed for the CO oxidation on Cu(110) (see Section VI.B.4), but since the reacting particles could not be resolved the same Langmuirian formulation as in macroscopic experiments had to be applied. As mentioned previously, STM opens the possibility to build ensemble averages from microscopic observations and thereby to arrive at macroscopic quantities that are not subject to averaging by a macroscopic technique.

This goal has been attained only for the previously discussed oxidation of CO on Pt(111) [Wintterlin *et al.* (*173*)]. Values for the reaction rates $-d\Theta(O_{ad})/dt$ were extracted from the data by determination of the fraction of (2×2) area removed in successive frames. Because of the phase separation into O_{ad} and CO_{ad} domains, the reaction rate was expected to be proportional to the length of the domain boundary. This was indeed the case, as follows from Fig. 32 (*173*), which shows the measured rate divided by the length of the boundary L vs time (squares in Fig. 32). It was found that this quantity remained almost constant during the reaction, i.e., $-d\Theta(O_{ad})/dt = kL$. The data were also used to test whether the time dependence of the rate could be described by the Langmuirian kinetics $-d\Theta(O_{ad})/dt = k\ \Theta(O_{ad})\Theta(CO_{ad})$, according to which a plot of the rate divided by $\Theta(O_{ad})\Theta(CO_{ad})$ should give a constant value. Figure 32 (crosses) shows that this was clearly not the case, emphasizing that such a description is inadequate under these conditions. However, Fig. 32 also gives an example showing that the reverse analysis—from the reaction rate, as a macroscopic quantity, to the mechanism—is not unique. If, in a macroscopic experiment, the rate had been measured only between ~200 and 1100 s, the Langmuirian kinetics (crosses) would have described the rate accurately, although the underlying physical picture is wrong.

In a second step an attempt was made to measure quantities that could be correlated with macroscopic experiments. For this purpose STM series such as that in Fig. 30 were taken over a range of temperatures between 237 and 274 K. An Arrhenius plot of the rates at equal boundary lengths is shown in Fig. 33. The data fall on a straight line, from which an activation energy of $E^* = 0.49$ eV per molecule (47 kJ/mol) and a preexponential factor of 3×10^{21} particles $cm^{-2}\ s^{-1}$ were obtained. Under the same low-temperature conditions, a molecular beam study revealed values of 0.51

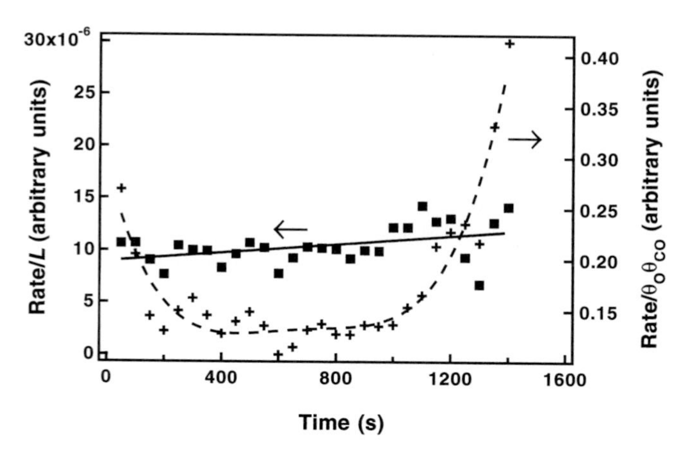

FIG. 32. Reaction rates of $CO_{ad} + O_{ad} \rightarrow CO_2 \uparrow$ determined from the change of the size of the $(2 \times 2)O_{ad}$ area between successive frames of the data of Fig. 30 vs time. ■, Rate divided by the length of the boundary between oxygen and CO areas; +, rate divided by $\Theta(O_{ad})\Theta(CO_{ad})$ [reprinted with permission from Wintterlin *et al.* (*173*). Copyright © 1997 American Association for the Advancement of Science].

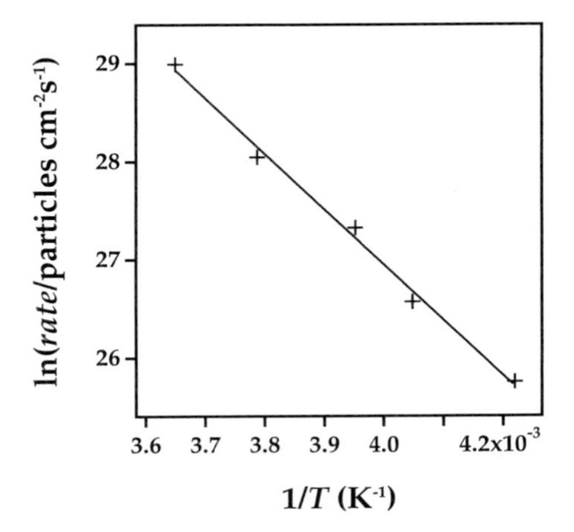

FIG. 33. Arrhenius plot of reaction rates at temperatures between 237 and 274 K. The CO pressure was always high enough to ensure that the reaction rate was determined by the surface processes [reprinted with permission from Wintterlin *et al.* (*173*). Copyright © 1997 American Association for the Advancement of Science].

eV for the activation energy and 10^{22}–10^{24} particles $cm^{-2}\,s^{-1}$ for the preexponential factor (*172*), both of which are in amazingly good agreement with the STM data. (To allow this comparison, the units of the molecular beam preexponential factor had to be converted). In this way, for the first time a direct connection between atomic events of a surface reaction and macroscopic quantities was established.

Another basis for quantitative analysis of STM data characterizing chemical reactions involves comparisons with kinetic MC simulations. MC simulations were the only previously available means for systematically testing non Langmuirian models of surface reactions (*177*), but their implementation was hampered because the entering parameters were usually not well-known, and only the resulting macroscopic quantities could be compared with the experimental results. In contrast, STM allows a comparison on the level of particle configurations.

Such a combined STM–MC study was also performed on the basis of the CO oxidation data on Pt(111) (*178*). The goals were to determine why the reaction was limited to the domain boundaries, although CO_{ad} molecules occupy sites between the O_{ad} atoms in the (2×2) structure (Fig. 31), and to explain the slight but reproducible increase in the rate/L values with time (Fig. 32). The configurations shown in Fig. 34 are from the model that finally revealed the best agreement with experiment (Fig. 30). To reproduce the relatively smooth domain boundaries in the experiments and

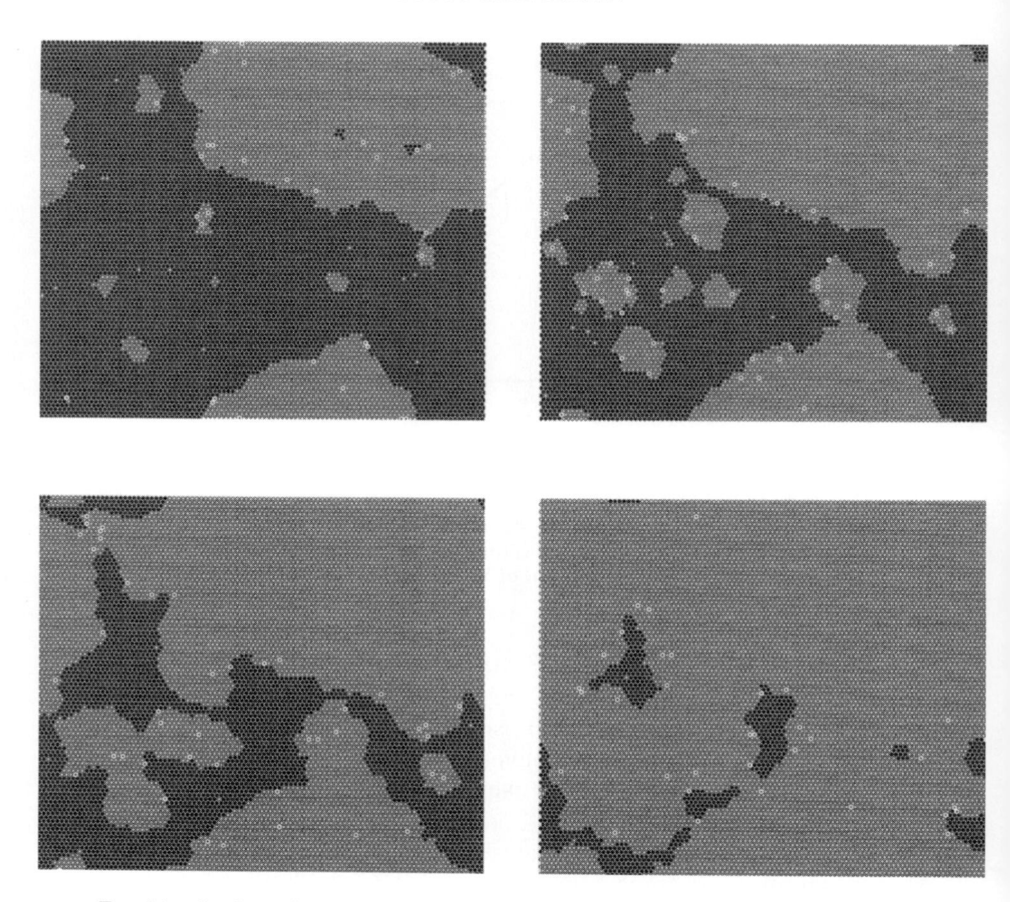

FIG. 34. Configurations from Monte Carlo simulations of the O_{ad} + CO_{ad} reaction on Pt(111). Dark gray circles are unit cells occupied by one O_{ad} atom and one CO_{ad} molecule (see structure model in Fig. 31); the few bright sites are empty sites; open circles are CO_{ad} molecules. CO was continuously adsorbing and CO_2 was leaving the surface. The O_{ad}–O_{ad} attraction is 25 meV; the CO_{ad}–O_{ad} repulsion is 25 meV; the activation energy for reaction is 0.49 eV per molecule (47 kJ/mol), expanded by an additional activation energy for oxygen coordinated O_{ad} atoms of 25 meV per O_{ad} neighbor [reprinted with permission from Völkening *et al.* (*178*)].

the reaction order of 0.5, the activation energy of 0.49 eV per molecule (47 kJ/mol) for the reaction between O_{ad} and CO_{ad} had to be expanded by contributions from the attractive interactions between the O_{ad} atoms (25 meV per neighbor atom). This made the reaction probability of O_{ad} atoms at the interface coordination dependent—O_{ad} atoms with few O_{ad}

neighbors are more reactive than highly coordinated atoms—and smoothed out the boundaries. Moreover, it explained the increasing rate/L values, equivalent to an increasing reaction probability of the interface O_{ad} atoms. Because the shapes of the domains changed during the course of the reactions [from small CO_{ad} islands within a (2×2)-covered surface to small (2×2) islands within a CO_{ad}-covered surface], the average coordination of the interface O_{ad} atoms decreased and hence their reaction probability increased. The reactivity of the interstitial CO_{ad} molecules had to be artificially suppressed. This could be understood from the structure (Fig. 31). The interstitial molecules occupy only on-top sites, whereas the CO_{ad} molecules in the $c(4 \times 2)$ structure occupy on-top and bridge sites. Configurations with short distances between O_{ad} and CO_{ad} therefore exist only at the $(2 \times 2)/c(4 \times 2)$ interface. Because of the O_{ad}/CO_{ad} repulsion, these are energetically unfavorable and most reactive.

G. Dynamic Pattern Formation: Oxidation of Hydrogen on Pt(111)

Several catalytic reactions display particularly interesting kinetics phenomena, rate oscillations, and dynamic concentration patterns on the catalyst surface (*179*). These can occur when the rate equations contain nonlinear terms. The spatial patterns, reaction fronts, target patterns, spiral waves, etc. have been made visible with various surface-imaging techniques, such as photoemission electron microscopy (*180*) and low-energy electron diffraction microscopy (*181*). With STM they were found, quite unexpectedly, in one case, demonstrating the capacity of STM to reveal information about reactions both on the atomic scale and on a mesoscopic scale.

Völkening *et al.* (*182*) investigated the Döbereiner reaction—the oxidation of hydrogen on platinum. Notwithstanding the long history of the reaction, essential parts of the mechanism, such as the nature of the reaction intermediate (*183–185*) and the reason for the dramatic increse in the activation energy with temperature (*186, 187*), remained obscure. Völkening's experiments were performed by exposing an O_{ad}-covered Pt(111) surface to hydrogen over a wide range of temperatures lower than 300 K:

$$H_2 \rightarrow 2\,H_{ad} \tag{15a}$$

$$2\,H_{ad} + O_{ad} \rightarrow H_2O_{ad}\ (\rightarrow H_2O\uparrow) \tag{15b}$$

Water is produced at temperatures higher than ~110 K, and below its desorption temperature of ~170 K water remains adsorbed.

STM series recorded during hydrogen dosing at temperatures <170 K showed a surprising behavior. The two large-scale images [Fig. 35 (*182*)], from an experiment at 111 K, were recorded after precovering the surface with oxygen under a constant hydrogen pressure. The area displayed several terraces, which were covered with many small bright dots and a bright ring that expanded in the time between the two frames. The ring appeared to travel at constant velocity and without changing its shape, both typical characteristics of reaction fronts in nonlinear reaction–diffusion systems.

The mechanism that could explain these data depended critically on the identification of the reaction intermediate. The expected OH_{ad} had previously not been clearly identified spectroscopically. A well-defined OH_{ad} layer, however, could be prepared by adsorption of H_2O on an O_{ad}-

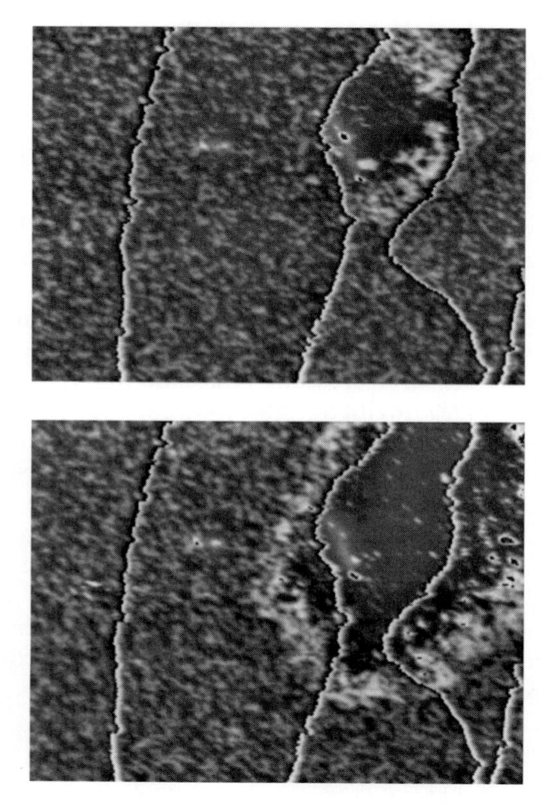

FIG. 35. STM images of the oxygen-covered Pt(111) surface during hydrogen adsorption at 111 K; time delay, 625 s. The bright ring consists of OH_{ad}; the outer area is covered by O_{ad} and small OH_{ad} islands (bright dots) and the inner area by H_2O_{ad} (2100 × 1760 Å) [reprinted with permission from Völkening *et al.* (*182*). Copyright © 1999 The American Physical Society].

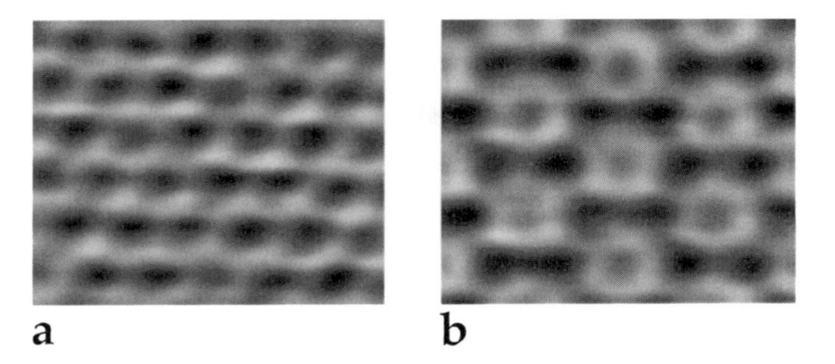

FIG. 36. STM images of OH_{ad} structures on Pt(111) formed by adsorption of water on the $(2 \times 2)O_{ad}$-covered surface at 123 K and annealing at 170 K. (a) $(\sqrt{3} \times \sqrt{3})R30°$ structure; (b) (3×3) structure (34 × 25 Å) [reprinted with permission from Bedürftig *et al.* (*188*)].

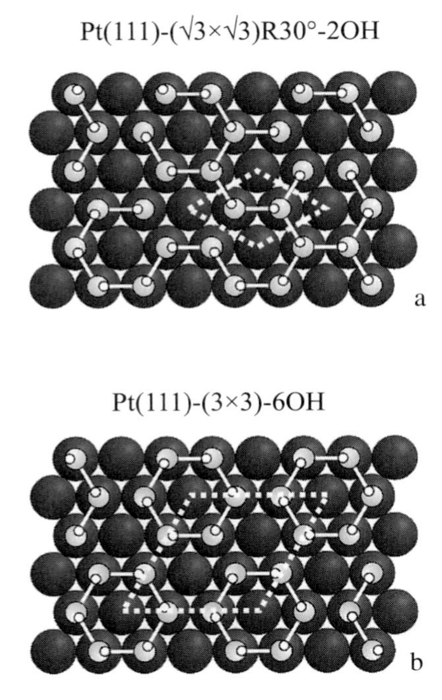

FIG. 37. Models of the OH_{ad} $(\sqrt{3} \times \sqrt{3})R30°$ structure (a) and the (3×3) structure (b). Bars indicate hydrogen bonds and the broken lines the unit cells [reprinted with permission from Bedürftig *et al.* (*188*)].

covered surface, leading to reaction to give 2 OH_{ad}. Figure 36 shows small-scale STM images of the two ordered phases that were observed on this surface. The hexagonal structure (Fig. 36a) corresponds to a $(\sqrt{3} \times \sqrt{3})R30°$ structure and the honeycomb pattern (Fig. 36b) to a (3×3) structure. Structure models (Fig. 37), which were proposed on the basis of combined HREELS and STM studies by Bedürftig *et al.* (*188*), show that the two phases differ only by the orientation of the hydrogen bridges between the OH_{ad} molecules.

Small-scale, atomically resolved data obtained during the hydrogen oxidation reaction [Reactions (15a) and (15b)] showed that the same structures also occurred under these conditions. The STM series of Fig. 38, recorded at 131 K during hydrogen adsorption on the oxygen-covered surface, shows in the beginning the $(2 \times 2)O_{ad}$ layer (O_{ad} atoms appear as dark dots), in which after 625 s the first bright islands had formed. These grew continuously with time, until they developed the hexagonal and honeycomb pat-

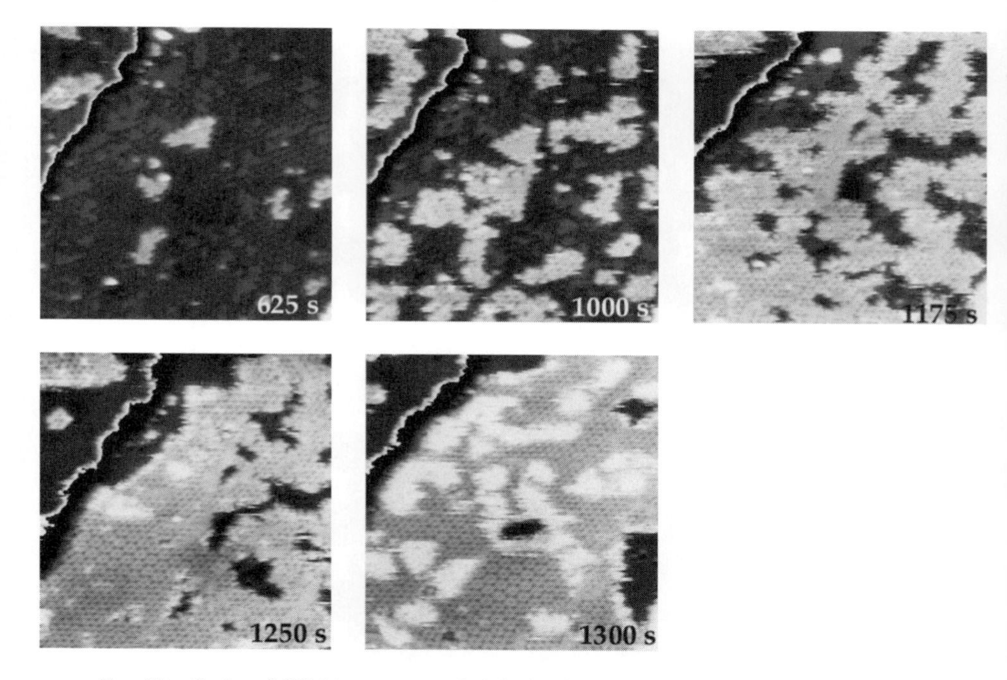

FIG. 38. Series of STM images recorded during hydrogen adsorption on the $(2 \times 2)O_{ad}$-covered Pt(111) surface at 131 K. The line at the left is an atomic step. Oxygen atoms appear dark, OH_{ad} islands bright, and H_2O_{ad} as white features in the last frame (170 × 170 Å) [reprinted with permission from Völkening *et al.* (*182*). Copyright © 1999 The American Physical Society].

terns of the OH_{ad} phases, clearly shown in the 1250-s frame. The additional white features appearing in the 1300-s frame were the first traces of H_2O_{ad}, into which the complete OH_{ad} layer was finally transformed. Since the reaction fronts were several hundred angstroms wide at ~130 K, much larger than the images of Fig. 38, the front did not become visible here. Rather, the series showed the structure at one location of the surface as the front passed this area. The induction period of ~600 s corresponds to the time before the front arrived. The small OH_{ad} islands that had formed at this stage explain the small bright dots seen on the area outside of the front in the large-scale view (Fig. 35). The front consisted mainly of OH_{ad} (bright ring in Fig. 35), and the area behind the front was covered by H_2O_{ad} (not resolved in Fig. 35 because of interactions with the tip). That most of the OH_{ad} was concentrated in the front, but had a small overall concentration, explains the difficulty encountered in resolving it as an intermediate in previous spectroscopic investigations.

The following mechanism was suggested to explain the fronts and the underlying nonlinearity:

$$O_{ad} + H_{ad} \rightarrow OH_{ad} \tag{16a}$$

$$OH_{ad} + H_{ad} \rightarrow H_2O_{ad} \tag{16b}$$

$$H_2O_{ad} + O_{ad} \rightarrow 2\, OH_{ad} \tag{16c}$$

H_{ad} reacts first with O_{ad} to give OH_{ad} and then H_2O_{ad} [Reactions (16a) and (16b)], but the water thus formed is mobile and can diffuse to areas with unreacted O_{ad}, where it comproportionates with oxygen to give OH_{ad} by Reaction (16c). Reaction (16c) is the process by which the OH_{ad} layer had been prepared (Fig. 36); it is a fast process that occurs at low temperatures. Together, Reactions (16b) and (16c) represent an autocatlytic process because each OH_{ad} particle that enters Reaction (16b) is transformed into two OH_{ad} by Reaction (16c) [which both enter Reaction (16b) again]. At low temperatures, the bulk of the water is therefore formed by cyclic repetition of the (fast) steps [Reactions (16b) and (16c)], whereas the slow step [Reaction (16a)] only initiates the process. At temperatures higher than 170 K, the lifetime of H_2O_{ad} on the surface is short so that the water can no longer sustain Reaction (16c). The autocatalytic process therefore breaks down at temperatures higher than the water desorption temperature. At higher temperatures, water has to be formed by Reactions (16a) and (16b), which is limited by the slow step [Reaction (16a)]. This explanation accounts for the finding of a lower rate at higher temperatures.

H. Reactions at High Pressures

Finally, the first studies at high pressures have been reported that attempt to expand the STM technique to practical catalytic conditions. McIntyre *et al.* (*189*) investigated the decomposition of propylene on Pt(111) in propylene–hydrogen mixtures at atmospheric pressure, a process that underlies coking. The processes are equivalent to those for ethylene (Fig. 27a). At the experimental temperatures of $T \geq 300$ K, the high adsorbate mobility prevented detection of molecular fragments such as propylidyne until, after annealing at higher temperatures, larger carbonaceous clusters C_xH_y became visible that were formed by subsequent decomposition and polymerization. Because most adsorbates are extremely mobile at the higher temperatures required for high-pressure reactions, the major challenge for STM experiments under these conditions is clear.

The investigations revealed an interesting tip effect (*190–192*). It was found that carbonaceous clusters, formed by annealing a layer of adsorbed propylidyne, could be removed again in an atmosphere of propylene–hydrogen mixtures or pure hydrogen from the area underneath the tunneling tip. The authors suggested that hydrogen dissociated on the Pt/Rh tip and that the hydrogen atoms were transferred to the sample surface, where they reacted with the C_xH_y clusters. The exact nature of the process is not clear. The finding shows that additional tip effects on surface reactions may come into play under such conditions; these have to be considered in high-pressure studies.

VII. Conclusions

STM has made it possible for the first time to view surface catalytic reactions on the atomic scale. This success is largely based on instrumental improvements in the past 10 years. In the initial experiments at room temperature, investigations of surface reactions were limited to systems with low particle mobility. Since then, instruments cooled by liquid helium or working at variable temperatures or at high recording speed have been developed, greatly expanding the range of conditions available for experiments. Although STM is in principle not restricted to vacuum, most investigations have been performed under usual surface science conditions (UHV and using single crystals), but the first results from high-pressure experiments have been reported.

The theory of STM imaging has developed to a degree (by exact calculation of the tunneling current and using realistic tip geometries) that allows one to reproduce the experimentally observed images of adsorbates. The results demonstrate that an interpretation of adsorbate images by simple inspection is not possible and that even the traditional understanding of the STM images as a LDOS at E_F contour is an oversimplification. Additional information from other techniques is thus required for particle identification.

Because of the electronic characteristics of most adsorbate–metal systems, STS has not found wide application for adsorbate identification. A promising development is IETS, which has recently provided vibrational spectra of individual molecules.

Information about the primary steps of catalytic reactions, adsorption from the gas phase, and surface dissociation has been obtained from the (static) surface distribution of the particles. The dissociation of O_2 and of NO on various surfaces has been studied in detail. The effects of hot atoms were observed, indicating nonadiabatic processes during dissociation, which could also be stimulated in a controlled way by the electrons from the tunneling tip. Atomic steps were found to act as active sites, and adsorbed particles were also found to be sites of enhanced dissociation probability.

Several STM investigations have provided values characterizing lateral interactions between adsorbates, parameters that are relevant for possible deviations from Langmuirian kinetics of surface reactions. These values were obtained by comparisons of experimental atomic configurations with those obtained from MC simulations and also directly from the experiments by means of the correlation function in dilute adsorbate layers.

Work on surface diffusion illustrated the capacity of STM for dynamic investigations. The motion of adsorbates such as oxygen, sulfur, and nitrogen atoms and ethylene and benzene molecules could be monitored, and the results allowed determination of hopping rates, activation barriers, and, in one case, the rotation rate of molecules. It was shown how much lateral interactions affect particle motions, which is required to assess the importance of surface diffusion at the nonzero coverages of surface reactions.

The central step of catalytic reactions, the reaction between the adsorbed reactants to give an intermediate or the final product, has been characterized in many STM investigations. Usually, the experiments were performed as titrations, by exposure of an adsorbate-covered surface to a second reactant, but the first measurements under steady-state reaction conditions have recently been reported as well.

Many investigations of the reaction step were performed with reconstructed surfaces, in particular with the added-row phases formed by oxygen on the(110) surfaces of copper, nickel, and silver. These reconstructions have the advantage of immobilizing the oxygen atoms, allowing STM experiments at room temperature.

Reactions of these phases with H_2S and with formic acid revealed structure rearrangements on a mesoscopic scale caused by the density changes (with respect to substrate atoms) resulting when the reconstructions are lifted as oxygen is reacted off. The altered surface topography can then affect the rate of subsequent structure-sensitive reactions. This work also demonstrated the inadequacy of the static checkerboard picture of catalytic surfaces, which STM shows may display drastic topography variations.

The added-row phases consist of long, one-atom-wide rows of metal and oxygen atoms, each held together by strong internal bonds but only weakly interacting with neighboring rows, and their structures significantly affect their reactivities. Reactions of these phases with methanol, formic acid, NH_3, CO, and hydrogen displayed pronounced anisotropies; the reactions were usually found to progress one-dimensionally along the rows but barely, if at all, from the sides. Because the reactions occur at the terminal atoms of the rows, the availability of open ends determines the reactivity of the surface. On well-ordered, added-row reconstructed surfaces, with only few open ends, or when the row ends become blocked by reaction products, reactivities can become quite small, which explains variations in reactive sticking coefficients. Atomic steps can be important in these cases because these are locations of open ends. Similar effects were discovered for the Rh(110) surface on which a different but also anisotropic reconstruction exists. For a reaction of this type, carbonate formation on Ag(110), the capacity of STM for determination of the stoichiometry of a chemical reaction was demonstrated.

Investigations of reactions on nonreconstructed surfaces have required reduced temperatures for STM monitoring of the highly mobile adsorbates. Examples include isomerization and cyclization reactions of hydrocarbons (ethylene and acetylene). These reactions showed nonrandom reactivities even in the absence of substrate defects, which were connected with domain boundaries in the adsorbate layer.

Such deviations from the assumptions of traditional Langmuirian kinetics were found to be quite pronounced for the oxidation of CO on Pt(111). At low temperatures the reaction was found to take place only at boundaries between oxygen and CO-covered domains. By means of comparison with MC simulations, the effect was traced to particularly reactive sites existing at the boundaries; hence, it is not caused by the spatial separation of CO_{ad} and O_{ad} alone. Statistical analyses of the data provided values of the reaction rates, which were found to be proportional to the boundary lengths. The activation energy and preexponential factor reproduced macroscopic results very well. Because these parameters are derived from atomic observations, they are not subject to the nonuniqueness problem occurring in traditional kinetics measurements because of averaging over macroscopic particle numbers.

Hydrogen oxidation on the same platinum surface was found to include an autocatalytic process at low temperatures, leading to the formation of reaction fronts traveling across the surface. Because dynamic pattern formation had previously been known only on larger scales, these results are the first demonstration that STM can provide evidence of such effects.

Recent reports of high-pressure STM experiments [characterizing the decomposition of propylene on Pt(111)] show that STM can be applied successfully under conditions similar to those of practical catalysis.

ACKNOWLEDGMENTS

I gratefully acknowledge F. Besenbacher, M. Bowker, W. Ho, N. D. Lang, F. M. Leibsle, R. J. Madix, T. Michely, M. Rose, P. Sautet, M. Salmeron, and P. S. Weiss, who kindly provided illustrations of their work for inclusion in this article. I also thank J. V. Barth, G. Ertl, W. Mahdi, S. Renisch, and R. Schuster for their helpful comments and W. Mahdi for help in preparing the figures.

REFERENCES

1. Binnig, G., Rohrer, H., Gerber, C., and Weibel, E., *Appl. Phys. Lett.* **40,** 178 (1982).
2. Binnig, G., Rohrer, H., Gerber, C., and Weibel, E., *Phys. Rev. Lett.* **49,** 57 (1982).
3. Murray, P. W., Besenbacher, F., and Stensgaard, I., *Isr. J. Chem.* **36,** 25 (1996).
4. Leibsle, F. M., Murray, P. W., Gordon, N. G., and Thornton, G., *J. Phys. D Appl. Phys.* **30,** 741 (1997).
5. Wintterlin, J., and Behm, R. J., *in* "Scanning Tunneling Microscopy I" (H.-J. Güntherodt and R. Wiesendanger, Eds.), Springer series in surface sciences, Vol. 20, p. 39. Springer, Berlin, 1992.
6. Chiang, S., *in* "Scanning Tunneling Microscopy I" (H.-J. Güntherodt and R. Wiesendanger, Eds.), Springer series in surface sciences, Vol. 20, p. 181. Springer, Berlin, 1992.
7. Besenbacher, F., and Stensgaard, I., *in* "The Chemical Physics of Solid Surfaces and Heterogeneous Catalysis" (D. A. King and D. P. Woodruff, Eds.), Vol. 7, p. 573. Elsevier, Amsterdam, 1993.
8. Besenbacher, F., and Nørskov, J. K., *Prog. Surf. Sci.* **44,** 5 (1993).
9. Wintterlin, J., *in* "Scanning Tunneling Microscopy I" (H.-J. Güntherodt and R. Wiesendanger, Eds.), 2nd ed., p. 245. Springer, Berlin, 1994.
10. Chiang, S., *in* "Scanning Tunneling Microscopy I" (H.-J. Güntherodt and R. Wiesendanger, Eds.), 2nd ed., p. 258. Springer, Berlin, 1994.
11. Besenbacher, F., *Rep. Prog. Phys.* **59,** 1737 (1996).
12. Gleich, B., Ruff, M., and Behm, R. J., *Surf. Sci.* **386,** 48 (1997).
13. Frank, M., Andersson, S., Libuda, J., Stempel, S., Sandell, A., Brena, B., Giertz, A., Bruhwiler, P. A., Bäumer, M., Martensson, N., and Freund, H. J., *Chem. Phys. Lett.* **279,** 92 (1997).
14. Besenbacher, F., Chorkendorff, I., Clausen, B. S., Hammer, B., Molenbroek, A. M., Nørskov, J. K., and Stensgaard, I., *Science* **279,** 1913 (1998).
15. Valden, M., Lai, X., and Goodman, D. W., *Science* **281,** 1647 (1998).
16. Eigler, D. M., and Schweizer, E. K., *Nature* **344,** 524 (1990).
17. Gaisch, R., Gimzewski, J. K., Reihl, B., Schlittler, R. R., and Tschudy, M., *Ultramicroscopy* **42–44,** 1621 (1992).
18. Kent, A. D., Renner, C., Niedermann, P., Bosch, J.-G., and Fischer, Ø., *Ultramicroscopy* **42–44,** 1632 (1992).
19. Stranick, S. J., Kamna, M. M., and Weiss, P. S., *Rev. Sci. Instrum.* **65,** 3211 (1994).
20. Rust, H.-P., Buisset, J., Schweizer, E. K., and Cramer, L., *Rev. Sci. Instrum.* **68,** 129 (1997).

21. Ferris, J. H., Kushmerick, J. G., Johnson, J. A., Youngquist, M. G. Y., Kessinger, R. B., Kingsbury, H. F., and Weiss, P. S., *Rev. Sci. Instrum.* **69,** 2691 (1998).
22. Wolkow, R. A., *Rev. Sci. Instrum.* **63,** 4049 (1992).
23. Schulz, R. R., and Rossel, C., *Rev. Sci. Instrum.* **65,** 1918 (1994).
24. Horch, S., Zeppenfeld, P., David, R., and Comsa, G., *Rev. Sci. Instrum.* **65,** 3204 (1994).
25. Bott, M., Michely, T., and Comsa, G., *Rev. Sci. Instrum.* **66,** 4135 (1995).
26. Crew, W. W., and Madix, R. J., *Rev. Sci. Instrum.* **66,** 4552 (1995).
27. Meyer, G., *Rev. Sci. Instrum.* **67,** 2960 (1996).
28. Behler, S., Rose, M. K., Dunphy, J. C., Ogletree, D. F., Salmeron, M., and Chapelier, C., *Rev. Sci. Instrum.* **68,** 2479 (1997).
29. Hoogeman, M. S., Vanloon, D. G., Loos, R. W. M., Ficke, H. G., Dehaas, E., Vanderlinden, J. J., Zeijlemaker, H., Kuipers, L., Cahn, M. F., Klik, M. A. J., and Frenken, J. W. M., *Rev. Sci. Instrum.* **69,** 2072 (1998).
30. Stipe, B. C., Rezaei, M. A., and Ho, W., *Rev. Sci. Instrum.* **70,** 137 (1999).
31. Ludwig, C., Gompf, B., Glatz, W., Petersen, J., Eisenmenger, W., Möbius, M., Zimmermann, U., and Karl, N., *Z. Phys. B Cond. Matter* **86,** 397 (1992).
32. Kuipers, L., Loos, R. W. M., Neerings, H., ter Horst, J., Ruwiel, G. J., de Jongh, A. P., and Frenken, J. W. M., *Rev. Sci. Instrum.* **66,** 4557 (1995).
33. Wintterlin, J., Trost, J., Renisch, S., Schuster, R., Zambelli, T., and Ertl, G., *Surf. Sci.* **394,** 159 (1997).
34. McIntyre, B. J., Salmeron, M., and Somorjai, G. A., *Rev. Sci. Instrum.* **64,** 687 (1992).
35. McIntyre, B. J., Salmeron, M., and Somorjai, G. A., *Catal. Lett.* **14,** 263 (1992).
36. Rasmussen, P. B., Hendriksen, B. L. M., Zeijlemaker, H., Ficke, H. G., and Frenken, J. W. M., *Rev. Sci. Instrum.* **69,** 3879 (1998).
37. Tersoff, J., and Hamann, D. R., *Phys. Rev. Lett.* **50,** 1998 (1983).
38. Tersoff, J., and Hamann, D. R., *Phys. Rev. B* **31,** 805 (1985).
39. Bardeen, J., *Phys. Rev. Lett.* **6,** 57 (1961).
40. Lang, N. D., *Phys. Rev. Lett.* **56,** 1164 (1986).
41. Sautet, P., Dunphy, J., Ogletree, D. F., and Salmeron, M., *Surf. Sci.* **295,** 347 (1993).
42. Wintterlin, J., Schuster, R., and Ertl, G., *Phys. Rev. Lett.* **77,** 123 (1996).
43. Doyen, G., Drakova, D., Barth, J. V., Schuster, R., Gritsch, T., Behm, R. J., and Ertl, G., *Phys. Rev. B* **48,** 1738 (1993).
44. Ruan, L., Besenbacher, F., Steensgaard, I., and Lægsgaard, E., *Phys. Rev. Lett.* **70,** 4079 (1993).
45. Tekman, E., and Ciraci, S., *Phys. Rev. B* **40,** 10286 (1989).
46. Doyen, G., Koetter, E., Vigneron, J. P., and Scheffler, M., *Appl. Phys. A* **51,** 281 (1990).
47. Chen, C. J., *J. Vac. Sci. Technol. A* **9,** 44 (1991).
48. Tsukada, M., Kobayashi, K., Isshiki, N., and Kageshima, H., *Surf. Sci. Rep.* **13,** 265 (1991).
49. Sacks, W., and Noguera, C., *Phys. Rev. B* **43,** 11612 (1991).
50. Sautet, P., and Joachim, C., *Chem. Phys. Lett.* **185,** 23 (1991).
51. Wiesendanger, R., and Güntherodt, H.-J. (Eds.), "Scanning Tunneling Microscopy III," Springer series in surface science, Vol. 29. Springer, Berlin, 1993.
52. Sautet, P., *Chem. Rev.* **97,** 1097 (1997).
53. Weiss, P. S., and Eigler, D. M., *Phys. Rev. Lett.* **71,** 3139 (1993).
54. Sautet, P., and Bocquet, M.-L., *Isr. J. Chem.* **36,** 63 (1996).
55. Sautet, P., and Bocquet, M.-L., *Phys. Rev. B* **53,** 4910 (1996).
56. Sautet, P., *Surf. Sci.* **374,** 406 (1997).
57. Lang, N. D., *Phys. Rev. B* **34,** 5947 (1986).
58. Feenstra, R. M., Stroscio, J. A., and Fein, A. P., *Surf. Sci.* **181,** 295 (1987).

59. Hamers, R. J., *in* "Scanning Tunneling Microscopy I" (H.-J. Güntherodt and R. Wiesendanger, Eds.), Springer series in surface science, Vol. 20, p. 83. Springer, Berlin, 1992.
60. Hasegawa, Y., and Avouris, P., *Phys. Rev. Lett.* **71,** 1071 (1993).
61. Crommie, M. F., Lutz, C. P., and Eigler, D. M., *Nature* **363,** 524 (1993).
62. Crommie, M. F., Lutz, C. P., and Eigler, D. M., *Phys. Rev. B* **48,** 2851 (1993).
63. Chua, F. M., Kuk, Y., and Silverman, P. J., *Phys. Rev. Lett.* **63,** 386 (1989).
64. Meyer, J. A., Kuk, Y., Estrup, P. J., and Silverman, P. J., *Phys. Rev. B* **44,** 9104 (1991).
65. Bartels, L., Meyer, G., Rieder, K.-H., Velic, D., Knoesel, E., Hotzel, A., Wolf, M., and Ertl, G., *Phys. Rev. Lett.* **80,** 2004 (1998).
66. Bartels, L., Meyer, G., and Rieder, K.-H., *Chem. Phys. Lett.* **297,** 287 (1998).
67. Hansma, P. K. (Ed.), "Tunneling Spectroscopy: Capabilities, Applications, and New Techniques." Plenum, New York, 1982.
68. Smith, D. P. E., Kirk, M. D., and Quate, C. F., *J. Chem. Phys.* **86,** 6034 (1987).
69. Stipe, B. C., Rezaei, M. A., and Ho, W., *Science* **280,** 1732 (1998).
70. Stipe, B. C., Rezaei, M. A., and Ho, W., *Phys. Rev. Lett.* **82,** 1724 (1999).
71. Weiss, P. S., and Eigler, D. M., *Phys. Rev. Lett.* **69,** 2240 (1992).
72. Barth, J. V., Zambelli, T., Wintterlin, J., and Ertl, G., *Chem. Phys. Lett.* **270,** 152 (1997).
73. Briner, B. G., Doering, M., Rust, H.-P., and Bradshaw, A. M., *Phys. Rev. Lett.* **78,** 1516 (1997).
74. Matsushima, T., *Surf. Sci.* **127,** 403 (1983).
75. Brune, H., Wintterlin, J., Behm, R. J., and Ertl, G., *Phys. Rev. Lett.* **68,** 624 (1992).
76. Zambelli, T., Barth, J. V., Wintterlin, J., and Ertl, G., *Nature* **390,** 495 (1997).
77. Steininger, H., Lehwald, S., and Ibach, H., *Surf. Sci.* **123,** 1 (1982).
78. Eigler, D. M., Lutz, C. P., and Rudge, W. E., *Nature* **352,** 600 (1991).
79. Stroscio, J. A., and Eigler, D. M., *Science* **254,** 1319 (1991).
80. Neu, B., Meyer, G., and Rieder, K.-H., *Mod. Phys. Lett. B* **9,** 963 (1995).
81. Bartels, L., Meyer, G., and Rieder, K.-H., *Appl. Phys. Lett.* **71,** 213 (1997).
82. Stipe, B. C., Rezaei, M. A., Ho, W., Gao, S., Persson, M., and Lundqvist, B. I., *Phys. Rev. Lett.* **78,** 4410 (1997).
83. Stipe, B. C., Rezaei, M. A., and Ho, W., *J. Chem. Phys.* **107,** 6443 (1997).
84. Winkler, A., Guo, X., Siddiqui, H. R., Hagans, P. L., and J. T. Yates, J., *Surf. Sci.* **201,** 419 (1988).
85. Zambelli, T., Wintterlin, J., Trost, J., and Ertl, G., *Science* **273,** 1688 (1996).
86. Piercy, P., De'Bell, K., and Pfnür, H., *Phys. Rev. B* **45,** 1869 (1992).
87. Kuk, Y., Chua, F. M., Silverman, P. J., and Meyer, J. A., *Phys. Rev. B* **41,** 12393 (1990).
88. Schuster, R., Barth, J. V., Ertl, G., and Behm, R. J., *Phys. Rev. B* **44,** 13689 (1991).
89. Salmeron, M., and Dunphy, J., *Faraday Discuss.* **105,** 151 (1996).
90. Tsong, T. T., *Rep. Prog. Phys.* **51,** 759 (1988).
91. Ehrlich, G., *Appl. Phys. A* **55,** 403 (1992).
92. Trost, J., Zambelli, T., Wintterlin, J., and Ertl, G., *Phys. Rev. B* **54,** 17850 (1996).
93. Kopatzki, E., and Behm, R. J., *Surf. Sci.* **245,** 255 (1991).
94. Besenbacher, F., Jensen, F., Lægsgaard, E., Mortensen, K., and Stensgaard, I., *J. Vac. Sci. Technol. B* **9,** 874 (1991).
95. Dunphy, J. C., Sautet, P., Ogletree, D. F., Dabbousi, O., and Salmeron, M. B., *Phys. Rev. B* **47,** 2320 (1993).
96. Dunphy, J. C., Sautet, P., Ogletree, D. F., and Salmeron, M. B., *J. Vac. Sci. Technol. A* **11,** 2145 (1993).
97. Stranick, S. J., Kamna, M. M., and Weiss, P. S., *Science* **266,** 99 (1994).
98. Stranick, S. J., Kamna, M. M., and Weiss, P. S., *Surf. Sci.* **338,** 41 (1995).

99. Österlund, L., Pedersen, M. Ø., Stensgaard, I., Lægsgaard, E., and Besenbacher, F., *Phys. Rev. Lett.* **83,** 4812 (1999).
100. Ertl, G., *in* "Molecular Processes on Solid Surfaces" (E. Drauglis, T. R. Gretz, and R. F. Jaffee, Eds.), Vol. p. 147. McGraw-Hill, New York, 1969.
101. Nagl, C., Schuster, R., Renisch, S., and Ertl, G., *Phys. Rev. Lett.* **81,** 3483 (1998).
102. Kern, K., Niehus, H., Schatz, A., Zeppenfeld, P., Goerge, J., and Comsa, G., *Phys. Rev. Lett.* **67,** 855 (1991).
103. Binnig, G., Fuchs, H., and Stoll, E., *Surf. Sci.* **169,** L295 (1986).
104. Trost, J., Brune, H., Wintterlin, J., Behm, R. J., and Ertl, G., *J. Chem. Phys.* **108,** 1740 (1998).
105. Zambelli, T., Trost, J., Wintterlin, J., and Ertl, G., *Phys. Rev. Lett.* **76,** 795 (1996).
106. Yoshinobu, J., Tanaka, H., Kawai, T., and Kawai, M., *Phys. Rev. B* **53,** 7492 (1996).
107. Barth, J. V., Zambelli, T., Wintterlin, J., Schuster, R., and Ertl, G., *Phys. Rev. B* **55,** 12902 (1997).
108. Briner, B. G., Doering, M., Rust, H.-P., and Bradshaw, A. M., *Science* **278,** 257 (1997).
109. Dunphy, J. C., Rose, M., Behler, S., Ogletree, D. F., Salmeron, M., and Sautet, P., *Phys. Rev. B* **57,** 12705 (1998).
110. Ichihara, S., Yoshinobu, J., Ogasawara, H., Nantoh, M., Kawai, M., and Domen, K., *J. Electron Spectrosc. Rel. Phen.* **88–91,** 1003 (1998).
111. Renisch, S., Schuster, R., Wintterlin, J., and Ertl, G., *Phys. Rev. Lett.* **82,** 3839 (1999).
112. Weckesser, J., Barth, J. V., and Kern, K., *J. Chem. Phys.* **110,** 5351 (1999).
113. Linderoth, T. R., Horch, S., Lægsgaard, E., Stensgaard, I., and Besenbacher, F., *Surf. Sci.* **402–404,** 308 (1998).
114. Hoffmann, H., Zaera, F., Ormerod, R. M., Lambert, R. M., Yao, J. M., Saldin, D. K., Wang, L. P., Bennett, D. W., and Tysoe, W. T., *Surf. Sci.* **268,** 1 (1992).
115. Zeppenfeld, P., Lutz, C. P., and Eigler, D. M., *Ultramicroscopy* **42–44,** 128 (1992).
116. Meyer, G., Neu, B., and Rieder, K.-H., *Appl. Phys. A* **60,** 343 (1995).
117. Meyer, G., Neu, B., and Rieder, K.-H., *Phys. Stat. Sol. B* **192,** 313 (1995).
118. Böhringer, M., Schneider, W.-D., and Berndt, R., *Surf. Sci.* **408,** 72 (1998).
119. Stipe, B. C., Rezaei, M. A., and Ho, W., *Science* **279,** 1907 (1998).
120. Stipe, B. C., Rezaei, M. A., and Ho, W., *Phys. Rev. Lett.* **81,** 1263 (1998).
121. Linderoth, T. T., Horch, S., Lægsgaard, E., Stensgaard, I., and Besenbacher, F., *Phys. Rev. Lett.* **78,** 4978 (1997).
122. Coulman, D. J., Wintterlin, J., Behm, R. J., and Ertl, G., *Phys. Rev. Lett.* **64,** 1761 (1990).
123. Jensen, F., Besenbacher, F., Lægsgaard, E., and Stensgaard, I., *Phys. Rev. B* **41,** 10233 (1990).
124. Hashizume, T., Taniguchi, M., Motai, K., Lu, H., Tanaka, K., and Sakurai, T., *Ultramicroscopy* **42–44,** 553 (1992).
125. Eierdal, L., Besenbacher, F., Lægsgaard, E., and Stensgaard, I., *Ultramicroscopy* **42–44,** 505 (1992).
126. Ruan, L., Besenbacher, F., Stensgaard, I., and Lægsgaard, E., *Phys. Rev. Lett.* **69,** 3523 (1992).
127. Besenbacher, F., Sprunger, P. T., Ruan, L., Olesen, L., Stensgaard, I., and Lægsgaard, E., *Topics Catal.* **1,** 325 (1994).
128. Leibsle, F. M., Haq, S., Frederick, B. G., Bowker, M., and Richardson, N. V., *Surf. Sci.* **343,** L1175 (1995).
129. Bowker, M., Rowbotham, E., Leibsle, F. M., and Haq, S., *Surf. Sci.* **349,** 97 (1996).
130. Haq, S., and Leibsle, F. M., *Surf. Sci.* **375,** 81 (1997).
131. Bowker, M., Bennett, R. A., Poulston, S., and Stone, P., *Catal. Lett.* **56,** 77 (1998).

132. Bowker, M., Poulston, S., Bennett, R. A., and Stone, P., *J. Phys. Condens. Matter* **10,** 7713 (1998).
133. Stone, P., Poulston, S., Bennett, R. A., Nicola, N. J., and Bowker, M., *Surf. Sci.* **418,** 71 (1998).
134. Bennett, R. A., Poulston, S., and Bowker, M., *J. Chem. Phys.* **108,** 6916 (1998).
135. Askgaard, T. S., Nørskov, J. K., Olesen, C. V., and Stoltze, P., *J. Catal.* **156,** 229 (1995).
136. Feidenhans'l, R., Grey, F., Nielsen, M., Besenbacher, F., Jensen, F., Lægsgaard, E., Stensgaard, I., Jacobsen, K. W., Nørskov, J. K., and Johnson, R. L., *Phys. Rev. Lett.* **65,** 2027 (1990).
137. Coulman, D., Wintterlin, J., Barth, J. V., Ertl, G., and Behm, R. J., *Surf. Sci.* **240,** 151 (1990).
138. Oertzen, A. v., Rotermund, H. H., and Nettesheim, S., *Surf. Sci.* **311,** 322 (1994).
139. Leibsle, F. M., Murray, P. W., Francis, S. M., Thornton, G., and Bowker, M., *Nature* **363,** 706 (1993).
140. Comelli, G., Dhanak, V. R., Kiskinova, M., Pangher, N., Paolucci, G., Prince, K. C., and Rosei, R., *Surf. Sci.* **260,** 7 (1992).
141. Leibsle, F. M., Francis, S. M., Davis, R., Xiang, N., Haq, S., and Bowker, M., *Phys. Rev. Lett.* **72,** 2569 (1994).
142. Leibsle, F. M., Francis, S. M., Haq, S., and Bowker, M., *Surf. Sci.* **318,** 46 (1994).
143. Francis, S. M., Leibsle, F. M., Haq, S., Xiang, N., and Bowker, M., *Surf. Sci.* **315,** 284 (1994).
144. Bowker, M., and Leibsle, F. M., *Catal. Lett.* **38,** 123 (1996).
145. Woodruff, D. P., McConville, C. F., Kilcoyne, A. L. D., Lindner, T., Somers, J., Surman, M., Paolucci, G., and Bradshaw, A. M., *Surf. Sci.* **201,** 228 (1988).
146. Poulston, S., Jones, A. H., Bennett, R. A., and Bowker, M., *J. Phys. Condens. Matter* **8,** L765 (1996).
147. Jones, A. H., Poulston, S., Bennett, R. A., and Bowker, M., *Surf. Sci.* **380,** 31 (1997).
148. Leibsle, F. M., *J. Phys. Condens. Matter* **9,** 8787 (1997).
149. Poulston, S., Jones, A. H., Bennett, R. A., and Bowker, M., *J. Phys. Condens. Matter* **9,** 8791 (1997).
150. Leibsle, F. M., *Surf. Sci.* **401,** 153 (1998).
151. Silva, S. L., Lemor, R. M., and Leibsle, F. M., *Surf. Sci.* **421,** 135 (1999).
152. Silva, S. L., Lemor, R. M., and Leibsle, F. M., *Surf. Sci.* **421,** 146 (1999).
153. Ruan, L., Stensgaard, I., Lægsgaard, E., and Besenbacher, F., *Surf. Sci.* **314,** L873 (1994).
154. Guo, X.-C., and Madix, R. J., *Surf. Sci.* **367,** L95 (1996).
155. Guo, X.-C., and Madix, R. J., *Farday Discuss.* **105,** 139 (1996).
156. Guo, X.-C., and Madix, R. J., *Surf. Sci.* **387,** 1 (1997).
157. Carley, A. F., Davies, P. R., and Roberts, M. W., *Chem. Commun.* **17,** 1793 (1998).
158. Kiskinova, M., Baraldi, A., Rosei, R., Dhanak, V. R., Thornton, G., Leibsle, F., and Bowker, M., *Phys. Rev. B* **52,** 1532 (1995).
159. Madey, T. W., and Benndorf, C., *Surf. Sci.* **152/153,** 587 (1985).
160. Crew, W. W., and Madix, R. J., *Surf. Sci.* **319,** L34 (1994).
161. Crew, W. W., and Madix, R. J., *Surf. Sci.* **349,** 275 (1996).
162. Crew, W. W., and Madix, R. J., *Surf. Sci.* **356,** 1 (1996).
163. Sprunger, P. T., Okawa, Y., Besenbacher, F., Stensgaard, I., and Tanaka, K., *Surf. Sci.* **344,** 98 (1995).
164. Villarubia, J. S., and Ho, W., *Surf. Sci.* **144,** 370 (1984).
165. van Santen, R. A., and Kuipers, H. P. C. E., *Adv. Catal.* **35,** 265 (1989).
166. Stensgaard, I., Lægsgaard, E., and Besenbacher, F., *J. Chem. Phys.* **103,** 9825 (1995).
167. Okawa, Y., and Tanaka, K., *Surf. Sci.* **344,** L1207 (1995).

168. Land, T. A., Michely, T., Behm, R. J., Hemminger, J. C., and Comsa, G., *Appl. Phys. A* **53,** 414 (1991).
169. Land, T. A., Michely, T., Behm, R. J., Hemminger, J. C., and Comsa, G., *J. Chem. Phys.* **97,** 6774 (1992).
170. Janssens, T. V. W., Völkening, S., Zambelli, T., and Wintterlin, J., *J. Phys. Chem. B* **102,** 6521 (1998).
171. Patterson, C. H., and Lambert, R. M., *J. Am. Chem. Soc.* **110,** 6871 (1988).
172. Campbell, C. T., Ertl, G., Kuipers, H., and Segner, J., *J. Chem. Phys.* **73,** 5862 (1980).
173. Wintterlin, J., Völkening, S., Janssens, T. V. W., Zambelli, T., and Ertl, G., *Science* **278,** 1931 (1997).
174. Gland, J. L., Sexton, B. A., and Fisher, G. B., *Surf. Sci.* **95,** 587 (1980).
175. Gland, J. L., and Kollin, E. B., *J. Chem. Phys.* **78,** 963 (1983).
176. Akhter, S., and White, J. M., *Surf. Sci.* **171,** 527 (1986).
177. Silverberg, M., Ben-Shaul, A., and Rebentrost, F., *J. Chem. Phys.* **83,** 6501 (1985).
178. Völkening, S., Wintterlin, J., and Ertl, G., to be published.
179. Imbihl, R., and Ertl, G., *Chem. Rev.* **95,** 697 (1995).
180. Rotermund, H. H., Engel, W., Kordesch, M., and Ertl, G., *Nature* **343,** 355 (1990).
181. Rose, K. C., Berton, B., Imbihl, R., Engel, W., and Bradshaw, A. M., *Phys. Rev. Lett.* **79,** 3427 (1997).
182. Völkening, S., Bedürftig, K., Jacobi, K., Wintterlin, J., and Ertl, G., *Phys. Rev. Lett.* **83,** 2672 (1999).
183. Fisher, G. B., Gland, J. L., and Schmieg, S. J., *J. Vac. Sci. Technol.* **20,** 518 (1982).
184. Mitchell, G. E., and White, J. M., *Chem. Phys. Lett.* **135,** 84 (1987).
185. Germer, T., and Ho, W., *Chem. Phys. Lett.* **163,** 449 (1989).
186. Ogle, K. M., and White, J. M., *Surf. Sci.* **139,** 43 (1984).
187. Smith, J. N., Jr., and Palmer, R. L., *J. Chem. Phys.* **56,** 13 (1972).
188. Bedürftig, K., Völkening, S., Wang, Y., Wintterlin, J., Jacobi, K., and Ertl, G., *J. Chem. Phys.* **111,** 11147 (1999).
189. McIntyre, B. J., Salmeron, M., and Somorjai, G. A., *J. Catal.* **164,** 184 (1996).
190. McIntyre, B. J., Salmeron, M., and Somorjai, G. A., *Science* **265,** 1415 (1994).
191. Schröder, U., McIntyre, B. J., Salmeron, M., and Somorjai, G. A., *Surf. Sci.* **331–333,** 337 (1995).
192. McIntyre, B. J., Salmeron, M., and Somorjai, G. A., *Catal. Lett.* **39,** 5 (1996).

Adsorption Energetics and Bonding from Femtomole Calorimetry and from First Principles Theory

QINGFENG GE, RICKMER KOSE,* AND DAVID A. KING

Department of Chemistry
University of Cambridge
Cambridge CB2 1EW, United Kingdom

The advent of an accurate, sensitive single-crystal adsorption calorimeter (SCAC) in 1991 meant that, for the first time, a general tool was available for determining the energetics of surface processes, both reversible and irreversible, under well-defined conditions. This is particularly valuable when the structure of the final state of the system, adsorbate and substrate, is well-characterized. Concurrently, first principles density functional theory (DFT) slab calculations were developed, particularly during the past six years, to the point where good comparisons can now be made with experimental results from the most reliable surface structure analyses. Results from the SCAC provide the most stringent benchmark currently available for these calculations: the total energy. Here, we provide a review of the current state of the art in both experimental and theoretical studies of the energetics of adsorption and surface reactivity, including an exhaustive comparison of data for which experimental SCAC data and DFT slab results are available for the same systems. We demonstrate how the current understanding of chemical bonding and reactivity at surfaces has been transformed through the use of these techniques. © 2000 Academic Press.

Abbreviations: AES, Auger electron spectroscopy; DFT, density functional theory; FPLAPW, full-potential linearized augmented plane wave; GGA, generalized gradient approximation; LDA, local density approximation; LEED, low energy electron diffraction; ML, monolayer; NEXAFS, near-edge X-ray absorption fine structure; PBE, Perdew-Burke-Ernzerhof form of GGA; PES, potential energy surface; PW91, Perdew-Wang form of GGA; RAIRS, reflection absorption infrared spectroscopy; rev PBE, revised PBE, or Zhang and Yang functional; $q_d(\Theta)$, differential heat of adsorption, kJ/mol [$= -\Delta H(\Theta)$]; $q_m(\Theta)$, calorimetric heat of adsorption, kJ/mol; $q_{int}(\Theta)$, integral heat of adsorption, kJ/mol; q_0, zero coverage heat at $\Theta = 0$, kJ/mol; s, sticking probability; SCAC, single crystal adsorption calorimetry; TPD, temperature programmed desorption; Θ, fractional surface coverage with respect to number of metal atoms in top layer.

* Present address: Sandia National Laboratories, MS-9161, Livermore, CA 94550.

0360-0564/00 $35.00

I. Introduction

The importance of understanding the nature of the surface chemical bond has never been more apparent than it is today. The enormous progress in technologies such as nano- and microchip fabrication is a direct outcome of the study of solid surfaces during the past 30 years (*1–4*). State-of-the-art surface diffraction and spectroscopic techniques (*5, 6*) are capable of providing detailed structural and electronic information; surface processes can be monitored with atomic resolution (*7*), and the dynamics of gas–solid scattering and the processes associated with chemisorption and surface reactions are under detailed investigation (*8–10*). In addition, continued advances in computing power and in theory have provided increasingly powerful tools to model surface properties with unprecedented reliability (*11, 12*). One of the most challenging tasks for computational surface science is the determination of the structure and the total energy of the ground state of an adsorbate, and single-crystal adsorption calorimetry (SCAC) uniquely provides important benchmarks for such calculations.

In this chapter, we review the accumulated experimental and theoretical results for adsorption and surface reaction energetics, particularly where a comparison is possible—be it directly, through the availability of both for identical adsorbate systems, or phenomenologically where suitable. The aim of this review is therefore to provide an overview and evaluation of the current state, quality, and accuracy of the theoretical and experimental investigations covering the broad spectrum of adsorbates on metal surfaces and also to assess the understanding of chemical bonding at surfaces developed through this rigorous methodology.

II. Adsorption Energetics: Current State of the Art

A. Experimental

One of the most challenging tasks of experimental surface science has been the determination of the total energy of the ground state of an adsorbate. This measurement not only provides important benchmarks for theoretical calculations but also allows for an evaluation of the chemical and electronic properties of a wide range of corresponding adsorption processes.

Since the first measurements by Tracy and Palmberg of CO on Pd{100} (*13*), Clausius–Clapeyron analyses of equilibrium adsorption isosteres have been widely used to determine adsorption heats on well-defined single-crystal surfaces. The main disadvantage of these techniques, however, is that they can only be used for studying adsorption processes that are fully reversible. The more indirect technique of temperature-programmed desorption suffers from the same problem (*14*). In general, the majority of processes, such as coadsorption, reactive adsorption, and oxidation, cannot be studied. However, through the development of a single-crystal adsorption calorimeter in Cambridge, it is now possible to measure accurately calorimetric adsorption and reaction heats on single-crystal surfaces as a function of coverage for a large variety of adsorption systems (*15–17*). A comprehensive review of the calorimetric data available up to 1998 has been published elsewhere (*18*).

1. *Measuring the Heat of Adsorption Calorimetrically*

A variety of approaches have been used to perform such measurements in the past, as reviewed elsewhere (*18, 19*), but these were all based on polycrystalline substrates. A general lack of consistent sample definition, preparation, and purity led to a considerable amount of irreproducibility of the experimental results, making a comparison of different experiments very difficult. Meaningful analyses require a knowledge of the structure of the final state of the adsorbate, and this is only available at known coverages on single-crystal surfaces.

The first attempt to perform heat of adsorption measurements on single crystals was initiated by Kyser and Masel (*20, 21*) for a Pt{111} crystal. However, the large thermal capacity of the crystal necessitated crystal temperature equilibration times of more than 10 h after crystal cleaning, by which time the sample cleanliness was notably affected.

At this point, the need for an experimental technique embracing these challenges, particularly the deployment of well-defined single crystals with a very significantly reduced thermal mass, was apparent—and so was its solution. In 1991, Borroni-Bird *et al.* (*15, 16*) presented an instrument, the Cambridge SCAC, that makes use of ultrathin single crystals with a thermal mass so small that the equilibration time after sample cleaning is negligible; in fact, the rapid thermal response of the thin crystal film was exploited to perform the actual heat of adsorption measurement in a contact-free manner.

The SCAC allows a precise, coverage-dependent measurement of differential heats of adsorption and sticking probabilities for the adsorption of a range of gases (such as CO, O_2, C_2H_2, and C_2H_4) on virtually any single-crystal surface [e.g., Ni{100}, Rh{100}, Pt{311}, and Pt{532}] at room temperature. An adaptation of the SCAC is an instrument developed by Campbell and coworkers (*22–25*) that allows for the measurement of metal-on-metal cohesion energies. Here, the temperature change at the crystal is not measured via blackbody radiation as in the Cambridge SCAC but by means of a pyroelectric ribbon that is brought into thermal contact with the sample during the adsorption experiments.

2. *The Single-Crystal Adsorption Calorimeter*

The Cambridge SCAC relies on a very simple mechanism (Fig. 1). A pulse of gas is rapidly adsorbed on the free-standing area of an ultrathin single crystal. As bond making and bond breaking occur, the overall excess bond energy is dissipated into the bulk, providing a temperature rise. Because the bulk is extremely small in volume, it equilibrates rapidly with its surroundings via blackbody radiation. This is detected by means of a remote infrared sensor (e.g., an MCT detector).

By using a pulsed molecular beam (50-ms pulse width and 2-s period), the adsorption process can be confined to a known surface area, and with known beam flux and sticking probability adsorbate coverages can be extracted. Furthermore, the pulsed nature of the beam eliminates the need to constantly monitor the crystal temperature, which would require an evaluation of ambient temperature fluctuations; instead, only a temperature differential is measured.

The sticking probability is measured by using the King and Wells technique (*26*), in which the pressure rise during adsorption of a gas pulse is compared with the pressure rise when the crystal is shielded by an inert gold flag (Fig. 2). The sticking probability at a coverage corresponding to the nth molecular beam pulse onto the sample is

$$s(n) = 1 - \frac{p_n}{p_0} \tag{1}$$

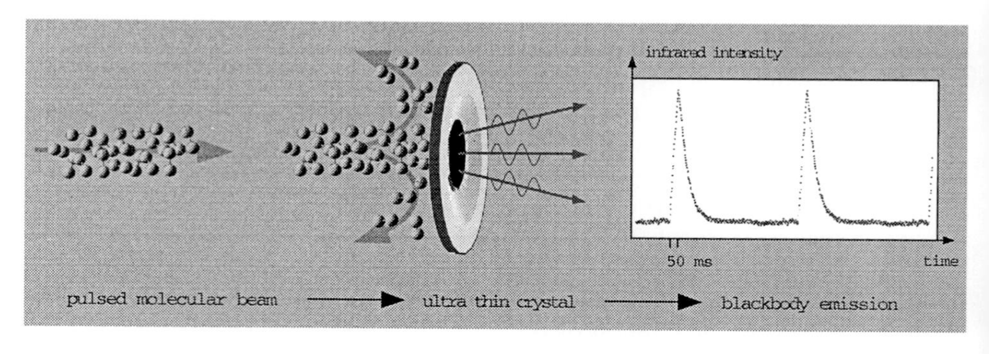

FIG. 1. The principal mechanism of the SCAC experiment. Using a pulsed molecular beam, a species is adsorbed onto an ultrathin single crystal. Cooling of the crystal occurs almost exclusively via blackbody radiation, which is detected remotely using an infrared detector.

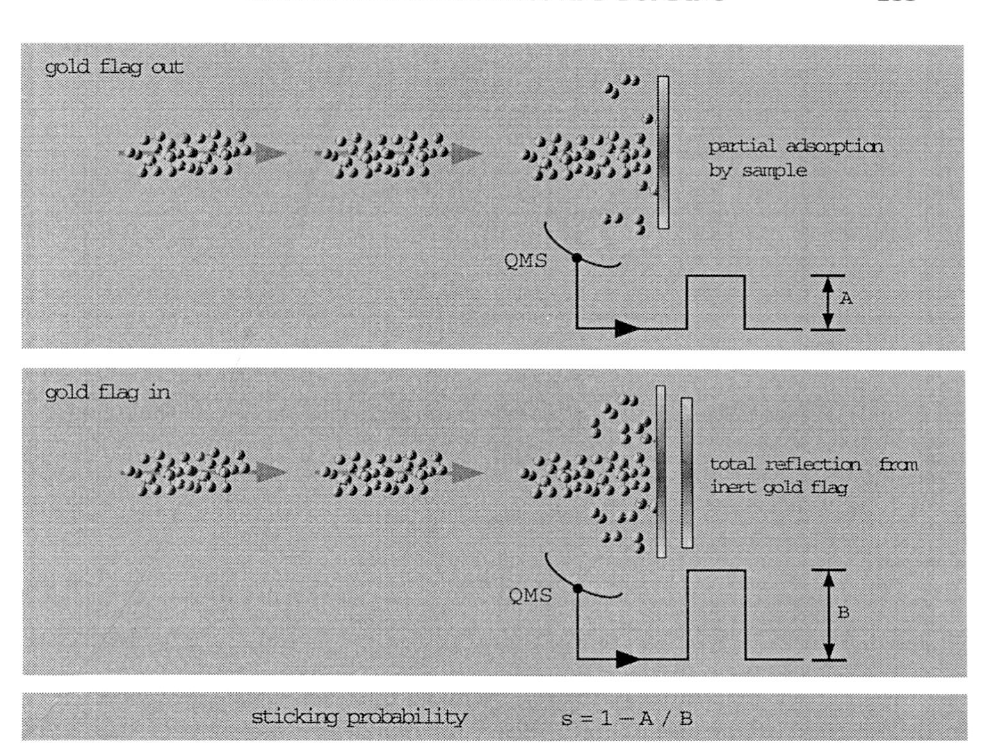

FIG. 2. Principle of the King and Wells technique (*26*) for measuring sticking probabilities for gaseous species in a system dosed by means of a molecular beam.

where p_n is the pressure rise for the nth pulse onto the sample and p_0 is the pressure rise when the sample is shielded by an inert surface (gold flag).

Because adsorption occurs pulse by pulse, at high coverages an adsorption regime is usually entered in which the adsorption process is governed by adsorbates with lifetimes longer than the timescale of the sticking probability measurement (~100 ms) but shorter than the pulse period (usually ~2 s). Species adsorbing and desorbing within this regime are adsorbed in terms of the sticking probability measurement but are actually desorbed before the next incoming molecular pulse. The sticking probability measurement is still correct because it refers to the instantaneous sticking probability, i.e., the sticking probability for the current molecular beam pulse hitting the surface. However, the determination of coverages by means of integration of subsequent sticking probabilities as a function of molecular pulse number inevitably leads to an apparent indefinite increase in the coverage. At coverages below saturation, this effect is negligible because the fraction of species leaving the surface within the pulse period compared with species remaining at the surface is effectively zero. At room temperature reversibility kicks in when the adsorption heat falls to ~80 kJ/mol.

To convert the measured data into differential heats (with units of kJ/mol), it is necessary to perform some calibrations. To determine the coverage after the *n*th pulse, the flux and the cross-sectional area of the molecular beam need to be known. The flux is measured in a beam calibration experiment. The cross-sectional beam area is determined by placing a photographic film into the sample position and shining a light through the beam path apertures (*17*). Knowledge of the surface orientation, lattice parameters, and unit cell allows the number of crystal surface atoms within the area sampled by the beam to be calculated.

A laser calibration is used to convert the amplitude of the collected infrared signal into the equivalent heat absorbed by the crystal. It is simply a simulation of the experiment, in which the molecular beam pulse is replaced by a HeNe laser beam pulse directed through the same apertures as the molecular beam (except the skimmer). The laser beam is turned on and off by means of a liquid crystal shutter, which has the same open/close characteristics as the molecular beam valve and is driven by the same pulse generator to ensure an exact replication of the temporal pulse shape. To measure the amount of energy contained in one laser pulse, the single crystal is replaced by an absolutely calibrated photodiode, and the signals are measured when the liquid crystal shutter is open and when it is closed. Finally, it is necessary to know the reflectivity of the crystal and the transmittance of the KBr window for this particular laser wavelength. The latter can be easily measured on a bench; the former is difficult to measure precisely and it was decided to use corresponding reflectivity values from the literature (*27*).

To perform beam flux measurements, a stagnation detector can be placed directly into the beam path. The molecular beam enters through a small pipe of fixed length and known conductance, and the pressure increase caused by beaming into the detector is measured by means of an absolute pressure gauge (spinning rotor gauge; SRG-2, MKS, Germany). This pressure change is directly proportional to the beam flux, which can thus be calculated easily.

The technique for producing the ultrathin single crystals was developed by J. Chevallier at Aarhus University in Denmark. A radiation-hardened NaCl single crystal is cut and polished to exhibit the desired crystal orientation needed for the metal single crystal to be grown. The metal crystal is then epitaxially grown by vapor deposition onto the rock salt crystal until the desired crystal thickness (usually ≃200 nm) is achieved. The salt is then dissolved away in distilled water until the metal crystal remains floating on top of the water surface. Using a polycrystalline ring of the same metal, which has previously been rinsed in hot acid and then flushed with distilled water to remove oxides, the metal film is then picked up, and the two are then gently pressed together to expel excess water and form a permanent cold weld. The resulting crystals were initially manufactured for use with ion channeling experiments, and it has been shown that they are indeed very good single crystals. A wide range of crystals and crystal plane orientations, including stepped and step-kink surfaces, have been prepared by this technique.

For the experiments, it is desired to achieve maximum blackbody intensity from the crystal back face. This can be increased by coating the crystal back face with a thin layer of amorphous carbon using a household candle. Notwithstanding this potential contaminant on the crystal back face, no diffusion of material from the back face to the experimental front face has been detected.

Before the experiments, the crystal front face is cleaned by means of a combination of Ar^+ sputtering, annealing, and chemical treatment. To prevent destruction of the extremely thin crystal, sputtering has to be very gentle, and it is estimated that each sputter cycle removes approximately 10–20 monolayers of substrate atoms. Typical sputter conditions are 10 μA drain current at 2 keV gun voltage and 10^{-5} mbar Ar background pressure for 5–10 min. Subsequent remote heating (for $\sim$30 s) by means of the projector lamp anneals out surface defects created by the sputtering procedure. Surface cleanliness and structure are checked by using Auger electron spectroscopy (AES) and low-energy electron diffraction (LEED), and when appropriate (as with platinum and rhodium crystals), any remaining residual surface oxygen is removed by the coadsorption of hydrogen and subsequent heating. Two or three cleaning cycles typically result in a clean surface which shows no contamination in AES and gives a good LEED pattern.

3. *Thermodynamic Considerations*

In a SCAC experiment, the temperature change at a crystal surface due to the adsorption of a pulse of molecules is monitored as a function of coverage. To explain the fundamental thermodynamics of the experiment, it is sufficient to examine the effect of a single molecular pulse on the evolution of crystal temperature as a function of time; the crystal temperature change is monitored by the radiative cooling of the crystal in the experiment.

It is important to note that conveniently (and intentionally) the following quantities are considered constant for the duration of an experimental run:

1. The crystal area where adsorption takes place
2. The quantity of gas impinging on the surface in one pulse

3. The surface temperature (T) at the moment when the gas pulse first impinges on the surface. Between pulses, the surface cools to this temperature, but we note that the temperature rise during the adsorption is $\leq$50 mK, which is negligible for the thermodynamic analysis.

We consider that the gas to be adsorbed is ideal:

$$H = U + pV \tag{2}$$

where H is the enthalpy of the gas, U is its internal energy, p is the pressure, and V is the volume. For the species in the gaseous state (indexed g) and in the adsorbed state (indexed a) the molar enthalpies are

$$H_g = U_g + p_g V_g \tag{3}$$

$$= U_g + RT \tag{4}$$

$$H_a = U_a \tag{5}$$

The term $p_a V_a$ is negligible.

Since adsorption processes are exothermic, we adopt the usual convention, assigning the differential molar adsorption heat, q, as the negative of the differential molar enthalpy change for the process.

In the experiment, the gas-to-adsorbate phase transition has an enthalpy change given by

$$\Delta H' = (H_a - H_g)$$

$$= \underbrace{U_a - U_g}_{-q_d} + \underbrace{RT}_{q_c}, \tag{6}$$

where q_d is the differential heat of adsorption and is the quantity extracted from the experiment as a function of adsorbate coverage. It describes the change in internal energy during adsorption along a path on which no work is done. The heat of compression, q_c, is a consequence of the transformation of the finite volume in the gas phase into the zero-volume adsorbate layer. Because the adsorption area is constant and the surface is considered thermodynamically inert, the heat of compression is the only relevant contribution to the measurement.

Finally, we can write the equation for calculating the coverage-dependent differential heat of adsorption as

$$-\Delta H(\Theta) = q_m(\Theta) = q_d(\Theta) + q_c(\Theta) \tag{7}$$

and thus

$$q_d(\Theta) = q_m(\Theta) - RT \tag{8}$$

where $q_d(\Theta)$ is the differential heat extracted from the results of the experiment that measures the calorimetric heat $q_m(\Theta)$ at a coverage Θ.

In practical terms, at 300 K RT is approximately 2.5 kJ mol^{-1}, two orders of magnitude smaller than typical heats of adsorption measured in the experiment. This is therefore generally a small number, but it cannot always be neglected. In simple terms it is acceptable to treat $q_d(\Theta)$, ΔH, and $q_m(\Theta)$ as being approximately equal and to make reasonable comparisons with the activation energy of desorption (E^*_{des}).

The differential heat of adsorption, $\Delta H(\Theta)$ [see Eq. (7)], describes the internal energy change of the adsorbed species and ideally refers to the adsorption of an infinitesimally small amount of gas. The differential heat is practically identical to the isosteric heat obtained at constant coverage for a system at equilibrium (*18, 28*).

a. *Integral Heat of Adsorption.* It is often of interest to consider the integral heat of adsorption, $\Delta H_{int}(\Theta)$. The integral heat accounts for the total amount of energy released when adsorbing from zero coverage up to a coverage Θ and is related to the differential heat of adsorption as follows:

$$\Delta H_{int}(\Theta) = \frac{\int_0^{\Theta} \Delta H \mathrm{d}\Theta}{\int_0^{\Theta} \mathrm{d}\Theta} \tag{9}$$

When the total energy is calculated theoretically for a given coverage, it is related to this quantity—the integral heat of adsorption.

b. *Isosteric Data.* For a fully reversible adsorption process, the heat measured in the SCAC experiment can be directly related to the isosteric heat, q_{st}, which is measured at constant-coverage pressure-temperature equilibrium:

$$q_{st} = -R\left(\frac{\partial \ln p}{\partial(1/T)}\right)_{\Theta} \tag{10}$$

Comparing the chemical potentials associated with an isosteric heat experiment (Gibbs–Duhem equation) with the SCAC experiment under constant coverage conditions, it can be shown that in this case the isosteric heat and the measured calorimetric heat equate to each other. Expansion of the total differential of the chemical potential of the adsorbed phase into partial differentials gives for the SCAC experiment on an inert surface

$$\mathrm{d}\mu_a = -S_a\,\mathrm{d}T + V_a\,\mathrm{d}p + \left(\frac{\partial \mu}{\partial A}\right)_{T,p,n_a}\mathrm{d}A + \left(\frac{\partial \mu}{\partial n_a}\right)_{T,p,A}\mathrm{d}n_a \tag{11}$$

where S_a is the differential entropy, and V_a the differential volume of the adsorbate with respect to the number of adsorbate species n_a and constant temperature T, pressure p, and adsorbate area A. Isosteric heats are measured at constant coverage, which implies that the last two terms of Eq. (11) are zero for a system in equilibrium. This means that Eq. (11) can be equated with the Gibbs–Duhem equation for the gas phase:

$$d\mu_g = -S_g\,dT + V_g\,dp \tag{12}$$

which gives

$$\Delta V_{g-a}\left(\frac{\partial p}{\partial T}\right)_{A,n_a} = \Delta S_{g-a} \tag{13}$$

where ΔV_{g-a} and ΔS_{g-a} are the volume difference and differential entropy, respectively. Using the ideal gas equation, and neglecting the volume change of the adsorbate covered substrate, gives

$$-R\left(\frac{\partial \ln p}{\partial(1/T)}\right)_{A,n_a} = T\Delta S_{g-a} \tag{14}$$

which is already close to the Clausius–Clapeyron-type equation displayed in Eq. (10). Equation (14) describes the released heat in the reversible adsorption–desorption phase change, corresponding to a pathway along which the only work done is through expansion of the gaseous phase at constant T and p, leading to

$$T\,\Delta S_{g-a} = q_d + RT = q_m \tag{15}$$

where q_d is the differential heat of adsorption and q_m the actually measured heat release. From Eqs. (10) and (14), we now conclude that

$$q_{st} = q_m \tag{16}$$

for the case in which all the constraints and conditions noted previously apply.

c. *Temperature-Programmed Desorption.* We now consider how SCAC data can be compared with results obtained from temperature-programmed desorption (TPD), particularly how the the calorimetric heat relates to the activation energy for desorption E^*_{des}. By making simple assumptions about the desorption kinetics, we will be able to directly compare it to the isosteric heat, introduced in Eq. (10).

The rate equation for the desorption process of xth order with a rate constant k for a species A that has a surface coverage Θ is

$$-\frac{d\Theta}{dt} = k\Theta^x \tag{17}$$

The temperature dependence of the rate constant k is described by the Arrhenius equation

$$k = k_0 \exp\left(-\frac{E^*_{\mathrm{des}}}{RT}\right) \tag{18}$$

In the language of surface science, k_0 translates into a frequency factor $\nu_{(x)}$ for xth order desorption. Combining Eqs. (17) and (18) gives

$$-\frac{\mathrm{d}\Theta}{\mathrm{d}t} = \nu_{(x)}\Theta^x \exp\left(-\frac{E^*_{\mathrm{des}}}{RT}\right) \tag{19}$$

which is the Polanyi–Wigner equation for the activated desorption process. Assuming first-order kinetics [$x = 1$ in Eq. (19)], then at equilibrium between adsorption and desorption

$$\nu\Theta \exp\left(-\frac{E^*_{\mathrm{des}}}{RT}\right) = \underbrace{\frac{p}{\sqrt{2\pi m k_{\mathrm{B}} T}}}_{\dagger} s(\Theta) \underbrace{\exp\left(-\frac{E^*_{\mathrm{ad}}}{RT}\right)}_{\ddagger} \tag{20}$$

Here, the term marked † expresses the rate of molecular impingement of a species with the molecular mass m during the adsorption process, and $s(\Theta)$ is the coverage-dependent sticking probability; k_{B} is the Boltzmann constant, and E^*_{ad} is the activation energy barrier for adsorption.

For nonactivated adsorption, the term marked ‡ in Eq. (20) is unity and thus, at fixed coverage, we can solve Eq. (20) for p and differentiate as follows:

$$\underbrace{-R\left(\frac{\partial \ln p}{\partial(1/T)}\right)_{\Theta}}_{q_{\mathrm{st}}} = E^*_{\mathrm{des}} \underbrace{-T\left(\frac{\partial E^*_{\mathrm{des}}}{\partial T}\right)_{\Theta}}_{\to 0 \text{ for } T = \text{const}} + \frac{1}{2}RT \underbrace{- R\left(\frac{\partial \ln(\nu/s)}{\partial(1/T)}\right)_{\Theta}}_{\to 0 \text{ for } T = \text{const}} \tag{21}$$

Therefore, the isosteric heat of adsorption q_{st} differs from the activation energy for desorption by $\frac{1}{2}RT$ even at constant temperature. Obviously, the assumption of constant temperature is not applicable to TPD. However, a comparison of SCAC data with activation energies for desorption often yields good agreement (*18*).

B. Theoretical: A First-Principles Approach to Chemisorption

A fundamental understanding of heterogeneous catalytic reactions relies on studies of the elementary surface processes. Chemisorption is one of the essential steps in heterogeneous catalysis and has been a central topic in surface science, both experimentally and theoretically. There are two different approaches to describe molecule–surface interactions. The first approach was derived from chemistry by considering the molecule–surface system as a big molecule. The "cluster" method derives from this approach (*29*). Another approach comes from solid-state physics, whereby translational symmetry is used to generate the infinite substrate. A range of empirical and semiempirical models have been developed to describe adsorption and surface reactions. Generally, these empirical methods require a model of the interactions between the atoms and the parameters of the model are usually determined by fitting the experimental results, in one way or another. In the past, they played important roles in understanding many surface processes. Readers are referred to van Santen (*30*) for a detailed description and application of these various models. However, the ultimate goal of chemisorption theory is to determine for any adsorbate–surface system the equilibrium structure, the ground state energetics, and other related properties by using only the values of the fundamental constants, i.e., from first principles. With the development of numerical techniques and computational power, the density functional theory (DFT)-based methods have become the most powerful tools for modeling chemical processes at surfaces. The preceding decade witnessed a boom in the application of *ab initio* methods to surface problems, advancing our understanding of chemisorption processes in an unprecedented manner. In this section, we give a brief introduction to the various methods, with emphasis on the state-of-the-art DFT slab method. More in-depth texts and reviews can be found in the literature (*31–33*).

The aim of *ab initio* methods is to solve the many-body Schrödinger equation for the system under study. The Born–Oppenheimer approximation (*34*) separates the electronic degrees of freedom from those of the nuclei on the basis of the fact that the electrons are much lighter than the nuclei but experience similar forces. We can therefore consider the energy of a given nuclear configuration to be that of the ground state of the electrons for that configuration. Instead of considering a coupled many-body nuclei and electrons problem, we solve the problem for the electronic variables at a given nuclear configuration.

The electronic Schrödinger equation can be written as follows:

$$\hat{H}\Psi = E\Psi \tag{22}$$

where Ψ is the many-body wavefunction for the N electronic eigenstates, an antisymmetric function of the electronic coordinates $\mathbf{r}$, and E is the total energy. The electronic Hamiltonian $\hat{H}$ is given by

$$\hat{H} = \sum_i -\frac{\hbar}{2m_e}\nabla_{\mathbf{r}}^2 + V_{\text{ext}}(\mathbf{R}_j) + V_{\text{e-e}}(\mathbf{r}) \tag{23}$$

where V_{ext} is the external potential due to nuclei at distance $\mathbf{R}_j$ and V_{e-e} is the Coulombic interaction between electrons at distance $\mathbf{r}$. The previous equation is complex and can be solved analytically only for the hydrogen atom, a single-electron system. Fortunately, the variational principle shows that the ground state energy can be found by minimizing the functional

$$E[\Psi] = \frac{\langle \Psi | \hat{H} | \Psi \rangle}{\langle \Psi | \Psi \rangle} \tag{24}$$

over all possible $\Psi(\mathbf{r})$. Quantum Monte Carlo techniques may be used to evaluate $E[\Psi]$ for a suitable trial wavefunction and to minimize $E[\Psi]$ over all possible configurations (*35, 36*). However, such calculations are still not practical for complex systems; the wavefunction Ψ becomes very complicated.

1. *Hartree–Fock Method*

The Hartree-Fock method starts from a consideration of each electron and its interactions with all other electrons and nuclei (*37–39*). The exact quantum mechanical Hamiltonian operator is used to describe explicitly the motion of each electron and its Coulomb interactions with all other charged particles in the system under consideration. The exact many-body Hamiltonian operator can be written, but the corresponding exact many-electron wavefunction is unknown. Hartree–Fock theory builds on the simplest possible approximation to such a many-electron wavefunction, namely, the product of one-electron wavefunctions, where each one-electron wavefunction corresponds to an individual electron. In Hartree–Fock theory, one additional aspect is included in the many-body wavefunction, i.e., the Pauli principle. Therefore, the exchange effect is taken into account in Hartree–Fock theory by using an antisymmetric wavefunction.

Ab initio quantum chemistry methods based on Hartree–Fock theory use an exact Hamiltonian and approximate many-body wavefunctions. The simplest of such wavefunctions is a single Slater determinant. More sophisticated many-electron wavefunctions lead to improved levels of accuracy. The essential task in the calculations is to solve the Hartree–Fock equations self-consistently in an iterative manner (*40*). Because of the mathematical nature of the Hartree–Fock approach, Gaussian-type functions are the most efficient and practical way to expand the one-electron wavefunctions. These Gaussian-type orbitals have dominated Hartree–Fock approaches since their introduction in the early 1950s (*41*). The Hartree–Fock methods have been implemented in computer programs using Gaussian basis functions and applied to simple molecular systems since the 1960s (*42*). In these programs, one starts with certain nuclear positions and selects a certain set of Gaussian basis functions. Solving the Hartree–Fock equations with this set of Gaussian basis functions will give a new set of one-electron wavefunctions. These new wavefunctions are then taken as input and the procedure is repeated until the input and output wavefunctions are the same within a certain numerical tolerance, i.e., until self-consistency has been achieved. Using the self-consistent wavefunctions, one can then evaluate the total energy of the system and thus the other related properties.

For simple molecules, the Hartree–Fock method can predict the molecular geometry within a few percent of the experimental results. Electron correlation effects beyond the single-determinant approximation can be included in a systematic way by expanding the many-electron wavefunction in a series of Slater determinants representing different electronic configurations. Formally, such a "configuration interaction" expansion converges to the exact many-body solution, although at a cost of substantially increased computational effort. Generally, these sophisticated quantum chemical methods can describe small molecules accurately but are computationally very demanding for large systems (*43*). In particular, we note that the quantum chemical cluster methods adapted for surface problems suffer from the fact that the properties of the extended system cannot be described properly. For example, it is known from a variety of experiments (*5*) that adsorption is often accompanied by relaxation and restructuring of the solid surface. This cannot be accounted for in the cluster approach.

2. *Density Functional Theory*

Density functional theory provides an alternative methodology which is both viable for systems with relatively large numbers of atoms and reliable, often to quantitative accuracy. In contrast to the Hartree–Fock method, DFT starts with a consideration of the entire electronic system. Hohenberg and Kohn (*44*) proved that the ground state energy of a system is a function of the electron density, i.e.,

$$E[\rho(\mathbf{r})] = F[\rho(\mathbf{r})] + \int [\rho(\mathbf{r})] V_{\text{ext}}(\mathbf{r})\, d^3\mathbf{r} \tag{25}$$

and the density $\rho(\mathbf{r})$ which minimizes the energy corresponds to the ground state density. This is a much simpler functional than the wavefunction $\Psi(\mathbf{r})$. Kohn and Sham (*45*) took a step further by formulating the density $\rho(\mathbf{r})$ in terms of a set of orthonormal noninteracting single-particle wavefunctions, $\psi_i(\mathbf{r})$:

$$\rho(\mathbf{r}) = \sum_i |\psi_i(\mathbf{r})|^2 \tag{26}$$

The functional $F[\rho(\mathbf{r})]$ is then written as

$$F[\rho(\mathbf{r})] = \sum_i \frac{\hbar^2}{2m_e} \int \psi_i^* \nabla \psi \, d^3\mathbf{r} + \frac{e^2}{2} \int \int \frac{\rho(\mathbf{r})\rho(\mathbf{r}')}{|\mathbf{r}' - \mathbf{r}|} d^3\mathbf{r} \, d^3\mathbf{r}' + E_{xc}[\rho(\mathbf{r})] \quad (27)$$

where $E_{xc}[\rho(\mathbf{r})]$ is the exchange-correlation energy, defined as the difference between the true F and the other terms. In reality, the true form of F is unknown and so is E_{xc}. However, many approximations to E_{xc} exist. In the original approximation of Kohn and Sham, the local density approximation (LDA) takes the form

$$E_{xc}[\rho(\mathbf{r})] = \int \varepsilon_{xc}[\rho(\mathbf{r})]\rho(\mathbf{r}) \, d^3\mathbf{r} \quad (28)$$

where $\varepsilon_{ex}[\rho(\mathbf{r})]$ is the exchange-correlation energy per unit volume of a homogeneous electron gas with a density of $\rho(\mathbf{r})$. Although LDA is valid only in the limit of slowly varying densities, it has been found to give good results for a range of systems, particularly for extended bulk systems. However, it was found that LDA tends to a high level of overbinding for molecular systems. To rectify this, many corrections, generally including the gradient of the electron density, were introduced. These are the so-called "generalized gradient approximations" (GGA) (*46, 47*). The inclusion of the density gradient does not have a significant effect on local properties such as bond lengths or vibrational frequencies, but it does improve the calculated bond strength of a molecule or that of a molecule on a surface (*48–50*). Current GGA calculations provide a basis for a systematic comparison with recent experimental results to evaluate accuracy. On the other hand, the search for more accurate exchange-correlation functionals continues.

Formally, only the electron density is a physically meaningful quantity in DFT. Practically, the eigenvalues of the Kohn–Sham equations provide good approximations to the one-electron energies. Analyses of these eigenstates often provide insightful understanding into the interactions among the constituent atoms (*51*).

As noted previously, only Gaussian-type basis sets are practical for Hartree–Fock calculations. However, with the DFT-based methodology, there are many different algorithmic implementations. Most of the modern quantum chemistry computational packages using Gaussian-type basis sets are equipped with DFT as a main alternative, particularly for large systems. The basis sets can be chosen as atomic orbitals, Gaussians, linear augmented plane waves, or plane waves, and various DFT methods are usually named after the particular type of basis function used.

Each basis set has its advantages and limitations. Atomic orbitals with Gaussian basis sets prove to be most appropriate for descriptions of free molecules and molecular systems. Many cluster quantum chemistry approaches implemented in DFT, as a natural extension of Hartree–Fock calculations, have used this type of basis set. In the full-potential linearized augmented plane wave method (*52*), the basis functions were chosen as linearized augmented plane waves. This approach treats the electrons in a scalar relativistic form and it includes all electrons self-consistently by using a full potential for the valence electrons. The method can be implemented for two- or three-dimensional systems. Although these calculations are very accurate, the computational demand for systems with open structures (e.g., surface problems) is prohibitively high.

Using plane waves as the basis set is a natural choice for three-dimensional periodic systems since we can use the Bloch theorem by writing

$$\psi_{i,\mathbf{k}} = \sum_{\mathbf{G}} c_{i,\mathbf{k}}(\mathbf{G}) e^{i(\mathbf{k}+\mathbf{G})\mathbf{r}} \tag{29}$$

where the sum is over reciprocal lattice vectors **G**, and **k** is a symmetry label within the first Brillouin zone. The convergence of the plane wave basis set can be systematically improved by increasing the cutoff energy—the maximum kinetic energy of the plane waves included in the reciprocal space.

If plane waves were used to represent the electrons close to the nucleus, the number of plane waves used would be prohibitively high; the theory would be impossible to use. This is where pseudopotentials come into play (*53*). On the basis of the facts that the core electrons of an element are almost independent of the environment and only valence electrons are important to the chemical bond, the core electrons can be combined with the nuclei to form frozen ions. In this so-called frozen core approximation, not only the electrostatic potential but also the quantum nature of the core electrons are experienced by the valence electrons. The pseudopotential also eliminates oscillations of valence electron wavefunctions in the core region. Modern *ab initio* pseudopotentials reproduce the potential of an atom exactly outside the core region defined by a radius r_c and are smooth inside the core region (*54*). The quality of a pseudopotential is judged by its transferability and softeness (*55*). A transferable pseudopotential should behave like the all-electron potential in a range of different chemical environments, whereas softness will allow the calculation to be done at a low cutoff energy, thus reducing computation cost. Pseudopotentials that reproduce the same charge inside the core region as that of the all-electron potentials are referred to as norm-conserving pseudopotentials. Recently, Vanderbilt proposed *ab initio* pseudopotentials that do not fulfill the condition of norm conservation and therefore can work at a greatly reduced number of plane waves. The calculation of forces within the pseudopotential plane wave approach is straightforward and efficient, thus making optimization of the ionic configuration and even dynamical simulation possible (*56, 57*).

In the plane wave pseudopotential method, the surface is modeled by the so-called supercell, in which a geometry is repeated in both vertical and lateral directions, as shown in Fig. 3. The lateral periodicity implies that the surface is modeled properly as an infinite slab of finite thickness, with finite coverage. A sufficient number of layers of metal atoms and a sufficiently large vacuum region between slabs are required to avoid interactions between the two surfaces through the slab or through the space. The adsorption energy calculated with this method is the integral energy at the corresponding coverage. We note one significant problem with this method: The size of the unit cell in the plane of the surface determines

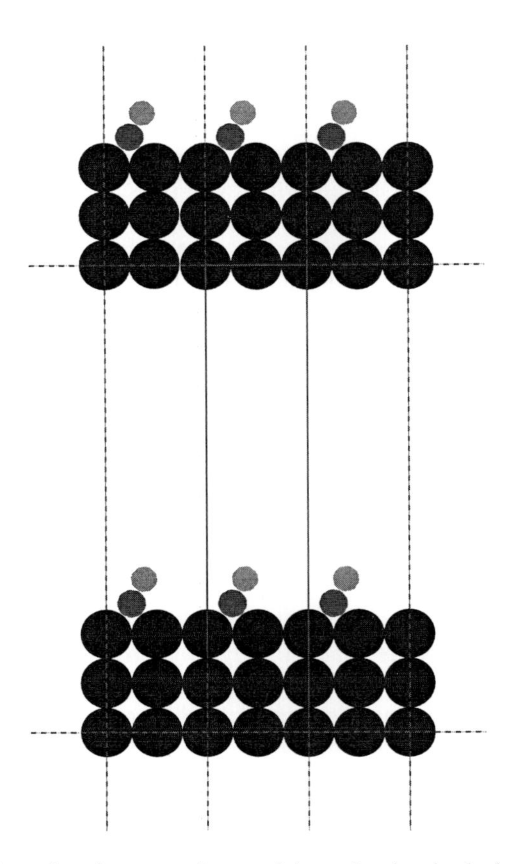

FIG. 3. Schematics of a supercell to model a molecule adsorbed on a surface.

the surface coverage. Thus, very low coverages imply large unit cells, which are in turn computationally expensive.

III. Survey of Calculated and Experimental Results

Here, we first compare experimental integral heats of adsorption with values computed by first-principles methods and draw conclusions from both sets of results. The full computation of an adsorption heat requires calculations to be made of the total energies of both the final state (the fully relaxed adsorbate–surface system) and the initial state (the gaseous molecule and the clean surface, also in their equilibrated states). Any relaxation of the metal surface which occurs on adsorption is therefore included in both experiment and theory. However, in several instances the experimental measurement has deliberately been made on both the metastable unrestructured clean surface and the stable restructured clean surface. This has led to an experimental SCAC measurement of the energy difference between clean metastable and stable phases on Pt{100} (*58*). Care has therefore been taken whenever possible to make the correct comparison with theory. Furthermore, the reported coverage dependencies of calorimetric adsorption heats refer to the differential heats; for the comparison here, we computed the integral heats from the SCAC data for coverages corresponding to those used in calculations.

Tables I–IV give an overview of the adsorbate systems that have been studied both experimentally and theoretically. For reference, the extrapolated zero coverage heat of adsorption as measured by experiment, q_0, is listed in addition to the integrated heats for particular coverages since the SCAC data were obtained at 300 K, at which temperature the adsorbate mobility is generally high.

It is in the nature of computational investigations that only simple adsorbate systems are within reach if the available computing power is limited, and only very recently, with the availability of parallel processors and fast workstations, has it become possible to study more complex systems. Many of these early studies were made for hydrogen on single-crystal surfaces. Hydrogen, however, poses experimental challenges in the SCAC setup in Cambridge, and only recent improvements have paved the way for accurate measurements of adsorption heats for hydrogen. To date, the optimum performance of SCAC has been for the adsorption of small gaseous species with sufficiently high vapor pressures at room temperature, such as O_2, NO, CO, ethene, and acetylene. Comparisons are made when results from both theoretical and experimental studies are available. In this section, we address the adsorption of CO, NO, and O_2 on the faces of platinum, palladium, nickel, and rhodium single crystals.

The column marked "*m*" in Tables I–IV indicates the percentage difference between theoretical and experimental integral heats of adsorption, giving a direct measure of the correlation between theory and experiment. A low *m* number clearly gives confidence in both. We note that parameters and approximations can be introduced into the theoretical calculations which may differ from experimental conditions, or it may be that bonding arrangements do not match the true (and possibly unknown) structure.

When theory and experiment differ significantly, we anticipate that the problems are attributable to the theoretical parameters used. However, we also note that although theory computes the ground state for the system, it is possible that the experimental measurement relates to a metastable state; the timescale of the SCAC experiment is 50 ms, and a slow transformation to the most stable state would therefore lead to an underestimate of the adsorption energy for that state. However, this is unlikely to be a significant factor except at the highest coverages.

A. CO Adsorption

1. *A Quantitative Comparison of Theory with Experiment*

Table I shows the compilation of theoretically and experimentally acquired integral heats for CO adsorption on nickel, palladium, platinum, and rhodium. The agreement between experimental and theoretical adsorption heats on palladium surfaces is very good. For adsorption on Pd{110}, Hu *et al.* (*50*) reported strong overbinding with LDA, but with GGA at one monolayer coverage, excellent agreement was attained, with the most stable site being the bridge, in agreement with conclusions from infrared spectroscopy. The calculation for CO on Pd{100} by Hammer *et al.* (*60*) also achieved excellent agreement with the experiment, when the revPBE (revised Perdew–Burke–Ernzerhof form of GGA) functional was implemented. The measured integral heat of adsorption at 0.25 monolayer (ML) coverage is 140 kJ/mol (*18, 59*) and the theoretical value for bridge site adsorption is 145 kJ/mol (*60*). Eichler and Hafner (*61*) also performed chemisorption energy calculations for CO adsorbed into bridge, atop, and fourfold sites on Pd{100}. They showed that the bridge site is energetically the most favorable. The atop site occupancy observed in experiments at low temperature (*62–65*) was attributed to the topology of the potential energy surface (PES) for CO at a large distance from the surface, which steers CO toward the energetically less favorable atop site (*61*).

The SCAC integral heat of CO adsorption on Pt{100}-(1 $\times$ 1) is 210 kJ/mol at 0.25 ML (*18, 58*), considerably higher than the isosteric heat of 111

TABLE I

Integral Heats from Theory and Experiment for the Adsorption of Carbon Monoxide on Single-Crystal Surfaces (in kJ/mol) Derived from Experimentally Published Differential Calorimetric Heats[a]

System	Adsite	Coverage (ML)	q_t (reference)	q_{exp} (reference)	q_0	m (%)
CO/Ni{100}	Hollow	0.25	152 (*60*)	123 (*28*)	123	19.6
CO/Ni{110}	Bridge	0.5	190 (*83*)	130 (*28*)	133	31.5
CO/Ni{111}	Atop	0.25	131 (*71*)	126 (*28*)	131	3.7
	Fcc	0.25	144 (*60*)	126 (*28*)		12.1
CO/Pd{100}	Bridge	0.25	145 (*60*)	140 (*18, 59*)	170	3.3
	Bridge	0.5	185 (*61*)	123 (*18, 59*)		33.9
CO/Pt{100}-(1 × 1)	Atop	0.25	150 (*66*)	209 (*18, 58*)	220	−39
CO/Pt{110}-(1 × 1)	Atop	0.50	209 (*179*)	175[b] (*80*)	220	16.6
	Atop	1	182 (*171*)	154[b] (*80*)		15.3
CO/Pt{110}-(1 × 2)	Bridge	0.25	204 (*79*)	185 (*80*)	197	9.0
CO/Pt{111}	Atop	0.25	140 (*71*)	173 (*18, 67*)	190	−23.4
	Atop	0.25	152 (*73*)	173 (*18, 67*)		−13.3
	Hcp	0.25	172 (*68*)	173 (*18, 67*)		−0.6
CO/Pt{211}	Bridge	0.25	211 (*66*)	165 (*82*)	185	21.9
	Atop	0.25	198 (*66*)	165 (*82*)		17.4
CO/Rh{100}	Bridge	0.25	178 (*60*)	117 (*91*)	117	34.2
	Bridge	0.25	215 (*78*)	117 (*91*)		45.7
	Bridge	0.5	198 (*78*)	115 (*91*)		43.1
	Bridge	0.75	176 (*78*)	109 (*91*)		37.9
	Bridge	1	170 (*78*)	102 (*91*)		39.8

[a] The column labeled "m" refers to the percentage match of measured and calculated integral adsorption energies at the corresponding coverage, calculated as $(q_t - q_{exp})/q_t \cdot 100$, with q_t being the theoretical value and q_{exp} the experimental value. [b] Reconstruction energy not included in the value.

kJ/mol measured for adsorption on Pt{100}-hex at very low coverage under conditions for which the hex reconstruction is not lifted. This large energy difference is responsible for driving the lifting of the reconstruction (*58*). Here, the agreement between theory and experiment is rather poor. For adsorption on the metastable (1 × 1) surface, Hammer *et al.* (*66*) found 150 kJ/mol, assuming CO bonded atop, and on the hex surface the most stable site yields 135 kJ/mol.

Excellent agreement, however, is achieved for CO on Pt{111}, for which the SCAC integral heat of adsorption for atop CO at 0.25 ML is 173 kJ/mol (*18, 67*). Lynch and Hu (*68*) calculated a value of 161 kJ/mol, but their results provide a discrepancy with experiment in predicting that the most stable site is hcp hollow. The calculated chemisorption energies for the high-symmetry sites (top, bridge, fcc hollow, and hcp hollow) are within a range of 11 kJ/mol (161, 164, 168, and 172 kJ/mol, respectively). Experiment clearly shows that the atop site is the most stable at low coverages (*18, 67*). Because the calculations are effectively performed for CO adsorption at zero temperature, the entropic contributions to the total free energy are ignored. At finite temperatures it has been demonstrated that vibrational entropy can cause site switching, favoring the atop and bridge sites over the threefold sites (*69, 70*). This effect does not fully compensate for the measured relative energies for CO on Pt{111}. The issue is unresolved.

Hammer *et al.* (*71*) also performed *ab initio* DFT calculations (*57*) for the same adsorbate system using LDA (*72*) for finding charge densities and GGA for the exchange-correlation potential to calculate total energy differences (*47*). Substrate relaxation was not permitted; the substrate was fixed with the bulk truncated geometries. Furthermore, the metal–carbon distance was fixed at 1.94 Å, and the carbon–oxygen distance for the adsorbed species was fixed at 1.14 Å. The effect of relaxation on the binding energy was claimed to contribute less than 5 kJ/mol to the total energy. The binding energy found for CO atop adsorption at 0.25 ML coverage was reported as 140 kJ/mol, which is significantly lower than the experimentally determined value of 172 kJ/mol and lower than any of the values found by Lynch and Hu (*68*). Since the different authors used essentially the same methods, the reason for the discrepancy probably lies in the absence of relaxation. For example, Hammer *et al.* (*71*) held the metal–carbon bond length fixed at 1.94 Å, whereas Lynch and Hu (*68*) found a value of 1.85 Å, matching the experimentally determined value of 1.85 $\pm$ 0.1 Å (*68*).

The adsorption of CO on Pt{111} was also investigated by Eichler and Hafner (*73*) as part of their studies of CO catalytic oxidation on this surface. For their calculations, they made use of a spin-polarized version of the Vienna *ab initio* simulation program (*74–76*) and a Perdew–Wang (PW91) form of the GGA for the exchange-correlation potential (*47*). They, too, neglected substrate relaxation, referring to their earlier calculations, according to which such relaxations showed little effect on the calculated energies (*61, 77, 78*). The chemisorption energy for CO on Pt{111} at 0.25 ML coverage was found to be 152 kJ/mol, which is higher than the value reported by Hammer *et al.* (*71*) but lower than the values calculated by Lynch and Hu (*68*) and found by the SCAC experiment (*18, 67*).

Good theory–experiment agreement is achieved between the theoretical heats reported by Ge and King (*79*) and the SCAC data for CO on metastable Pt{110}-(1 $\times$ 1), at both 0.5 and 1.0 ML. In the experiment the adsorption of CO on the stable (1 $\times$ 2) reconstructed surface was accompanied by a lifting of the reconstruction at room temperature at 0.25 ML.

For the adsorption of CO on the stable Pt{110}-(1 × 2) surface without lifting the reconstruction, the theory–experiment agreement is even more acceptable; in this case the experimental and theoretical values deviate from each other by less than 10%. The chemisorption energy was calculated to be 204 kJ/mol at 0.25 ML coverage, which is comparable to the experimentally determined heat of adsorption of 185 kJ/mol (*80*).

Strong overbinding with respect to the experimental results was found for CO adsorption on the stepped Pt{211} surface by Hammer *et al.* (*66*). For on-top adsorption and a coverage of 0.25 ML, they found a chemisorption heat of 200 kJ/mol. For adsorption onto bridge sites, the value is even higher (211 kJ/mol), indicating the most stable bond, whereas experimental results clearly show that on-top adsorption is favored at low coverages (*81*). The integral SCAC heat of adsorption is 165 kJ/mol at 0.25 ML (*82*), which is significantly lower than the energies calculated by Hammer *et al.*, even though in this case substrate relaxation effects were considered in the calculation.

We note a potential problem in calculations for adsorption on the surface of a ferromagnetic material such as nickel, for which inclusion of a correction term for spin polarization could be essential. In support of this, it was found that for CO adsorption on Ni{110} the inclusion of spin polarization dramatically decreases the otherwise strong overbinding (*83*). With spin polarization, and including a nonlinear core correction (*84*), the heat of adsorption at 0.5 ML coverage is computed to be 190 kJ/mol, which is a considerable reduction of the value of 247 kJ/mol found when the spin effect is excluded. The resulting overbinding from the non-spin-polarized calculations is mainly due to an overestimation of the total energy of the bare surface, which has a significantly higher magnetic moment than the adsorbate-covered surface.

Hammer *et al.* (*60, 71*) investigated the adsorption of CO on the {100} and {111} faces of nickel. For CO/Ni{100} and CO/Ni{111}, Table I shows the calculated energies under implementation of the Zhang and Yang functional (*85*) (rev PBE), which generally yield better agreement with the experimental results than do other functionals (*60*). Although the preferred adsorption site for CO on Ni{100} at low coverages is the atop site (*86, 87*), the agreement of the theoretical value for hollow site adsorption of 152 kJ/mol with the experimental value of 123 kJ/mol at 0.25 ML (*28*) is remarkable, even more so if one assumes that the atop site, similar to CO/Ni{111}, yields a lower calculated chemisorption energy.

The nature of the CO adsorption site on Ni{111} at low coverages has been subject to examination in many experiments, and it is now agreed that threefold hcp and fcc sites are occupied at low coverages (0.25 ML) (*88*), identical to those covered when a c(4 × 2) structure forms at 0.5 ML

(*89*). Atop sites are occupied only at higher coverages (*90*). These results are supported by the findings of Hammer *et al.* (*60, 71*), who found the hollow site [with a chemisorption energy of 144 kJ/mol when using the revPBE functional (*60*)] to be more stable than the atop site [131 kJ/mol (*71*)], yielding a chemisorption value that is in quite close agreement with the experimentally determined heat of adsorption of 126 kJ/mol (*28*).

One of the parameters that could persistently lie behind some of the disagreement between experiment and theory is the sample temperature. It is 300 K for the referenced SCAC experiments, whereas it is effectively 0 K for the theoretical calculations. The inclusion of relaxation of the adsorbate system due to the effective electronic potential is also clearly important to account correctly for the adsorption process. For example, there are very significant differences between the experimental results (*91*) and the theoretically obtained adsorption energies of CO on Rh{100}, both from Hammer *et al.* (*60*) and Eichler and Hafner (*78*). The discrepancy may possibly be attributable to the adsorbates being trapped in a metastable state that is not readily accessible in the calculations, which consequently results in a lower measured heat of adsorption than theoretically calculated. However, we believe that the disagreement probably arises from the theoretical model. Hammer *et al.* (*60*) showed that although there is an appreciable improvement of the widely used PW91 or PBE forms of GGA compared with LDA, these do not perform equally well for all systems, and the authors demonstrated that the performance of the PBE functional can be improved by introducing one additional parameter. A further possible source of the discrepancy may be the pseudopotentials used. This is currently an open question.

2. *The Bonding of CO to Transition Metal Surfaces*

Theoretical investigations of CO chemisorption on metals have produced a clear understanding of the nature of the surface chemical bond. Hu *et al.* (*51*) examined the individual eigenstates of the CO/Pd{110} system, for which agreement between experiment and theory is good, and they examined the charge density distribution of each Bloch state for the adsorption system and, where appropriate, compared these distributions with the Kohn–Sham orbitals computed for a raft of CO molecules without the metal slab. The comparison (Fig. 4) demonstrates quite spectacularly the mixing of the molecular CO 4σ, 1π, and 5σ states with d states of the palladium surfaces and also the presence of metal-derived bonds showing d–2π bonding and antibonding states below and above the Fermi level. These Kohn–Sham orbitals show a very close resemblance to the anticipated charge density distributions for these states. The schematic orbital

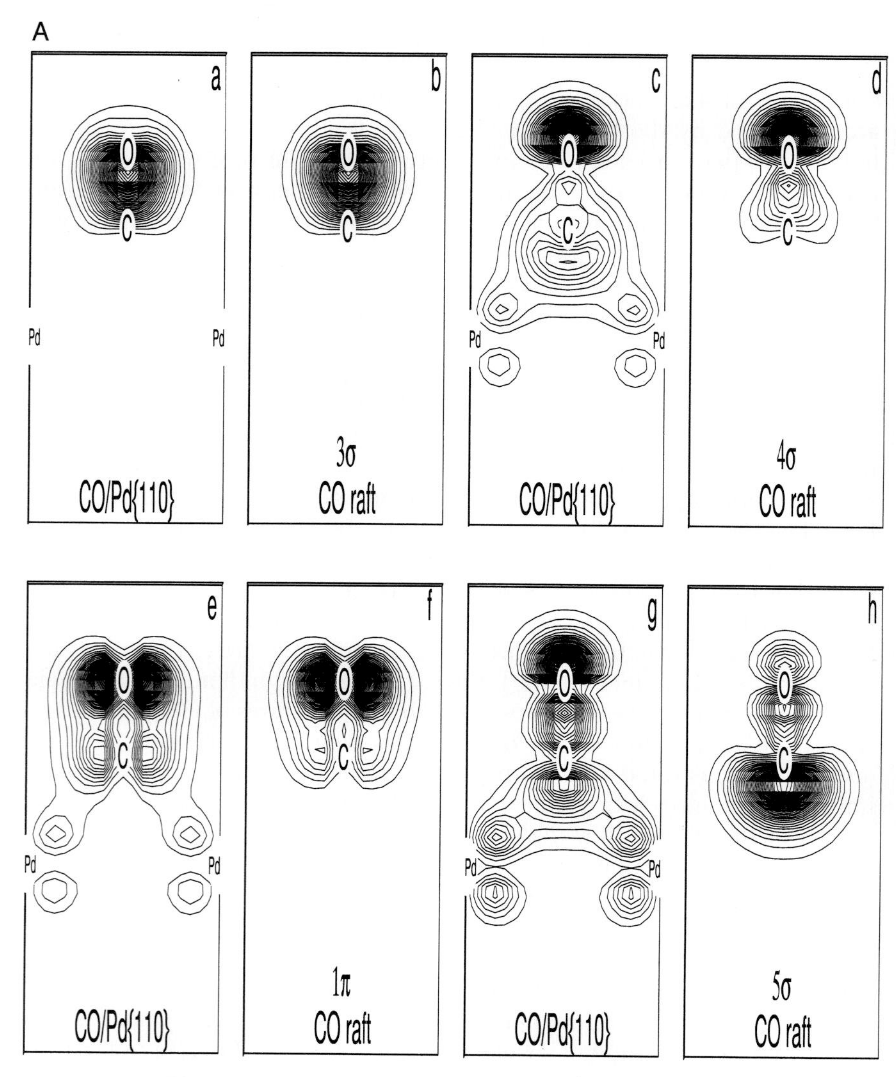

FIG. 4. (A) A charge density comparison of 3σ-, 4σ-, 1π-, and 5σ-derived orbitals between CO/Pd{110} and CO monolayer without metal; (a), (c), (e) and (g) are 2-D contours of charge densities of 3σ-, 4σ-, 1π-, and 5σ-derived orbitals, respectively, cut through the C–O and C–Pd bond axes, which correspond to $\rho_{i_{\text{band}}}$, $\mathbf{k}_j$ from CO/Pd{110} at $\mathbf{k}_j = \Gamma$, $i_{\text{band}} = 2$ (−23.86 eV), 4 (−9.23eV), 10 (−5.41eV), and 7 (−6.99eV), respectively, where E_F is defined as zero. Oxygen, carbon, and top-layer palladium atom positions are labeled; (b), (d), (f), and (h) are the same but from a CO monolayer without metal, which correspond to $\rho_{i_{\text{band}}}$, $\mathbf{k}_j$ at $\mathbf{k}_j = \Gamma$, $i_{\text{band}} = 2$, 4, 8, and 10 respectively.

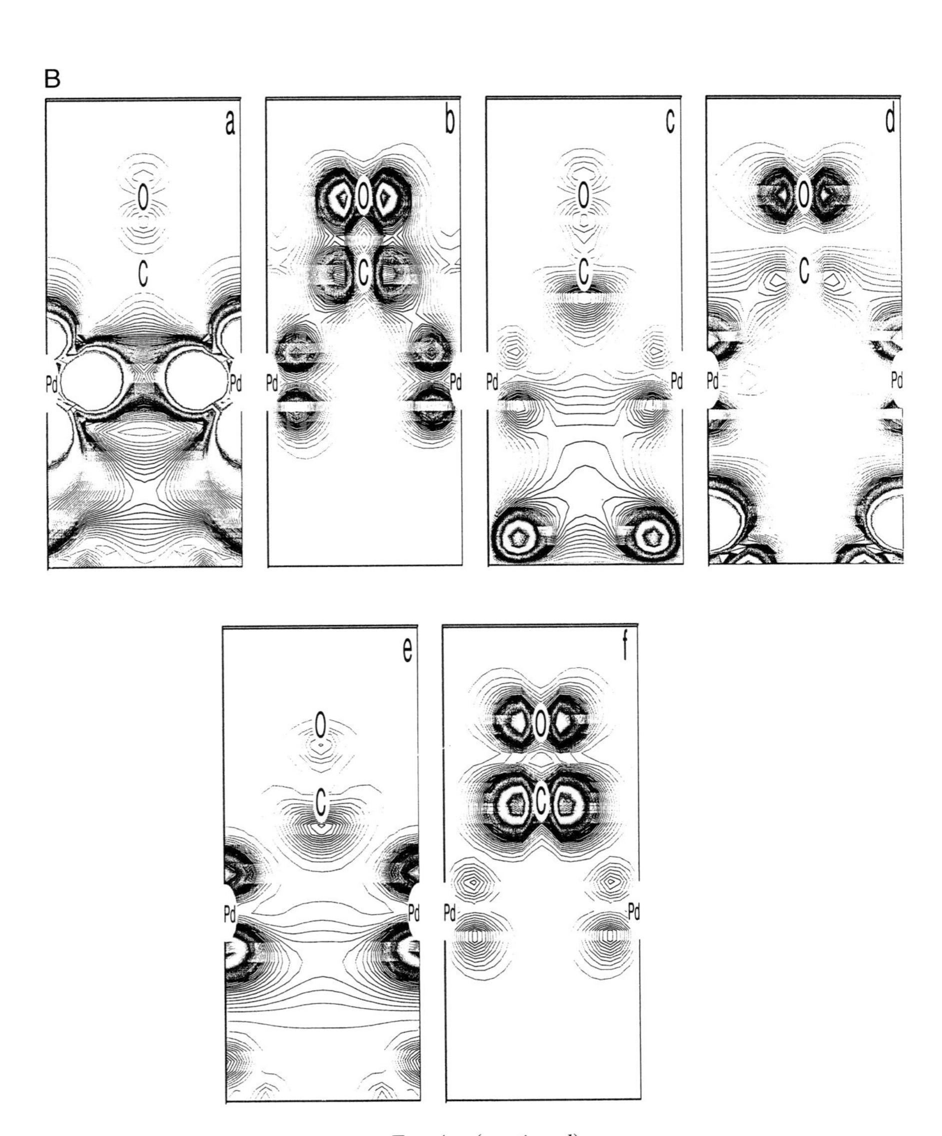

FIG. 4. *(continued)*

(B) Illustration of 4σ, 1π, 5σ, and 2π characters in metal-derived bands and metal–5σ and metal–2π antibonding states; a–d correspond to charge density $\rho_{i_{\text{band}}}$; $\mathbf{k}_j$ from CO/Pd{110} at $\mathbf{k}_j = \Gamma$; $i_{\text{band}} = 11$ (-4.72 eV), 12 (-4.30 eV), 16 (-3.97 eV), and 34 (-0.85 eV), respectively, where E_F is defined as zero, cutting through the C–O and C–Pd bond axes. Oxygen, carbon, and top-layer palladium atom positions are labeled; e and f are the same cuts but from $\rho_{i_{\text{band}}}$ at $\mathbf{k}_j = \Gamma$, $i_{\text{band}} = 42$ (1.29 eV) and 50 (3.71 eV), respectively, which are above the Fermi level, showing metal–5σ and metal–2π antibonding states, respectively [after Hu *et al.* (*51*)].

mixing model, showing bonding and antibonding states in CO/Pd{110}, is shown in Fig. 5.

From this work, Hu *et al.* (*51*) reached an important general conclusion. In crossing the periodic table from right to left (e.g., from copper to iron), the d band energy levels shift up. An upward shift in the d band energy levels results in an increased overlap in population of the mixed 2π–d states and hence an increase in the CO–metal adsorption heat. These states, as exemplified by the mixed orbital shown in Fig. 4B(d), are bonding with respect to CO adsorption since the equivalent antibonding states are above the Fermi level and hence unoccupied [Fig. 4B(f)], but their occupancy weakens the C–O bond. Thus, as concluded from a semiempirical analysis by Hoffmann *et al.* (*92, 93*), shifting far enough to the left in the periodic table results in C–O bond dissociation. We note, for example, that CO readily dissociates on tungsten surfaces.

Hammer *et al.* (*71*) developed the conclusion from Hu and King's work into a general principle. Focusing on the "d band center" as a measure of the position of the d band, and accounting for the crystal-plane-dependent narrowing of the d band in the surface metal layer arising from a lowering of the metal atom coordination, they demonstrated a clear correlation between the energy of the d band center and the calculated heat of adsorption for CO in atop sites on a range of single-crystal transition metal surfaces.

Although the d band center principle is clearly a useful didactic tool, it is by no means the only factor in determining CO adsorption heats. In moving to the left beyond tungsten and molybdenum, for which the d band is half occupied, the d band occupancy is reduced, and the bonding strength between metal and adsorbate also begins to diminish.

We also note the following important point: The adsorption properties of the stable surface Pt{100}, which has a hexagonal close-packed top layer, are quite different from the properties of Pt{111}, on which again the top layer is hexagonal. On the other hand, the metastable, square

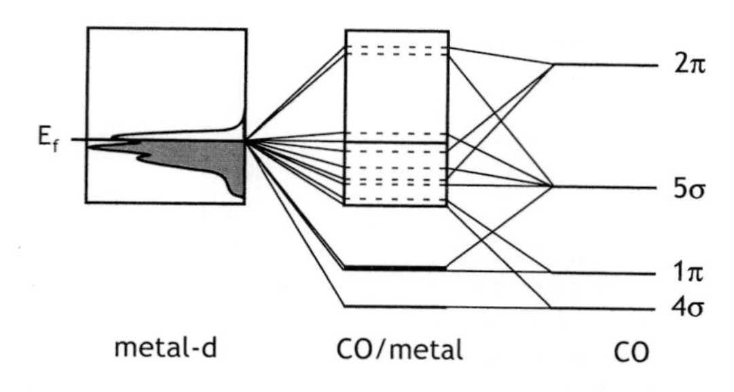

FIG. 5. Schematic illustration of orbital mixing model for CO on palladium [after Hu *et al.* (*51*)].

Pt{100}-(1 × 1) surface shows some adsorption properties which are similar to those of hexagonal Pt{111}. This is contrary to an expectation based on the top-layer metal atom coordination. Thus, the dissociative sticking probabilities for O_2 on Pt{100} hex, {100}-(1 × 1), and {111} surfaces are, respectively, 10^{-4} (*94*), 0.3 (*95*), and 0.2 (*96*). Relevant to the current consideration, the adsorption heat of CO on Pt{100} hex is considerably smaller [115 kJ/mol (*97*)] than those on {100}-(1 × 1) [220 kJ/mol (*58*)] and on {111} [198 kJ/mol (*67*)]. We conclude that the detailed electronic structure of the strain-relieved stable surface of Pt{100} is quite different from those of the unrestructured {111} and {100} surfaces, which is not accounted for in the position of the d band center.

An additional important new result for CO chemisorption deserves coverage in this brief overview. Much work has been done on the diffusion of adsorbates on surfaces, whereby it has been almost universally assumed that the lateral potential energy surface for the diffusion of an adsorbate species across a single-crystal surface can be described by a simple cosine-like function with single hops from one stable high-symmetry site to the next (*98*). However, DFT slab calculations for CO on both Pt{110}-(1 × 2) and -(1 × 1) surfaces by Ge and King (*80*) show that both bridge and atop sites along the atom rows are local potential minima; the diffusion barrier is situated at a low symmetry point between these high-symmetry sites. The diffusional PES for CO along {$1\bar{1}0$} for CO on Pt{110}-(1 × 2) at 0.25 ML is shown in Fig. 6.

There is clearly a deep trapping well at the bridge site, between atop sites, requiring a double hop between high-symmetry sites. In this case, the bridge site is marginally more stable than the atop site at 0 K, whereas on Pt{110}-(1 × 1) the calculations demonstrate a small preference for the atop site at this temperature (*79*).

B. NO Adsorption

Table II shows the theoretically and experimentally acquired integral heats of adsorption for NO, which are discussed in this section.

In contrast to CO adsorption, only a few NO adsorption systems have been studied with both SCAC and theory. This lack is not because NO is less important than CO. Rather, the adsorption of NO on metal surfaces is of considerable scientific and technological interest. For example, the reduction of NO_x species by automobile exhaust catalysis has a great influence on the environment. Brown and King (*99*) recently reviewed studies of NO adsorption on metal surfaces and suggested that some of the traditional views toward NO adsorption need to be altered.

NO is a paramagnetic molecule; there is an unpaired electron in its

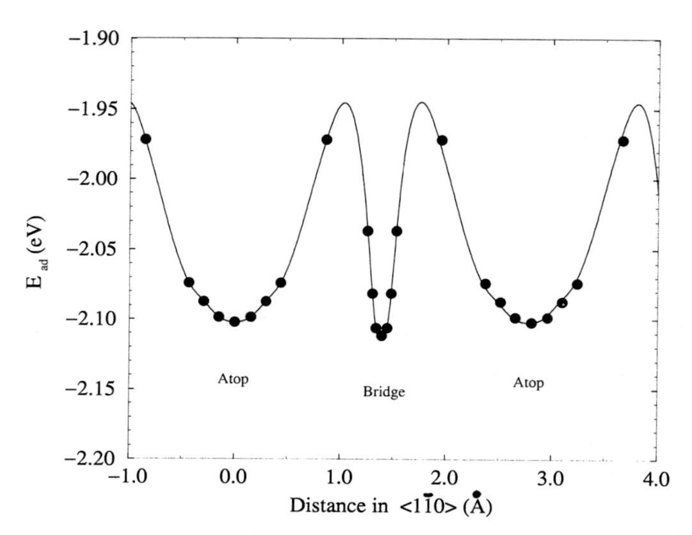

FIG. 6. Diffusional potential energy surface for CO on Pt{110}-(1 × 2) along the {1$\bar{1}$0} direction [after Ge and King (*79*)].

antibonding 2π orbital. Although it is generally considered to be similar to CO when adsorbed, results show that the adsorption process can be much more complicated. Recent DFT studies show that NO loses its spin identity when adsorbed on transition metal surfaces (*100, 101*), whereas it essentially retains its spin density on noble metal surfaces (*102*). This residual spin of NO when adsorbed on silver and on copper stabilizes adsorbed $(NO)_2$ and leads to the rich chemistry of NO on these surfaces at very low temperatures (*103, 104*).

TABLE II

Integral Heats from Theory and Experiment for the Adsorption of Nitric Oxide on Single-Crystal Surfaces (in kJ/mol)[a]

System	Adsite	Coverage (ML)	q_t (reference)	q_{exp} (reference)	q_0	m (%)
NO/Ni{100}	Dissoc.	0.25	366 (*60*)	228 (*18, 105, 106*)	400	37.7
NO/Pd{100}	Hollow	0.25	157 (*60*)	104 (*18, 59*)	111	33.7
	Bridge	0.50	149 (*172*)	96 (*18, 59*)		35.1
NO/Pt{100}-(1 × 2)		0.25	210 (*107*)	161 (*80*)	164	23.4
No/Pt{110}-(1 × 1)		0.50	212 (*107*)	151[b] (*80*)		28.6

[a] The column labeled "m" refers to the percentage match of measured and calculated integral adsorption energies as in Table I. [b] Reconstruction energy not included in the value.

A comparison of theoretically determined chemisorption heats with experimental results is generally quite disappointing, with significant overbinding in the calculations. However, it may be relevant that for each of the underlying substrates—nickel, palladium, and platinum—the discrepancy is almost a constant fraction. The theoretical integral heats of adsorption at 0.25 ML of NO on Ni{100} (dissociative, 357 kJ/mol) and Pd{100} (157 kJ/mol) calculated by Hammer *et al.* (*60*) differ by approximately 35% from the experimentally determined values, which are 228 kJ/mol for NO/Ni{100} (*18, 105, 106*) and 104 kJ/mol for NO/Pd{100} (*18, 59*). The difference between the measured heat and the calculated value for NO on Ni{100} may be in part due to incomplete NO dissociation at 0.25 ML in the experiment. In support of this suggestion, we note that the initial heat, when NO dissociation is complete, is in reasonable agreement with the theoretical value. For the adsorption of molecular NO on the (1×1) and (1×2) phases of Pt{110}, the calculations by Ge *et al.* also yield significant disagreement. Calculated values at 0.25 ML coverage are 212 kJ/mol for NO/Pt{110}-(1×1) (*107*), compared with the experimental value of 151 kJ/mol (*80*), and 210 kJ/mol for NO/Pt{110}-(1×2) (*107*), compared with the experimental value of 161 kJ/mol (*80*).

Notwithstanding this level of disagreement between theory and experiment in absolute adsorption energies, the combined results of theoretical and experimental studies have greatly advanced our understanding of NO adsorption on surfaces (*99*). The golden rule regarding adsorbate site occupancy based on infrared spectra for metal nitrosyl complexes has had to be modified (*99*). Quite large overlaps between the N–O stretch frequencies for different geometries can still make the site assignment ambiguous in some cases.

The hollow site occupancy found by LEED (*108–110*) and supported later by a DFT calculation (*101*) suggested a breakdown of the vibrational frequency-based site assignment. It is clear that the best way for NO adsorption site assignments to be made, and generally for adsorption structure determination, is by a combination of techniques, including theory (*111*). A combination of DFT and reflection adsorption infrared spectroscopy studies of NO adsorption on Pt{110} has shown very promising results (*107, 112, 113*). The lateral potential energy surface calculated with DFT helped to identify the metastable states populated by NO at very low temperatures (30 K). Local minima on the lateral PES indicate where NO can be trapped when thermal diffusion is suppressed, and the vibrational bands observed are assigned to these trapped species at these sites in metastable states.

Both molecular and dissociative adsorption occur for NO on Ni{100}, depending on coverage and temperature. At room temperature, NO dissociates initially at low coverages. With the building up of the N and O adatom coverage on the surface, the strong repulsive interactions between the adatoms drive the adsorption from dissociative to molecular. This is clearly shown in the SCAC-measured differential heats of adsorption (*105, 106*) as shown in Fig. 7.

The initial heat of adsorption on the clean surface is 385 kJ/mol, and it decreases smoothly with coverage up to 0.16 ML. At this point, there is a large change in the slope of the differential heat with coverage; beyond this critical coverage the heat is relatively constant, as the molecular NO state is populated on the surface up to 0.4 ML. When the surface is precovered with oxygen adatoms, the initial heat of adsorption is lower, and the additional coverage of dissociated NO needed to reach the molecular state is much lower than that on the clean surface. [The SCAC-measured heats of NO dissociative adsorption can be used to calculate the heat of dissociative adsorption of N_2, which cannot be measured directly due to a very high activation barrier and therefore a very low dissociative sticking probability for N_2 on Ni (18).]

Strong lateral interactions between the adatoms force the differential heat to decrease rapidly with increasing coverage, and at $\theta > 0.16$ ML the adsorption process becomes predominantly molecular. This switch of

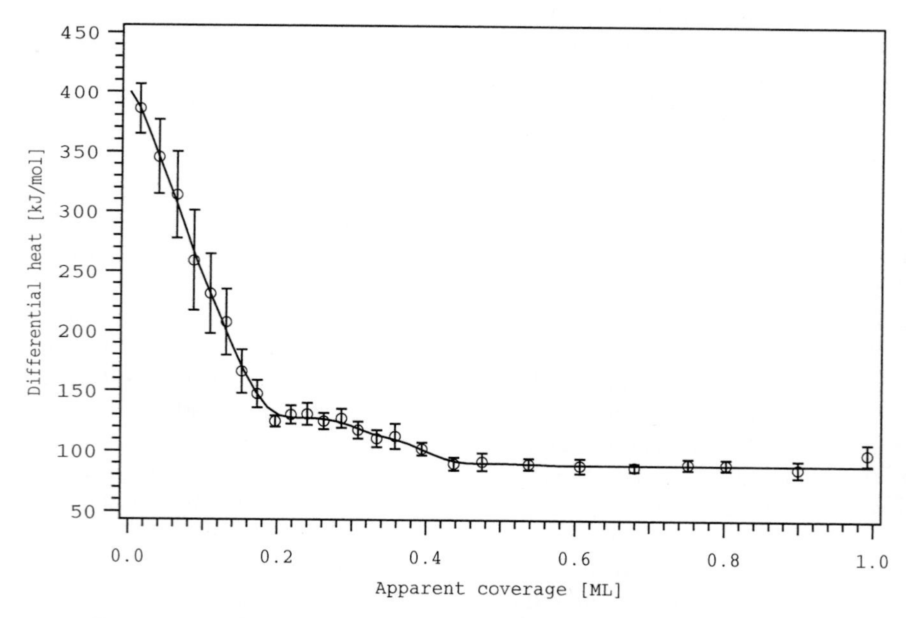

FIG. 7. Heat of adsorption for NO on Ni{100} [after Vattuone *et al.* (*106*)].

processes is driven by these lateral interactions, which actually drive the stability of $N_a + O_a$ to be lower than that of NO_a. This can be critically important in catalytic processes, leading to a control of reaction selectivity. Thus, a molecular beam investigation of ammonia oxidation showed that the preadsorption of oxygen on Pt{100} above the critical coverage of 0.16 ML can switch the product of ammonia oxidation from N_2 at low O coverages to NO at high coverages with 100% efficiency even at temperatures as low as 500 K (*114, 115*). Any NO formed on the surface at these O adatom coverages cannot dissociate; it can only desorb. At lower adatom coverages the NO formed in the surface reaction dissociates, yielding N_2 by N atom recombination and switching the product of the reaction.

C. Oxygen Adsorption

1. *Comparison of Theory with Experiment*

Table III shows a compilation of the theoretically and experimentally determined integral heats of adsorption for oxygen on metal single crystals. Oxygen is an important reactant in many heterogeneous reactions. In the Langmuir–Hinshelwood mechanism, the reaction is governed by surface oxygen adatoms. The study of the energetics and structure of oxygen adsorption is clearly of great importance in many catalytic reactions.

On most transition metal surfaces, oxygen undergoes irreversible dissociative adsorption. Siegbahn and Wahlgren (*116*), using a cluster model,

TABLE III

Integral Heats from Theory and Experiment for the Adsorption of Oxygen on Single-Crystal Surfaces (in kJ/mol)[a]

System	Adsite	Coverage (ML)	q_t (reference)	q_{exp} (reference)	q_0	m (%)
O_2/Ni{100}	Hollow	0.25	541 (*60*)	515 (*117*)	564	4.8
	Hollow	0.0	544 (*116*)	564 (*117*)		−3.7
O_2/Ni{111}	Fcc	0.25	487 (*60*)	461 (*117*)	420	5.3
	Hollow	0.0	481 (*116*)	420 (*117*)		−12.8
O_2/Pt{111}	Fcc	0.25	259 (*33*)	212 (*18, 67*)	335	17.9
	Fcc	0.25	304 (*118*)	212 (*18, 67*)		30.2
	Fcc	0.25	261 (*68*)	212 (*18, 67*)		18.8
O_2/Rh{100}	Hollow	0.25	479 (*60*)	357 (*119*)	384	25.6
	Hollow	0.25	425 (*122*)	357 (*119*)		15.9
	Hollow[b]	0.50	377 (*122*)	315 (*119*)		16.6

[a] Values given are per O_2 molecule, based on a calculated dissociation energy of 550 kJ/mol. The column labeled "m" refers to the percentage match of measured and calculated integral adsorption energies as in Table I. [b] The substrate is clock reconstructed.

calculated the isolated adatom (zero coverage) chemisorption energies of atomic oxygen on Ni{100} and Ni{111}. By electronic excitation of the cluster, it is placed into a state suitable for bonding. The calculation produced adsorption energies that match the experimental results well. For oxygen adsorbed on Ni{100}, the experimentally determined adsorption heat is 564 kJ/mol in the zero-coverage limit (*117*). The calculations by Siegbahn and Wahlgren yielded 544 kJ/mol (*116*). For adsorption on Ni{111}, the extrapolated zero-coverage adsorption energy is 420 kJ/mol (*117*), whereas Siegbahn and Wahlgren (*116*) calculated it to be 481 kJ/mol. The underlying argument is that excitations to higher (bonding) states are part of the cluster preparation. For an infinite surface, the excitation energy is close to zero, but for a cluster the required excitation energy should be taken into account in the calculated adsorption energy.

Hammer *et al.* (*60*) performed DFT-slab calculations for the dissociative adsorption of oxygen on the {100} and {111} faces of Ni at 0.25 ML oxygen atom coverage. The agreement of the calculated adsorption energy with experiment is very good. The calculated chemisorption energies are 541 kJ/mol on Ni{100} (hollow) and 487 kJ/mol on Ni{111} (fcc) (*60*) compared with the experimental integral heats of 515 and 461 kJ/mol (*117*), respectively.

Although the theoretical results for oxygen adsorption on Pt{111} from Hammer and Nørskov (*33*) and Lynch and Hu (*68*) show comparably good agreement with the experimental results, the results obtained by Bleakley and Hu (*118*) show considerable overbinding. For fcc adsorption at 0.25 ML coverage, Hammer and Nørskov calculated an energy value of 259 kJ/mol, whereas Lynch and Hu calculated an energy of 304 kJ/mol, compared with an experimental integral heat of adsorption of 212 kJ/mol (*18, 67*).

Comparison of the calculated chemisorption energies (hollow site) by Hammer *et al.* (*60*) with the experimental results for 0.25 ML oxygen on Rh{100} (*119*) also showed strong overbinding; the theoretical value is 480 kJ/mol, which is significantly larger than the experimental value of 357 kJ/mol. Upon oxygen adsorption, the substrate undergoes a phase transition to form a "clock" reconstructed (2×2)p4g structure (*120, 121*). However, this structure is unlikely to be present at 0.25 ML coverage, and there should be no contribution to the chemisorption heat due to the clock reconstruction at higher coverages. The discrepancy between experiment and theory is much smaller in the calculations of Ge and King (*122*), at both 0.25 and 0.5 ML coverages. Even in this example, the comparison suggests overbinding from theory. A possible reason for this discrepancy is the choice of the pseudopotential, as was discussed by Walter and Rappe (*123*). They suggested the use of pseudopotentials with different core sizes for the adsorbate and for the free O_2 molecule to warrant accuracy and to gain enhanced calculational efficiency.

2. *The Bonding of O Adatoms to Transition Metal Surfaces*

Coverage-dependent measurements and calculations of heats for dissociative adsorption of oxygen on metals have in recent years provided a wealth of new information. Experimental results for O_2 adsorption on the principal planes of Ni(*117*) revealed a very sharp decrease in adsorption heat at low coverages due to repulsive interactions between adatoms. Data are shown in Fig. 8 for the {100}, {111}, and {110} crystal planes.

On Ni{100}, for example, the differential adsorption heat decreases from 550 kJ/mol at zero coverage to 180 kJ/mol at 0.5 ML. This represents a massive tuning down of the O–Ni lattice bond energy, a factor of critical importance in heterogeneous catalysis. Evaluation of the substrate temperature-dependent calorimetric heat data obtained by use of a Monte Carlo approach provided pairwise lateral interaction energies between adatoms. There is a large second nearest-neighbor repulsive pairwise interaction of 30 ± 5 kJ/mol which is responsible for lowering the adsorption heat at

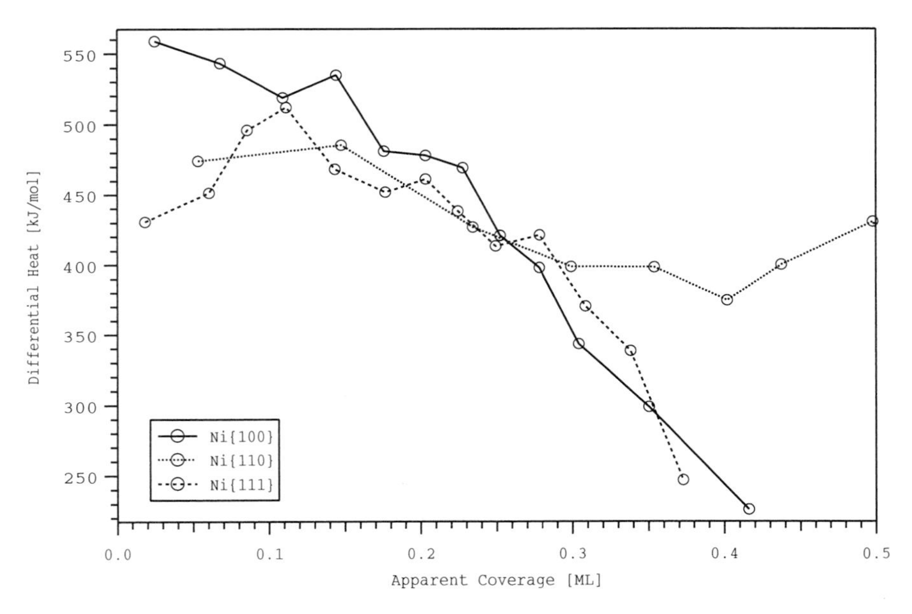

FIG. 8. Heat of adsorption of O_2 as a function of coverage at 300 K for all three nickel low-index surfaces [after Stuckless *et al.* (*117*)].

coverages exceeding 0.25 ML. The origin of this repulsive force is clarified by DFT calculations, as discussed later.

The calorimetric heat data for O_2 on nickel surfaces provide a simple explanation for the onset of oxide film formation at coverages between 0.4 and 0.5 ML on Ni{100} and Ni{111} and at coverages close to 1 ML on Ni{110}. In Fig. 8, the onset of oxide film growth is marked by an upturn in the differential heat of adsorption. The strong repulsive adatom–adatom second-neighbor interactions eventually force the heat of adsorption of the chemisorbed layer of O atoms down, until it is equal to the heat of formation of an oxide film. Oxide growth is therefore initiated at or close to the crossover of the integral heat of chemisorption curve with that of the formation of an oxide layer. This is therefore a thermodynamic basis for the onset of oxide film growth. The kinetic requirement for dissociation is a complex eight- or nine-site empty array, which drives the dissociative sticking probability sharply down as the coverage increases; however, the onset of oxide formation, which is accompanied by an increase in the dissociative sticking probability, is thermodynamically determined (*117*).

First-principles DFT calculations for O_2 on Pt{100}-(1 $\times$ 1) not only reveal the origin of the second-neighbor pairwise repulsive interactions but also demonstrate that these interactions can lead to a switching of the most stable site as the coverage is increased (*11*). At 0.25 ML the fourfold hollow site is found to be the most stable, with an integral heat of 180 kJ/mol. However, at 0.5 ML the integral heat in this site is only 56 kJ/mol, whereas the bridge site can be occupied at this coverage with a heat of 182 kJ/mol. Clearly, the fourfold site becomes destabilized with increasing coverage, in favor of the bridge site. The origin of the repulsive interactions is shown in Fig. 9, which displays the Kohn–Sham orbitals for the occupied p_x and p_y orbitals on the O atoms, approximately 5.7 eV below Fermi level, for both sites. The repulsive Pauli interaction between these p states on next-nearest-neighbor O adatoms in the fourfold hollow site is clearly demonstrated. Interestingly, although the fourfold hollow site is the most stable site on Rh{100}, it does induce a clock reconstruction, whereas on Ir{100} the bridge site is the most stable for O adatoms at coverages of 0.25 and 0.5 ML (*122*).

IV. Energy Changes in Surface Reactions

A. CO Oxidation

The oxidation of CO on metal surfaces is a textbook example of a catalytic reaction and has been extensively investigated (*124–127*) because of its roles in environment catalysis and in fundamental research. With the recent

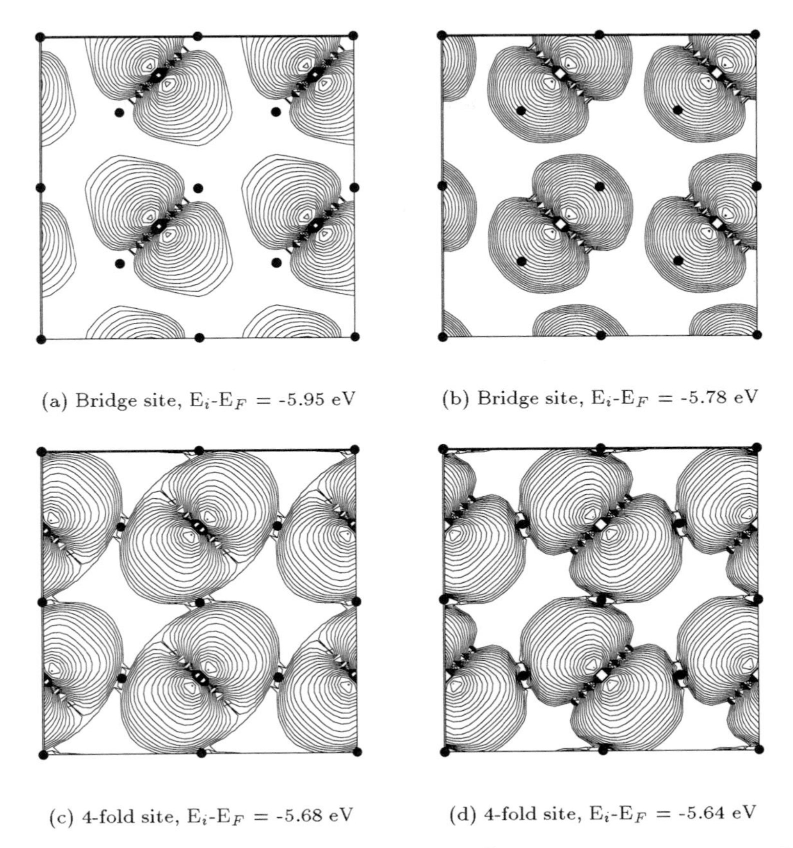

FIG. 9. Logarithmic charge density contour plots at $\bar{\Gamma}$ for oxygen chemisorbed on Pt{100}-(1 × 1) at 0.5 ML, cut through the oxygen atom center parallel to the surface. The black dots show the projected center of platinum atoms in the surface plane [after Ge *et al.* (*11*)].

advances in theory (and in computer technology) described previously, this reaction has now also been studied on the basis of first principles (*73, 128*).

The heat released during CO oxidation to give gaseous CO_2 on Pt{111} (*18, 67*) and on Pt{110} (*18, 129*) has been measured with SCAC. The heats of the reactions

$$CO(g) + O_{(Pt)} \rightarrow CO_2(g); \tag{30}$$

and

$$\tfrac{1}{2} O_2(g) + CO_{(Pt)} \rightarrow CO_2(g); \tag{31}$$

can be calculated, assuming that the product CO_2 is fully thermalized with the crystal temperature, by using the known heats of formation of gaseous CO and CO_2 and, from SCAC data, of O_{Pt} and CO_{Pt}. Thus, any difference between the measured heat change and that calculated for thermalized CO_2 is attributable to excess internal and translational energy carried away by the product CO_2. In this way, for Pt{110} Wartnaby *et al.* (*129*) reported that when CO is dosed onto a saturated O adatom layer [Reaction (30)], the product CO_2 has no measurable excess internal energy; it is close to being equilibrated. However, when O_2 is beamed at predosed CO [Reaction (31)], the excess CO_2 internal energy is 49 ± 21 kJ/mol. It was suggested that in the latter case the adsorbed CO reacts with "hot" O adatoms produced in the exothermic process of O_2 dissociative adsorption (*130, 131*). However, it is also possible that the reaction occurs via molecularly chemisorbed O_2, with the reaction proceeding via an $[O_2CO]^*$-activated complex which is unstable with respect to $CO_2(g)$. Impulsive desorption from the complex would produce hot CO_2.

Only Reaction (30) on Pt{111} could be investigated because the O_2 sticking probability on CO-precovered Pt{111} could not be measured. The SCAC measurements showed a relatively small excess CO_2 energy of 12 ± 8 kJ/mol (*18, 67*).

These calorimetric results are quite consistent with earlier data, both from chemiluminescence measurements of vibrational and rotational excitation in product CO_2 at steady state (*132*) and from temperature-programmed reaction data for CO with preadsorbed O_2 (*133*). In the latter case, the production of highly excited molecules was attributed to hot O adatom production, but the route may again be via an $[O_2CO]^*$ complex.

On the basis of DFT slab calculations, a low-energy pathway for CO oxidation by O adatoms on Pt{111} was identified by Alavi *et al.* (*128*) as shown in Fig. 10. The transition state along this pathway was found with the O adatom displaced from its most stable hollow site into a bridge site and CO displaced from the atop site. This pathway was later shown by Eichler and Hafner (*73*) to be lower in energy than other pathways. These authors also calculated the overall energetics of CO oxidation on the surface. The surface reaction barrier was found to be 101 kJ/mol by Alavi *et al.* and 72 kJ/mol by Eichler and Hafner. The heat of formation for gaseous CO_2 from gaseous CO and adsorbed O adatoms is 234 kJ/mol (*73*), higher than the experimental value of 185 kJ/mol. We note, however, that the calculated formation energy for CO_2 in the gas phase shows overbinding by 45 kJ/mol (*73*). The results of these DFT calculations are consistent with the SCAC observation of near-thermalized CO_2 for Reaction (30).

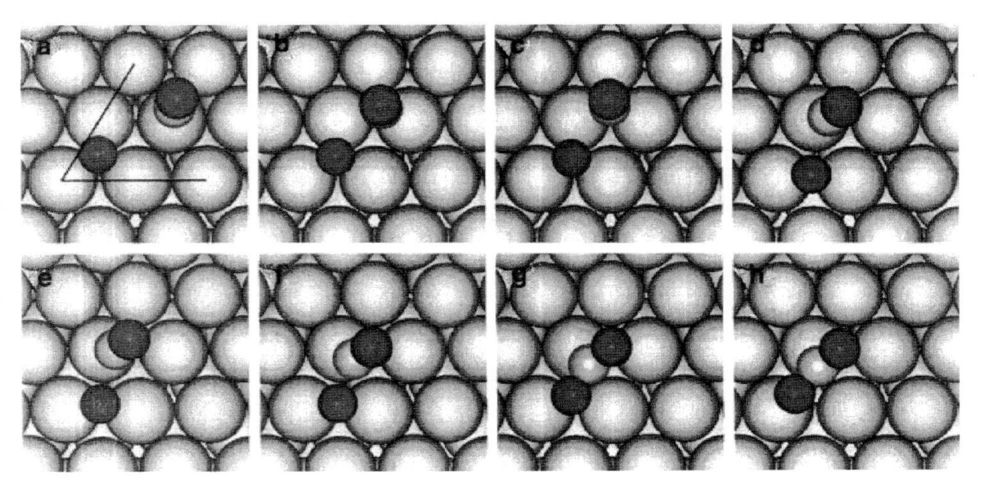

FIG. 10. Snapshots of the reaction pathway between CO and O on Pt{111} from the initial state (a) to the final state (h). The lattice vectors are indicated in a [after Alavi *et al.* (*128*)].

B. HYDROCARBONS

Hydrocarbon reactions are important in many industrial catalytic processes, and an understanding of the energetics of the adsorption pathways is essential to unveil the mechanism of the reaction steps involved. The surface reactions accompanying hydrocarbon adsorption often result in complex product distributions. In a calorimetric measurement it is the overall heat change in these processes that is measured, from gaseous molecule to the final species formed by reaction. The irreversible nature of the processes determines that no adsorption heat can be measured directly. On the basis of the possible products formed on the surface, carbon–metal bond energies can be calculated from the measured heats of reaction (*18, 134–136*). On the other hand, once the nature of the intermediates is known, theoretical results can help to establish the reaction pathway on the basis of calculated energetics (*137*).

1. *Ethene Adsorption*

Table IV is a compilation of theoretically and experimentally acquired integral heats of adsorption on Pt{111}, and Table V is a summary of values of average metal–carbon bond energies for individual surfaces. The adsorption and reactions of ethene on metal surfaces under well-defined conditions are often used as a prototype for studies of alkene hydrogenation and other reactions. Today, ethene on Pt{111} is an experimentally well-

TABLE IV

Integral Heat of Dissociative Adsorption from Theory and Experiment for the Adsorption of Ethene on Pt{111} (in kJ/mol)[a]

State	Adsite	Coverage (ML)	q_t (reference)	q_{exp} (reference)	q_0	m (%)
Molecular	Bridge	0.25	122 (*140*)	116 (*136*)		4.8
	Bridge	0.0	151 (*148*)			
Dissociative		0.25	154 (*140*)	163 (*18, 136*)	202	−5.6
		0.0	229 (*148*)	202 (*18, 136*)		11.8

[a] The column labeled "m" refers to the percentage match of measured and calculated integral adsorption energies as in Table I.

characterized system. At temperatures below 90 K an atop-bonded species is formed, which is converted to bridge-bonded ethene at higher temperatures. Warming to room temperature results in the desorption of a fraction of the adlayer and a conversion of the remaining surface species into ethylidyne ($\equiv CCH_3$) and adsorbed H adatoms. This adlayer forms the only known ordered structure [with a (2×2) LEED pattern] corresponding to 0.25 ML of ethylidyne. The surface crystallography of (2×2) ethylidyne has been derived from a tensor LEED analysis (*138*). Ethylidyne adsorbs in fcc threefold hollow sites, with the C–C bond axis perpendicular to the surface.

a. *Nondissociative Adsorption.* A diffuse LEED study of the nondissociative disordered "di-σ" adlayer formed at low temperatures (*139*) showed that the bonding takes place in fcc threefold hollow sites with the

TABLE V

Average Metal–Carbon Bond Energies for the Individual Surfaces Studied with SCAC (in kJ/mol)[a]

Surface	Energy	Average	Theory (reference)
Pt{111}	244		230 (*140*), 218 (*148*)
Pt{110}	242		
Pt{100}	240	242	
Pd{100}	171	171	
Ni{110}	205		
Ni{100}	191	198	
Rh{100}	273	273	

[a] Based on Brown *et al.* (*18*) and Ge and King (*171*). Available theoretical values are also listed.

C–C bond axis tilted up slightly from the plane of the surface. However, this study was based on a small database, and the large error bars do not give a high level of confidence in the results. In contrast, Ge and King (*140*) demonstrated that for molecular ethene adsorption, for which four different configurations were examined in detail, the twofold bridge site, with the C–C axis lying along the bridge, was found to be energetically most favorable, with an adsorption energy of 122 kJ/mol. The calculated result is in close accord with results of a near-edge X-ray absorption fine structure study (*141*) which showed that the C–C bond axis lies in the plane of the surface, with the C–C bond length stretched to 1.49 Å (gas-phase ethene has a C=C bond length of 1.34 Å). The alternative site favored in the DLEED analysis showed an adsorption energy of only 69 kJ/mol (*140*). In this configuration, the C–C bond length is 1.46 Å, and its axis is angled away from the surface. The calculation for ethene adsorbed in the favored bridge site yields a C–C bond length of 1.48 Å, with the bond axis lying in the plane of the surface. Interestingly, the bonding is shown to be dominated by π and π^* mixing with metal d states, as in CO adsorption. The C_2H_4 π bond is still intact in the chemisorbed molecule. Charge density distributions in these mixed π–d eigenstates are shown in Fig. 11, in which the comparison with the π orbitals of a raft of ethene molecules in the distorted configuration of the chemisorbed species but deviod of the platinum slab clearly demonstrates the π character of the chemisorption mixed orbitals.

The atop site, with an adsorption energy of 53 kJ/mol, is the least stable site. The two hollow-site structures (fcc and hcp) represent local potential minima. Although the C–C bond length for the two hollow sites is significantly stretched, the bond energies are much closer to those of the atop site than to those of the bridge site. This comparison indicates that the two hollow sites do not provide optimal bonding conditions for ethene adsorption.

b. *Dissociative Adsorption.* Using SCAC, Yeo *et al.* (*18, 136*) determined the reactive heat of adsorption for ethene on Pt{111} as a function of surface coverage at 300 K. The initial adsorption heat (in the zero-coverage limit) was found to be 173 kJ/mol and corresponds to the formation of ethylidyne and coadsorbed hydrogen. As the coverage increases to 0.2 ML, the heat decreases to 124 kJ/mol, with the formation of the (2×2) ethylidyne structure, which was understood to be accompanied by the endothermic process of H_2 desorption. In a subsequent molecular beam analysis of this system, Haq and King (*142*) confirmed that initial exposure of Pt{111} to ethene yields no H_2 evolution, but as the surface coverage moves toward 0.2 ML, ethene adsorption is accompanied by H_2 evolution.

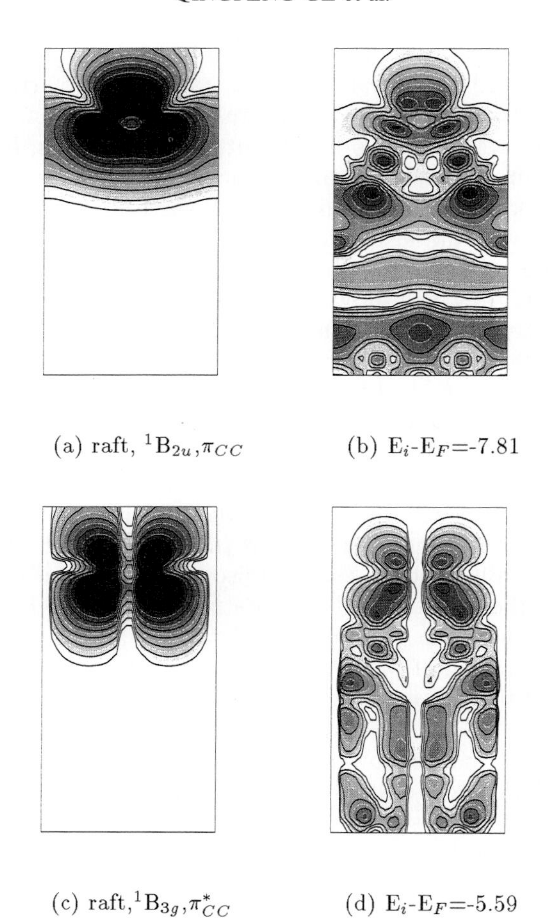

FIG. 11. Logarithmic partial charge density contour plots at $\bar{\Gamma}$ for C_2H_4 on Pt{111} on the bridge site, with a contour spacing of 0.5, compared with similar plots for a raft of C_2H_4 with the structure distorted, as in the chemisorbed state. The similarity of charge distributions in the vicinity of the C_2H_4 molecule between a and b and between c and d provides a ready identification of the orbital character of the bonding states in the adsorbed state [after Ge and King (*140*)].

At a coverage of 0.25 ML a surprising increase in the adsorption heat was observed, to 184 kJ/mol, which was attributed to the formation of ethylidene ($=CHCH_3$) in this coverage range. Cremer *et al.* (*143*) obtained evidence from a sum frequency generation (SFG) study for the presence of ethylidene on Pt{111}.

On the basis of the total energies obtained in the DFT slab calculations of Ge and King (*140*), an energy diagram was constructed (Fig. 12). The experimentally measured heat for the dissociative adsorption of ethene to form coadsorbed ethylidyne and a hydrogen adatom at low coverage is 164 kJ/mol (*136*) compared with the calculated value of 154 kJ/mol, and the experimental estimate for the production of ethylidyne and gaseous H_2 is 123 kJ/mol compared with 122 kJ/mol from theory. The calculated energy difference between ethylidyne-surface and quartet ethylidyne plus bare slab of 692 kJ/mol yields a C–Pt single bond energy of 231 kJ/mol, which is in excellent agreement with the calorimetrically measured value of 255 kJ/mol (*18, 136*). When this C–Pt bond energy was used with the energy difference between C–C and C=C of 334 kJ/mol, the bond energy of the di-σ-bonded ethene on the surface was estimated to be 116 kJ/mol, in good agreement with the directly calculated bond energy of the bridge-bonded ethene. However, note that the calculated bond energy is in direct contrast to the "binding energy" (203 kJ/mol) for the bridge-bonded C_2H_4 on Pt{111}, which was estimated on the basis of the threshold energy from collision-induced desorption data (*144*). This value is considerably larger than the experimental desorption energy [71 kJ/mol (*145*)] or our calculated value of 116 kJ/mol. The authors argued that the activation energy determined from thermal desorption spectroscopy is less than the adsorption energy because the adsorption process is precursor mediated and due to the nonequilibrium nature of the thermal desorption experiment. However, π-bonded (or atop) ethene, which is considered to be formed as a precursor

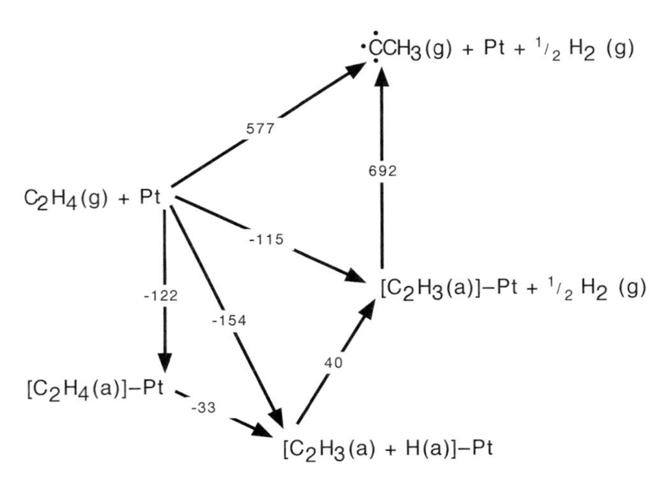

FIG. 12. Schematic of the energetics for ethene adsorption and dissociation on Pt{111} (energy units, kJ/mol) [after Ge and King (*140*)].

state in the adsorption process, starts to convert to di-σ (or bridge)-bonded species at temperatures as low as 52 K (*146*). Thus, the barrier between the precursor state and the chemisorption state is very low and the desorption energy should approximate the bond energy. If the C–Pt σ bond energy is only 125 kJ/mol, as the authors quoted, the di-σ-bonded species would not be formed on Pt{111}. In this case, the energy gain in forming the two di-σ bonds with the surface cannot compensate the energy cost (~347 kJ/mol) to break one of the C–C bonds in ethene, and the adsorption process would be endothermic. Therefore, we believe that even at threshold the collision-induced desorption process produces vibrationally excited ethene in the gas phase so that the measured threshold energy (*144*) is significantly higher than the heat of adsorption.

Ethene adsorption on Pt{111} was also investigated by Kua and Goddard (*147, 148*), who used an interstitial electron surface model. The surface was modeled with a cluster of 10 platinum atoms. The effective electronic configuration was chosen to be similar to that of an infinite surface. The calculation produced binding energies of 150 and 36 kJ/mol for the bridge-bonded and atop-bonded ethene, respectively. The results for the ethene adsorption energy and the corresponding C–Pt σ bond energy are listed in Tables IV and V, respectively. Use of a cluster model effectively implies the zero-coverage adsorption limit. As shown in Table IV, the calculated adsorption energy for ethene dissociative adsorption and consequent formation of ethylidyne and coadsorbed hydrogen agrees well with the measured initial heat of adsorption. The exclusive use of a nonlocal density functional without a density gradient also probably contributes to the observed overbinding. In addition, as mentioned earlier, cluster calculations cannot allow for substrate relaxation.

Various surface species have been postulated to be formed as intermediates in the dissociation and transformation of adsorbed ethene to ethylidyne, but none of these intermediates has been clearly identified experimentally, and the exact reaction mechanism for ethene to ethylidyne conversion on Pt{111} is debatable (*149*). Different reaction mechanisms for the conversion proposed in the past have been critically reviewed by Carter and Koel (*150*). They calculated the heats of adsorption for a large variety of adsorbed hydrocarbon species on a Pt{111} surface. From their related gas-phase values, heats of formation for possible products, and energies for many surface reactions, they estimated the relative stabilities of many surface intermediates. On the basis of experimental and theoretical considerations, they ruled out many previously proposed mechanisms and discussed the two most likely reaction mechanisms in more detail: the isomerization of ethene to ethylidene ($HCCH_3$), followed by dehydrogenation of ethene to ethylidyne and hydrogen adatoms, and the dehydrogenation of ethene to vinyl ($HCCH_2$) and hydrogen adatoms, followed by isomerization of vinyl to ethylidyne. These mechanisms are the simplest models, and they encompass two essential features in ethylidyne formation: the breaking of a C–H bond and the transfer of a second hydrogen from one carbon atom to the other. On the basis of their calculations, the authors gave preference to the isomerization of ethene to ethylidene as the first elementary step of the overall reaction on Pt{111}. Consequently, the second step is the dehydrogenation of ethylidene to ethylidyne. A recent SFG study of ethene to ethylidyne conversion provides support for this mechanism (*143*). A peak at 2957 cm^{-1} was assigned to the CH_3 antisymmetric stretch of ethylidene and/or ethyl. The existence of the ethylidene intermediate has been reported on potassium-modified Pt{111} (*151*).

Both mechanisms have been questioned by Zaera (*149*). He argued that the transformation of both vinyl and ethylidene groups into surface ethene prior to ethylidyne formation made the reverse conversion of ethene to ethylidyne via either vinyl or ethylidene quite unlikely. However, if the second step (i.e., the conversion of either vinyl or ethylidene to ethylidyne) is much faster than the formation of these intermediates, both mechanisms seem viable. Kua and Goddard (*148*) summarized the conversion of ethene to ethylidyne and classified four different pathways (Fig. 13) on the basis of previous findings (*145, 152–154*). Their calculated energetics ruled out the Kang–Anderson pathway (*154*) since the conversion of $CHCH_2$ to CCH_2 is almost 84 kJ/mol uphill. It seems more likely that $CHCH_2$ isomerizes to form ethylidyne, in accordance with the pathway proposed by Zaera

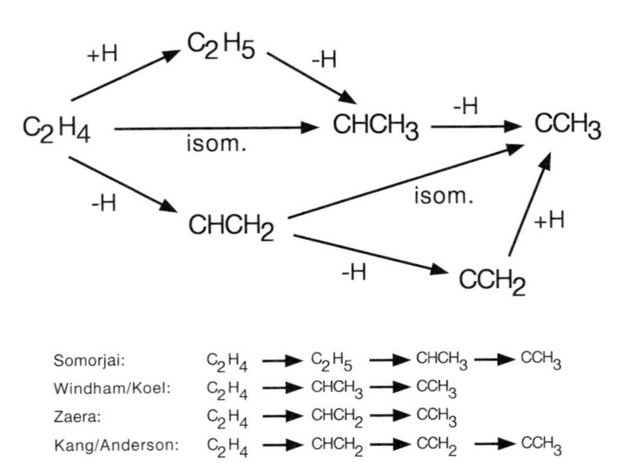

FIG. 13. Four pathways for ethene conversion to ethylidyne [adapted from Kua and Goddard (*148*)].

(*153*). However, the calculated energetics only give an indication of the thermodynamic viability of a reaction; kinetics may play an important role in determining the actual reaction mechanism.

2. *Methane Adsorption*

As a primary source for both energy and synthetic materials, methane plays a crucial role in industrial catalysis. Examples include the Fischer–Tropsch process (*155–159*), methane partial oxidation (*160, 161*), and water–gas shift reactions (*162, 163*). Conversion of methane to enable easier handling has been an important subject in carbon chemistry research. Although the heats of adsorption and reactions of methane have not been measured with SCAC, there are a few systems that have been studied theoretically and provide energetics for the reaction processes involved.

Kua and Goddard (*148*) investigated the chemistry of CH_x on a Pt_8 cluster and found that the preferred binding site is the one that allows carbon to form four σ bonds. Therefore, the most stable structures are CH_3 in the atop site, CH_2 in the bridge site, and CH and C in the hollow site. The calculated C–Pt σ-bond energy is approximately 225 kJ/mol (averages of the bond energies for CH_3, CH_2, and CH). This value is in good accord with the value calculated for ethene adsorption (*148*). Michaelides and Hu (*164*) calculated the energetics along the CH_4 dehydrogenation path. Using constraint search, they were able to locate the transition state and could thus calculate the energy barrier for hydrogenation/dehydrogenation processes. On the basis of their results, we constructed the energy diagram for CH_4 conversion shown in Fig. 14. The two paths, lower and upper, correspond to hydrogen from the dissociation remaining on and desorbing from the surface, respectively. Energetically, methane dissociation needs to overcome a barrier of 64 kJ/mol. CH is the most stable species on the surface. Recent molecular beam experiments characterizing methane dehydrogenation on Pt{110} by Watson and King (*165*) provide the first experimental confirmation that the most stable species formed at temperature between 300 and 400 K is CH, in agreement with the energy profile shown. At higher temperatures this species is converted to adsorbed C as the CH dissociates and recombinative desorption occurs to give H_2. This process is driven by the entropy contribution of gaseous H_2, as in a straightforward (endothermic) desorption process.

C. Coadsorption and Promoter Action

Alkali metals and their oxides have long been known to act as promoters for many catalytic reactions. Recent SCAC, quantitative LEED, and DFT

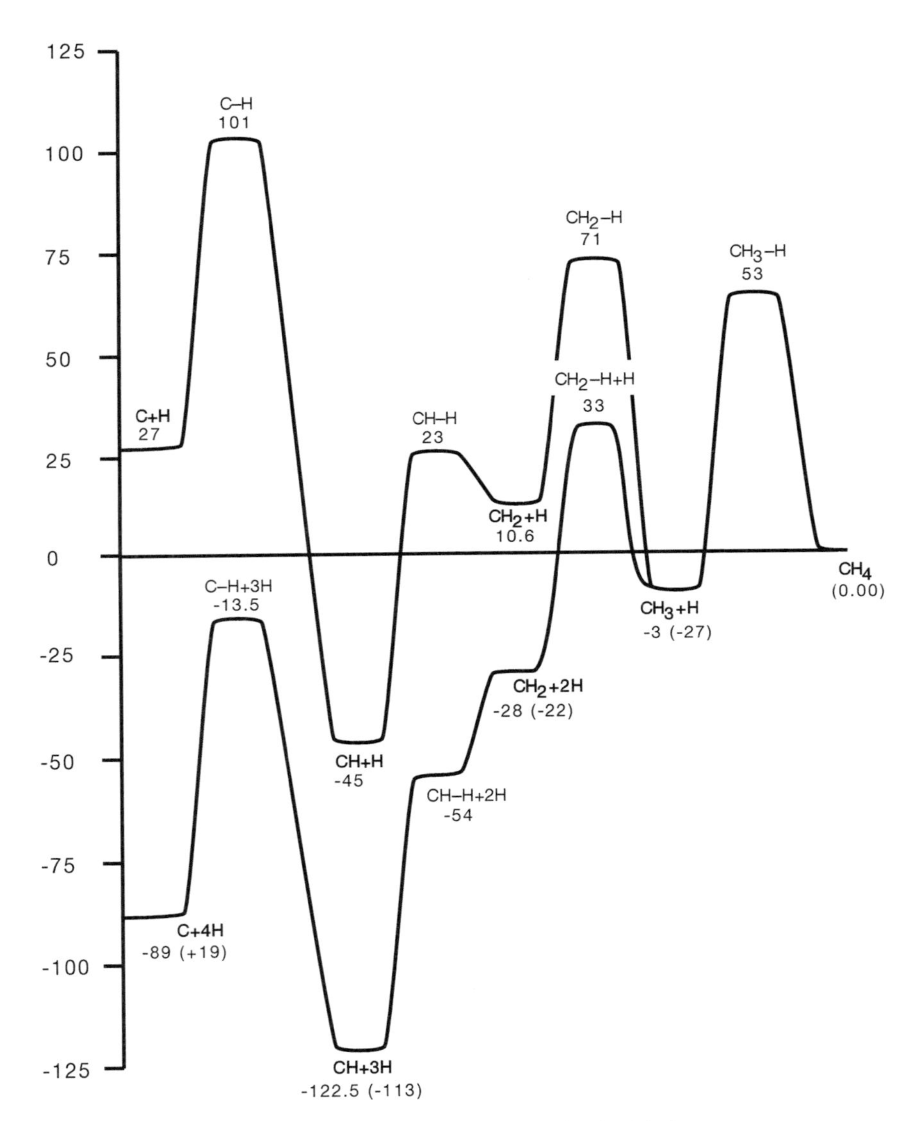

FIG. 14. Energetics of methane dehydrogenation on Pt{111}, with the upper route accompanied by hydrogen desorption. Data are based on Michaelides and Hu (*164*) (energy units, kJ/mol). The numbers in parentheses are from Kua and Goddard (*148*).

slab calculations throw new light on this process through studies of CO coadsorption with potassium on transition metal surfaces.

Calorimetric adsorption heats for CO adsorption on Ni{100} were measured by Al-Sarraf *et al.* (*166*) for a range of potassium precoverages. The influence of preadsorbed potassium was found to be unexpectedly large. The initial CO adsorption heat increases from 124 kJ/mol on the clean surface to 210 kJ/mol with 0.03 ML potassium preadsorbed and up to 310 kJ/mol with 0.2 ML potassium preadsorbed. At low potassium coverages charge transfer to the metal surface leaves an ionic species on the surface (*3*). Following the model of Nørskov *et al.* (*167*), this was attributed to a rigid downward shift in the CO molecular orbital binding energies in the presence of the electric field generated by the ion and its image charge, leading to an increased population in the CO $2\pi^*$ orbital [or, more correctly (as shown in Fig. 4), the 2π–d mixed state] and an increase in the heat of adsorption. At higher potassium precoverages (e.g., 0.2 ML), mutual depolarization in the potassium adlayer strongly decreases the charge transfer, generating a predominantly metallic bond. When CO is adsorbed, the same $K^+CO^-K^+$ final state is generated as at lower coverages, and the large increase in q, from 210 kJ/mol at $\Theta_K = 0.03$ to 310 kJ/mol at $\Theta_K = 0.2$, was attributed to the return of the promoting potassium atom to its ionic state. We will show that recent DFT slab calculations alter this interpretation significantly.

The discovery of an ordered phase of CO coadsorbed with K, Co{10$\bar{1}$0}-C(2 × 2)-(K + CO), at half monolayer coverages of both K and CO led Kaukasoina *et al.* (*168*) to make a quantitative determination of the surface structure using LEED intensity analysis. The resulting structure (Fig. 15)

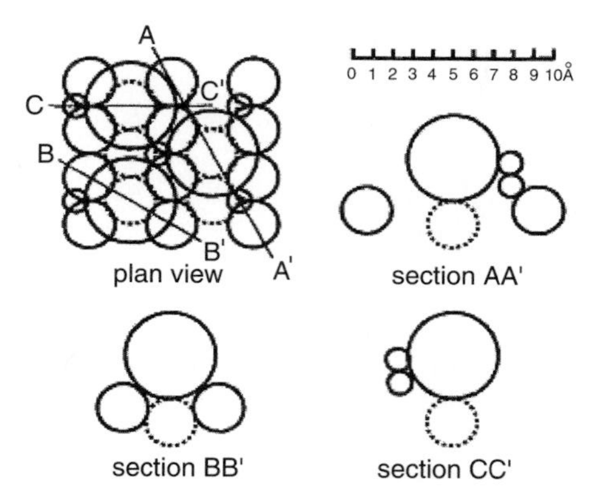

FIG. 15. Top and side views of the favored geometric structure for CO coadsorbed with K on Co{10$\bar{1}$0} [after Kaukasoina, *et al.* (*168*)].

was unexpected. Although the plan view of the structure is not surprising, with each K atom in the fourfold coordinate hollow site and with three surrounding CO molecules somewhat pushed off the favored bridge site, the cross section demonstrates that the K atom is displaced upwards by 0.4 Å compared with the c(2 × 2)-K structure (*169*) in the absence of CO. This displacement leaves the center of the K atom 0.23 Å above the oxygen atoms in the CO molecule. Each K atom sits on the tripod of CO molecules. This unusual structure, with no analogy in three dimensions, indicated a missing link in the understanding of the valence charge reorganization for these systems (*168*).

DFT slab calculations by Jenkins and King (*170*) of the Co{10$\bar{1}$0}-c(2 × 2)-(K + CO) structure have recently been completed. The experimental structure (*168*) is reproduced with extraordinary precision, particularly the upward shift of the K atom. However, more interesting is the analysis of the net charge and dipole moment of CO in the coadsorbed layer. In the absence of potassium, the net charge on the adsorbed CO was estimated to be 0.24e, with a molecular dipole moment of 0.37 D. However, the addition of potassium enhances the polarization of the CO molecule to a remarkable 0.90 D, with the net charge on the CO remaining effectively unchanged at 0.71 e. This result is clearly not in accord with the previously accepted model based on effective medium theory (*167*). The major influence of potassium coadsorption on CO is therefore the enormous polarization of the CO bond, a strong negative charge being generated on the O atom. The significance of this result is a new paradigm for promoter action—molecular polarization. Polar bonds are highly reactive and can be subject to both electrophilic and nucleophilic attack. Of secondary importance, this result provides the rationale for the observed K + CO structure.

V. Summary and Conclusions

The advent of a precise and general method for measuring calorimetric heats at single-crystal surfaces has provided a broad range of new information on the energetics of both reversible and irreversible chemisorption and catalytic processes at single-crystal surfaces. In addition to this new range of information, the single-crystal adsorption calorimeter has provided an important benchmark against which state-of-the-art first-principles calculations of the structure, energetics, and reaction pathways at surfaces can be judged. In particular, developments in density functional theory slab calculation procedures for surface studies during the past 6 years have led to a remarkable degree of reliability in structural analyses, even for quite complex systems. The current review comprehensively and critically demonstrates that agreement between experiment and theory in the total adsorption energy varies considerably from one study to another. Part of this variation is due to factors under the control of the theoretician. For example, the gradient density correction procedures are approximate and are always subject to improvement. There is also a choice of pseudopotentials, and inclusion of spin has been shown to be important for magnetic materials, such as nickel and cobalt. Finally, full convergence in slab and vacuum layer thickness and in *k*-point sampling has not always been checked. Relaxation of the adsorbate and the top few layers of the metal slab can also be critical.

Progress has been quite remarkable, and we now have the capability of providing reliable structures and detailed descriptions of chemical bonding at surfaces. This combined approach of quantitative experiment and first-principles theory of surface processes can be used with considerable reliance in the design of catalysts and even in chemical reactor technology. This will be increasingly important in developing new catalytic technologies during the next few decades.

Acknowledgments

We acknowledge financial support from EPSRC (Q. G.) and the Oppenheimer Trust (R. K.) and the contributions of many in our group at Cambridge (mentioned in the Reference section) whose work made this review possible.

References

1. King, D. A., and Woodruff, D. P. (Eds.), "The Chemical Physics of Solid Surfaces and Heterogeneous Catalysis. Vol. 7: Phase Transitions and Adsorbate Restructuring at Metal Surfaces." Elsevier, Amsterdam, 1994.
2. Somorjai, G. A. "Introduction to Surface Chemistry and Catalysis." Wiley, New York, 1994.
3. Zangwill, A., "Physics at Surfaces." Cambridge Univ. Press, Cambridge, United Kingdom, 1992.
4. Masel, R. I., "Principles of Adsorption and Reaction on Solid Surfaces." Wiley, New York, 1996.
5. Titmuss, S., Wander, A., and King, D. A., *Chem. Rev.* **96,** 1291 (1996).
6. Duke, C. B. (Ed.), "Surface Science—The First Thirty Years." Elsevier, Amsterdam 1994; *Surf. Sci.* **299/300.**
7. Besenbacher, F., *Rep. Prog. Phys.* **59,** 1737 (1996).
8. Rettner, C. T., and Ashfold, M. N. R. (Eds.), "Dynamics of Gas–Surface Interactions." Royal Society of Chemistry, Cambridge, United Kingdom, 1991.
9. Rettner, C. T., Auerbach, D. J., Tully, J. C., and Kleyn, A. W., *J. Phys. Chem.* **100,** 13021 (1996).
10. Gruyters, M., and King, D. A., *J. Chem. Soc. Faraday Trans.* **93,** 2947 (1997).
11. Ge, Q., Hu, P., King, D. A., Lee, M.-H., White, J. A., and Payne, M. C., *J. Chem. Phys.* **106,** 1210 (1997).

12. Gross, A., and Scheffler, M., *Prog. Surf. Sci.* **53,** 187 (1996).
13. Tracy, J. C., and Palmberg, P. W., *Surf. Sci.* **14,** 274 (1965).
14. King, D. A., *Surf. Sci.* **47,** 384 (1975).
15. Borroni-Bird, C. E., and King, D. A., *Rev. Sci. Instrum.* **62,** 2177 (1991).
16. Borroni-Bird, C. E., Al-Sarraf, N., Andersson, S., and King, D. A., *Chem. Phys. Lett.* **183,** 516 (1991).
17. Stuck, A., Wartnaby, C. E., Yeo, Y. Y., Stuckless, J. T., Al-Sarraf, N., and King, D. A., *Surf. Sci.* **349,** 229 (1996).
18. Brown, W. A., Kose, R., and King, D. A., *Chem. Rev.* **98,** 797 (1998).
19. Černý, S., *Surf. Sci. Rep.* **26,** 1 (1996).
20. Kyser, D. A., and Masel, R. I., *J. Vac. Sci. Technol. A* **4,** 1431 (1986).
21. Kyser, D. A., and Masel, R. I., *Rev. Sci. Instrum.* **58,** 2141 (1987).
22. Stuckless, J. T., Starr, D. E., Bald, D. J., and Campbell, C. T., *J. Chem. Phys.* **107,** 5547 (1997).
23. Stuckless, J. T., Starr, D. E., Bald, D. J., and Campbell, C. T., *Phys. Rev. B* **56,** 13496 (1997).
24. Stuckless, J. T., Frei, N. A., and Campbell, C. T., *Rev. Sci. Instrum.* **69,** 2427 (1998).
25. Campbell, C. T., *Curr. Opin. Solid State Mater. Sci.* **3,** 439 (1998).
26. King, D. A., and Wells, M. G., *Surf. Sci.* **29,** 454 (1972).
27. Lide, D. R. (Ed.), "CRC Handbook of Chemistry and Physics," 75th ed. CRC Press, Boca Raton, FL, 1997.
28. Stuckless, J. T., Al-Sarraf, N., Wartnaby, C. E., and King, D. A., *J. Chem. Phys.* **99,** 2202 (1993).
29. Pacchioni, G., Bagus, P. S., and Parmigiani, F. (Eds.), "Cluster Models for Surface and Bulk Phenomena," Vol. 283, NATO ASI Series B. Plenum, New York, 1991.
30. van Santen, R. A., "Theoretical Heterogeneous Catalysis." World Scientific, Singapore, 1991.
31. van Santen, R. A., and Neurock, M., *Catal. Rev. Sci. Eng.* **37,** 557 (1995).
32. Whitten, J., and Yang, H., *Surf. Sci. Rep.* **24,** 55 (1996).
33. Hammer, B., and Nørskov, J. K., "Chemisorption and Reactivity on Supported Clusters and Thin Films: Towards on Understanding of Microscopic Processes in Catalysis," Vol. 331, NATO ASI Series E, p. 285. Kluwer Academic, Dordrecht, 1997.
34. Born, M., and Oppenheimer, J. R., *Ann. Phys.* **79,** 361 (1927).
35. Ceperley, D. M., "Monte Carlo and Molecular Dynamics of Condensed Matter Systems," p. 443. Italian Physical Society, Bologna, 1996.
36. Hammond, B. L., Jr., Lester, W. A., and Reynolds, P. J., "Monte Carlo Methods in ab Initio Quantum Chemistry." World Scientific, Singapore, 1994.
37. Hartree, D. R., *Proc. Cambridge Phil. Soc.* **24,** 89 (1928).
38. Fock, V., *Z. Phys.* **61,** 126 (1930).
39. Fock, V., *Z. Phys.* **62,** 795 (1930).
40. Szabo, A., and Ostlund, N. S., "Modern Quantum Chemistry: Introduction to Advanced Electronic Structure Theory." McMillan, New York, 1989.
41. Boys, F., *Proc. R. Soc. London A* **200,** 542 (1950).
42. Hehre, W. J., Lathan, R. A., Ditchfield, R., Newton, M. D., and Pople, J. A., Gaussian 70, Program No. 237. Quantum Chemistry Program Exchange, 1970.
43. Bauschlicher, C. W., Langhoff, S. R., and Taylor, P., *Adv. Chem. Phys.* **77,** 103 (1990).
44. Hohenberg, P., and Kohn, W., *Phys. Rev.* **136,** B864 (1964).
45. Kohn, W., and Sham, L. J., *Phys. Rev.* **140,** A1133 (1965).
46. Becke, A. D., *Phys. Rev. A* **38,** 3098 (1988).
47. Perdew, J. P., Chaevary, J. A., Vosko, S. H., Jackson, K. A., Pederson, M. R., Singh, D. J., and Fiolhais, C., *Phys. Rev. B* **46,** 6671 (1992).

48. Hammer, B., Jacobsen, K. W., and Nørskov, J. K., *Phys. Rev. Lett.* **70,** 3971 (1993).
49. Philipsen, P. H. T., te Velde, G., and Baerends, E. J., *Chem. Phys. Lett.* **226,** 583 (1994).
50. Hu, P., King, D. A., Crampin, S., Lee, M.-H., and Payne, M. C., *Chem. Phys. Lett.* **230,** 501 (1994).
51. Hu, P., King, D. A., Lee, M.-H., and Payne, M. C., *Chem. Phys. Lett.* **246,** 73 (1995).
52. Wimmer, E., Krakauer, H., and Freeman, A. J., "Electronics and Electron Physics," Vol. 65, p. 357. Academic Press, Orlando, FL, 1985.
53. Bachelet, G. B., Hamann, D. R., and Schlüter, M., *Phys. Rev. B* **26,** 4199 (1982).
54. Pickett, W. E., *Comp. Phys. Rep.* **9,** 115 (1989).
55. Lee, M.-H., Advanced pesudopotentials. PhD thesis, University of Cambridge, Cambridge, UK, 1995.
56. Car, R., and Parrinello, M., *Phys. Rev. Lett.* **55,** 2471 (1985).
57. Payne, M. C., Teter, M. P., Allan, D. C., Arias, T. A., and Joannopoulos, J. D., *Rev. Mod. Phys.* **64,** 1045 (1992).
58. Yeo, Y. Y., Wartnaby, C. E., and King, D. A., *Science* **268,** 1731 (1995).
59. Yeo, Y. Y., Vattuone, L., and King, D. A., *J. Chem. Phys.* **106,** 1990 (1997).
60. Hammer, B., Hansen, L. B., and Nørskov, J. K., *Phys. Rev. B* **59,** 7413 (1999).
61. Eichler, A. and Hafner, J., *Phys. Rev. B* **57,** 10110 (1998).
62. Bradshaw, A. M., and Hoffmann, F. M., *Surf. Sci.* **72,** 513 (1978).
63. Hoffmann, F. M., *Surf. Sci. Rep.* **3,** 107 (1983).
64. Biberian, J. P., and Van Hove, M. A., *Surf. Sci.* **118,** 443 (1982).
65. Uvdal, P., Karlsson, P.-A., Nyberg, C., Andersson, S., and Richardson, N. V., *Surf. Sci.* **202,** 167 (1988).
66. Hammer, B., Nielsen, O. H., and Nørskov, J. K., *Catal. Lett.* **46,** 31 (1997).
67. Yeo, Y. Y., Vattuone, L., and King, D. A., *J. Chem. Phys.* **106,** 392 (1997).
68. Lynch, M., and Hu, P., submitted for publication (1999).
69. Gu, J., Sim, W. S., and King, D. A., *J. Chem. Phys.* **107,** 5613 (1997).
70. Schweizer, E., Persson, B. N. J., Tüshaus, M., Hoge, D., and Bradshaw, A. M., *Surf. Sci.* **213,** 49 (1989).
71. Hammer, B., Morikawa, Y., and Nørskov, J. K., *Phys. Rev. Lett.* **76,** 2141 (1996).
72. Ceperley, D. M., and Alder, B. J., *Phys. Rev. Lett.* **45,** 56 (1980).
73. Eichler, A., and Hafner, J., *Phys. Rev. B* **59,** 5960 (1999).
74. Kresse, G., and Hafner, J., *Phys. Rev. B* **47,** R558 (1993).
75. Kresse, G., and Furthmüller, J., *Phys. Rev. B* **54,** 11169 (1996).
76. Kresse, G., and Furthmüller, J., *Comput. Mater. Sci.* **6,** 15 (1996).
77. Eichler, A., and Hafner, J., *Phys. Rev. Lett.* **79,** 4481 (1997).
78. Eichler, A., and Hafner, J., *J. Chem. Phys.* **109,** 5585 (1998).
79. Ge, Q., and King, D. A., *J. Chem. Phys.* **111,** in press (1999).
80. Wartnaby, C. E., Stuck, A., Yeo, Y. Y., and King, D. A., *J. Phys. Chem.* **100,** 12483 (1996).
81. Hayden, B. E., Kretzschmar, K., Bradshaw, A. M., and Greenler, R. G., *Surf. Sci.* **149,** 394 (1985).
82. Kose, R., Karmazyn, A., and King, D. A., manuscript in preparation.
83. Ge, Q., and King, D. A., manuscript in preparation.
84. Louie, S. G., Froyen, S., and Cohen, M. L., *Phys. Rev. B* **26,** 1738 (1982).
85. Zhang, Y., and Yang, W., *Phys. Rev. Lett.* **80,** 890 (1998).
86. Yoshinobu, J., and Kawai, M., *Surf. Sci.* **363,** 105 (1996).
87. Grossmann, A., Erley, W., and Ibach, H. *Surf. Sci.* **313,** 209 (1994).

88. Davis, R., Woodruff, D. P., Hofmann, P., Schaff, O., Fernandez, V., Schindler, K. M., Fritzsche, V., and Bradshaw, A. M., *J. Phys.* **8,** 1367 (1996).
89. Mapledoram, L. D., Bessent, M. P., Wander, A., and King, D. A., *Chem. Phys. Lett.* **228,** 527 (1994).
90. Held, G., Schuler, J., Sklarek, V., and Steinrück, H. P., *Surf. Sci.* **398,** 154 (1998).
91. Kose, R., Brown, W. A., and King, D. A., *J. Phys. Chem.* **103,** 8722 (1999).
92. Sung, S.-S., and Hoffmann, R., *J. Am. Chem. Soc.* **107,** 578 (1985).
93. Glassey, W. V., Papoian, G. A., and Hoffmann, R., *J. Chem. Phys.* **111,** 893 (1999).
94. Guo, X.-C., Bradley, J. M., Hopkinson, A., and King, D. A., *Surf. Sci.* **301,** 163 (1994).
95. Bradley, J. M., Guo, X.-C., Hopkinson, A., and King, D. A., *J. Chem. Phys.* **104,** 4283 (1996).
96. Luntz, A. C., Brown, J. K., and Williams, M. D., *J. Chem. Phys.* **93,** 5240 (1990).
97. Hopkinson, A., Bradley, J. M., Guo, X.-C., and King, D. A., *Phys. Rev. Lett.* **71,** 1597 (1993).
98. King, D. A., *J. Vac. Sci. Technol.* **17,** 241 (1980).
99. Brown, W. A., and King, D. A., *J. Phys. Chem.,* in press.
100. Hass, K. C., Tsai, M.-H., and Kasowski, R. V., *Phys. Rev. B* **53,** 44 (1996).
101. Ge, Q., and King, D. A., *Chem. Phys. Lett.* **285,** 15 (1998).
102. Pérez Jigato, M., King, D. A., and Yoshimori, A., *Chem. Phys. Lett.* **300,** 639 (1999).
103. Brown, W. A., Gardner, P., Jigato, M. P., and King, D. A., *J. Chem. Phys.* **102,** 7277 (1995).
104. Brown, W. A., Sharma, R. K., King, D. A., and Haq, S., *J. Phys. Chem.* **100,** 12559 (1996).
105. Vattuone, L., Yeo, Y. Y., and King, D. A., *Catal. Lett.* **41,** 119 (1996).
106. Vattuone, L., Yeo, Y. Y., and King, D. A., *J. Chem. Phys.* **104,** 8096 (1996).
107. Ge, Q., Brown, W. A., Sharma, R. K., and King, D. A., *J. Chem. Phys.* **110,** 12082 (1999).
108. Mapledoram, L. D., Wander, A., and King, D. A., *Chem. Phys. Lett.* **208,** 409 (1993).
109. Mapledoram, L. D., Wander, A., and King, D. A., *Surf. Sci.* **312,** 54 (1993).
110. Materer, N., Barbieri, A., Gardin, D., Starke, U., Batteas, J. D., Van Hove, M. A., and Somorjai, G. A., *Surf. Sci.* **303,** 319 (1994).
111. Titmuss, S., A new approach to surface structure determination by low energy electron diffraction. PhD thesis, Cambridge University, Cambridge, UK (1999).
112. Brown, W. A., Ge, Q., Sharma, R. K., and King, D. A., *Chem. Phys. Lett.* **299,** 253 (1999).
113. Brown, W. A., Ge, Q., Sharma, R. K., and King, D. A., *Phys. Chem. Chem. Phys.* **1,** 1995 (1999).
114. King, D. A., *Stud. Surf. Sci. Catal.* **109,** 79 (1997).
115. Bradley, J. M., Hopkinson, A., and King, D. A., *J. Chem. Phys.* **99,** 17032 (1995).
116. Siegbahn, P. E. M., and Wahlgren, U., *Int. J. Quantum Chem.* **42,** 1149 (1992).
117. Stuckless, J. T., Wartnaby, C. E., Al-Sarraf, N., Dixon-Warren, S. J. B., Kovar, M., and King, D. A., *J. Chem. Phys.* **106,** 2012 (1997).
118. Bleakley, K., and Hu, P., *J. Am. Chem. Soc.* **121,** 7644 (1999).
119. Kose, R., Brown, W. A., and King, D. A., *Surf. Sci.* **402–404,** 856 (1998).
120. Mercer, J. R., Finetti, P., Leibsle, F. M., McGrath, R., Dhanak, V. R., Baraldi, A., Prince, K. C., and Rosei, R., *Surf. Sci.* **352,** 173 (1996).
121. Baraldi, A., Dhanak, V. R., Comelli, G., Prince, K. C., and Rosei, R., *Phys. Rev. B* **53,** 4073 (1996).
122. Johnson, K., Ge, Q., Titmuss, S., and King, D. A., *J. Chem. Phys.,* in press, 2000.
123. Walter, E. J., and Rappe, A. M., *Surf. Sci.* **427/428,** 11 (1999).
124. Campbell, C. T., Ertl, G., Kuipers, H., and Segner, J., *J. Chem. Phys.* **73,** 5862 (1980).
125. Coulston, G. W., and Haller, G. L., *J. Chem. Phys.* **95,** 6932 (1991).

126. Mullins, C. B., Rettner, C. T., and Auerbach, D. J., *J. Chem. Phys.* **95,** 8649 (1995).
127. Ertl, G., *Surf. Sci.* **299/300,** 742 (1994).
128. Alavi, A., Hu, P., Deutch, T., and Silvestrelli, P. L., and Hutter, J., *Phys. Rev. Lett.* **80,** 3650 (1998).
129. Wartnaby, C. E., Stuck, A., Yeo, Y. Y., and King, D. A., *J. Chem. Phys.* **102,** 1855 (1995).
130. Akerlund, C., Zoric, I., and Kasemo, B., *J. Chem. Phys.* **104,** 7359 (1996).
131. Wintterlin, J., Schuster, R., and Ertl, G., *Phys. Rev. Lett.* **77,** 123 (1996).
132. Mantell, D. A., Kunimori, K., Ryali, S. B., Haller, G. L., and Fenn, J. B., *Surf. Sci.* **172,** 281 (1986).
133. Allers, K.-H., Pfnür, H., Fenlner, P., and Menzel, D., *J. Chem. Phys.* **100,** 3985 (1994).
134. Yeo, Y. Y., Stuck, A., Wartnaby, C. E., Kose, R., and King, D. A., *J. Mol. Catal. A* **131,** 31 (1998).
135. Stuck, A., Wartnaby, C. E., Yeo, Y. Y., and King, D. A., *Phys. Rev. Lett.* **74,** 578 (1995).
136. Yeo, Y. Y., Stuck, A., Wartnaby, C. E., and King, D. A., *Chem. Phys. Lett.* **259,** 28 (1996).
137. Kose, R., Brown, W. A., and King, D. A., *J. Am. Chem. Soc.* **121,** 4845 (1999).
138. Starke, U., Barbieri, A., Materer, N., Van Hove, M. A., and Somorjai, G. A., *Surf. Sci.* **286,** 1 (1993).
139. Doll, R., Gerken, C. A., Van Hove, M. A., and Somorjai, G. A., *Surf. Sci.* **374,** 151 (1997).
140. Ge, Q., and King, D. A., *J. Chem. Phys.* **110,** 4699 (1999).
141. Stohr, J., Sette, F., and Johnson, A. L., *Phys. Rev. Lett.* **53,** 1684 (1984).
142. Haq, S., and King, D. A., unpublished data (1998).
143. Cremer, P., Stanners, C., Niemantsverdriet, J. W., Shen, Y. R., and Somorjai, G., *Surf. Sci.* **328,** 111 (1995).
144. Szulczewski, G., and Levis, R. J., *J. Am. Chem. Soc.* **118,** 3521 (1996).
145. Windham, R. G., Bartram, M. E., and Koel, B. E., *J. Phys. Chem.* **92,** 2862 (1988).
146. Hugenschmidt, M. B., Dolle, P., Jupille, J., and Cassuto, A., *J. Vac. Sci. Technol.* **7,** 3312 (1989).
147. Kua, J., and Goddard, W. A., III, *J. Phys. Chem. B* **102,** 9481 (1998).
148. Kua, J., and Goddard, W. A., III, *J. Phys. Chem. B* **102,** 9492 (1998).
149. Zaera, F., *Langmuir* **12,** 88 (1996).
150. Carter, E. A., and Koel, B. E., *Surf. Sci.* **226,** 339 (1990).
151. Zhou, X. L., Zhu, X. Y., and White, J. M., *Surf. Sci.* **193,** 387 (1988).
152. Somorjai, G. A., van Hove, M. A., and Bent, B. E., *J. Phys. Chem.* **92,** 973 (1988).
153. Zaera, F., *J. Phys. Chem.* **94,** 5090 (1990).
154. Kang, D. B., and Anderson, A. B., *Surf. Sci.* **155,** 639 (1985).
155. Fischer, F., and Tropsch, H., *Brennst. Chem.* **7,** 97 (1926).
156. Fischer, F., and Tropsch, H., *Chem. Ber.* **59,** 830 (1926).
157. Biloen, P., and Sachtler, W. M. H., *Adv. Catal.* **30,** 165 (1981).
158. Herrmann, W. A. *Angew. Chem. Int. Ed.* **21,** 117 (1982).
159. Keim, W. (Ed.), "Catalysis in C_1 Chemistry." Reidel, Dordrecht, 1983.
160. Bergman, R. G., *Science* **223,** 902 (1984).
161. Arndtsen, B. A., Bergman, R. G., Mobley, T. A., and Peterson, T. H., *Acc. Chem. Res.* **28,** 154 (1995).
162. Rostrup-Nielson, J. R., "Catalysis—Science and Technology," Vol. 5, p. 1. Springer, Heidelberg, 1984.

163. Besenbacher, F., Chorkendorff, I., Clausen, B. S., Hammer, B., Molenbroek, A. M., Norskov, J. K., and Stensgaard, I., *Science* **279,** 1913 (1998).
164. Michaelides, A., and Hu, P., to be submitted (1999).
165. Watson, D. T. P., Titmuss, S., and King, D. A., *J. Am. Chem. Soc.,* submitted (2000).
166. Al-Sarraf, N., Stuckless, J. T., and King, D. A., *Nature* **160,** 243 (1992).
167. Nørskov, J. K., Lang, N. D., and Holloway, S., *Surf. Sci.* **137,** 65 (1984).
168. Kaukasoina, P., Lindroos, M., Hu, P., King, D. A., and Barnes, C. J., *Phys. Rev. B* **51,** 17063 (1993).
169. Barnes, C. J., Hu, P., Lindroos, M., and King, D. A., *Surf. Sci.* **251/252,** 561 (1991).
170. Jenkins, S. J., and King, D. A., submitted for publication (1999).
171. Ge, Q., and King, D. A., manuscript in preparation (1999).
172. Loffreda, D., Simon, D., and Sautet, P., *J. Chem. Phys.* **108,** 6447 (1998).
173. Kose, R., Brown, W. A., and King, D. A., *Chem. Phys. Lett.* **311,** 109 (1999).
174. Ogletree, D., Van Hove, M. A., and Somorjai, G. A., *Surf. Sci.* **173,** 351 (1986).

Active Sites on Oxides: From Single Crystals to Catalysts

HICHAM IDRISS

Department of Chemistry
University of Auckland
Auckland, New Zealand

AND

MARK A. BARTEAU

Center for Catalytic Science and Technology
Department of Chemical Engineering
University of Delaware
Newark, DE 19716

Metal oxides are ubiquitous in heterogeneous catalysis, serving as catalysts, as catalyst supports, and as modifiers and promoters, among other roles. The surface science approach to understanding the structure and reactivity of metal oxide surfaces has blossomed during the preceding decade. We consider here important concepts drawn from catalysis by metal oxides and their connections to reaction pathways and principles of oxide surface reactivity revealed by experimental and theoretical studies of well-defined oxide surfaces. The applications of reaction mechanisms and site requirements from surface science studies of carboxylic acid chemistry, in particular, to emerging catalysts for dehydration, coupling, and reduction reactions provide important examples of contributions to heterogeneous catalysis from oxide surface science. We also consider

Abbreviations: AES, Auger electron spectroscopy; AFM, atomic force microscopy; DFT, density functional theory; $E_{(M\text{-}O)}$, metal—oxygen bond energy; E_{ads}, adsorption energy; HOMO, highest occupied molecular orbital; HREELS, high resolution electron energy loss spectroscopy; INDO, intermediate neglect of differential overlap; LEED, low energy electron diffraction; MA, maleic anhydride; MIEC, metastable impact electron spectroscopy; NEXAFS, near edge X-ray absorption fine structure; NMR, nuclear magnetic resonance; Q, metal oxide heat of formation; q, partial charge; q_M, partial charge on metal ion; q_O, partial charge on oxygen ion; RAIRS, reflection absorption infrared spectroscopy; r_e, equilibrium metal—oxygen interatomic distance; RHEED, reflection high energy electron diffraction; SCF, self-consistent field; STM, scanning tunneling microscopy; TPD, temperature programmed desorption; UHV, ultra-high vacuum; UPS, ultra-violet photoelectron spectroscopy; $V_{Mad}(M)$, Madelung potential at metal ion; $V_{Mad}(O)$, Madelung potential at oxygen ion; V_{Mad}, Madelung potential; ΔV_{Mad}, V_{Mad} (M)-V_{Mad}(O); XPS, X-ray photoelectron spectroscopy; α, Madelung constant; α_O, electronic polarizability of oxygen ion; α_{Mz+}, electronic polarizability of metal ion; α_{OX}, oxide polarizability; θ, fractional surface coverage; χ, electronegativity; χ_O, electronegativity of oxygen ion; χ_M, electronegativity of metal ion; Δ_χ, χ_O-χ_M.

0360-0564/00 $35.00

relationships between oxide surface reactivity and the physical and electronic properties of oxides as represented by characteristics such as electronegativity, bond energies, ionicity, Madelung potential, and polarizability. This analysis highlights the need for reaction data on well-defined single-crystal surfaces in order to distinguish structural and electronic effects when comparing patterns of catalytic activity and selectivity among different metal oxides. This review of the connections between metal oxide single-crystal surfaces and high surface area catalysts (or their components) demonstrates the potential of surface science approaches to elucidate the chemical and physical bases for catalysis by metal oxides, in pursuit of the goal of catalyst design.

I. Introduction

Metal oxides are key components of a wide range of solid catalysts. Whether as supports or modifiers of metal catalysts, as precursors of other components (e.g., metal sulfides), or as reactive components that participate directly in the transformation of reactants into products, metal oxides play a variety of roles. Often one tends to dismiss them as inert, perhaps thinking first of their widespread utilization as supports in maintaining the dispersion of small metal crystallites in most supported metal catalysts. An earlier review of surface science studies of single-crystal oxide surfaces likewise suggested that "stoichiometric low-index faces of oxides are often rather unreactive" (*1*).

However, even in supported metal catalysts, the nominally inert oxide support may not be nearly as unreactive as is often taken for granted. Oxides in supported metal catalysts facilitate well-established phenomena such as hydrogen spillover (*2, 3*) and the transport of organic reactants between the metal and the oxide support. Examples include the classical demonstrations of vapor-phase transport of unsaturated hydrocarbons between metal and acidic functions of platinum-on-alumina reforming catalysts (*4*). In this case, the reactivity of the oxide, which provides the acid sites, is essential to the performance of the catalyst. Oxides have been suggested to facilitate the spillover of organic intermediates between adjacent metal and oxide regions of a variety of catalysts (*5–7*). They may also function as catalyst promoters, occupying a portion of the metal surface and in some cases interacting directly with adsorbates at the metal–support interface (*8–11*). Even in cases in which the physical and chemical characteristics of oxide supports have been chosen to minimize the activity of the support, oxide-catalyzed reactions may still affect catalyst performance. For example, alkali metal promoters in supported silver catalysts for selective oxidation of ethylene to ethylene oxide (*12–14*) have been suggested to function as poisons for residual acidic sites on the support. Such sites may otherwise catalyze undesirable isomerization and degradation reactions, lowering the selectivity to ethylene oxide. Given the high metal loading and low surface area of the relatively inert support material in this catalyst, one might expect the interactions of the support with reactants and products to be much less important than the interactions of the support with reactants in more conventional supported metal catalysts. Because these interactions are significant even in such a low surface area catalyst with a nominally inert support, clearly the assumption that oxides are inert components in catalysts must be treated in general with considerable caution.

What sorts of reactions do oxides carry out when they function as catalysts in their own right, without the presence of zero-valent metals? Kung (*15*) suggested that the most important applications of transition metal oxides as catalysts are in processes such as selective oxidation, ammoxidation, and selective dehydrogenation of organic reactants. Well-known examples include selective oxidation of butane with phosphorus-containing vanadium oxide catalysts, ammoxidation of propylene to give acrylonitrile with molybdate-based catalysts, and oxidative dehydrogenation of methanol to formaldehyde with ferric molybdate catalysts (*15–21*). Nonoxidative dehydrogenations of ethylbenzene and of isobutane to give styrene and isobutylene, respectively, are also important processes catalyzed by metal oxides (*21, 22*).

All of the catalytic processes mentioned previously exemplify one of the critical roles common to oxide catalysts: "activation" of the organic reactant. They involve bond-scission processes (C–H and O–H in the previous examples) whether or not oxygen (or nitrogen in the case of ammoxidation) is inserted. This observation suggests that oxides may catalyze other processes involving such scission or rearrangements, whether or not these catalysts make available labile components for insertion. This suggestion is borne out by numerous examples. For instance, oxides frequently give rise to acid-base chemistry at their surfaces. Acid–base characteristics arise from the distribution of charge in these materials: The electronegative oxygen atoms accumulate electron density and are typically described as anions, with the metal constituents described as cations. This statement is not meant to imply that these anion and cation centers bear actual charges corresponding to their formal oxidation states but rather that some charge separation occurs whether the oxide is viewed as covalent [e.g., SiO_2 (*23*)] or ionic [e.g., MgO (*1*)].

The issue of charge distribution and electronic properties is considered in greater detail later. Now, however, it is sufficient to recognize that the anions and cations at oxide surfaces may be described as Lewis base and acid sites. The strengths of these sites vary with the identity and structure of the oxide, and especially with the formation of mixed oxides, which can lead to significant increases in basicity or acidity (*15, 24–26*).

Silica-alumina is a classic example. Surface basicity and acidity can be of either the Brønsted or Lewis type. Because Brønsted acid catalysis is a whole other topic, and because it is less amenable to study by single-crystal surface science, which is one of the emphases of this review, the principal focus on surface acid–base properties will be on those of the Lewis type. It should not be forgotten, however, that many important Brønsted acid catalysts (e.g., zeolites) can be regarded as oxides.

The literature of heterogeneous catalysis by metal oxides provides important concepts that serve to link the physical, chemical, and electronic properties of these materials with their function and performance as catalysts. Perhaps the oldest of these is the application of acid–base reaction mechanisms to describe oxide surface chemistry, as noted previously.

Although this approach has indeed proven fruitful, one must be careful about uncritical extension to surfaces of concepts derived from solution chemistry. For example, how does one quantify the strength of acid and base sites on a surface in the absence of solvent? Measures such as pH are inherently solvent dependent and may lose their meaning in the absence of solvent, i.e., in the gas phase or on solid surfaces. Reaction mechanisms, even for realtively simple organic transformations, may also differ between solutions and surfaces. For example, one of the classical probes of surface acid-base chemistry is the dehydration/dehydrogenation selectivity for conversion of alcohols (*25–30*). The assumption built into the use of this probe, by analogy with acid-catalyzed dehydration reactions in the liquid phase, is that dehydration is catalyzed by surface Brønsted acid sites, whereas dehydrogenation activity may be taken as a measure of surface base sites or acid–base site pairs. However, it has also become clear that oxide surfaces that are completely devoid of Brønsted acid sites (such as oxide single crystals investigated in ultrahigh vacuum surface science experiments) can also dehydrate alcohols (*31, 32*). The mechanism must be different in such cases from that of Brønsted acid catalysis in solution.

In general, the determination of surface properties on the basis of incorrect mechanistic assumptions is problematical. There is a clear need to establish the mechanisms of chemical reactions on oxide surfaces at a fundamental level before the results of probe reactions such as alcohol dehydration are used as diagnostics or design inputs for catalyst performance.

The second important concept, again carried over from another catalytic realm, is the dependence of reactivity on solid structure, either surface or bulk. The concept of "structure sensitivity" has emerged as a powerful tool in catalysis by metals (*33*). The idea that surface reactions may require a metal atom ensemble of specific size or structure has provided a foundation for building structure–property relationships in both single-crystal surface science and catalysis by metals. This concept serves to explain why certain classes of reactions (e.g., ethane hydrogenolysis) exhibit rates that depend strongly on particle size/surface structure and on catalyst composition (*33*). This connection between structure sensitivity and chemical specificity of reactions catalyzed by metals has played an important role in the design of catalysts for industrial processes such as catalytic reforming of hydrocarbons (*34*).

Although much of the focus has been on physical structure–reactivity relationships, this concept also provides a basis for developing electronic structure–property relationships. For example, as one reduces the size of metal crystallites to the nanometer scale, their electronic levels (band structures) begin to deviate from those of the corresponding bulk metal (*35*). One might expect that chemical properties would be altered as well. There is indeed evidence for reactivity differences of small particles that may be attributable to such "quantum size effects" (*36*). It is also well-known that the ionization potential changes with increasing cluster size in a nonlinear manner, indicating that structural effects are most likely the cause (*37*).

However, it is difficult to distinguish clearly the effects of (undefined) surface structural variations from those of electronic structure modification for particles on this size scale. Intriguing evidence for the alteration of surface chemical properties of metals by confining one dimension to quantum length scales has emerged from investigations of thin films and structures produced lithographically (*38*).

What about structure sensitivity in oxide catalysis? As discussed in greater detail later, reactions on oxide surfaces also depend on surface crystallographic structure. One must be careful, however, to disconnect "structure sensitivity" and "specificity" to particular materials in translating these concepts to oxides. As noted previously, reactions as simple as Brønsted acid dissociation may require sites present on some crystal planes but not on others for a given oxide (*39*). Notwithstanding this dependence on surface structure, such reactions may be common to many oxides. In other words, one may find reactions on oxides that are structure sensitive but not specific to a particular oxide. This has led some authors to suggest that one should speak of "crystal-face anisotropy" rather than structure sensitivity in catalysis by oxides (*40*). It is important to recognize that not all the phenomena associated with structure sensitivity in catalysis by metals carry over to oxides.

There is a small but growing literature of novel effects in catalysis by nanometer-scale oxide particles (*41, 42*). Once again, one faces the challenge of distinguishing electronic from structural effects on chemical and catalytic properties in order to demonstrate conclusively that a particular example is in fact a quantum size effect.

However, it is also worth noting that metal oxides are likely to provide fertile ground for exploring quantum size effects in catalysis. First, nanometer-scale particles of oxides are easier to generate by many routes than are such particles of metals. Unlike some finely divided metals, metal oxides are nonpyrophoric and need not be supported to avoid sintering. Because they are semiconductors and insulators, metal oxides allow relatively straightforward monitoring of quantum size effects on electronic properties (e.g., changes in the band gap), and these effects might be expected to have a dramatic influence on redox and photochemical processes that involve promotion of electrons across the gap. The increasing number of studies of catalysis by oxide nanoparticles may lead to a new focus on the electronic properties of oxide catalysts.

Because of the compositional variability and more localized electronic structure of oxides than of metals, one might expect defects to play a more important role in oxide surface chemistry and catalysis than in metal surface chemistry and catalysis. For example, if one removes an atom from a metal surface, to a first approximation one simply alters the coordination number of neighboring atoms. In contrast, if one removes an atom from an oxide surface, one alters the local composition (depending on whether a metal or oxygen center is removed) and local charge as well as the coordination number of its neighbors. Since average coordination numbers of both metal and oxygen atoms are lower in oxides than coordination numbers in metals, creation of coordination vacancies on oxide surfaces may be expected to have a more dramatic effect on the local coordination environment for oxides than for metals.

Moreover, on materials such as oxides, with their localized electronic properties, one must also consider the charge associated with defects. Removal of ions from the surface may leave behind electrons or holes at defect centers, for example. It is not our intent to review the solid-state properties of defects in oxides, but we stress that the range of defects (and their potential chemical properties) is much richer for oxides than for metals. Although there is a long history of ascribing catalytic phenomena to defects on oxides, an important challenge from a fundamental viewpoint is the generation of model systems with controllable populations of well-defined defects that permit one to test directly the chemical properties assigned to these defects. As discussed later, surface science approaches are perhaps best suited to addressing such issues, and they remain challenging.

The importance of oxides as oxidation catalysts has led to concepts specific to these materials and processes, without particular connections to catalysis by metals. Perhaps foremost among these concepts is the Mars–van Krevelen mechanism for oxidation by reducible metal oxides (*16, 43*). In outline, this mechanism involves the oxidation of reactants (e.g., hydrocarbons) by their reaction with oxygen atoms at the surface of the oxide lattice. The catalytic cycle is completed by resupply of oxygen atoms by dissociation of dioxygen from the fluid phase. The sites of hydrocarbon oxidation and oxygen activation need not be the same. Net transport of oxygen across the surface, within the near-surface region, or even to and from the bulk oxide lattice by the full range of solid-state diffusion mechanisms provides communication between these spatially and chemically distinct sites.

A key to selective oxidation is the delivery of a controlled amount of oxygen to specific positions on the target molecule. For reactants such as light hydrocarbons, which constitute feedstocks for the most important commercial selective oxidation processes, the number of potential points of attack is limited, and issues regarding regioselectivity of oxidation can often be neglected. Therefore, the principal issue is the selection/design of catalytic materials that effect limited oxidation of the target. The foremost criterion is the redox behavior of the oxide. In effect, this needs to be tuned to the task at hand. One wants oxidation activity sufficient to oxidize the reactant to the desired product, but overly aggressive oxidation behavior will produce undesirable further oxidation and ultimately combustion of the desired product.

One can recognize the illustration of this concept in the classifications of catalysts developed for commercial hydrocarbon oxidation processes. Catalysts for selective oxidative functionalization of olefins (e.g., of propylene to give acrolein or acrylonitrile) are typically based on oxides of molybdenum (*16, 17*). Catalysts for oxidation of more refractory hydrocarbons (e.g., aromatics and alkanes) typically contain vanadium oxides (*16, 19*). The latter are more aggressive oxidizing agents and tend to be less selective when used for oxidation of reactants that can undergo less highly activated C–H scission processes. Conversely, the molybdate catalysts optimized for olefin oxidations are relatively unreactive for alkane conversion. Again, one must match the redox properties of the oxide catalyst to the reactivity of the feedstock and the target products.

The translation of concepts from catalysis by metals to catalysis by metal oxides was discussed previously. There is also an important example of translation in the opposite direction, from oxides to metals—the concept of site isolation. Simply stated, if one wants to effect limited oxidation of the reactant, one needs to limit the supply of oxygen from the catalyst. For noble metal catalysts, chemisorbed oxygen atoms are the reactive species of interest, and the principal mechanism for controlling their population is by control of the oxygen partial pressure or reactant : oxygen ratio in the gas phase. For metal oxides operating by the Mars–van Krevelen mechanism, the oxide lattice is the oxygen source that reacts directly with the organic reactant, and one can also effect spatial control of this source. The concept of site isolation suggests that by control of surface composition and structure, one can control the number of adjacent sites on the surface which can oxidize an adsorbed organic molecule. Conceptually, one isolates in a less reactive matrix the site, or a small ensemble of sites, required for selective oxidation.

This concept and its critical importance in the development of oxide catalysts for selective oxidation and ammoxidation of propylene have been reviewed comprehensively by Grasselli *et al.* (*17, 44*). It has also found application in the development of supported organometallic catalysts (*45–47*) as well as zeolites and redox-active molecular sieves (*48, 49*). How one accomplishes the spatial separation of active sites on the catalyst represents a challenging problem in materials design and synthesis that is beyond the scope of this review. In the case of oxide catalysts, the most common approaches include manipulation of solid-state structure and composition, control of crystal habit, surface modification by promoter addition, and creation of dispersed oxide sites and phases on a nominally inert support.

These key concepts from the literature of heterogeneous catalysis by metal oxides—acid–base properties, surface structural dependence of reactions, electronic theories of catalysis, Mars–van Krevelen mechanisms, redox properties, and site isolation—provide a framework for considering the impact of surface science on the field. The surface science approach, utilizing structurally and compositionally well-defined surfaces to examine individual reaction steps and intermediates under ultrahigh vacuum conditions in which powerful spectroscopic techniques can be applied, may serve to test, refine, and extend these concepts. Ideally, one would wish to develop a rigorous description of the principles of surface reaction processes and to apply these for the purpose of catalyst design and invention. Although the field of metal oxide surface science has generally lagged behind that of metals, it has progressed dramatically within the preceding decade. We consider here the advances made in the surface science of oxides, areas of oxide reactivity that may require other approaches, and connections to heterogeneous catalysis.

II. Surface Science of Metal Oxides

A. Experimental Considerations

As noted previously, the surface science of metal oxides emerged much later than that of metals. The principal historical barrier was the low electrical conductivity of metal oxides. These materials are typically semiconductors or insulators and can develop electrostatic charge when charged particle spectroscopies (both electron and ion spectroscopies) are applied. Charging and band-bending phenomena can skew or shift the apparent energy distribution of charged particles emitted from the surface, e.g., low-energy electrons that are the backbone of most surface-sensitive spectroscopic techniques. However, uniform materials such as single crystals tend to exhibit uniform charging and band-bending effects. Thus, the problem of applying photoelectron spectroscopies to metal oxide single crystals, even those with large band gaps, becomes one of appropriate energy referencing rather than one of compensating for differential charging across the surface. In addition, many of the oxides chosen for single-crystal surface science studies to date have sufficiently small band gaps (or exhibit sufficiently large conductivity when the bulk composition is slightly below the stoichiometric oxygen : metal ratio) that the application of electron spectroscopies to them is little different from that on metals. Examples include TiO_2 (rutile), ZnO, NiO, and others (*1, 50*).

We previously reviewed the spectroscopic techniques that can be applied to characterize adsorbed intermediates on model oxide surfaces (*50*). Suffice it to say that during the past two decades, the full arsenal of surface science techniques has been brought to bear on oxide surfaces.

X-ray photoelectron spectroscopy (XPS) was one of the first techniques used to distinguish adsorbed intermediates such as alkoxides and carboxylates (*51*), and it remains a valuable tool for adsorbate identification. Ion scattering spectroscopy (ISS) has been used by several groups to demonstrate that the conjugate base ligands deposited by Brønsted acid dissociation are indeed anchored to surface cations. Studies on model MgO (*52*) and NiO (*53*) surfaces have shown that the ion scattering signal from the metal cations is attenuated disproportionately more than that from oxygen anions upon Brønsted acid dissociation. This phenomenon is due to the screening of the surface cations by the bulky organic ligands bound to them.

Vibrational spectroscopies have long been applied to characterize adsorbates on polycrystalline oxide powders. Vibrational spectroscopies applied to single-crystal and thin-film oxides include high-resolution electron energy loss spectroscopy (HREELS) (*51–61*) and reflection absorption infrared spectroscopy (*61–64*). Both techniques have been successfully applied to identify and characterize adsorbed intermediates, such as carboxylates, alkoxides, and acetylides, on model oxide surfaces under ultrahigh vacuum (UHV) conditions. The application of both of these techniques to oxides is more challenging than the application to metals. HREELS experiments must be carried out under conditions that minimize surface charging; furthermore, the loss spectra from semiconductors and insulators are dominated by the phonon modes of the lattice (*65*).

Signals from phonon modes at higher frequency than the fundamental can be removed by relatively straightforward deconvolution techniques (*66–70*), permitting the adsorbate modes to be isolated. Even so, a complication that arises is the apparent relaxation of the surface dipole selection rules. On metals, adsorbate vibrational modes with dynamic dipoles parallel to the surface plane are screened by the image charge in the conductive surface, and thus they are not detected. On semiconductors this screening is incomplete, and "surface dipole forbidden" modes may still be detected, adding uncertainty to attempts to deduce adsorbate orientation (*71, 72*).

Reflection infrared techniques can also be more difficult to apply to oxides than to metals, owing to the lower reflectivity of most oxides. An advantage of using thin oxide films deposited on metals as model systems for such experiments is that the metal surface supporting the oxide may help to reflect infrared radiation. An elegant example of the application of RAIRS to oxides is the work of Hayden *et al.* (*64*), who used this technique to detect formate intermediates on $TiO_2(110)$, including both a symmetric bidentate configuration (perpendicular to the surface plane) and a nonsymmetrically bound monodentate configuration.

Although electron and ion spectroscopies on oxide single crystals turned out to be much more feasible than once thought, experimental difficulties resulting from the differences between the physical and mechanical properties of oxides and metals were perhaps underestimated. Simple operations in vacuum, such as heating, cooling, and temperature measurement, are more difficult for nonmetals (which do not permit spot welding) than for metals. Thus, some early surface science investigations of oxide single crystals reflected artifacts in temperature measurement (*73, 74*). However, experiments characterizing oxides and other non-metals have become much more routine over the preceding decade.

An alternative to bulk oxide single crystals is provided by crystalline oxide films on metal single-crystal surfaces. Experimentation with such samples has grown in parallel with investigations of oxide single crystals. By depositing a thin crystalline oxide film on the surface of a refractory metal, one can obtain most of the benefits of the ease of working with metallic samples, including ease of mounting, heating, temperature measurement, and electrical conductivity (*75*). Ideally, one would like to prepare thin oxide crystal films with the orientation one chooses, and with minimal deviation of the structural, chemical, and electronic properties of the thin film from those of the bulk oxide. The principal control variables are the substrate composition, orientation, and temperature, combined with the method and conditions of oxide film growth. The details of crystalline oxide film growth are outside the scope of this review (and are addressed in a separate chapter in this volume). Suffice it to say that a variety of crystalline oxide films have been demonstrated in the literature (*75–85*), from simple cubic systems such as MgO and NiO to more complex oxides of titanium, iron, aluminum, etc.

Critical issues common to both single-crystal and thin-film studies are the choice, control, and characterization of the crystallographic structure of the exposed surface to be investigated. A disadvantage of the thin-film approach is that one may not be able to induce crystal growth along desired axes to produce a surface of the desired orientation. In contrast, by the use of bulk single crystals, the problems of crystal growth and surface orientation are decoupled—one simply cuts a previously grown crystal to the desired orientation.

That does not mean, however, that the surface will in fact be a simple truncation of the bulk lattice (even allowing for minor relaxation). Oxide surfaces are prone to thermal faceting and rearrangement. One can rationalize this tendency in terms of the lower coordination numbers of the individual metal centers (and oxygen) in oxide lattices than in bulk metals. Creation of surfaces, especially those that are not close packed, may have a much greater relative effect in decreasing the average coordination number of surface atoms in oxides than those in metals. In addition, one must consider the issue of charge balancing for polar and ionic oxides. Surfaces on which the formal charges of exposed anions and cations are not balanced may also be thermodynamically unfavorable or unstable relative to surfaces on which charges are balanced. The net result is that even low Miller index planes of oxides undergo reconstruction and faceting. Although this reconstruction may produce new and interesting reactivity patterns, as discussed for some examples later, it also produces a greater challenge in surface characterization—to define surface structure and to connect it to reactivity.

The advent of scanning tunneling microscopy (STM) methods for the characterization of oxide surfaces in the preceding decade (*86*) has begun to provide a great deal of atomic-level detail about the structure of oxide surfaces, but it has also demonstrated considerable complexity that earlier methods [e.g., low-energy electron diffraction (LEED)] had only incompletely revealed (*86–94*).

The strategy of surface science, after all, is to use surfaces with a limited number of well-defined sites or structures to develop relationships between structure and reactivity—and thereby to minimize the ambiguity that is unavoidable with structurally complex practical catalysts. It is essential that the features of model systems be amenable to definition rather than adding to the complexity of constructing structure–property relationships.

We later consider results of surface science investigations conducted in the preceding two decades and the principles of oxide surface reactivity that emerge from them. For the most part, these principles have been developed without the benefit of surface structural determinations by STM. One can argue that, if these principles are not derived from completely incorrect descriptions of surface structure, they are at the appropriate level of detail for catalyst description and design.

STM has not yet been widely applied to most oxides, and the use of this technique to characterize adsorbates and surface reactions has until now been limited to simple systems. STM, however, is clearly the wave of the future. We therefore consider the knowledge gained from "classical" surface science studies of oxide surfaces and their reactivity as well as recent contributions, refinements, and amplifications based on STM studies. The pace of progress in the field is perhaps best illustrated by the adjectives applied to it. From a field declared to be in its "infancy" in 1985 (*95*), we now speak of "classical" surface science studies of oxides.

B. Adsorption and Reaction on Different Oxide Surface Structures

Because of the great variety of bulk crystal structures represented among metal oxides, the surfaces of these materials present a plethora of different geometric arrangements of cations and oxide anions. As noted previously (*96*), cataloging oxide surface reactivity in terms of surface crystallographic structure is a nearly impossible task. Nonetheless, here we review results for oxides of four basic structure classes: wurtzite, rock salt, fluorite, and rutile.

Fortunately, and not unexpectedly, the adsorption and surface reaction behavior of small molecules is often dominated by local site characteristics, and the geometric arrangement of these sites with respect to each other can often be neglected. That is not to say that cooperative effects do not exist in oxide surface chemistry, but beginning with the simpler, localized picture of oxide surface reactivity may help to delineate these.

The most important characteristic of chemically reactive oxide surfaces is the presence of accessible, coordinatively unsaturated cation centers. In other words, if one wishes to do coordination chemistry on oxide surfaces, vacant sites in the coordination spheres must be available to bind molecules from the fluid phase (*96, 97*). The parallel with metal complex chemistry in solution is immediately apparent. As in organometallic chemistry, metal centers with easily displaced ligands are expected to be reactive. On oxides, surface hydroxyl groups can sometimes act as "placeholders" (*50*), in effect occupying coordination sites on surface metal cations until more strongly interacting species are introduced. That detail aside, what is perhaps surprising is that all oxide surfaces do not necessarily present coordinatively unsaturated cation sites to the external environment. After all, creation of surfaces must involve bond breaking and creation of coordination vacancies, and it is textbook material that atoms exposed at transition metal surfaces exhibit multiple coordination vacancies or "dangling bonds."

Certain oxide structures, however, may be terminated only by anions at the outermost layer. Foremost among these are materials with layered bulk structures such as MoO_3. The bulk structure of this material can be described as layers of octahedrally coordinated Mo centers, bounded on both sides by a hexagonal layer of oxide anions. The interactions between the terminal oxo groups that face each other between adjacent layers are weak, and thus the material can be cleaved between layers, much as materials such as graphite or compound analogs such as mica or MoS_2 can be cleaved.

The basal plane surface, designated (010), exposed by cleavage between layers thus presents a hexagonally packed anion array. Investigations of MoO_3 crystallites with different aspect ratios reported in the catalysis literature have fueled debate about whether this surface structure possesses any intrinsic reactivity—there are claims on both sides of the issue (*98–101*). Recent studies of MoO_3(010) single-crystal surfaces by Smith and Rohrer (*102–104*), who used atomic force microscopy (AFM), suggested that reaction of weak acids does indeed require defect sites on this surface. They observed that exposure to vapor-phase water or methanol at elevated temperature led to the formation of pits and step edges on the (010) surface, producing dramatic increases in surface reactivity. These observations are consistent with the relative lack of reactivity of MoO_3 and MoS_2 basal planes demonstrated in earlier surface science investigations (*105, 106*) and with the observation that increasing exposure of edge planes increases reactivity. These observations also serve to reconcile some of the conflicting results in the catalysis literature (noted previously) by demonstrating the occurrence of dynamic restructuring of these surfaces under catalytic reaction conditions, which leads to changing populations of active sites.

1. *ZnO Single Crystals (Wurtzite)*

Zinc oxide crystallizes in the wurtzite structure. This is a noncentrosymmetric crystal structure, cleavage of which perpendicular to the *c*-axis produces polar surfaces that are respectively anion and cation terminated [see Henrich and Cox (*1*) and Barteau (*96*) for structural schematics]. The ideal (0001) surface consists of an outermost layer of zinc cations in a hexagonal arrangement. Each of these is bonded to three oxygen anions in the plane immediately below; thus, each surface cation has a single coordination vacancy relative to the tetrahedrally coordinated cations in the bulk structure. On the $(000\bar{1})$ surface the situation is reversed: The surface is terminated with a hexagonal array of oxygen anions, and the zinc cations below are coordinatively saturated.

The polar surfaces of ZnO have served as important model systems to demonstrate the crucial role of cation coordination vacancies in oxide surface reactivity. Studies by Kung *et al.* (*73, 107–110*), Jacobi *et al.* (*111, 112*), Zhang *et al.* (*113*), and Vohs *et al.* (*31, 51, 54–56, 114–117*) demonstrated that the zinc polar (0001) surface is reactive for adsorption and dissociation of a wide variety of Brønsted acids, including alcohols, thiols (*118*), carboxylic acids, and even alkynes. In contrast, the oxygen polar $(000\bar{1})$ surface is unreactive. These observations have been explained in terms of the requirement that the surface present cations with coordination vacancies. These act as Lewis acid sites to coordinate the acid molecules and then to provide binding sites for the conjugate base anions formed by acid dissociation. Coordinatively unsaturated surface cations are active sites for other adsorption and reaction phenomena as well. For example, aldehydes are adsorbed at these Lewis acid centers on the ZnO(0001) surface and subsequently may undergo nucleophilic attack by surface oxygen atoms to produce stable carboxylate intermediates (*119–122*). This chemistry does not occur on the $(000\bar{1})$ surface.

A few reactions have been reported to occur on the $(000\bar{1})$ surface. These include nonselective oxidation of aromatic compounds (*122*) and the formation of carbonates from CO_2 (*123*). It is not clear, however, to what extent these reactions, especially carbonate formation, may depend on the presence of defects on the oxygen polar surface. Although atomic resolution STM images of the $ZnO(000\bar{1})$ surface have not been obtained (*124–126*), the best images of this surface to date show evidence for a high density of triangular pits and protrusions after annealing of the sample to 900 K in oxygen (*126*). The authors interpreted their images as evidence for $(10\bar{1}0)$ facets, the edges of which may contain coordinatively unsaturated Zn cations.

Thus, heavily faceted $ZnO(000\bar{1})$ surfaces may exhibit some features of the reactivity of the zinc polar (0001) plane. Such effects may explain occasional observations of Lewis and/or Brønsted acid–base chemistry on the oxygen polar surface. The clear reactivity difference between the two polar planes demonstrates the requirement of coordinatively unsaturated surface cations in surface acid–base reactions; the oxygen polar surface exhibits reactivity only when defects exposing such sites are present. These results also point out the need for methods such as STM to verify actual surface structures, in addition to more traditional techniques such as LEED that provide information about the average periodicity of surface structure.

2. *MgO and CaO Single Crystals (Rock Salt)*

Well-defined crystals of MgO have received considerable attention from both experimental and theoretical researchers in the preceding decade.

Among the reasons for this attention are the relative stability of these materials, the ease of studying them as single crystals, their highly ionic character, and their simple structure (rock salt). Interest in MgO has also been stimulated by the fact that this material acts as a catalyst for base-catalyzed reactions such as β aldolization (*127, 128*), olefin isomerization (*129*), Cannizzaro-type disproportionation (*130*), methane coupling (*131–133*), ketonization (*134*), and Tishchenko reactions (*135*). In contrast, well-defined CaO surfaces have received only sporadic attention; in most cases, the motivation has been to make comparisons with MgO.

Depending on the preparation route, MgO can be obtained either as a high-surface-area material in which the building blocks of irregular particles are very small cubes or as a low-surface-area oxide of nearly perfect cubelets (*136*). Figure 1 (*137*) shows an idealized structure of MgO. The (100) face contains fivefold coordinated Mg cations (designated Mg^{2+}_{5c}), whereas on the edges Mg cations are fourfold coordinated; one can also see Mg^{2+}_{3c} on the corners. Measurements by diffuse reflectance spectroscopy have shown the three electronic transitions corresponding to surface states involving O^{2-} in three-, four-, and fivefold coordination environments (*138*). O^{2-}, unstable in the gas phase (in which it decomposes to give O^-), is stabilized by

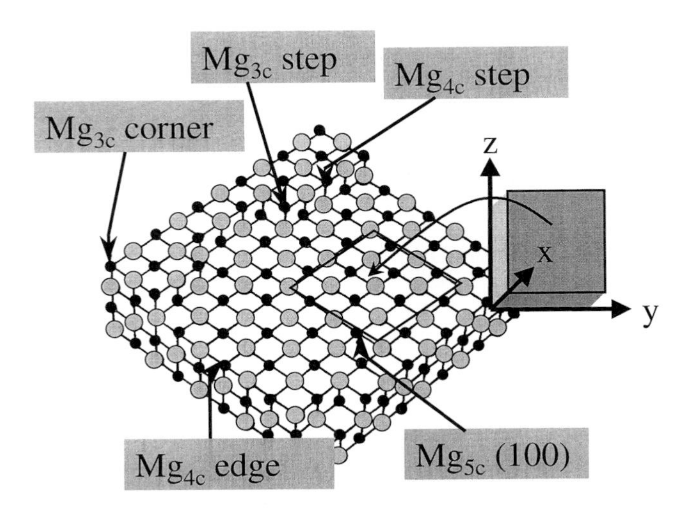

FIG. 1. Schematic representation of the MgO(100) surface showing three-, four-, and fivefold coordinated ions, designated Mg_{nc}.

the potential created by the surrounding anions and cations (the Madelung potential). Thus, the lower the coordination number, the less stable (i.e., the more reactive with adsorbates) is the site. Table I (*139, 140*) shows the decrease of the Madelung potential of oxygen anions (due to the decrease in the Madelung constant) from O^{2-}_{6c} (bulk) to O^{2-}_{3c} (corner). The energy required to remove a neutral O atom, leaving two electrons at the vacancy (i.e., to form an F center), decreases from the bulk to the surface and decreases further with decreasing coordination number of the anion. Also presented for comparison in Table I are the potentials for the corresponding sites in CaO. Clearly, the main factor affecting the difference in stability between MgO and CaO is α/r_e, where α is the corresponding Madelung constant and r_e is the equilibrium Mg–O or Ca–O bond length in the lattice.

The roles of these different sites in adsorption of simple molecules have been extensively investigated, as described elsewhere (*141, 142*). In particular, more detailed data characterizing adsorption of CO, CO_2, and H_2 are given in a review by Sauer *et al.* (*137*). Self-consistent field calculations representing CO adsorption on $(MgO_4)^{6-}$ and on $(MgO_3)^{6-}$, each embedded in a large array of point charges, have provided quantitative values for the increase in adsorption energy with decreasing coordination number. The calculated binding energies of CO on the various sites are as follows: on nondefective (*100*), 22 kJ/mol; on the edge, 32 kJ/mol; and on the corner, 57 kJ/mol. The calculated binding energy of CO on a free Mg^{2+} ion is much higher—161 kJ/mol.

Two reactions on CaO(100) single crystals—the decomposition of N_2O to give N_2 + O(a) [where O(a) denotes adsorbed atomic oxygen] (*143*) and the reaction of CO with O(a) (*144*)—have been investigated by *ab initio* electronic structure calculations. These reactions are relevant to the industrial removal of SO_2 with calcite under fluidized bed combustion

TABLE I

The Madelung Constant and the O^{2-} Madelung Potential in MgO and CaO at Different Coordination Sites [from Pacchioni et al. (139)]; also Shown Is the Energy of Formation of F Centers in MgO [from Pacchioni and Pescarmona (140)]

Site	Madelung constant	MgO Madelung potential eV (r = 2.106 Å)[a]	CaO Madelung potential eV (r = 2.399 Å)[a]	MgO, energy of formation of F centers (eV)
O^{2-}_{6c} bulk	1.747	23.89	20.97	9.67
O^{2-}_{5c} surface (100)	1.681	22.98	20.18	8.06
O^{2-}_{4c} step	1.591	21.76	19.1	6.85
O^{2-}_{3c} corner	1.344	18.38	16.13	5.58

[a] r, metal–oxygen interatomic distance.

conditions (*143*) since removal of surface oxygen by parallel reactions is required to reactivate the surface. The barrier height for the reaction $N_2O \rightarrow N_2 + O(a)$ was found to be 113 kJ/mol (*144*). In this reaction, the O atom is transferred to the surface in a concerted mechanism whereby the N_2 to O σ-bond is broken, the N_2 to O σ-bond is weakened, and the O is converted into O^{2-}_{surf}.

The reactions of CO_2 on the surfaces of MgO and of CaO have been investigated by both theoretical (*139, 145*) and experimental methods (*146–148*). The experimental work was conducted with thin ($\simeq$10 nm) films of CaO and MgO grown on Si(111) and utilized metastable impact electron spectroscopy, ultraviolet photoelectron spectroscopy (UPS, He I), and XPS. The amount of chemisorbed CO_2 differed dramatically from one oxide thin film to the other. Although in both cases CO_2 adsorption led to the formation of carbonates (CO_3^{2-}), CaO was two orders of magnitude more reactive than MgO. Theoretical workers had in fact predicted this behavior (*139*), providing one of the rare examples in which theory has taken the lead in the field. They demonstrated that on a cation that is fivefold coordinated to oxygen anions [i.e., the exposed cations at the (100) surface of a rock salt structure] CO_2 adsorption is more favorable on CaO than on MgO. More precisely, the CO_2/OMg_{5c} complex is unstable with respect to dissociation by 0.8 eV, whereas the CO_2/OCa_{5c} complex is stable by approximately 1 eV.

The difference in stability between the two chemisorbed species is mainly a consequence of the different efficiencies of the charge donation mechanism in MgO, on the one hand, and in CaO, on the other hand, as follows: First, the electron cloud of an O^{2-} ion in CaO is spatially more diffuse than that in MgO and can overlap more efficiently than that in MgO with the unoccupied π_u^* orbital of CO_2. Second, the highest occupied molecular orbital donor orbital, O^{2-}, lies at ~11 eV in MgO but at only 8.5 eV in CaO. Thus, charge can be donated to the adsorbing CO_2 molecule at a lower cost in CaO than in MgO. Table I shows that the Madelung potential at a regular O^{2-}_{5c} site in CaO is smaller than that of O^{2-}_{5c} in MgO; indeed, it is smaller than that of O^{2-}_{4c} in MgO. Thus, the oxygen anions on CaO surfaces are less stable (more reactive) than those on MgO surfaces.

Numerous experimental and theoretical studies have been devoted to investigating more complex molecules than the di- and triatomics discussed previously on well-defined MgO surfaces. The list includes acetic acid (*149*), formaldehyde (*130*), ammonia (*150*), acetylene (*151*), formic acid (*152*), and methane (*153, 154*). Calculations based on density functional theory (DFT) and the pseudopotential method were carried out to determine the optimum mode of adsorption of HCOOH on a "nondefective" MgO(100) surface (*152*). In all cases, dissociative adsorption is favorable. This result is consistent with direct experimental results. XPS showed that the acetic acid, for example, is dissociated at room temperature to give acetate species (*149*) on MgO [as well as on other oxide surfaces such as CeO_2(111) (*155*), TiO_2(001) (*156*), Cu_2O(100) (*157*), and ZnO(0001) (*115*)]. Three configurations of adsorbed carboxylates were considered: monodentate (D1), bidentate (D2), and bridging (D3). In all cases, the bidentate mode was not stable but was transformed into monodentate or bridging species (*152*). Moreover, the bridging configuration was found to be far more stable than the monodentate configuration. At a surface coverage of one-half monolayer ($\theta = 0.5$), the energy of adsorption of the form D1 was calculated to be almost four times less than that of D3 (0.24 vs 0.88 eV; values reported are for calculations using the generalized gradient approximation). Decreasing the coverage to $\theta = 0.25$ increased the adsorption energies (to 0.52 and 1.07 eV for D1 and D3 forms, respectively) in both cases. This latter result indicates that lateral interactions may influence the energy of adsorption.

The adsorption of methane on well-defined MgO crystals has been investigated by both theoretical and experimental methods. It is beyond the scope of this review to present the oxidative coupling reaction of methane [other reviews have been devoted to this subject (*158, 159*)]. It is sufficient to note that surface structural effects on the adsorption and reaction modes were clearly observed. In fact, the coordination geometry of Mg_{nc} and O_{nc} ions at the chemisorption sites (*n* is the coordination number) was shown to be crucial for the adsorption and reaction process (*153, 154*). Again, as in the case of CO_2 adsorption, the chemisorption energy follows the trend Mg^{2+}_{3c} (corner) > Mg^{2+}_{4c} (edge) > Mg^{2+}_{5c} (face).

For dissociative adsorption of weak Brønsted acids such as hydrocarbons, sites with low coordination numbers may be required. An example that ties together results characterizing high-surface-area powders with those of surface science experiments, and that also ties together experimental and theoretical results, is the dissociative adsorption of acetylene on MgO to form acetylides. On the basis of spectroscopic investigations of MgO powders, Stone *et al.* (*160*) concluded that the dissociation of strong Brønsted acids such as carboxylic acids occurred on five-coordinate Mg–O site pairs, i.e., surface sites at which the coordination numbers of both the anion and cation were one less than in the bulk material (Fig. 1). The weakest acids, acetylenes, required four-coordinate Mg–O sites. Peng and Barteau (*52, 161*) examined thin-film and single-crystal MgO surfaces, concluding that, although carboxylic acids dissociated to give essentially monolayer coverage on high coordination sites [i.e., the five-coordinate MgO sites of the (100) surface], alcohols and water were dissociated to a lesser extent, and acetylide coverages were even lower (<0.16 monolayer).

These observations are qualitatively consistent with those of Stone *et al.* (*160*) as well as with results of single-crystal investigations by Onishi *et al.* (*162*). These workers compared adsorption of formic acid, methanol, and water on (100) and faceted (111) surfaces. They reported that formate coverages resulting from formic acid were comparable on the two surfaces, but coverages of the dissociation products of methanol and water on the (100) surface were less than one-third of those on the (111) surface. This difference is consistent with the greater population of sites of lower coordination number on the faceted (111) surface.

The requirement of sites of low coordination number to dissociate acetylene has been confirmed by Nicholas *et al.* (*163*), who used a combination of experimental and theoretical approaches. These workers measured nuclear magnetic resonance spectra of polycrystalline MgO samples after exposure to acetylene; the resulting lines were most consistent with acetylides bound at three-coordinate Mg cation sites. Quantum chemical calculations showed that the formation of acetylides by dissociative adsorption of acetylene was thermodynamically favorable on low coordination sites (three- and four-coordinate sites), whereas this reaction is endothermic on sites of higher coordination number.

3. *CeO_2 and UO_2 Single Crystals (Fluorite)*

Both CeO_2 and UO_2 have the fluorite structure, with the bulk cations eightfold coordinated to oxygen anions; oxygen anions are fourfold coordinated to the respective cations. Interestingly, UO_2 single crystals were among the first oxides investigated in UHV systems, with the first reported study dating back to 1968 (*164*). The obvious motivation for these early studies was the importance of UO_2 as a nuclear fuel. Consequently, none of these investigations were devoted to surface reactions relevant to catalysis. In contrast, CeO_2 single crystals have only recently received significant attention—motivated by the use of CeO_2 in catalytic converters in automobiles.

A discussion of UO_2 warrants mention of several pioneering works. Ellis *et al.* (*165–168*) investigated in detail the low-index surfaces of UO_2 single crystals. With LEED, they observed ordering/disordering of terraces, ledges, and kinks on the three surfaces vicinal to UO_2(111). They also found that upon annealing of these surfaces at high temperatures (700–900 K), the ledges rotated and decomposed irreversibly into (553) and (311) microfacets (*165*). ISS investigations were also conducted with the planar (111) and stepped (553) surfaces of single-crystal UO_2 (*166*). The outermost layer of each surface consists of oxygen occupying the lattice position expected from a simple termination of the bulk structure. LEED and ISS investigations have also been conducted to characterize the (100) surface (*167*), which consists of a monolayer of oxygen atoms arranged in distorted bridge-bonded, zigzag chains along [100] directions.

Other important contributions include the work of Tasker (*169*), who reported the first surface energy calculations for low-index UO_2 as an example of a fluorite structure, as discussed later. Hubbard *et al.* (*170–173*) provided some of the most complete optical and electronic studies of the UO_2 system.

Table II shows the surface energies of UO_2 and a comparison of these to the values characterizing the corresponding CeO_2 surfaces. Calculations for both were based on the Born model of potentials, including a long-range term to account for Coulombic interaction and a short-range term to model the repulsions and van der Waals attractions between neighboring charge clouds. The short-range interactions between interacting ions were described in Buckingham form: $V(r) = A \exp(-r/\rho) - C/r^6$. The parameters A, ρ, and C for anion–anion and cation–anion interactions can be found in Sayle *et al.* (*174*) and Tasker (*169*) for CeO_2 and UO_2, respectively; the cation–cation interaction was set equal to zero.

The most stable surface of each oxide (i.e., the surface having the lowest surface energy) is the (111), followed by the (110). The (111) surface is an oxygen-terminated surface, whereas the (110) contains alternating cations and anions. Relaxation occurs on the surfaces of both oxides. On the (111) surface, O anions are relaxed inward by 0.14 Å. Relaxation of the (110) surface also occurs. As shown in Table II, the effect of relaxation is dramatic, a consequence of the highly ionic character of each oxide.

Recent STM studies of $UO_2(111)$ (*175, 176*) and $UO_2(110)$ (*177*) single crystals have confirmed that the surfaces are almost ideal terminations of the bulk. The (001) surface of UO_2 has also been imaged by STM (*178*). The images reveal a maze-like structure with atomic periodicities corre-

TABLE II

Surface and Defect Energies of CeO_2 and UO_2 Crystal Planes

Surface	$CeO_2(111)$	$CeO_2(110)$	$CeO_2(310)$	$UO_2(111)$	$UO_2(110)$
Surface energy (J/m^2)					
Relaxed	1.195[a]	1.575[a]	2.475[a]	1.069[b]	1.539[b]
Unrelaxed	1.707	3.597	11.57	1.479	3.148
Defect equilibrium (eV)					
Reaction *c*	2.71	−0.47	−6.25	n.a.	n.a.
Reaction *d*	−0.22	−3.40	9.18		

[a] From Sayle *et al.* (*174*). [b] From Tasker (*169*). [c] $O_l^{2-} + 2\,Ce_l^{4+} \rightarrow \frac{1}{2}O_2(g) + V_O + 2\,Ce^{3+}$.
[d] $CO(g) + CeO_2 \rightarrow CO_2(g) + V_O + 2\,Ce^{3+}$.

sponding to a half monolayer of oxygen arranged in (1 × 1) out-of-phase domains along the [110] and [$1\bar{1}0$] directions. The CeO_2(111) surface has also been examined recently by LEED (*155*) and STM (*179*); both techniques show that the surface exhibits a (1 × 1) termination, with oxygen anions being the features most likely imaged by STM.

Bulk properties of both CeO_2 and UO_2 are presented in Table III. The Madelung potentials of O^{2-} for the two oxides are nearly the same, as are those of the M^{4+} ions in each. These similarities arise because both materials are composed of cations with similar sizes that are arranged in an identical structure. In contrast to the situation for MgO and CaO, data characterizing the change of the Madelung potential with surface structure are not available (the values are expected to change similarly for the two fluorite-structure oxides). Although both oxides have the same electronegativity difference, $\Delta\chi$, the values of the average partial charge of oxygen anions in the bulk of each material, q_O, appear to indicate that CeO_2 is more ionic than UO_2.

There have been relatively few experimental investigations of chemical reactions on CeO_2 and UO_2 single crystals. Moreover, we are aware of only one computational investigation devoted to the surface reactions of CeO_2 (*174*). The oxidation of CO to CO_2 has been investigated on three low-index surfaces of CeO_2: (111), (110), and (310). As shown in Table II, the (310) surface is expected on thermodynamic grounds to be far more

TABLE III

Selected Common Bulk Properties of CeO_2 and UO_2

Oxide	Bulk CeO_2	Bulk UO_2
Defect equilibrium (eV)	6.58[a]	9.8[g]
	3.65[b]	10.0[h]
Bulk $E_{(M^{4+}-O^{2-})}$(kJ/mol [O])	960[c]	930[c]
R_{av} (Å)	2.343[d]	2.370[i]
M^{4+} radius (Å)	0.92[e]	0.93[e]
$V_{Mad}(O^{2-})$ (eV)	21.7[f]	21.4[j]
$V_{Mad}(M^{4+})$ (eV)	−40.3[f]	−39.9[j]
$\Delta\chi(Ev)$	1.84	1.83
q_O	−1.16	−0.57

[a] $O_l^{2-} + 2\,Ce_l^{4+} \rightarrow \frac{1}{2}O_2(g) + V_O + 2\,Ce^{3+}$. [b] $CO(g) + CeO_2 \rightarrow CO_2(g) + V_O + 2\,Ce^{3+}$. [c] Calculated following the equation in Rethwisch and Dumesic (*180*). [d] Average nearest-neighbor distance around a given ion from Broughton and Bagus (*181*). [e] From Hausner (*182*). [f] From Broughton and Bagus (*181*). [g] From Catlow (*184*) (semiempirical calculations). [h] From Pettit *et al.* (*185*) (*ab initio*). [i] From Jollet *et al.* (*183*). [j] From Catlow (*184*).

active than the (111) surface, and the (110) surface should show intermediate activity. The (310) surface is a stepped surface on which Ce^{4+} and O^{2-} ions alternate along the steps. Thus, both ions are characterized by lower coordination numbers than those in the bulk or on a flat surface. The low coordination number of Ce^{4+} on these steps may result in a higher binding energy for CO than on the corresponding Ce^{4+} on terraces and drive the reaction with neighboring oxygen anions as depicted schematically in Fig. 2. Since there are no experimental investigations of the (310) surface of CeO_2, it is not possible to compare these predictions with experimental results. However, it is intuitively expected that low coordination cation sites on oxide surfaces will be more reactive than high coordination cation sites. As discussed previously, similar results were obtained for MgO surfaces (both experimentally and theoretically).

The reactions of ethanol on the $UO_2(111)$ surface have been investigated by temperature-programmed desorption (TPD) (*186*). Ethanol is converted by both dehydration and dehydrogenation pathways, forming ethylene and acetaldehyde, respectively. The driving force for dehydrogenation is (in addition to structural requirements) the ability to break one U–O bond (in other words, to make UO_{2-x} from UO_2). One would therefore expect that the stabilities of various surface defects would be an important issue determining the selectivities for aldehyde formation on a series of oxides. The reaction of ethanol has also been studied on three other single-crystal oxide surfaces: $TiO_2(001)$ (*32*), $TiO_2(110)$ (*187*), and ZnO(0001) (*116*). On $TiO_2(110)$, small (but not quantified) amounts of acetaldehyde were observed (*187*).

Table IV shows the relation of the Madelung potentials of the metal cations and the oxygen anions for these oxides with their selectivities for alcohol reactions in TPD experiments. The stated values of the Madelung potential are those of the bulk and serve only as a proxy for surface

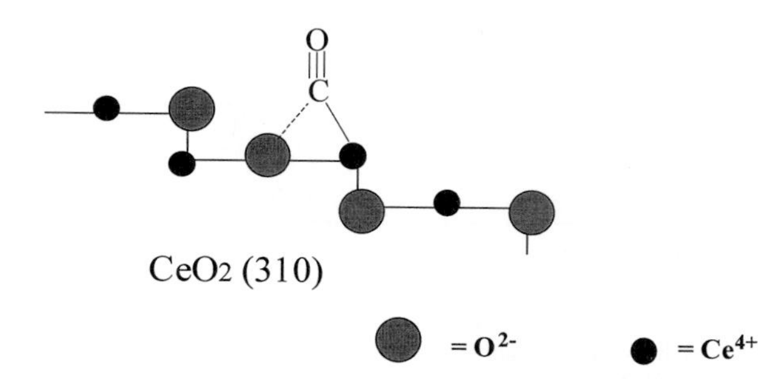

FIG. 2. Side view of $CeO_2(310)$ surface. Preferential CO adsorption on Ce^{4+} ions with low coordination number (step) is shown.

TABLE IV

Dehydrogenation/Dehydration Selectivities of Alcohols Determined from TPD Experiments with Oxide Single Crystals: Comparison with Bulk Madelung Potentials

Oxide surface	Reactant	$CH_3CHO/CH_2{=}CH_2$[a]	$V_{Mad}(O^{2-})$	$V_{Mad}(M^{x+})$	ΔV_{Mad}
$TiO_2(001)$[b]	CH_3CH_2OH	0.2	25.9	−44.7	−70.6
$TiO_2(110)$[c]	CH_3CH_2OH	0[c]	25.9	−44.7	−70.6
$UO_2(111)$[d]	CH_3CH_2OH	0.8	21.4	−39.9	−61.3
$ZnO(0001)$[e]	CH_3CH_2OH	3.2	24.0	−24.0	−48.0
$Cu_2O(100)$[f]	$CH_3CH_2CH_2OH$	4.6[g]	21.8	−12.8	−34.6

[a] CH_3CH_2OH to CH_3CHO and $CH_2{=}CH_2$. [b] From Kim and Barteau (*32*). [c] Trace acetaldehyde production not quantified [Gamble *et al.* (*187*)]. [d] From Chen *et al.* (*186*). [e] From Vohs and Barteau (*116*). [f] From Schulz and Cox (*157*). [g] $CH_3CH_2CH_2OH$ to CH_3CH_3CHO and $CH_3CH{=}CH_2$.

properties. As illustrated in Fig. 3, the more negative the Madelung potential difference between anions and cations, the lower is the selectivity for the alcohol dehydrogenation pathway. Conversely, the less negative the Madelung potential difference, the higher is the dehydrogenation selectiv-

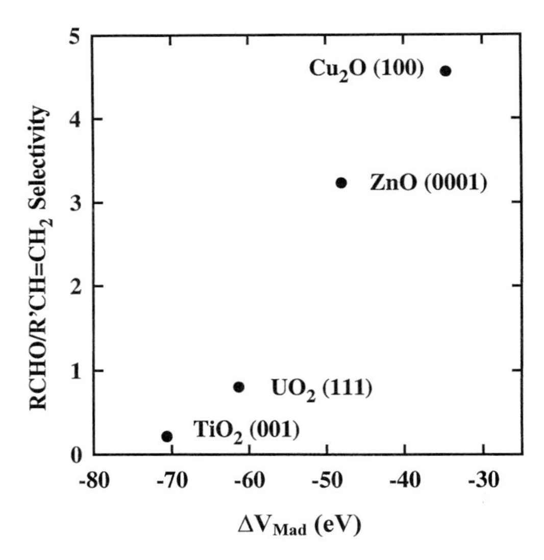

FIG. 3. $RCHO/R'CH{=}CH_2$ selectivity from alcohol TPD from a series of oxide single crystals as a function of ΔV_{Mad}. Selectivity data are for ethanol TPD from $TiO_2(001)$ (*32*), $UO_2(111)$ (*186*), and $ZnO(0001)$ (*116*) and for 1-propanol TPD from $Cu_2O(100)$ (*188*).

ity. Correlations of reactivity with physical and electronic characteristics of oxides are considered in greater detail in Section III.

The reactions of HCOOH and CH_3COOH on the $CeO_2(111)$ surface have also been investigated by TPD and HREELS (*155*), and the reaction of CH_3COOH on $UO_2(111)$ has been examined by TPD (*186*). On $CeO_2(111)$, adsorbed formates and acetates, produced by dissociative adsorption of formic and acetic acids, respectively, decompose at temperatures near 600 K to produce CO and CH_2CO, respectively, via dehydration. Both formates and acetates were found in a monodentate configuration, as evidenced by the wide separation (220 cm^{-1}) between the symmetric and asymmetric OCO stretching modes observed in HREELS. TPD results indicated that the ketene yield from acetic acid was almost the same (~0.8) for both $CeO_2(111)$ and $UO_2(111)$ surfaces. Although this result is perhaps not surprising given the similar structural and electronic properties of these surfaces (discussed previously), it is emphasized that the demonstration of these physical and chemical similarities has been made possible by the application of surface science techniques and approaches.

4. *TiO_2 and SnO_2 Single Crystals (Rutile)*

TiO_2 single-crystal surfaces are among the most thoroughly investigated in surface science and catalysis by both experimental and theoretical methods. More important, rutile TiO_2 single crystals have been found to be reactive for a rich variety of reaction pathways, some of which have not been encountered on any other well-defined surface (oxide or metal) to data. Examples of reactions observed on TiO_2 surfaces include reduction of aldehydes to give alcohols (*189, 190*), cyclization of alkynes (*191, 192*), dehydration of alcohols to give olefins (*32*) and of carboxylic acids to give ketenes (*156, 193*), dehydrogenation of alcohols to give aldehydes (*32, 194*), condensation of aldehydes to give unsaturated higher aldehydes (*190, 195*), coupling of carboxylic acids to form ketones (*156, 193*), and reductive coupling of aldehydes and of ketones to form symmetric olefins (*196–198*). All these reactions, with the exception of cyclization and reductive coupling (which are specific to surfaces containing defects), have been observed on the reconstructed surfaces of $TiO_2(001)$ single crystals.

On the other hand, SnO_2 single crystals have received only a small fraction of the effort devoted to TiO_2 surfaces. Both oxides are wide band gap semiconductors and both are most stable in the rutile structure. The study of chemical reactions on TiO_2 has been motivated by its catalytic activity, but that of reactions on SnO_2 is mainly due to its gas-sensing activity and its use in solar-cell devices. Table V shows some of the bulk electronic properties of these two oxides.

TABLE V
Selected Common Bulk Properties of TiO_2 and SnO_2

Oxide	TiO_2	SnO_2
$\Delta\chi$ (eV)	1.74	1.44
q_O	−0.76	−0.86
$E_{(M-O)}$ (eV)	20.7	15.5
$V_{Mad}(M)$ (eV)	−44.74	−42.85
$V_{Mad}(O)$ (eV)	25.88	24.64
Band gap (eV)	3.1–3.2	3.5–3.6

This section presents the surface structure and properties of the two oxides and links them, where appropriate, to chemical reactions. Theoretical work related to the surface characteristics, as well as to the modes of adsorption of surface species, is also discussed. Because of the relatively large volume of work on various surfaces of TiO_2, the structure and properties of individual crystal planes are considered separately.

TiO_2 is composed of chains of TiO_6 octahedra. Stoichiometric TiO_2 is a d^0 compound, and reduction can lead to a wide range of intermediate or suboxide structures between Ti_2O_3 (corundum—Ti^{+3}/d^1) and Ti_nO_{2n-1}, where $n = 3$–10. These intermediate compounds may have variable compositions or a repeating lattice structure, as in the case of Magnéli phases. Strong reduction may produce the unstable rock salt structure TiO_x, where $0.6 < x < 1.28$ (*1*).

Three bulk structures exist—rutile, anatase, and brookite—and in these the octahedra are differently distorted and the chains differently assembled (*199*). Rhombohedral brookite (*200*) is rather unstable and rarely encountered in chemical applications. Anatase is the most commonly used in powder studies, because it has the highest surface area and is easily obtainable. The TiO_6 octahedra in this material share four edges. However, it is very difficult to produce single crystals of anatase of an adequate size for surface science study because anatase is metastable and transforms into rutile at elevated temperatures (>1073 K) (*201*). Consequently, rutile has been the form of choice for single-crystal studies.

In the bulk rutile structure of TiO_2, each titanium cation is surrounded by a distorted octahedron of six oxygen anions, and each oxygen is in a threefold coordination environment, with the lattice having an overall tetragonal symmetry. The TiO_6 octahedra share only two edges. Figure 4 illustrates the rutile unit cell. Cleaving a bulk lattice at different orientations yields different surface geometries. The actual surface structure will differ from that produced by simple cleavage of the bulk structure. Typically, the top several layers will relax by moving in or out a fraction of an angstrom

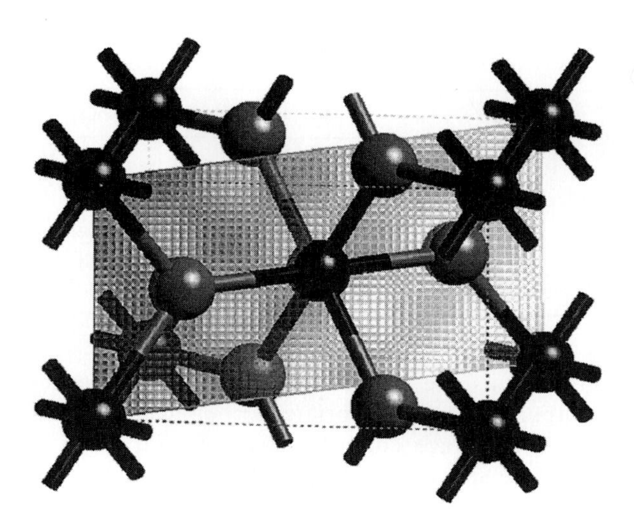

FIG. 4. TiO_2 rutile unit cell, with the (110) plane shaded. Physical properties of rutile: $\Delta G_f^o = -889.5$ kJ/mol, Ti–O bond length = 1.95 Å, and Ti–O dimer bond strength = 644 kJ/mol [data from Linsebigler *et al.* (*199*) and Lide (*204*)].

unit to minimize the surface energy (*202*). In some cases, heating to higher temperatures may produce rearrangement of the surface to a thermodynamically more stable structure (*203*).

a. *TiO_2 and SnO_2 (110).* A major contribution to the stability of a surface is the degree of coordinative unsaturation of the surface layer cations, with greater unsaturation giving less stability. The (110) face is the one most commonly found in both nature and synthetic rutile crystals because it gives the maximum coordination of the surface titanium ions—half remain six-coordinate, and half become five-coordinate.

Figure 5 shows a model of the bulk terminated (110) surface (ionic radii in Figs. 5–9 are not to scale). Three-coordinate O^{-2} and six-coordinate Ti^{+4} have ionic radii of approximately 1.36 and 0.61 Å, respectively (*204*). Rows of bridging oxygen atoms alternate with rows of fivefold coordinated Ti atoms. Surface X-ray diffraction (*205*), ion scattering (*206*), and theoretical calculations (*202*) have shown a small relaxation of the surface. The surface fivefold coordinated Ti cations move in toward the bulk (*202*) (slightly out of the plane of the four oxygen atoms) as do the bridging oxygen atoms. Several STM investigations (*207–210*) have enabled near-atomic level visualization of the surface, further confirming the structure. Reconstruction by high-temperature annealing to (1 × 2) has also been observed (*211*).

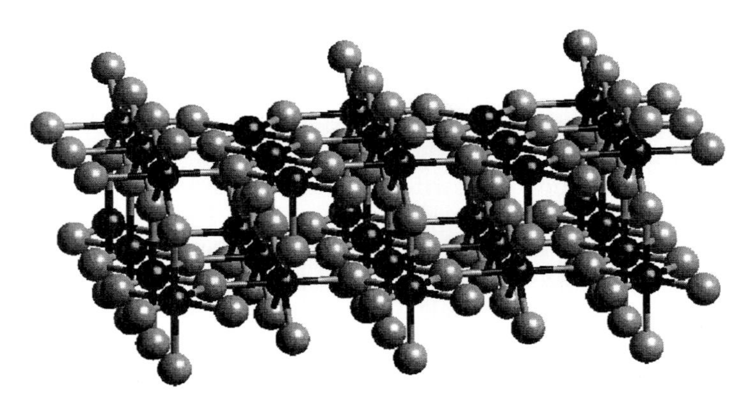

FIG. 5. Structure of the unrelaxed $TiO_2(110)$ surface. Black and gray spheres are Ti and O atoms, respectively.

There are three interpretations for this reconstruction: the missing-row model (*211*), the added-row model (*212*), and the added-Ti_2O_3-row model (*210*). The missing-row model is generally discounted in light of electron-stimulated desorption ion angular distribution (ESDIAD) experiments (*213*). Which of the added-row models is the better representation of the surface continues to be debated (*214, 215*).

The reactions of a variety of molecules on the surfaces of stoichiometric and defect-containing $TiO_2(110)$ single crystals have been investigated, although the reactivity of the stoichiometric surface is low, partly due to the presence of out-of-plane bridging oxygen anions and the fact that 50% of Ti^{+4} ions are sixfold coordinated and are thus, in principle, unreactive. The alternating rows of these sixfold coordinated cations force the reaction to occur along the fivefold coordinated cation rows, unless the molecule is large enough to cross the sixfold coordinated barrier row (see the example of bi-isonicotinic acid). The following reactions, among others, on the stoichiometric as well as reduced surfaces have been investigated by a variety of methods: reactions of alcohols (*187, 216*), hydrocarbons (*217*), halogenated hydrocarbons (*218, 219*), carboxylic acids (*220–223*), CO (*224–226*), CO_2 (*227*), formaldehyde (*217, 228*), H_2O (*229–231*), 2-propanol (*232, 233*), pyridine (*234, 235*), and benzene (*234*). Selected reactions are discussed in the following sections.

i. Reactions of Primary Alcohols. Studies of methanol adsorption by DFT (*216*) and *ab initio* methods (*236*) and of ethanol reaction by TPD (*187*) have been carried out on the $TiO_2(110)$ surface. In the case of methanol, different adsorption modes were considered. In the unsymmetric molecular (UM) mode, the oxygen atom of methanol is bound to Ti^{+4}_{5c} and the hydrogen atom is coordinated to a bridging oxygen anion; in the unsymmetric dissociated (UD) mode, the proton is bound to a bridging oxygen ion. Bates *et al.* (*216*) have also performed the analysis at two coverages—$\theta = 1.0$ and 0.5. At each coverage, the adsorption energy (E_{ads}) increased only slightly from UM to UD.

A similar investigation was carried out to characterize $SnO_2(110)$ (which has the same unrelaxed structure as in Fig. 5) by DFT as well as Hartree–Fock pseudopotential methods (*237*). Table VI shows the DFT results as a function of coverage; also included in the table are the energies of the relaxed (110) surface of each oxide. Both molecular and dissociative adsorption of methanol on $SnO_2(110)$ are more favorable than on $TiO_2(110)$. In addition, in contrast to the adsorption on $TiO_2(110)$, methanol prefers to adsorb dissociatively on $SnO_2(110)$. The authors (*237*) concluded that, on $SnO_2(110)$, dissociative adsorption via C–O bond breaking is more favored than that involving O–H dissociation. Experimental evidence for alcohol activation by initial C–O scission on clean, stoichiometric oxide surfaces is lacking, however.

The binding of ethoxide groups, formed by dissociative adsorption of ethanol, on the $TiO_2(110)$ surface was also shown to involve two adsorption

TABLE VI

Surface and Binding Energies of Adsorbates on $TiO_2(110)$ and $SnO_2(110)$[a]

Oxide surface	$TiO_2(110)$	$SnO_2(110)$
Surface energy (J/m^2)		
Relaxed	1.78	2.06
Unrelaxed	2.05[b]	2.16
E_{ads} of adsorbed CH_3OH ($kJ\ mol^{-1}$)		
$\theta = 0.5$	100.8	232.6
$\theta = 1$	70.4	110.4
E_{ads} of adsorbed CH_3O^- ($kJ\ mol^{-1}$)		
$\theta = 0.5$	119.1	191.7
$\theta = 1$	74.3	114.5
E_{ads} of adsorbed HCOOH ($kJ\ mol^{-1}$)		n.a.
$\theta = 0.5$	62.7	
$\theta = 1$	58.9	
E_{ads}[c] of adsorbed $HCOO^-$ ($kJ\ mol^{-1}$)		n.a.
$\theta = 0.5$	134.1	
$\theta = 1$	118.7	
E_{ads}[d] of adsorbed $HCOO^-$ ($kJ\ mol^{-1}$)		n.a.
$\theta = 0.5$	187.2	

[a] Also shown is the effect of surface coverage on binding energy [from Bates *et al.* (*216*) and Calatayud *et al.* (*237*)]. [b] From Oliver *et al.* (*202*). [c] Unidentate formates with the C=O pointing away from the surface. [d] Bidentate formates.

states (*187*). Ethoxide groups bound to fivefold coordinated titanium atoms desorbed as ethanol at low temperatures along with water. This water is formed by recombination of bridging hydroxyl groups, leaving bridging oxygen anion vacancies. These may be filled by migration of adsorbed ethoxide groups, which are then much more tightly bound at these sites and do not react further until a temperature of 650 K is reached. At this temperature, some of these ethoxide species undergo dehydration and desorb as ethylene (*187*).

ii. Reactions of Carboxylic Acids. Carboxylic acids on the $TiO_2(110)$ surface have been examined in detail. There is also one reported investigation of a dicarboxylic acid (*220*). Formic acid (*221*), acetic acid (*222*), and benzoic acid (*223*) (and most likely higher carboxylic acids) adsorb dissociatively on $TiO_2(110)$. Table VI shows calculated adsorption energies (E_{ads}) for dissociative and nondissociative modes of formic acid adsorption. Dissociative adsorption is favored, in full agreement with experiments. Moreover, the formate species are predicted to be most stable in a bidentate configuration.

However, in this regard experiments have shown the situation to be considerably more complex. It has been observed by LEED and ESDIAD that adsorption of formic acid (*221*), acetic acid, (*222*), and benzoic acid (*223*) results in a reconstruction of the surface; a (2×2) structure has also been reported for the last case (*238*). It is proposed that the carboxylates adsorb as a (2×1) overlayer, bridge bonding with the fivefold coordinated Ti^{+4} ions. The O–O distance of acetates is 2.3 Å, which is less than the Ti^{+4}–Ti^{+4} distance of 2.9 Å (a lattice parameter of the bulk crystal). This comparison suggests contraction of the Ti^{+4} ion pairs, thus creating the reconstruction. Overall, a saturation coverage of not more than 0.5 monolayers is therefore expected.

Absorption infrared spectroscopy of formates on $TiO_2(110)$ has further enhanced the understanding of this mode of adsorption (*64*). Hayden *et al.* (*64*) found that saturation occurs at 0.59 monolayer, higher than that expected from LEED or ESDIAD experiments. Their interpretation is as follows: There are two formate species. In the first, with C_{2v} symmetry, the OCO plane is aligned in the (001) direction; the second has C_s symmetry, with the OCO plane in the (110) direction. The first species (observed at $\theta = 0.4$) is responsible for the (2×1) reconstruction. The second is an adsorbed formate having one of the oxygen atoms incorporated in the $TiO_2(110)$ oxygen rows. This species is most likely adsorbed on oxygen vacancies created via dissociative adsorption of formic acid to produce the first species, followed by reaction of two hydroxyl groups to produce water.

The reactions of formic acid on $SnO_2(110)$ crystals have also been investigated. Regarding TiO_2, the (110) surface of SnO_2 reconstructs. There are three reconstructions: (4×1), (1×1), and (1×2). They are induced by annealing at 900, 950, and 1075 K, respectively (*239*). The main reaction products formed from formic acid are CO_2, CO, and H_2O (*240*). Irwin *et al.* (*241*) investigated, by angle-resolved photoemission, Auger electron spectroscopy (AES), and LEED, the modifications of the surface electronic structure and symmetry of SnO_2(110)-(1 × 1) and -(1 × 2) surfaces upon formic acid adsorption/reaction. LEED analysis showed that although the surface order of the (1 × 1) surface is not modified, the half-order beams from the (1 × 2) surface are completely removed, i.e., transformation to the (1 × 1) surface occurs. As a consequence of the uncertainty about the exact structure of the reconstructed SnO_2(110) surface, it was inferred that the (1 × 2) surface either contains a missing row [i.e., is more reduced than the (1 × 1)] or extra bridging oxygen atoms [i.e., is more oxidized than the (1 × 1)]. Irwin *et al.* (*241*) suggested two possible mechanisms for this transformation. They attributed it either to the removal of the bridging oxygen rows thought to be present on the (1 × 2) surface (*242*) or to the deposition of oxygen atoms, converting the surface into a fully oxidized (1 × 1).

The adsorption of bi-isonicotinic acid (BINA; 2,2′-bipyridine-4,4′-dicarboxylic acid) on the TiO_2(110) surface has been investigated by XPS, near edge X-ray absorption fine structure (NEXAFS), and quantum chemical [intermediate neglect of differential overlap (INDO)] calculations by Patthey *et al.* (*220*). BINA is a ligand of important organometallic dye molecules (*243*) and has been intensively investigated for solar cell and related applications. On the basis of O (1s) XP spectra, the authors claimed that the two carboxylic groups were deprotonated—no chemical shift and/or double peak structure was observed in the O (1s) signal of the adsorbed molecule. Hence, the two oxygen atoms of each carboxyl group are chemically equivalent and both groups are involved in bonding to the surface. This inference is reasonable in light of the formic, acetic, and benzoic acid data discussed previously. Moreover, no adsorption occurs through the N atom. This inference is also reasonable since adsorption of pyridine has shown a weak interaction between the p orbitals of N and Ti_{5c}^{+4} ions; E_{ads} of pyridine was computed to be 84 kJ mol^{-1} [compared with 187 kJ mol^{-1} for bidentate formate species (*216*)]. The angular dependences of the π resonance intensities for the two azimuthal orientations (measured by NEXAFS) indicated that BINA is bound at an angle of 44° with respect to the (001) direction and is tilted by 25° with respect to the surface normal (*220*). INDO calculations show that the most stable configuration is bidentate, with a Ti–O bond distance of 2.0 Å. Clearly, these results show that BINA is bound to four Ti_{5c}^{4+} ions in two different rows. Moreover, the most stable bidentate configuration is that in which Ti_{5c}^{4+} ions at opposite ends are shifted by one crystallographic unit cell in the (001) direction.

b. *TiO_2(100).* The (100) surface is less stable than the (110) surface because all surface cations are fivefold coordinated. Figure 6 shows the structure of the (100)-(1 × 1) surface. Rows of bridging oxygen atoms are shown as for the (110) surface, but the exposed Ti atoms are no longer arranged symmetrically between them with the coordination vacancy oriented perpendicular to the surface; instead, they are arranged such that the "dangling bonds" are tilted. LEED measurements indicate that the (100) surface reconstructs to give successively (1 × 3) [with three distinct Ti sites (*244*)], (1 × 5), or (1 × 7) patterns (*245*) as the annealing temperature is increased. These probably arise from the creation of (110) facets, forming deep troughs which have been clearly visualized by STM (*246*).

The difference between the (110) and (100) surfaces affects the mode of adsorption of alcohols, as has been investigated both experimentally and theoretically (*236*). On the (100) surface, CH_3OH molecules adsorbed on the Ti atoms are able to form hydrogen bonds to bridging oxygen atoms, $O_{(br)}$, and thus dissociate, giving the following:

$$Ti - HOCH_3 + O_{(br)} \rightarrow Ti - OCH_3 + H - O_{(br)} \qquad (1)$$

The adsorption energies of methanol on a $Ti_6O_{15}H_{20}$ cluster simulating the (100) surface have been calculated by *ab initio* electronic structure calculations. The adsorption energy of the undissociated form was calculated to be 61.1 kJ mol^{-1}, and the adsorption energy of the dissociated form was found to be greater by ~73 kJ mol^{-1} (*236*). Thus, dissociative adsorption is clearly the thermodynamically preferred bonding mode for alcohols on this surface.

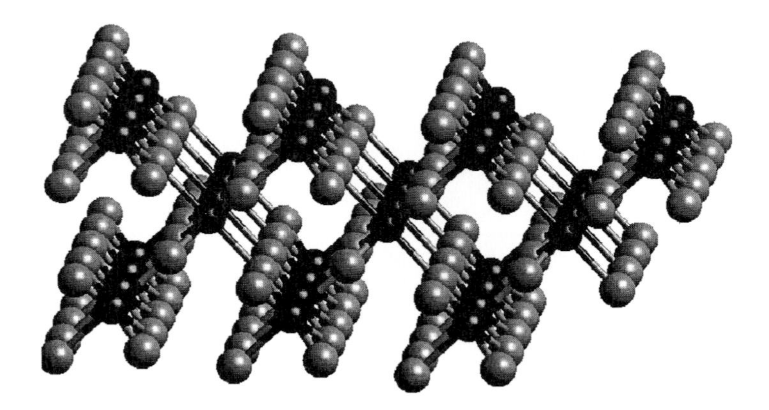

FIG. 6. Structure of the unreconstructed and unrelaxed TiO_2(100) surface. Black and gray spheres are Ti and O atoms, respectively.

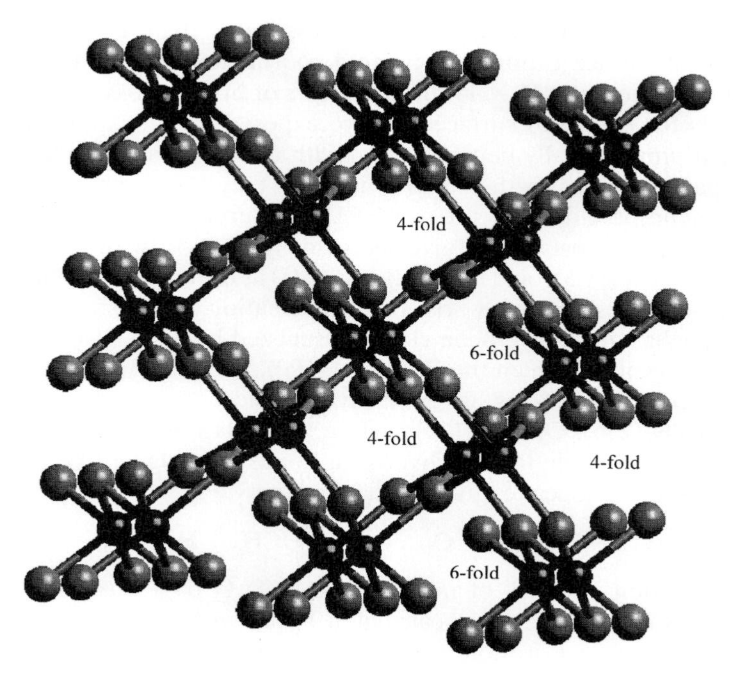

FIG. 7. Structure of the unreconstructed $TiO_2(001)$ surface. Black and gray spheres are Ti and O atoms, respectively.

c. *$TiO_2(001)$*. On the bulk terminated (001) structure, all surface Ti atoms are only fourfold coordinated (Fig. 7); thus, this surface is quite unstable. *Ab initio* calculated surface energies for TiO_2 are shown in Table VII.

The formation of a low-temperature "phase I" and a subsequent high-temperature "phase II" structure of $TiO_2(001)$ upon annealing was ob-

TABLE VII
Energies of TiO_2 Surfaces[a]

Surface	Surface Ti coordination	Unrelaxed surface energy $[meV/(a.u.)^2]$	Relaxed surface energy $[meV/(a.u.)^2]$
(110)	5, 6	30.7	15.6
(100)	5	33.8	19.6
(001)	4	51.4	28.9
(011)	5	36.9	24.4

[a] From Ramamoorthy *et al.* (*203*).

served and characterized with LEED, AES, EELS, and UPS (*247, 248*) in the late 1970s. Comparison of theoretically calculated LEED patterns for different structures with those observed allowed the determination of the facets present (*248*). Annealing to 750–800 K produced the (011)-faceted structure (Fig. 8), in which all surface Ti atoms are fivefold coordinated, reducing the free energy of the surface. Relaxation of the (011) surface has been modeled (*202*), with the results indicating a movement of the surface oxygen and Ti atoms 0.02 and 0.01 Å out of the surface, respectively, while the second-layer oxygens move 0.01 Å inward.

Annealing to temperatures higher than 900 K produces the (114)-faceted structure (Fig. 9). The (114)-faceted surface is particularly interesting in that, although the average coordination number of the surface Ti atoms is still five, in fact one-third of Ti cations are sixfold coordinated, one-third fivefold coordinated, and one-third only fourfold coordinated. The latter have two vacant coordination sites, leading to some unique reaction possibilities resulting from the coordination of two reactant species to the same cation center. Recently, a third reconstruction has been observed by STM and reflection high-energy electron diffraction, after heating of the crystal to 1473 K followed by quenching at a rate of 100 K s^{-1} (*249*). This structure

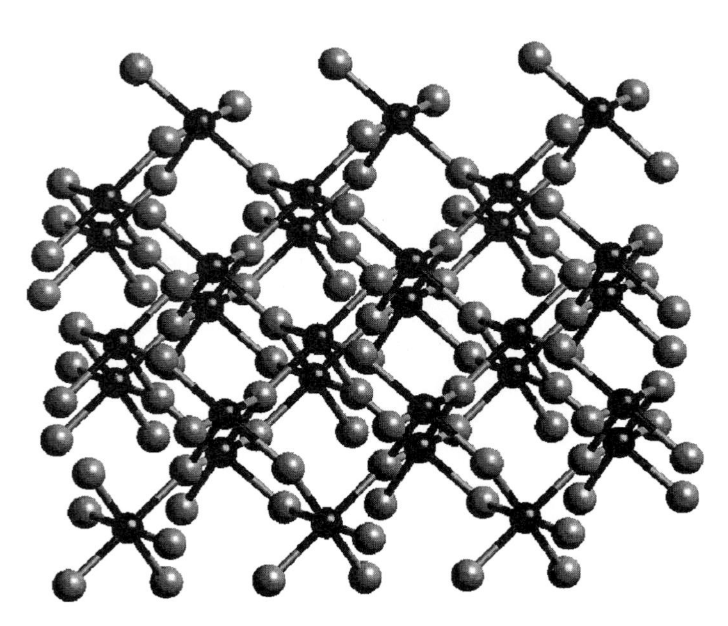

FIG. 8. Structure of the unrelaxed $TiO_2(011)$ surface. Black and gray spheres are Ti and O atoms, respectively.

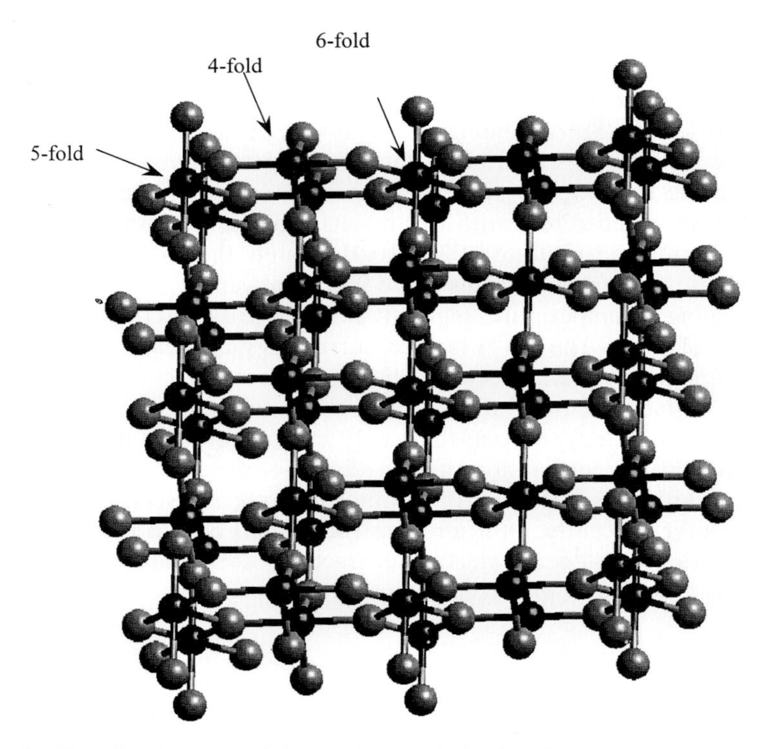

FIG. 9. Top view structure of the unrelaxed $TiO_2(114)$ surface. Black and gray spheres are Ti and O atoms, respectively.

has a two-domain network and is composed of ridges running in the (110) and $(\bar{1}10)$ directions at a spacing seven times the lattice constant ($a_{o(110)}$ = 0.65 nm). The reconstruction appears to be a $(7\sqrt{2} \times \sqrt{2})R45°$ (*249*). Ion sputtered (001) as well as the (011)- and (114)-faceted surfaces have been characterized during the preceding decade with a variety of techniques, including XPS (*156, 194, 250*), AES (*156*), UPS (*251*), NEXAFS (*252*), AFM (*253*), and STM (*254, 255*). These latter two techniques provide a more local view of faceting.

The (001) surface is especially advantageous for experimental work in that simply altering the annealing temperature may produce quite different surface structures. Comparisons of reaction chemistry occurring on singly coordinatively unsaturated Ti cations with that on doubly coordinatively unsaturated Ti cations can be readily made. This surface has also served as an important system for studying effects of surface reduction. In actuality, any surface, no matter how carefully prepared, will contain many defects, which may play a large role in determining its catalytic or chemical properties (*256*). In some cases disorder or defects may be deliberately introduced (*207, 257–259*), for example, by sputtering, high-temperature annealing, or annealing in the presence of hydrogen. These methods all reduce the surface and produce oxygen anion vacancies (i.e., suboxide compounds, TiO_{2-x}). The reduced state may be clearly detected with UPS, which shows the presence of filled electronic levels within the normal band gap region (*247, 251*). XPS is particularly useful in characterizing the degree of reduction of a TiO_2 surface, as the Ti ($2p_{3/2,\,1/2}$) peaks shift with oxidation state (*156, 194, 250, 251*). Thus, the relative populations of individual Ti oxidation states in the near-surface region may be determined from the areas of these peaks. Reannealing reoxidizes and reorders the surface.

Isolated defects such as oxygen anion vacancies or interstitial atoms are known as point defects. Extended defects include steps and crystallographic shear (CS) planes formed by shearing to eliminate oxygen vacancies lying on the same plane (*256, 260*). If such a CS plane occurs at every *n*th lattice plane throughout the crystal, a Magnéli phase may form with a stoichiometry of Ti_nO_{2n-1} (*1*). The inevitable presence of steps means that metal ions of lower coordination than the rest of the surface will be present. Kinks in steps will give rise to even lower coordination at the outside corner, but large inward relaxation of these exposed cations will reduce their instability (*1*).

i. Reactions of Alcohols, Aldehydes, and Ketones. Kim and Barteau (*32, 194*) used TPD to investigate adsorption and reaction of C_1–C_3 alcohols on both rutile (001) single crystals and anatase powders (*261*). In general, molecularly adsorbed alcohols were found to desorb when the temperature was increased to 300–350 K, and at 350–400 K some alkoxide species were found to recombine with surface hydroxyl groups and desorb. Low-temperature desorption of water (by recombination of hydroxyl groups) competes with ethanol formation (*262*). At higher temperatures (500 K or higher), two alternative decomposition pathways compete: dehydrogenation, giving aldehydes or ketones, and dehydration, giving alkenes, with smaller alkoxides favoring dehydrogenation.

Similar chemistry has been observed to occur on rutile powder (*262*). On the TiO_2(001)-(114)-faceted surface, where two alcohol molecules can coordinate to the same metal center, coupling of alcohols to give ethers occurs (*194*). Isopropanol, however, is not coupled to give diisopropyl ether because the bulk of the side chain appears to sterically hinder coordination of multiple isopropoxide ligands (*263*) to surface Ti centers.

Formaldehyde (*189*) and acetaldehyde reactions have been investigated on both (001) single crystals (*190, 195*) and powders (*190*). On the (011)-faceted surface, acetaldehyde and ethanol (a reduction product) formed from acetaldehyde desorbed first at ~400 K. Bimolecular reaction resulted in crotonaldehyde and crotyl alcohol by β aldolization at ~450 K, with small amounts of butene and butadiene (formed by reductive coupling) desorbing at ~550 K. The (114)-faceted surface gave similar chemistry, but it was more selective for aldolization. On stoichiometric TiO_2 powder, in the presence of gas-phase O_2, aldehydes were oxidized to give carboxylates (*264*). However, in UHV the Ti–O bond is too strong (644 kJ/mol for a Ti–O dimer and 2000 kJ/mol of oxygen for the bulk) to be broken.

ii. Reactions of Carboxylic Acids. The decomposition pathways of monocarboxylic acids (C_1–C_3), on both anatase powder (*265*) and rutile $TiO_2(001)$ surfaces, have been characterized extensively by TPD (*156, 266*). Three main decomposition pathways for acetic acid were observed with increasing annealing temperature after sputtering, namely, unselective deoxygenation, dehydration, and ketonization (*156*). From the reduced surface, desorption of water and acetic acid occurred at low temperature (390 K), followed by desorption of products, including CO, CH_4, ketene, and water, at higher temperatures. Deoxygenation of acetate species to fill oxygen vacancies also led to deposition of carbon on the surface.

Similar reaction channels for acetic acid were observed for the (011)-faceted surface, with the major pathway being unimolecular dehydration at 610 K, giving water and ketene [>70% selectivity (*156*)]. Some CO and traces of H_2 were also produced. TPD of CH_3COOD gave an initial desorption peak of CH_3COOD and D_2O, and the second peak gave CH_3COOH and H_2O, indicating that acetate, not acetic acid, is the adsorbed species that reacts at higher temperatures (as was further confirmed by XPS). Hydroxyl groups produced by dissociative adsorption were removed at low temperatures.

On the (114)-faceted surface, a new pathway for ketonization also opened up (*156*). Two acetate molecules bound to the same Ti center formed acetone at ~580 K. Propanoic acid reacted similarly, giving 3-pentanone. Similar chemistry was observed to occur on anatase powders (*265*). TPD of acrylic acid (*193*), an unsaturated carboxylic acid, on the (011)-faceted $TiO_2(001)$ crystal yielded ethylene from decarboxylation, acrolein from reduction, and butene and butadiene. On the (114)-faceted surface, divinyl ketone was also formed by ketonization.

iii. Reactions of Maleic Anhydride. The reaction of maleic anhydride (MA) on the (011)-faceted surface and on defect-containing $TiO_2(001)$ surfaces was investigated by TPD (*267*). (Strictly speaking, MA is not a dicarboxylic acid, although it may behave similarly to one.) Unreacted MA desorbed from the (011)-faceted surface at 380 K, followed by a complex set of desorption products observed at 550 K, including ketene, ethylene, acetylene, vinylacetylene, butene, CO, and CO_2. Small amounts of benzene were also observed at ~620 K. The yield of the coupling products (vinyl-acetylene, butene, and benzene) formed on the argon ion-sputtered (reduced) surface increased with prolonged sputtering. This result is expected in light of the reactivity of reduced surfaces for alkyne coupling. Moreover, evidence of dehydration from both ends of the molecule, giving rise to O=C=C=C=C=O (m/e 80) was proposed. Although m/e 80 was not observed with the mass spectrometer, a signal at m/e 40, attributed to a O=C=C= fragment, was observed. This complex set of products may be explained as follows:

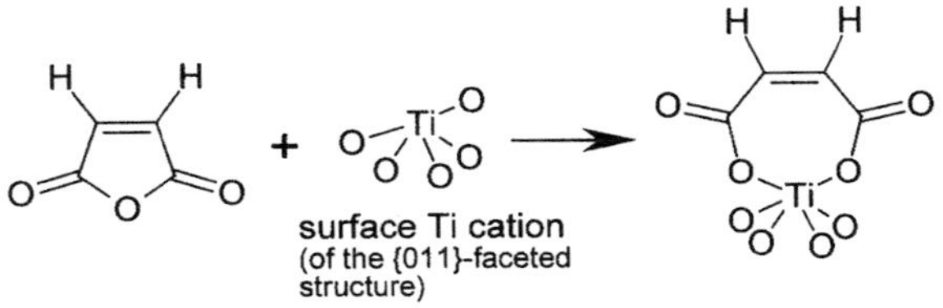

SCHEME 1

For adsorption on surface Ti^{4+}_{5c} ions to take place, electron donation from one surface oxygen atom to one of the two carbonyl groups of MA, followed by electron donation from the σ-bonded oxygen atom of MA to Ti^{4+}, must occur, as shown in Scheme 1. Decarboxylation of this intermediate from both ends results in acetylene formation at high temperature. Acetylene may also be hydrogenated to give ethylene. The high observed ratio of acetylene to ethylene is simply a consequence of kinetics limitations (i.e., the small concentration of adsorbed hydrogen present from total decomposition of a small fraction of MA). Cyclotrimerization of three molecules of acetylene on certain defect sites yields benzene. Ketene is formed via C–C bond breaking.

To achieve this step, the adsorbed molecule likely bonds through the olefin center. Such bonding has been observed by HREELS following MA adsorption and rehybridization of its surface-bound carbon atoms to sp^3 hybridization (*268*) on Pd(111) surfaces. The (011)-faceted surface structure may allow this adsorption and hybridization because (as shown in Fig. 8) there is no limitation to further interactions on adjacent sites [in contrast to the situation on the (110) surface, for example]. The other product, carbon oxide (O=C=C=C=C=O), belongs to a class of molecules of which the simplest is O=C=O. The stability of these molecules varies, with molecules containing an odd number of carbon atoms being stable and those containing an even number being unstable. However, carbon oxide was successfully synthesized and identified in the early 1990s (*269*), even though it is extremely unstable. This instability may be the reason

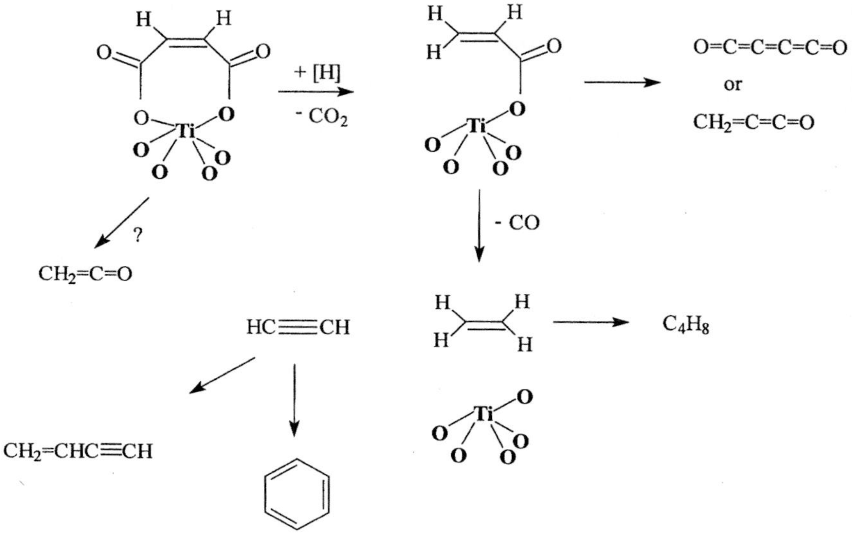

SCHEME 2

for the rapid decomposition and the absence of m/e 80 in the TPD spectrum. No mass spectrometer fragmentation patterns are available for this expected product. Scheme 2 summarizes the different reaction products.

iv. Carbon–Carbon Bond-Forming Reactions. Carbon–carbon bond-forming reactions are of particular importance to the synthetic chemist because they allow the building up of the carbon "backbone" of a molecule. Four different carbon–carbon bond-forming reactions have been demonstrated to occur on metal oxides under UHV conditions, and each was first discovered to occur on $TiO_2(001)$ single-crystal surfaces. These reactions were presented in a previous review (*96*) and are discussed only briefly here.

β-Aldolization. This reaction occurs analogously to the aldol condensation of two C_n aldehydes to give a C_{2n} product in solution. An aldehyde molecule coordinates to a Ti center via the carbonyl oxygen atom, and deprotonation occurs, giving an enolate-type species (Scheme 3).

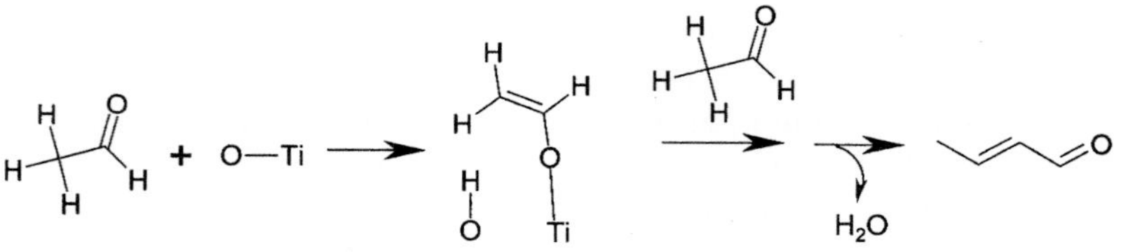

SCHEME 3

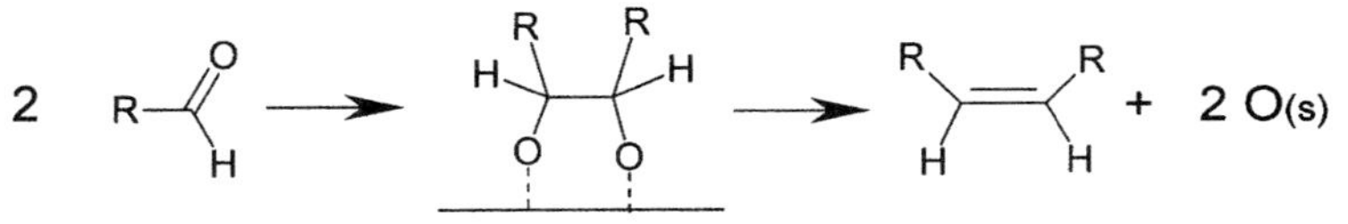

SCHEME 4

This enolate species can then react as a nucleophile at the carbonyl carbon atom of another adsorbed aldehyde. In solution this may give either an intermediate β-hydroxy aldehyde or the dehydration product, an α,β-unsaturated aldehyde. Only the latter product has been observed as a result of the reaction on TiO_2, however. This reaction is structure insensitive—it can occur on all TiO_2 surfaces, including both faceted structures of the (001) surface, sputtered and disordered surfaces, and powder surfaces.

Reductive coupling. This reaction has been known for two decades as the stoichiometric, liquid-phase McMurry reaction, useful for the production of a symmetric C_{2n} olefin from two C_n aldehydes or ketones (*270*). Carbonyl groups of two reactant species coordinate to adjacent Ti atoms on the surface, undergoing a reductive dimerization to give a pinacolate intermediate which is then further deoxygenated to give the olefin shown in Scheme 4.

The surface gains two oxygen atoms in the process. The formation of the strong metal–oxygen bonds provides the thermodynamic driving force for the reaction, and an ensemble of reduced Ti cations [Ti^{+x} ($0 < x < 4$)], able to undergo a four-electron reduction is required. Spectroscopic results have shown that reduced surface defects are necessary for reaction, but metallic, zero-valent titanium is not (*197*), contrary to what had been proposed for the liquid-phase McMurry reaction (*270*).

Butene and stilbene, among other products, have been produced from acetaldehyde (*195*) and benzaldehyde (*197*), respectively, on TiO_2(001) surfaces reduced by argon ion bombardment. The yield of reductively coupled product was found to vary monotonically with the degree of reduction measured by XPS. Furthermore, *p*-benzoquinone was coupled to give both biphenyl and terphenyl (*198*). The use of such multifunctional reactant species offers the possibility of the formation of conducting polymers on the surface (Scheme 5).

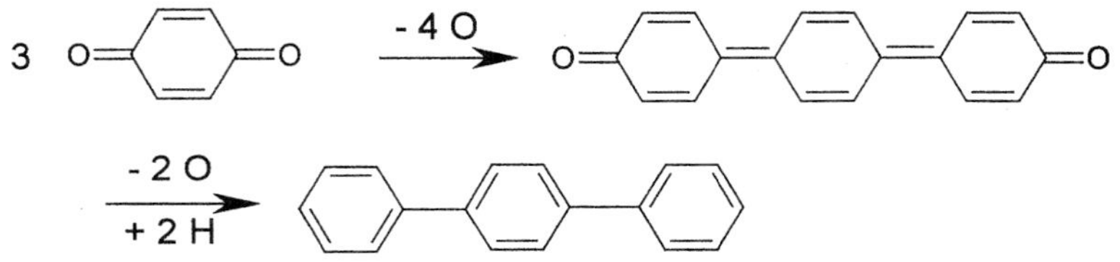

SCHEME 5

Lu *et al.* (*228*) also reported the formation of ethylene from formaldehyde on the reduced $TiO_2(110)$ surface. Thermal annealing to high temperature produced bridging oxygen vacancies and associated Ti^{3+} sites. Adsorbed $^{13}CH_2O$ (or D_2O or ^{15}NO) was reduced at these sites to give $^{13}C_2H_4$ (or D_2 or $^{15}N_2O$, respectively), and the vacancies were filled with oxygen atoms abstracted from the reactant.

Cyclotrimerization. Cyclotrimerization of alkynes to give aromatics has been demonstrated to occur on soluble transition metal complexes as well as on metal catalyst surfaces, including metal single crystals in UHV [e.g., Pd(111)] (*271–274*). In a manner analogous to Ziegler–Natta catalysis, three alkynes on TiO_2 are coupled to give an aromatic ring, via a metallacyclopentadiene intermediate, which may also give rise to a butadiene species (Scheme 6).

This reaction requires a Ti^{2+} center with two coordination vacancies that is able to undergo a two-electron oxidation; thus, it occurs primarily on reduced, defective surfaces such as the sputtered surface. The extent of reaction correlates well with the concentration of Ti^{2+} ions as determined by XPS (*191*). This reaction may be the first example of a truly catalytic carbon–carbon bond formation on a metal oxide surface in UHV (*275*). Additional examples include the production of trimethyl- and hexamethylbenzene from propyne and 2-butyne, respectively (*192*).

Ketonization. This reaction involves the dissociative adsorption of two carboxylic acid molecules and their coordination to the same Ti center, from which the two C_n carboxylates couple, with the elimination of CO_2, to give a C_{2n-1} ketone. Thus, the reaction is structure sensitive—it requires doubly coordinatively unsaturated Ti^{4+} centers. It occurs only on the (114)-faceted $TiO_2(001)$ surface, and not on the (011)-faceted surface, because the latter exposes no fourfold coordinate Ti cations.

The reaction of formic acid produces formaldehyde (*266*), acetic acid produces acetone (*156*), and acrylic acid produces divinyl ketone (*193*).

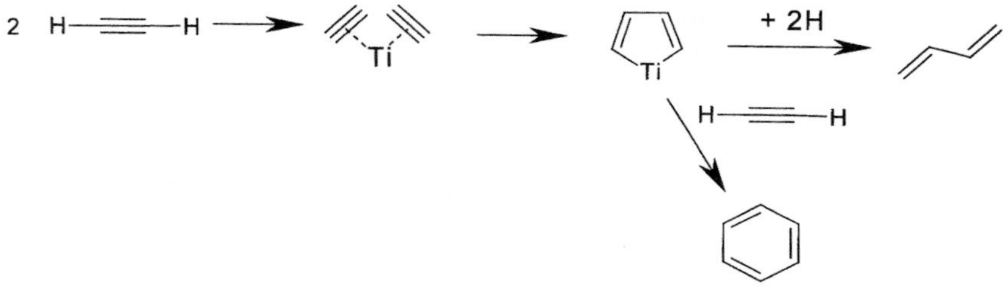

SCHEME 6

Because the ketonization reaction occurs only on the (114)-faceted (001) surface, the presence of these products can be used as an indication of the presence of such structures when LEED or other spectroscopic techniques are not available for confirmation. The increase of the yield of ketonization products with increasing annealing temperature [producing greater formation of (114) facets] has been well characterized (Fig. 10).

In the (114)-faceted structure (Fig. 9), each row is composed of alternating six-, four-, and fivefold coordinated Ti^{4+} ions, and each column contains alternating four- and sixfold coordinated Ti^{4+} ions in this stepped structure. Adjacent fourfold Ti^{4+} ions are present in the (110) directions. This arrangement results in zigzag lines incorporating only fourfold cations. Consequently, one might except that ketonization of dicarboxylic acids

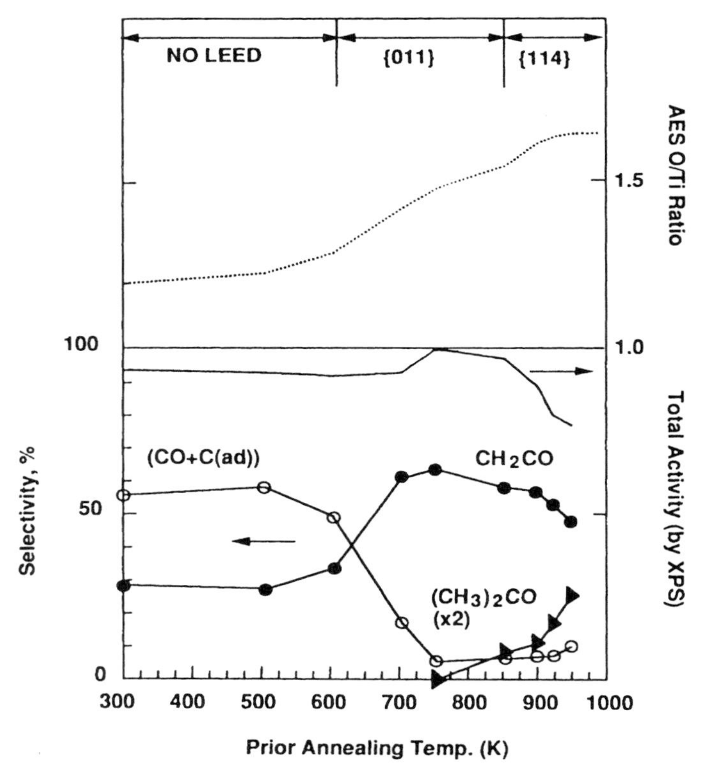

FIG. 10. Selectivity of acetic acid decomposition on $TiO_2(001)$ surfaces as a function of prior annealing temperature. The characteristic surface structures and compositions are depicted at the top [reproduced with permission from Kim and Barteau (*156*)].

along these lines could yield polymeric species that map these sites; no work addressing this possibility has been reported.

C. Principles and Applications of Oxide Surface Science

The picture of adsorption and reaction of organic compounds that has emerged from the preceding two decades of surface science of oxides, summarized in the previous section, complements very well the usual depiction of adsorption on oxide catalysts. Surface cations represent localized Lewis or Brønsted acid sites, and surface oxygen anions are the corresponding base or nucleophilic sites. Perhaps the most important conclusion is that although defects are indeed more reactive than other surface sites on oxides, they are not required for most oxide surfaces to exhibit reactivity. Coordination chemistry on oxide surfaces is rich indeed, and most oxide surfaces provide sites with the necessary coordination vacancies for the acid–base reactions described previously.

The real advantage of surface science approaches to the understanding of oxides is thus perhaps not so much in determining how and where adsorbed intermediates are formed but in understanding the properties of the surface that control the reactions of these to give different products. Surface science experiments permit one to work with uniform surfaces and thus to determine, for example, whether different product channels arise from different surface sites or whether instead they represent competing reactions of common intermediates on common sites. These experiments permit one to uncouple surface structure from the materials properties, e.g., surface area or crystal habit that may dominate the behavior of polycrystalline powder samples. They permit one to determine properties such as surface stoichiometry or oxidation state and to examine in a controlled fashion the effects of these properties on surface reactivity.

Most of the chemistry observed in oxide surface science investigations can be understood in terms of the coordination environment of surface cations, the oxidation state of the surface cations, or the redox properties of the oxide surface (*96*). For example, if one considers the chemistry of carboxylic acids on oxides, one can deduce the following reactivity principles from the reactions observed on different oxide surfaces (*96, 276*):

1. Coordinatively unsaturated surface cations are required for the initial dissociation of the acid. Although such sites are present on most oxide surfaces, surfaces lacking these sites, such as $ZnO(000\bar{1})$, are inactive for Brønsted acid dissociation.
2. Carboxylate ligands formed at cations originally having a single coordination vacancy undergo dehydration preferentially. Representative exam-

ples include carboxylate decomposition on ZnO(0001), MgO(100), and TiO_2(110) and -(001)-(011)-faceted surfaces.

3. Sites that can bind multiple carboxylate ligands (i.e., sites with multiple coordination vacancies initially) can couple pairs of carboxylates to form ketones. The clearest examples from surface science experiments are carboxylate coupling reactions observed to occur on the TiO_2(001)-(114) faceted surface.

4. Reduction of carboxylates to give aldehydes can occur on surfaces that are reduced. Such reactions reoxidize surface cations and are thus favored by the energetics of metal–oxygen bond formation.

5. Decarboxylation of carboxylic acids is favored on surfaces that are relatively easily reduced. This chemistry can be related to the ease of removal of oxygen atoms from the surface lattice. The production of H_2O by acid–base reactions at the surface, followed by CO_2 liberations as a result of decomposition of the carboxylate, involves net consumption of oxygen from the surface. On reducible oxides (e.g., ZnO), this reaction is favorable. On nonreducible oxides (e.g., MgO), net dehydration of the carboxylic acid (which does not require net consumption of lattice oxygen) is favored.

The operation and indeed the application of these principles is evident in several new examples of oxide-catalyzed reactions. For example, the catalytic dehydration of acetic acid and of higher carboxylic acids to give ketenes has been demonstrated to occur on high-surface-area silica (*276, 277*). In this case, surface hydroxyl groups act as placeholders; these are titrated from the surface by the adsorbing carboxylic acids to form water, resulting in binding of isolated carboxylate ligands to the surface. These undergo dehydration selectively to give the corresponding ketenes. The overall catalytic cycle has been depicted as follows (*276, 277*):

$$CH_3CHOOH + OH(a) \longrightarrow CH_3COO(a) + H_2O \tag{2}$$

$$CH_3COO(a) \longrightarrow H_2C = C = O + OH(a) \tag{3}$$

Silica is a nonreducible oxide with limited coordination sites. High-temperature dehydration of silica produces stable Si–O–Si linkages that are not easily opened by adsorption of Brønsted acids from the vapor phase. Thus, the role of OH groups in this case is to maintain sites at which isolated carboxylates can be formed; results of both spectroscopic and gravimetric experiments have shown a 1:1 correspondence between the population of isolated OH groups on this material and its capacity for carboxylate formation.

The performance of these silica catalysts for ketene production is striking. Recent experiments have demonstrated per-pass yields of 80% for ketene production from acetic acid and 95% for dimethyl ketene from isobutyric acid, provided that the silica catalyst is operated in a short contact time (<10 ms) mode (*278*).

The formation of both ketenes and ketones from carboxylic acids on oxides has been known for decades (*279*). Issues in ketene synthesis are primarily related to selectivity; in ketone synthesis, they also include catalyst deactivation. The surface science results described previously and the principles deduced from them provide an explanation for selectivity differences between different oxides in terms of surface structure, coordination environment, and redox properties, thus providing guidance in the selection of potential catalysts. The silica catalyst for ketene synthesis clearly exemplifies this approach.

A second example has recently emerged. Dooley *et al.* (*280, 281*) reported a family of ceria-on-alumina catalysts that show good performance for the synthesis of ketones from carboxylic acids. Of particular interest is the formation of nonsymmetric ketones by cross-coupling of mixtures of two carboxylic acids. For example, these authors reported the synthesis of cyclopropyl methyl ketone by cross-coupling of cyclopropyl carboxylic acid and acetic acid:

$$\triangleright\!\!-\!COOH + CH_3COOH \longrightarrow \triangleright\!\!-\!\overset{\overset{\displaystyle O}{\|}}{C}\!-\!CH_3 + CO_2 + H_2O \qquad (4)$$

As expected on the basis of surface science results for TiO_2 single crystals, as well as previous results from the catalysis literature, TiO_2 is also an active catalyst for carboxylic acid ketonization. It has been suggested that CeO_2-based catalysts are advantageous owing to a lesser tendency to deactivate.

In the past few years, Ponec *et al.* (*282–284*) reported a series of studies of carboxylic acid reactions on oxide catalysts that are in good accord with the reactivity patterns emerging from single-crystal surface science experiments. They observed that iron oxide catalysts were active and selective for the hydrogenation of acetic acid to give acetaldehyde when the catalyst was maintained in a partially reduced state. This state could be enhanced by the addition of platinum to the catalyst, which served to dissociate hydrogen, which in turn spilled over to the iron oxide. This hydrogen served two functions: maintenance of the iron oxide in its active, partially reduced state and reaction with acetic acid. Ketonization reactions, observed on an oxidized catalyst, were suppressed by metal addition and reduction of the iron oxide. The authors concluded that the best oxide catalysts for selective reduction of carboxylic acids to give aldehydes are those with intermediate metal–oxygen bond strengths because these can be operated in a partially reduced state. Oxides that are substantially more difficult to reduce generate the ketene and ketone products observed on stoichiometric single-crystal surfaces discussed previously.

In each of the previous examples, the results of oxide surface science have helped to define important properties of surface sites for different organic transformations, in principle permitting researchers to narrow considerably the scope of their search for active and selective catalysts. Such examples represent significant contributions to catalysis from oxide surface science—a field that, as noted previously, barely existed 15 years ago.

One may anticipate that other such examples will emerge, but we emphasize that "catalyst design" from first principles remains a difficult challenge. Although one can utilize the principles of oxide surface reactivity to narrow the materials selection process, it is difficult to specify the optimum material *a priori.* In effect, greater understanding of reactivity trends among oxides, particularly the separation of "electronic effects" from "structural effects," is needed. By serving to delineate the structural side, oxide surface science has helped to cast less well-developed aspects in sharper relief.

Electronic theories of catalysis have waxed and waned in popularity throughout the decades. In the following section, we consider characteristics of the local electronic structure (e.g., electronegativity and Madelung potential) that may serve to distinguish differences between oxides and thereby contribute to oxide catalyst design. Connections with oxide reactivity can be constructed both on the basis of classical studies carried out with polycrystalline oxide powders and on the basis of single-crystal surface science investigations. Thus, although the principal contribution of oxide surface science may be elucidation of structural requirements of oxide surface reactions, one can also extract important concepts regarding electronic aspects of oxide reactivity and catalysis from such studies, as discussed in the following section.

III. Trends in Oxide Surface Reactivity

A. Trends in the Physical and Electronic Properties of Oxides

Although catalysis by solids is a surface phenomenon, bulk properties of oxides require considerable attention if one is to understand the chemistry of heterogeneous catalysis. Bulk properties govern several important aspects of surface reactions, including the nature and distribution of defects, and the stabilization of surface structures. Several bulk properties have been commonly discussed in connection with catalytic reactions, including electronegativity of metal cations (*285*), Madelung potential (*139*), optical basicity (*286*), metal–oxygen bond strength (*180*), basicity (*135*), and ionicity (*287, 288*). This section is an examination of some of these concepts (or properties) for oxide materials and their consequences for two representative catalytic reactions.

The oxides considered here include the alkali-metal oxides Li_2O to Cs_2O, the alkaline earths MgO to BaO, oxides of aluminum, silicon, transition metals, lanthanides, and one actinide. The objective is simply to test the limits of the physical characteristics enumerated previously in classifying oxide catalytic materials. Among others, the following questions are considered:

1. To what extent can one consider the metal–oxygen bond strength as a measure of the covalent–ionic character?
2. Is there a direct relationship between the partial charge on the oxygen anion and its Madelung potential?
3. Does the electronegativity scale also apply to oxide materials in a useful fashion?

1. *Fundamental Properties*

We first define key concepts.

a. *Partial Charge, q_O.* In a simple diatomic gas-phase molecule, the degree of ionic character of the bond reflects the degree of charge separation. In a molecule M–A formed by a metal atom M and nonmetal atom A, the ionic character may be quantified by the partial charge, q, on either of the atoms. For example, a partial charge of zero of A implies pure covalent bonding, whereas $q = -1$ implies relatively strong ionic character on A with a formal negative charge equivalent to one electron. For an oxide, a partial charge on the oxygen, q_O, of -2 will result in destabilization of the compound; in the Born–Haber cycle for oxides, the negative electron affinity of the O^{2-} ion contributes to instability (*289*).

b. *Electronegativity, χ.* This term describes the force exerted by the nuclear charge on the electrons at the periphery of an atom. It indicates the tendency to become charged and not the state of being charged. Many different scales have been used to quantify the electronegativity χ (*290*), but regardless of the scale used it is generally the electronegativity difference $\Delta\chi$ between two atoms that governs bonding character (*291*). $\Delta\chi$ represents the difference in the atomic electronegativities χ_O of the neutral oxygen atom and χ_M of the neutral metal in their actual lattice positions: $\Delta\chi = \chi_O - \chi_M$. The difference can be calculated by using the heat of formation Q for the oxide of the formula MO_x according to the following equation (*289*):

$$\Delta\chi = [(Q + 1.13x)/2x]^{1/2} \tag{5}$$

where Q is in eV. The term 1.13 appears in the equation to account for double bonding in the O_2 molecule.

Duffy (*289*) compiled computed values of the partial charge of oxygen anions in oxides and compared them to χ_O. He obtained the following relationship:

$$q_O = \chi_O - 4.1 \tag{6}$$

Equation (6) requires use of a value for χ_O that is characteristic of a given oxide. To calculate χ_O one can employ the relationship $\chi_O = \chi_M + \Delta\chi$, together with a value of $\Delta\chi$ computed from Eq. (5) and conventional literature values for χ_M on a Pauling scale. The use of conventional literature values for χ_M represents an approximation that may not reflect the actual value of χ_M in an oxide lattice. It is important that q_O, and therefore the ionic character that it describes, is semiempirical.

c. *Metal–Oxygen Bond Strength $E_{(M-O)}$*. Calculation of $E_{(M-O)}$ begins with the determination of the heat of formation of an ion gas from the following equation (*180*):

$$y\,\Delta H_{M-O}(\text{ionic}) = -\Delta H_f(M_xO_y) + \Delta H_{sub}(M) + (y/2)\,\Delta H_{dis}(O_2) - y\sum \text{EA(O)} + x\sum \text{IE(M)} \tag{7}$$

where $\Delta H_f(M_xO_y)$ denotes the heat of formation of the oxide (*204*), $\Delta H_{sub}(M)$ is the heat of sublimation of the metal (*292*), $\Delta H_{dis}(O_2)$ is the dissociation energy of diatomic oxygen (*204*), ΣEA(O) the sum of the first and second electron affinities for atomic oxygen (*204*), and ΣIE(M) is the sum of the first through $2y/x$ ionization energies for the metal (*204*). $E_{(M-O)}$ can then be estimated by dividing $y\Delta H_{(M-O)}$ by the coordination number C_O of oxygen in the oxide. For example, C_O is equal to 3, 4, 3, and 2 for Fe_2O_3, CaO, TiO_2, and SiO_2, respectively.

d. *Madelung Potential, V_{Mad}*. This term quantifies the collective electrostatic interaction of the entire lattice with the type of atom in question. In an oxide, a Madelung potential can be defined for both the metal {[$V_{Mad}(M)$]} and the oxygen [$V_{Mad}(O)$] atoms. For lattice oxygen ions $V_{Mad}(O)$ is a function of the interatomic equilibrium metal–oxygen distance (r_e), the metal cation charge (q_M), and the Madelung constant (α) and is given by $\alpha q_M/r_e$. We have seen, for example, that $V_{Mad}(O)$ of CeO_2 and UO_2 are almost the same since both oxides have the same structure and both cations have the same oxidation state as well as nearly the same ionic sizes. In oxides, the values of $V_{Mad}(O)$ vary from ~15 to ~32 eV, whereas

TABLE VIII
Physical Properties of Oxides

Oxide	Q (eV)	$\Delta\chi$ (eV)	$E_{(M-O)}$ (eV)	χ_M (eV)	χ_O (eV)	q_O
Rutile systems						
TiO_2	9.8	1.74	20.7	1.5	3.24	−0.76
SnO_2	6.0	1.44	15.5	1.8	3.24	−0.86
MoO_2	6.1	1.45	16.7	1.8	3.25	−0.85
WO_2	6.1	1.45	16.9	1.7	3.15	−0.95
IrO_2	2.8	1.13	15.7	2.2	3.33	−0.77
RuO_2	3.2	1.16	15.4	2.2	3.36	−0.74
PbO_2	2.9	1.13	15.4	1.8	2.93	−1.17
Cr_2O_3 systems						
V_2O_3	12.6	1.63	12.3	1.6	3.23	−0.87
Fe_2O_3	8.5	1.41	16.6	1.8	3.21	−0.89
Al_2O_3	17.5	1.86	13.5	1.5	3.36	−0.70
Fe_3O_4	11.6	1.42	11.0	1.8	3.22	−0.88
Rh_2O_3	3.6	1.08	9.5	2.2	3.28	−0.82
Wurtzite						
ZnO	3.6	1.54	10.4	1.6	3.14	−0.96
La_2O_3 systems						
La_2O_3	18.6	1.91	7.0	1.1	3.01	−1.09
Sm_2O_3	18.9	1.92	9.9	1.1	3.02	−1.08
Fluorite systems						
CeO_2	11.3	1.84	10.0	1.1	2.94	−1.16
UO_2	11.2	1.83	9.7	1.7	3.53	−0.57
NaCl systems	6.3	1.92	4.3	1.2	3.12	−0.98
MgO	6.6	1.97	3.6	1.0	2.97	−1.13
CaO	6.1	1.91	3.3	1.0	2.91	−1.19
SrO	5.8	1.86	3.0	0.9	2.76	−1.34
BaO						
Na_2O systems	6.2	1.91	2.0	1.0	2.91	−1.19
Li_2O	4.3	1.65	1.5	0.9	2.55	−1.55
Na_2O	3.7	1.56	1.2	0.8	2.36	−1.74
K_2O	3.5	1.52	1.1	0.8	2.31	−1.79
Rb_2O	3.3	1.49	1.0	0.7	2.19	−1.91
Cs_2O						
Cu_2O system						
Ag_2O	0.3	0.85	4.2	1.9	2.75	−1.35
Other systems						
PdO	0.9	1.0	6.9	2.2	3.20	−0.90
UO_3	12.7	1.64	13.9	1.7	3.36	−0.76
U_3O_8	37.1	1.70	24.7	1.7	3.37	−0.74
PtO_2	0.9	0.88	16.4	2.2	3.08	−1.01
OsO_4	4.1	1.04	26.3	2.2	3.24	−0.86
WO_3	8.7	1.42	36.6	1.7	3.12	−0.98
CuO	1.6	1.17	7.1	1.9	3.07	−1.03
ZrO_2	11.4	1.77	14.2	1.4	3.17	−0.93
SiO_2	34.2	1.67	34.2	1.8	3.47	−0.60

those of $V_{Mad}(M)$ may have values of -20 to -60 eV. In highly ionic oxides, the Madelung potential is relatively small (in Na_2O, $V_{Mad}(O) = 19.6$ eV; $V_{Mad}(M) = -9.1$ eV), whereas in more covalent oxides the Madelung potential rises significantly (in β-quartz SiO_2, $V_{Mad}(O) = 30.5$ eV; $V_{Mad}(M) = -47.9$ eV).

e. *Oxygen Electronic Polarizability.* The electronic polarizability α_{OX} of an oxide describes the tendency of the bonding electrons to remain near their ion cores in the presence of an electric field. Because most of the valence electrons in an oxide are associated with the oxygen anion, the net material polarizability is dominated by the polarizability α_O of that oxygen. Indeed, materials scientists have found several interesting relationships between α_{OX} [as well as a related quantity: the change in the Auger parameter (*293*)] and the bulk properties of binary and ternary oxide materials (*294–296*).

The oxygen electronic polarizability α_O of a binary oxide M_xO_y may be calculated from the oxide polarizability α_{OX} and the solid-phase cation polarizability α_{Mz+} by use of the following equations:

$$\alpha_{OX} = (3V_m/4\pi N_a)(\kappa - 1)/(\kappa + 2) = (3V_m/4\pi N_a)(n^2 - 1)/(n^2 + 2) \quad (8)$$

and

$$\alpha_O = 1/y(\alpha_{OX} - x\alpha_{Mz+}) \quad (9)$$

where κ is the dielectric constant, V_m is the molar volume, N_a is Avogadro's number, and n is the refractive index. Equation (8) is taken from Duffy (*289*).

2. *Relationships between Fundamental Properties*

a. *Electronegativity Difference, M–O Bond Strength, and Partial Charge.* Table VIII and Fig. 11 show the calculated values for $\Delta\chi$, q_O, and $E_{(M-O)}$ for many oxides belonging to several families. Within a given family, close relationships exist. For example, for both the Na_2O and NaCl systems the decrease of $\Delta\chi$ is paralleled by a decrease in $E_{(M-O)}$ and an increase in q_O. However, it is also clear that there is no overall trend among all the oxides in Table VIII. As shown in Fig. 11, hardly any relationship between $\Delta\chi$ and $E_{(M-O)}$ is evident. Moreover, for $E_{(M-O)}$ values exceeding ~10 eV, almost all the oxides considered have similar oxygen partial charges [the linear relation at smaller $E_{(M-O)}$ values is due to the Na_2O and NaCl systems; see Table VIII].

Figure 12 presents a plot of q_O as a function of $\Delta\chi$. There are indeed correlations between $\Delta\chi$ and q_O within certain subsets of oxides. For exam-

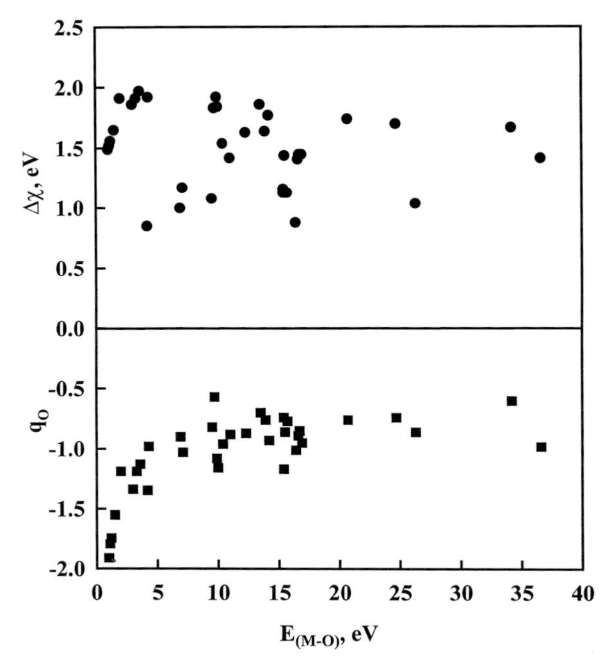

FIG. 11. Electronegativity difference $\Delta\chi$ and partial charge q_O of the oxides in Table VIII as a function of their corresponding metal cation–oxygen anion bond energy $E_{(M-O)}$.

ple, the alkali metal, alkaline earth, and lanthanide oxides cluster together along a rough line extending through the lower right portion of the diagram. The noble metal oxides form a nearly straight line at the upper left of Fig. 12. The remaining d block oxides, together with PbO_2, Al_2O_3, SiO_2, and uranium oxides, cluster loosely along a line running through the upper center of Fig. 12. The three lines for the oxides have distinctly different slopes and intercepts, showing conclusively that in the solid phase additional factors affect $\Delta\chi$ or q_O or both. Thus, except within carefully defined limits, it is not valid to use $\Delta\chi$ as a proxy for q_O. For example, SiO_2 and Na_2O have essentially the same $\Delta\chi$, but the former has a partial charge that is only 40% of the latter.

b. *Madelung Potential and Partial Charge.* A few workers have discussed the relationships between V_{Mad} and the properties of oxide materials. Broughton and Bagus (*181*) compiled extensive data for oxides in an attempt to correlate variations in V_{Mad}, as well as in X-ray photoelectron spectra, to ionicity. However, these workers could find no consistent quanti-

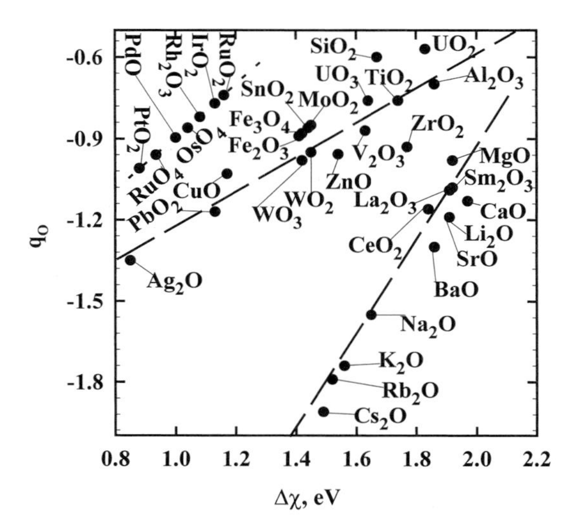

FIG. 12. Variation of the partial charge q_O as a function of the electronegativity difference $\Delta\chi$ (in eV) for several families of oxides.

tative relationship between ionicity and any of the quantities $V_{Mad}(M)$, $V_{Mad}(O)$, or ΔV_{Mad}. Pacchioni *et al.* (*139, 297*) showed that the direct relationship between $V_{Mad}(O)$ of a small series of oxides (MgO to BaO) may track the ionicity (basicity) as determined by Tanabe and Saito (*135*) for these same oxides. We have also previously shown that ΔV_{Mad} could also be used to correlate successfully alcohol dehydrogenation/dehydration selectivities on four different oxide single crystals.

The full possibilities for using V_{Mad} as a tool to predict catalytic properties for a wide range of oxides remain only incompletely explored. Figure 13 shows q_O as a function of $V_{Mad}(O)$. [$V_{Mad}(O)$ values were taken from Broughton and Bagus (*181*) for all but the alkaline earth series MgO to BaO, and for these $V_{Mad}(O)$ values were drawn from Pacchioni *et al.* (*139*).] Figure 13 exhibits a clear trend that can be expressed by an empirical equation of quadratic form:

$$q_O = -4.57 + 0.23\, V_{Mad}(O) - 3.3 \times 10^{-3}\, V^2_{Mad}(O) \qquad (10)$$

Equation (10) fits the entire data set with a correlation coefficient of 0.84. Equally important, within two families of oxides linear relationships are observed. $q_O = -2.28 \pm 0.05\, V_{Mad}(O)$ for the series MgO, CaO, SrO, and BaO; and $q_O = -3.1 \pm 0.08\, V_{Mad}(O)$ for the series Li_2O, Na_2O, K_2O, and Rb_2O. These two linear relationships indicate that one can estimate with satisfactory accuracy either quantity if the other is known.

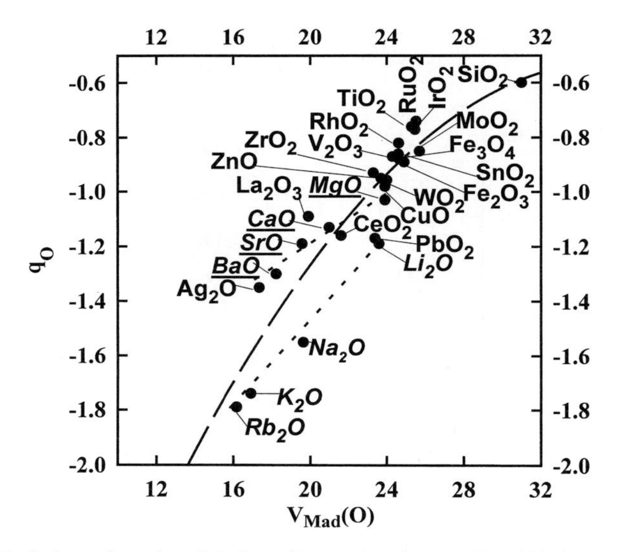

FIG. 13. Variation of q_O (partial charge) as a function of $V_{Mad}(O)$ for several families of oxides.

Why the partial charge on the oxygen atom decreases as the Madelung potential increases can be understood as follows. An O^{2-} ion in the gas phase dissociates readily into $O^- + e^-$. Similarly, an O^{2-} ion in a solid is nominally less stable than O^-. Dissociation of the O^{2-} ion in a solid such as Na_2O is prevented by the electrostatic potential exerted by all the surrounding ions. Nevertheless, the Madelung potential represents a good metric for materials stability; higher magnitudes of V_{Mad} correspond to greater stability. Thus, the relative stability implied by a larger value of V_{Mad} should correlate with a decrease the magnitude in q_O away from -2. Figure 13 shows exactly this trend.

In summary, it is safe to state the following:

1. No general relationship exists between the electronegativity difference and the partial charge on the oxygen anions of more than 30 oxides investigated.
2. No general relationship exists between the bond strengths of transition metal oxides and the electronegativity difference. This result indicates that most likely neither quantity alone can be used to classify transition metal oxides.
3. To some approximation, there is a reasonable relationship between the ionicity and the Madelung potential of oxygen anions among the oxides investigated (including transition metal oxides). In particular, this relationship holds extremely well for alkali metal oxides.

B. Correlations with Oxide Reactivity

In the previous section, we discussed the different fundamental oxide properties and their relationships to each other. In this section, we analyze data for two simple reactions in the light of these relationships:

1. The esterification (also called the Tishchenko reaction) of benzaldehyde to benzylbenzoate on MgO, CaO, SrO, and BaO (*135*)
2. The dehydrogenation of ethanol to give acetaldehyde on CeO_2 (*298*), CuO (*299*), CaO and SiO_2, Fe_3O_4, Fe_2O_3, and TiO_2 (*300*)

As was previously shown, these oxides exhibit great diversity in their properties. The oxides of the alkaline earth series MgO to BaO represent typical examples of ionic materials having a rock salt structure, in which both bulk anions and cations are sixfold coordinated. SiO_2, in contrast, is a prototypical covalent oxide. TiO_2 (anatase) is an amphoteric material that exhibits both ionic and covalent character. Fe_3O_4 is a mixed oxide composed of Fe^{2+} and Fe^{3+} ions in an inverse spinel structure, and Fe_2O_3 (hematite) is largely ionic.

In principle, rates of acid–base reactions should be expected to scale with Lewis acid–base strength, which in turn is affected by electronegativity (*1*). Correlations between the electronegativity of the metal cations and the rate of CO and CO_2 hydrogenation have been proposed (*285*). On the other hand, as indicated previously, oxidation reactions are often limited by a step in which oxygen is removed from the lattice to leave a vacancy [although several counterexamples can be identified (*301–303*)]. In such cases, the reaction rate would be expected to correlate with M–O bond strength. Indeed, a decrease in the rate of CO oxidation to give CO_2 has been well correlated with an increase in M–O bond strength for this reason (*180, 304*).

The titration of basic sites of a series of oxides by CO_2 adsorption at room temperature was recently carried out with a series of metal oxides (*300*). CO_2 adsorption leads to carbonate formation on many metal oxides, as shown in Scheme 7 (*305*).

M–O bond breaking, or at least considerable weakening, clearly takes place. Thus, in the absence of significant surface structural effects, adsorp-

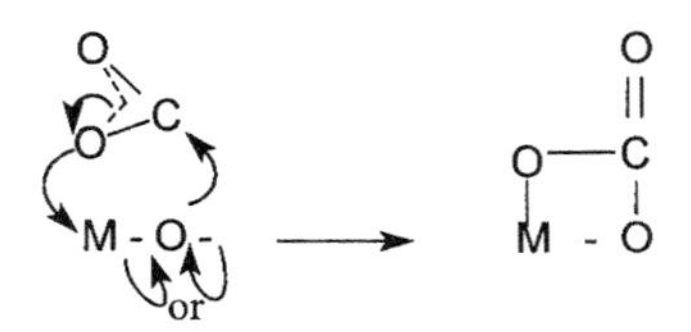

Scheme 7

tion capacity should vary inversely with $E_{(M-O)}$. The data for CO_2 uptake and $E_{(M-O)}$ shown in Table IX (*306*) confirm this trend. This correlation suggests that carbonate formation does not merely reflect the titration of surface Lewis base sites, as is often assumed, but is also sensitive to the ability of the surface to share or donate oxygen to adsorbates.

Thus, some caution must be exercised in using this probe to count basic sites at which different acids react. For example, if one normalizes the rate of a reaction, the oxidative dehydrogenation of ethanol to give acetaldehyde, according to CO_2 uptakes for the limited set of oxides in Table IX a reasonable correlation is obtained (*300*). The main exception is CaO (which exhibits unusual CO_2 adsorption capacity, as discussed in Section II.) The failure of CaO to fit the simple pattern for M–O bond strength indicates a need to separate mere counting of surface sites from measures of their reactivity or catalytic activity. This remains an important challenge in heterogeneous catalysis.

The Tishchenko reaction provides one of the best available examples of correlations between reaction rate and oxide electronic properties for polycrystalline materials. This reaction involves the conversion of two aldehyde molecules to form an ester. Tanabe and Saito (*135*) investigated the effects of the basic properties of alkaline–earth oxides (BeO, MgO, CaO, SrO, and BaO) on the reaction of benzaldehyde to give benzyl benzoate. The Tishchenko reaction requires hydrogen transfer from one adsorbed aldehyde, which is oxidized, to another adsorbed aldehyde, which is reduced to alkoxide. This process may involve a transition state that requires participation of a surface oxygen atom (Scheme 8) in which an adsorbed aldehyde molecule, with the participation of a surface oxygen anion, transfers a hydride to another adsorbed aldehyde molecule. This process results in a carboxylate species in proximity to an alkoxide species. The role of surface oxygen here is that of a nucleophile.

TABLE IX

CO_2 Uptake[a] (Molecules/m^2) and $E_{(M-O)}$ (eV) for a Series of Oxides

Oxide	*CO_2 uptake (molecules/m^2)*	$E_{(M-O)}$ (eV)
SiO_2	0	34.2
TiO_2	1.4×10^{17}	20.7
Fe_2O_3	5.0×10^{17}	13.6
Fe_3O_4	6.5×10^{17}	11.0
CaO	30.2×10^{17}	3.6

[a] From Idriss and Seebauer (*300*).

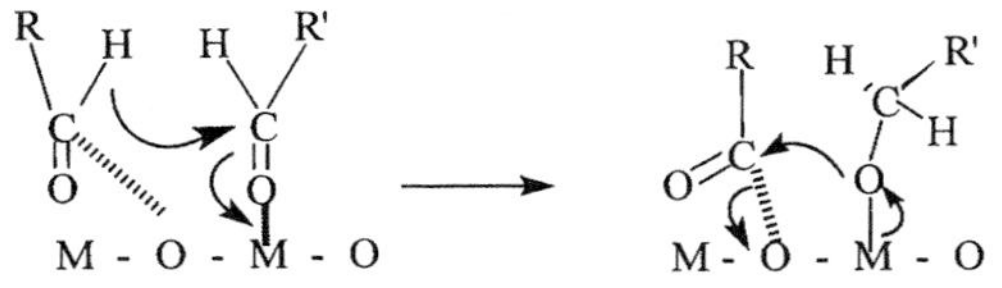

→ RC(O)OCH2R'

SCHEME 8

The rates of reaction on the surfaces of the alkaline earth oxides do indeed scale with those physical properties that provide a measure of surface basicity. Table X includes values of the logarithm of the rate constant for the reaction [from Tanabe and Saito (*135*)] and q_O, $V_{Mad}(O)$, and α_O for the series MgO to BaO. Each of the three properties scales very well with the rate constant of the reaction. The more ionic and more polarizable the oxygen, the greater the rate of the reaction.

Although the correlation of activity with oxide electronic properties for this example is impressive, it is not clear that this correlation can be extended straightforwardly to a broader range of oxides, including those of transition metals. Furthermore, no single-crystal surface studies of this reaction have been carried out, and thus effects of the surface structure have not been probed.

We therefore return to the example of alcohol dehydrogenation and dehydration to address some of these issues. As noted previously, the dehydrogenation/dehydration selectivity of alcohol reactions is often viewed as a criterion for classifying oxide materials. The conventional wisdom is that basic oxides tend toward dehydrogenation, whereas acidic oxides (which are less ionic) tend toward dehydration (*24–30, 297, 307,*

TABLE X

Comparison of the Electronic Properties of the Oxides in the Series from MgO to BaO, $V_{Mad}(O)$, and the Partial Charge q_O with the Reaction Rate Constant for the Reaction of Benzaldehyde to Benzyl Benzoate[a]

Oxide	ln k_{453K} (k in $min^{-1}m^{-2}$)	α_O	q_O	$V_{Mad}(O)$ (eV)
MgO	−5.7	1.81	−0.98	23.9
CaO	−4.4	2.96	−1.13	20.98
SrO	−2.1	3.93	−1.19	19.57
BaO	1.6[b]	5.25	−1.3	18.22

[a] From Tanabe and Saito (*135*).

[b] Calculated from Tanabe and Saito (*135*).

308). Although this classification may be valid under restricted conditions, it is at best an oversimplified picture.

Table XI includes values of the logarithm of the rate constant at 473 K for the oxidative dehydrogenation of ethanol to give acetaldehyde on several unrelated oxides [the data were taken from three sources (*298–300*)]. The comparison between these rate constants and the corresponding values of $\Delta\chi$, $E_{(M-O)}$, q_O, and $V_{Mad}(O)$ from Table VIII and Figs. 11–13 suggests the following:

1. $\Delta\chi$ and $E_{(M-O)}$ do not correlate with the relative reaction rates.
2. Correlations with q_O and $V_{Mad}(O)$ show imperfect trends; there is a good linear relationship between q_O [or $V_{Mad}(O)$] for six of the eight oxides, but CaO and SiO_2 are exceptions, exhibiting activities that are nearly two orders of magnitude lower than predicted by the correlation with q_O or $V_{Mad}(O)$.

To proceed with further analysis, it is helpful to examine more closely the dehydrogenation mechanism of alcohols. The dehydrogenation reaction of primary alcohols requires several steps that might be viewed as follows.

First, upon adsorption of the alcohol, the O–H bond dissociates heterolytically to yield an ethoxide and a proton, as shown in Scheme 9. The amount of ethoxide does not necessarily correlate with the surface catalytic activity. For example, considerable amounts of alkoxy species were observed at room temperature on oxides such as SiO_2 (*309*), although SiO_2 exhibits little activity for oxidative dehydrogenation (*300*). Instead, the amount of ethoxide correlates with the number of metal cations that exhibit coordinative unsaturation. Furthermore, the amount of surface ethoxide may be influenced by the electronic charge distributions around the oxygen anions that tend to abstract the hydrogen in the form of protons.

TABLE XI

Catalytic Activities for Ethanol Oxidative Dehydrogenation to Acetaldehyde Catalyzed by a Series of Polycrystalline Oxides

Oxide	ln k (k in ml/m^2 catalyst · s)	Reference
SiO_2	−14.6	Idriss and Seebauer (*300*)
TiO_2	−7.4	Idriss and Seebauer (*300*)
Fe_2O_3	−5.6	Idriss and Seebauer (*300*)
Fe_3O_4	−6.5	Idriss and Seebauer (*300*)
CaO	−6.2	Idriss and Seebauer (*300*)
CuO	−2.6	Tu *et al.* (*299*)
CeO_2	−1.2	Yee *et al.* (*298*)

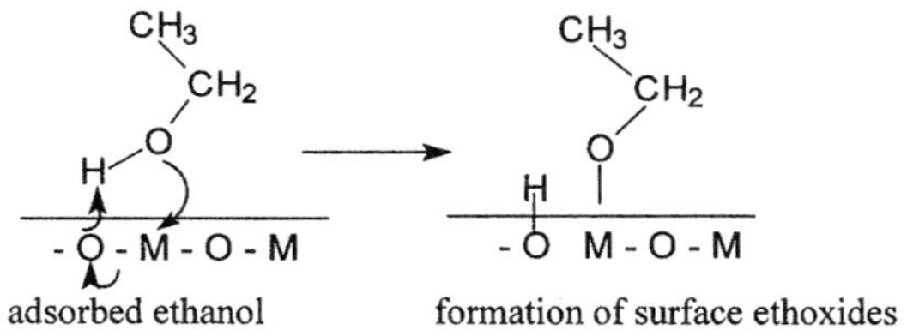

adsorbed ethanol

formation of surface ethoxides

(step 1)

SCHEME 9

Ethoxides undergo dehydrogenation with subsequent electron donation to the cation, yielding acetaldehyde, as shown in Scheme 10. This picture shows why surfaces that promote dehydrogenation reactions tend to be those containing reducible (and reoxidizable) cations.

Second, the hydrogen atoms released by this reaction are removed from the surface primarily either as H_2O or as H_2. H_2O can originate from the recombination of adsorbed OH and M–H. This product then desorbs, leaving an oxygen vacancy and a partially reduced metal atom. Two adsorbed OH species can also combine to make one molecule of H_2O, one oxygen vacancy, and one restored oxygen anion site. On the other hand, two M–H species may combine to make H_2.

Finally, the key characteristic of a surface promoting the alkoxide dehydrogenation reaction is its ability to break one M^{+x}–O^{2-} bond. Indeed, the stability of surface point defects is important in determining the selectivity to acetaldehyde.

Surface science investigations of well-defined oxide surfaces have helped to demonstrate the complexity of correlating or predicting such reactivity trends. A clear example is the following: The (011)-faceted structure of $TiO_2(001)$, containing only Ti^{+4}_{5c} ions, is active for methanol conversion (*194*). The dehydration route ultimately yields methane. (This path is a

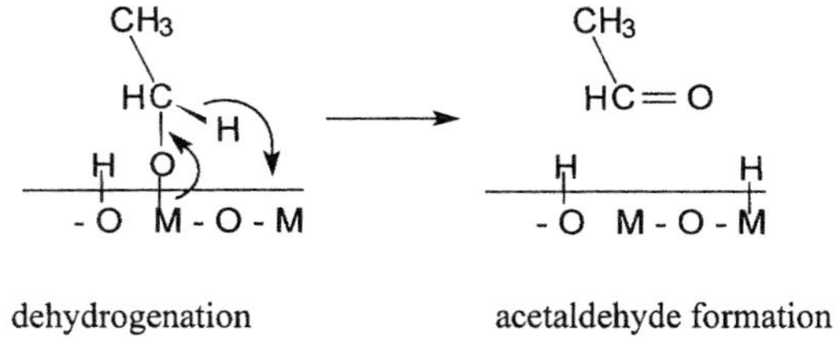

dehydrogenation

acetaldehyde formation

(step 2)

SCHEME 10

combined dehydration and hydrogenation, but for the sake of simplicity it may still be considered to track dehydration.) Dehydrogenation, in contrast, gives formaldehyde. The relative activities of the surface for these two reactions can be given by the HCHO : CH_4 ratio, which was observed to be 0.22 (*194*).

However, if one transforms the (011)-faceted structure into the (114)-faceted structure, where one-third of the surface Ti^{+4} ions are now 4-fold coordinated to O^{2-} (i.e., have two coordination vacancies), a second dehydration route is observed. This new route yields dimethyl ether (with 25% selectivity). The formation of the ether is accompanied by an increase of the HCHO : CH_4 ratio to 3.5 (i.e., a 16-fold increase).

This change in reaction selectivity is obviously not related to a change in the Ti^{4+}–O^{2-} bond strength. The most likely explanation is that accommodation of two methoxides by individual Ti^{+4} ions having the two coordination vacancies leads to dimethyl ether. In other words, the dominant effect is structural.

Results characterizing ethanol dehydrogenation are given in Fig. 14, which is a plot showing the relationship between α_O and the rate constant k for the series of polycrystalline oxide samples referred to in Table XI. The fit is relatively good, although CeO_2 deviates slightly from the line.

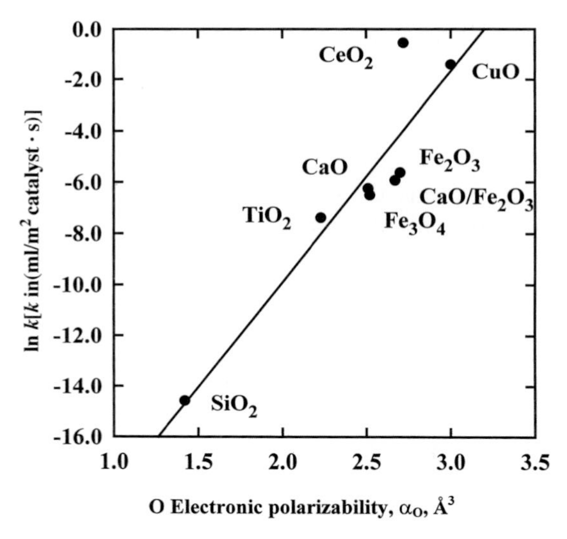

FIG. 14. Catalytic activity, represented as ln k (k in ml/m^2 of catalyst · s), as a function of α_O (Å^3) for the oxidative dehydrogenation reaction of ethanol to acetaldehyde catalyzed by a series of oxides [adapted from Idriss and Seebauer (*306*)].

The deviations might be attributed to several causes. There may be significant errors in the values used for the metal cation polarizabilities. More important, variations in α_O originating from structural effects almost certainly occur, and these may be not be taken into account when polycrystalline materials are used.

Thus, there is a need for single-crystal data. Although surface science studies of oxides have been carried out for more than two decades, there is still only a limited set of reaction examples that allow testing of general trends. For example, the reaction of ethanol has been conducted in a quantitative manner on only four well-defined oxide surfaces: TiO_2(001)-(011) faceted (*32*), TiO_2(110) (*187*), ZnO(0001) (*116*), and UO_2(111) (*186*). On each of these surfaces, alcohols undergo both dehydration and dehydrogenation reactions. Each of these surfaces incorporates cations having one coordination vacancy. The extent of reaction may therefore be followed by one or more of the intrinsic characteristics of these oxides.

Table XII shows the relationships between the reaction selectivity and the optical properties of oxides, the partial charge q_O, the bond strength $E_{(M-O)}$, and ΔV_{Mad}. Again, several trends are evident. Oxides with high values of q_O (strong ionic character) are more selective for the dehydrogenation reaction than the others. ΔV_{Mad} values also vary in the same order as the reaction selectivity, as illustrated in Fig. 3. A poorer correlation with α_O is apparent because of the value for ZnO(0001). Deviations due to changes of the electronic properties as a function of crystallographic orientation (not considered in Table XII) are probably most important in the case of ZnO polar surfaces. The polarizability of the zinc polar face of a ZnO single crystal is certainly far from those of the oxygen polar face or of nonpolar faces, and thus from the average oxygen polarizability. These results again demonstrate the need to account explicitly for surface structure in attempting to develop physical property–activity correlations for oxide catalysts.

TABLE XII

Comparison of the Optical Properties of Oxides, the Bond Strength, $E_{(M-O)}$, and ΔV_{Mad} with the Reaction Selectivity for Alcohol Dehydrogenation/Dehydration on Oxide Single-Crystal Surfaces

Oxide	q_O	$E_{(M-O)}$ (eV)	α_O	ΔV_{Mad} (eV)	Aldehyde/olefin selectivity
TiO_2(001)	−0.76	20.73	2.23	−70.6	0.2 from ethanol
UO_2(111)	−0.57	9.67	2.56	−66	0.8 from ethanol
ZnO(0001)	−0.9	10.36	2.40	−48	3.2 from ethanol
Cu_2O(100)	−1.10	4.91	5.5	−34.6	4.6 from *n*-propanol

IV. Summary

We conclude with a brief evaluation of the accomplishments and future expectations of oxide surface science in the advancement of heterogeneous catalysis. First and foremost among these must be the clear examples provided by surface science experiments (and to a lesser extent theory) of the site requirements for adsorption and various chemical reactions on oxide surfaces. These have been considered in detail and in previous publications. The principles of oxide surface reactivity developed from investigations of well-defined oxide surfaces represent the foundation on which a more complete understanding of oxide catalysis may be built.

For example, the understanding of surface site requirements for dissociative adsorption and unimolecular dehydration on oxide single-crystal surfaces has led directly to the demonstration of a new oxide catalyst for the synthesis of ketenes. Studies of the surface reactivity of titanium oxide single crystals have demonstrated dramatic changes with variation of cation oxidation state and coordination environment. The ability to define these physical characteristics of the surface has permitted correlation of rates of hydrocarbon and oxygenate reactions with them, providing both useful diagnostic probes and new catalytic syntheses (e.g., of functionalized olefins by reductive carbonyl coupling). Such studies have also provided valuable insights into organic reaction mechanisms on oxide surfaces, serving to delineate examples in which competing reaction channels do or do not proceed via common surface intermediates or sites. For example, UHV experiments with oxide single crystals and thin films have demonstrated that dehydration and dehydrogenation reactions of oxygenates occur via common intermediates on oxide surfaces. One need not invoke different sites or different descriptions of oxide surface reaction mechanisms to account for the different reaction channels observed.

However, the "catalog" of examples of reactions on well-defined surfaces, and of chemical principles that one might expect would emerge from these, is still far from complete. As noted previously, if one wishes to compare a common reaction on differnt single-crystal oxide surfaces, it is frequently difficult to construct a database from the current literature that consists of more than three or four different materials or surfaces. This situation will undoubtedly change in the future, permitting more complete tests of principles of oxide surface reactivity such as those advanced here.

Perhaps of equal importance will be the observation of exceptions to these principles and the elucidation of the origins of such exceptions. As noted previously, surface defects remain challenging targets for studies of model surfaces. The creation of "well-defined" defects—or at least the definition of the physical characteristics of the defects present—is only part of the challenge. Understanding the reactivity of what may be a small minority of such sites in the presence of more abundant reactive sites is indeed a formidable task.

By establishing for a broader range of materials the "baseline" behavior of various probe reagents on well-defined surfaces tending toward minimum defect populations, one can hope to distinguish the chemistry of defect sites as their population and variety are deliberately manipulated. Experiments with partially reduced single-crystal surfaces represent but a first step in this direction; the sites on such surfaces are often well defined in terms of their oxidation states but not necessarily in terms of their structures or coordination environments.

STM holds considerable promise for producing structural definition of oxide surface defects and for probing the occupancy of these (as well as some characteristics of the surface–adsorbate interaction) under various conditions. The difficulty (and time requirements) of the necessary experiments should not be underestimated, however.

At the beginning of this review, we enumerated important concepts derived from or applied to heterogeneous catalysis by metal oxides. All of these, from acid–base concepts to site isolation, predate the growth of the field of oxide surface science, and one may well ask what oxide surface science has contributed to testing and refining these concepts or to creating new ones. Clearly, the most direct contributions have been in demonstrating the dependence of many reactions on local site structure and in providing both unequivocal examples and explanations of surface structural anisotropy in surface reactions. Indeed, in the case of TiO_2, we have been able to show that the reactions of small oxygenates depend dramatically on the coordination environment of individual surface cations, but the influence of bulk crystallographic structure (i.e., the spatial relationships between cations) is largely unimportant once one accounts for differences in surface site density between different bulk structures.

The major contributions of oxide surface science to surface acid–base chemistry have been twofold. First, these techniques, both experimental and theoretical, have confirmed the classical picture of Brønsted acid–base reactions on surface cation–anion site pairs. Surface oxygen anions are base sites that bind protons; metal cations are Lewis acid sites that coordinate the conjugate base anions of dissociatively adsorbed Brønsted acids. Second, experiments with well-defined surfaces have helped to provide new insights into the mechanisms of surface acid–base reactions.

As discussed previously, alcohol dehydration and dehydrogenation reactions are common probes of surface acid–base properties. Single-crystal experiments have shown that both reactions can occur on oxide surfaces that are devoid of Brønsted acid sites, and they involve a common surface intermediate—the alkoxide. Although the selectivity of alkoxide reactions on oxides can be correlated with characteristics (e.g., the Madelung potentials of surface anions and cations and oxygen polarizability) that can be related to surface acidity and basicity, surface science studies have demonstrated that the mechanistic assumptions underlying the use of such probe reactions need to be reconsidered. Significant alcohol dehydration can occur in the absence of Brønsted acid sites via reactions in which the initial step is proton donation to basic surface anions.

Single-crystal experiments have provided an opportunity to resolve effects of surface crystallographic structure, stoichiometry, surface cation coordination environment, etc. on alcohol reactions and thus to begin to address electronic effects in oxide catalysis. We previously discussed correlations of oxide surface reactivity with metal–oxygen bond strength, ionicity, Madelung potential, etc.; clearly, more data representing well-defined surfaces are needed to provide more comprehensive tests of such correlations.

It is perhaps ironic that the concepts most closely connected with selective oxidation catalysis, Mars–van Krevelen mechanisms and site isolation, have been the least impacted by oxide surface science. The key to Mars–van Krevelen schemes, of course, is the redox behavior of the oxide surface. Organic reactants reduce the surface; oxygen adsorption must therefore occur on partially reduced surfaces. Surface science studies of oxygen adsorption on fully oxidized surfaces may have little to offer in this regard because stoichiometric surfaces frequently exhibit little affinity for further oxygen uptake, especially under UHV conditions. However, strongly reduced surfaces, such as those created by ion bombardment, may be able to provide little insight into structural aspects of oxygen adsorption and dissociation and may be well outside the surface composition range actually sampled during redox catalysis.

There is a need for creation of model systems containing structurally well-defined defects, as discussed previously, or at least ordered suboxide phases at their surfaces. The demonstration by STM of the formation of crystallographic shear planes at the surfaces of molybdenum oxides and titanium oxides represents an important example of structural characterization of reduced sites on oxides. STM may also be capable of providing information about the transport of oxygen between surface sites and cooperative rearrangements of the surface or near-surface regions that may be involved in such processes. Again, the key is the creation of structurally well-defined surfaces on which one can examine the site requirements of oxygen adsorption and transport.

In principle, the same needs apply to surface science experiments that might bolster the concept of site isolation in reactions on oxide surfaces.

In the literature of single-crystal experiments to date, however, it is difficult to discern clear examples that support this concept. Several issues may be at play. First, many of the oxides that have been used in surface science studies are not particularly good oxygen donors; thus, the influence of limiting the access of adsorbates to surface oxygen may be difficult to determine from the available data. Second, TPD experiments at low pressure (in UHV) may also tend to decrease the importance of interactions of adsorbates with neighboring sites. In such experiments, products formed by reactions occurring during the temperature ramp may desorb without interacting with other surface sites, and readsorption may be negligible. Furthermore, as discussed previously, much of the chemistry of small molecules examined on oxide surfaces can be understood in terms of the local coordination environment of surface cations; evidence for cooperative effects has been minimal. This is not to suggest that the concept of site isolation is not valid; rather, it has not been validated by oxide surface science studies. Again, a significant challenge exists in the fabrication of model surfaces that expose isolated sites of varying, but well-controlled, size and structure.

In conclusion, what oxide surface science has to offer oxide catalysis is structural definition of surfaces. The consequences of this are twofold. First, there is the opportunity, illustrated by the examples discussed previously, to connect the catalytic activities of oxides to specific surface structural features, i.e., to define the site requirements for reactions. By making such connections, one can establish a basis for comparison of the activities of the surfaces of different metal oxides in terms of sites of like geometric structure or coordination environment. By making such comparisons on the basis of comparable structures, one can identify and perhaps quantify other characteristics (e.g., electronic properties) that contribute to differences in activity and selectivity among oxide catalysts. Thus, the second consequence of the ability to characterize reactions on structurally well-defined oxide surfaces is a greater opportunity to understand and predict electronic effects in oxide catalysis. Oxide surface science promises to grow in importance as a tool for elucidating the physical and chemical bases for catalysis by metal oxides and for advancing the field toward the goal of catalyst design.

References

1. Henrich, V. E., and Cox, P. A., "The Surface Science of Metal Oxides." Cambridge Univ. Press, Cambridge, UK, 1994.
2. Khoobiar, S., *J. Phys. Chem.* **68,** 411 (1964).
3. Conner, W. C., *Stud. Surf. Sci. Catal.* **77,** 61 (1993)
4. Weisz, P. B., and Swegler, E. W., *Science* **126,** 31 (1957).

5. Conner, W. C., and Falconer, J. L., *Chem. Rev.* **95,** 7–59 (1995).
6. Robbins, J. L., and Marucchisoos, E., *J. Phys. Chem.* **93,** 2885 (1987).
7. Fu, S. S., and Somorjai, G. A., *Appl. Surf. Sci.* **48,** 93 (1991).
8. Borer, A. L., Prins, R., Goodwin, J. G., Dejong, K. P., Bell, A. T., Solymosi, F., Gonzalez, R. D., Koningsberger, D. C., Pinna, F., Johnston, P., Ponec, V., Waugh, K. C., Coenen, J. W. E., Ichikawa, M., Klier, K., Schmal, M., and Sachtler, W. M. H., *Stud. Surf. Sci. Catal.* **75** (Part A), 765 (1993).
9. Idriss, H., Diagne, C., Hindermann, J. P., Kiennemann, A., and Barteau, M. A., *J. Catal.* **155,** 219 (1995).
10. Vannice, M. A., and Sudhakar, C., *J. Phys. Chem.* **88,** 2429 (1984).
11. Dandekar, A., and Vannice, M. A., *J. Catal.* **183,** 344 (1999).
12. Minahan, D. M., Hoflund, G. B., Epling, W. S., and Schoenfeld, D. W., *J. Catal.* **168,** 393 (1997).
13. Yong, Y. S., Kennedy, E. M., and Cant, N. W., *Appl. Catal.* **76,** 31 (1991).
14. Mao, C. F., and Vannice, M. A., *Appl. Catal. A General* **122,** 61 (1995).
15. Kung, H. H., "Transition Metal Oxides: Surface Chemistry and Catalysis." Elsevier, Amsterdam, 1989.
16. Gates, B. C., Katzer, J. R., and Schuit, G. C. A., "Chemistry of Catalytic Processes." McGraw-Hill, New York, 1979.
17. Grasselli, R. K., and Burrington, J. D., *Adv. Catal.* **30,** 133 (1981).
18. Gates, B. C., "Catalytic Chemistry." Wiley, New York, 1992.
19. Ebner, J. R., Franchetti, V., Centi, G., and Trifiro, F., *Chem. Rev.* **88,** 55 (1988).
20. Muhler, M., *in* "Handbook of Heterogeneous Catalysis" (G. Ertl, H. Knözinger, and J. Weitkamp, Eds.), Vol. 5, p. 2274. Wiley–VCH, Weinheim, 1997.
21. Haber, J., *in* "Handbook of Heterogeneous Catalysis," (G. Ertl, H. Knözinger, and J. Weitkamp, Eds.), Vol. V, p. 2253. Wiley-VCH, Weinheim, 1997.
22. Weissermel, K., and Arpe, H.-J., "Industrial Organic Chemistry," 2nd ed. VCH, Weinheim, 1993.
23. Segall, R. L., Smart, R. St. C., and Turner, P. S., *in* "Surface and Near-Surface Chemistry of Oxide Materials" (J. Nowotny and L.-C. Dufour, Eds.), p. 257. Elsevier, Amsterdam, 1988.
24. Tanabe, K., "Solid Acids and Bases." Academic Press, New York, 1970.
25. Krylov, O. V., "Catalysis by Non-Metals." Academic Press, New York, 1970.
26. Pines, H., and Manassen, J., *Adv. Catal.* **16,** 49 (1966).
27. Winterbottom, J. M., *in* "Catalysis" (C. Kemball and D. A. Dowden, Eds.), Vol. 4, p. 141. Royal Society of Chemists, London, 1981.
28. Ai, M., *Bull. Chem. Soc. Jpn.* **50,** 2579 (1977).
29. Ai, M., and Suzuki, S., *J. Catal.* **30,** 362 (1973).
30. Mars, P., *in* "The Mechanism of Heterogeneous Catalysis" (J. H. de Boer, Ed.), p. 49. Elsevier, Amsterdam, 1959.
31. Vohs, J. M., and Barteau, M. A., *J. Phys. Chem.* **95,** 297 (1991).
32. Kim, K. S., and Barteau, M. A., *J. Mol. Catal.* **63,** 103 (1990).
33. Boudart, M., and Djega-Mariadassou, G., "Kinetics of Heterogeneous Catalytic Reactions." Princeton Univ. Press, Princeton, NJ, 1984.
34. Sinfelt, J. H., "Bimetallic Catalysts, Discoveries, Concepts, and Applications." Wiley, New York, 1983.
35. Burdett, J. K., "Chemical Bonding in Solids." Oxford Univ. Press, New York, 1995.
36. Schmid, G., Baumle, M., Geerkens, M., Helen, I., Osemann, C., and Sawitowski, T., *Chem. Soc. Rev.* **28,** 179 (1999).
37. Whitten, R. L., Cox, D. M., Trevor, D. J., and Kaldor, A., *Surf. Sci.* **156,** 8 (1985).

38. Zuburtikudis, I., and Saltsburg, H., *Science* **258,** 1337 (1992).
39. Barteau, M. A., and Vohs, J. M., *in* "Successful Design of Catalysts" (T. Inui, Ed.), p. 89. Elsevier, Amsterdam, 1988.
40. Oyama, S. T., *Bull. Chem. Soc. Jpn.* **61,** 2585 (1988).
41. Ying, J. Y., and Sun, T., *J. Electroceramics* **1,** 219 (1997).
42. Cao, L., Spiess, F.-J., Huang, A., Suib, S. L., Obee, T. N., Hay, S. O., and Freihaut, J. D., *J. Phys. Chem. B* **103,** 2912 (1999).
43. Mars, P., and van Krevelen, D. W., *Chem. Eng. Sci. Suppl.* **3,** 41 (1954).
44. Callahan, J. L., and Grasselli, R. K., *AIChE J.* **6,** 755 (1963).
45. Didillon, B., Houtman, C., Shay, T., Candy, J. P., and Basset, J.-M., *J. Am. Chem. Soc.* **115,** 9380 (1993).
46. Humblot, F., Candy, J. P., Le Peltier, F., Didillon, B., and Basset, J.-M., *J. Catal.* **179,** 459 (1998).
47. Annis, D. A., and Jacobsen, E. N., *J. Am. Chem. Soc.* **121,** 4147 (1999).
48. Sheldon, R. A., *Topics Curr. Chem.* **164,** 21 (1993).
49. Bein, T., *Curr. Opin. Solid State Mater. Sci.* **4,** 85 (1999).
50. Barteau, M. A., and Vohs, J. M., *in* "Handbook of Heterogeneous Catalysis" (G. Ertl, H. Knözinger, and J. Weitkamp, Eds.), Vol. 2, p. 873. Wiley–VCH, Weinheim, 1997.
51. Vohs, J. M., and Barteau, M. A., *Surf. Sci.* **176,** 91 (1986).
52. Peng, X. D., and Barteau, M. A., *Langmuir* **7,** 1426 (1991).
53. Gleason, N. R., and Zaera, F., *Surf. Sci.* **385,** 294 (1997).
54. Petrie, W. T., and Vohs, J. M., *Surf. Sci.* **245,** 315 (1991).
55. Petrie, W. T., and Vohs, J. M., *Surf. Sci.* **259,** L750 (1991).
56. Petrie, W. T., and Vohs, J. M., *J. Vac. Sci. Technol. A* **11,** 2169 (1993).
57. Dilara, P. A., and Vohs, J. M., *J. Phys. Chem.* **97,** 12919 (1993).
58. Wu, M. C., and Goodman, D. W., *Catal. Lett.* **15,** 1 (1992).
59. Truong, C. M., Wu, M. C., and Goodman, D. W., *J. Chem. Phys.* **97,** 9447 (1992).
60. Wu, M. C., Truong, C. M., and Goodman, D. W., *J. Phys. Chem.* **97,** 9425 (1993).
61. Freund, H. J., Dillman, B., Seiferth, O., Klivenyi, G., Bender, M., Erlich, D., Hemmerich, I., and Coppins, D., *Catal. Today* **32,** 1 (1996).
62. Xu, C., and Koel, B. E., *J. Chem. Phys.* **102,** 8158 (1995).
63. Evans, J., Hayden, B. E., and Lu, G., *J. Chem. Soc. Faraday Trans.* **92,** 4733 (1996).
64. Hayden, B. E., King, A., and Newton, M. A., *J. Phys. Chem. B* **103,** 203 (1999).
65. Ibach, H., *Phys. Rev. Lett.* **24,** 1416 (1970).
66. Pireaux, J. J., Thiry, P. A., Sporken, R., and Caudano, R., *Surf. Interface Anal.* **15,** 189 (1990).
67. Thiry, P. A., Liehr, M., Pireaux, J. J., and Caudano, R., *J. Elec. Spectrosc. Rel. Phenom.* **39,** 69 (1986).
68. Cox, P. A., Flavell, W. R., Williams, A. A., and Egdell, R. G., *Surf. Sci.* **152/153,** 784 (1985).
69. Egdell, R. G., *in* "Adsorption and Catalysis on Oxide Surfaces" (M. Che and G. C. Bond, Eds.), p. 173. Elsevier, Amsterdam, 1985.
70. Frederick, B. G., Nyberg, G. L., and Richardson, N. V., *J. Elec. Spectrosc. Rel. Phenom.* **64/65,** 825 (1993).
71. Jupille, J., and Richardson, N. V., *Surf. Sci.* **274,** L501 (1992).
72. Petrie, W. T., and Vohs, J. M., *Surf. Sci.* **274,** L503 (1992).
73. Akhter, S., Lui, K., and Kung, H. H., *J. Phys. Chem.* **89,** 1958 (1984).
74. Firment, L. E., *Surf. Sci.* **116,** 205 (1982).
75. Xu, C., and Goodman, D. W., *in* "Handbook of Heterogeneous Catalysis" (G. Ertl, H. Knözinger, and J. Weitkamp, Eds.), Vol. 2, p. 827. Wiley–VCH, Weinheim, 1997.
76. Goodman, D. W., *J. Vac. Sci. Technol. A* **14,** 1526 (1996).

77. Street, S. C., Xu, C., and Goodman, D. W., *Annu. Rev. Phys. Chem.* **48,** 43 (1997).
78. Freund, H.-J., Kuhlenbeck, H., and Staemmler, V., *Rep. Prog. Phys.* **59,** 283 (1996).
79. Weiss, W., Boffa, A. B., Dumphy, J. C., Galloway, H. C., Salmeron, M. B., and Somorjai, G. A., *in* "Adsorption on Ordered Surfaces of Ionic Solids and Thin Films" (H.-J. Freund and E. Umbach, Eds.). Springer, Berlin, 1993.
80. Ritter, M., Ranke, W., and Weiss, W., *Phys. Rev. B* **57,** 7240 (1998).
81. Freund, H.-J., *Angew. Chem. Int. Ed. England* **36,** 452 (1997).
82. Kim, Y. J., Gao, Y., and Chambers, S. A., *Surf. Sci.* **371,** 358 (1997).
83. Oh, W. S., Xu, C., Kim, D. Y., and Goodman, D. W., *J. Vac. Sci. Technol. A* **15,** 1710 (1997).
84. Bandara, A., Kubota, J., Wada, A., Domen, K., and Hirose, C., *Surf. Sci.* **364,** L580 (1996).
85. Chambers, S. A., Tran, T. T., Hileman, T. A., and Jurgens, T. A., *Surf. Sci.* **320,** L81 (1994).
86. Bonnell, D. A., *Prog. Surf. Sci.* **3,** 187 (1998).
87. Sander, M., and Engel, T., *Surf. Sci.* **302,** L263 (1994).
88. Szabo, A., and Engel, T., *Surf. Sci.* **329,** 241 (1995).
89. Onishi, H., and Iwasawa, Y., *Surf. Sci.* **358,** 773 (1996).
90. Iwasawa, Y., *Surf. Sci.* **402,** 8 (1998).
91. Murray, P. W., Leibsle, F. M., Muryn, C. A., Fisher, H. J., Flipse, C. F. J., and Thornton, G., *Surf. Sci.* **321,** 217 (1994).
92. Lennie, A. R., Condon, N. G., Leibsle, F. M., Murray, P. W., Thornton, G., and Vaughan, D. J., *Phys. Rev. B* **53,** 10244 (1996).
93. Li, M., Hebenstreit, W., and Diebold, U., *Surf. Sci.* **414,** L951 (1998).
94. Shaikhutdinov, Sh. K., and Weiss, W., *Surf. Sci.* **432,** L627 (1999).
95. Henrich, V. E., *Rep. Prog. Phys.* **48,** 1481 (1985).
96. Barteau, M. A., *Chem. Rev.* **96,** 1413 (1996).
97. Barteau, M. A., *J. Vac. Sci. Technol. A* **11,** 2162 (1993).
98. Machiels, C. J., Cheng, W. H., Chowdhry, U., Farneth, W. E., Hong, F., McCarron, E. M., and Sleight, A. W., *Appl. Catal.* **25,** 249 (1986).
99. Farneth, W. E., Staley, R. H., and Sleight, A. W., *J. Am. Chem. Soc.* **108,** 2327 (1986).
100. Tatibouet, J. M., and Germain, J. E., *J. Catal.* **72,** 375 (1981).
101. Tatibouet, J. M., Germain, J. E., and Volta, J. C., *J. Catal.* **82,** 240 (1983).
102. Smith, R. L., and Rohrer, G. S., *J. Catal.* **163,** 12 (1996).
103. Smith, R. L., and Rohrer, G. S., *J. Catal.* **180,** 270 (1998).
104. Smith, R. L., and Rohrer, G. S., *J. Catal.* **184,** 49 (1999).
105. Chowdhry, U., Ferretti, A., Firment, L. E., Machiels, C. J., Ohuchi, F., and Sleight, A. W., *Appl. Surf. Sci.* **19,** 360 (1984).
106. Roxlo, C. B., Deckman, H. W., Gland, J., Cameron, S. D., and Chianelli, R. R., *Science* **235,** 1629 (1987).
107. Lui, K., Vest, M., Berlowitz, P., Akhter, S., and Kung, H. H., *J. Phys. Chem.* **90,** 3183 (1986).
108. Cheng, W. H., Akhter, S., and Kung, H. H., *J. Catal.* **82,** 341 (1983).
109. Akhter, S., Cheng, W. H., Lui, K., and Kung, H. H., *J. Catal.* **85,** 437 (1984).
110. Berlowitz, P., and Kung, H. H., *J. Am. Chem. Soc.* **108,** 3532 (1986).
111. Zwicker, G., Jacobi, K., and Cunningham, J., *Int. J. Mass Spectrosc. Ion. Proc.* **60,** 213 (1984).
112. Jacobi, K., *in* "Adsorption on Ordered Surfaces of Ionic Solids and Thin Films" (E. Umbach and H.-J. Freund, Eds.), p. 103. Springer, Berlin, 1993.
113. Zhang, R., Ludviksson, A., and Campbell, C. T., *Catal. Lett.* **25,** 277 (1994).
114. Vohs, J. M., and Barteau, M. A., *J. Phys. Chem.* **91,** 4766 (1987).
115. Vohs, J. M., and Barteau, M. A., *Surf. Sci.* **201,** 418 (1988).
116. Vohs, J. M., and Barteau, M. A., *Surf. Sci.* **221,** 590 (1989).

117. Vohs, J. M., and Barteau, M. A., *J. Phys. Chem.* **95,** 297 (1991).
118. Casarin, M., Favero, G., Glisenti, A., Granozzi, G., Maccato, C., Tabacchi, G., and Vittadini, A., *J. Chem. Soc. Faraday Trans.* **92,** 3247 (1996).
119. Vohs, J. M., and Barteau, M. A., *Surf. Sci.* **197,** 109 (1988).
120. Vohs, J. M., and Barteau, M. A., *J. Catal.* **113,** 497 (1988).
121. Vohs, J. M., and Barteau, M. A., *Langmuir* **5,** 965 (1989).
122. Vohs, J. M., and Barteau, M. A., *J. Phys. Chem.* **93,** 8343 (1989).
123. Au, C. T., Hirsch, W., and Hirschwald, W., *Surf. Sci.* **197,** 391 (1988).
124. Rohrer, G. S., and Bonnell, D. A., *Surf. Sci.* **247,** L195 (1991).
125. Thibado, P. M., Rohrer, G. S., and Bonnell, D. A., *Surf. Sci.* **318,** 379 (1994).
126. Parker, T. M., Condon, N. G., Lindsay, R., Leibsle, F. M., and Thornton, G., *Surf. Sci.* **415,** L1046 (1998).
127. Discosimo, J. I., and Apestegua C. R., *J. Mol. Catal. A Chem.* **130,** 177 (1998).
128. Palomares, A. E., Eder-Mirth, G., Rep, M., and Lercher, J. A., *J. Catal.* **180,** 56 (1998).
129. Hattori, H., Tanaka, Y., and Tanabe, K., *Chem. Lett.,* 659 (1975).
130. Peng, X. D., and Barteau, M. A., *Langmuir* **5,** 1051 (1989).
131. Sokolvski, V. D., *React. Kinet. Catal. Lett.* **35,** 337 (1987).
132. Ito, T., and Lunsford, J. H., *Nature* (*London*) **314,** 721 (1985).
133. Lunsford, J. H., *Langmuir* **5,** 12 (1989).
134. Sugiyama, S., Sato, K., Yamasaki, S., Kawashiro, K., and Hayashi, H., *Catal. Lett.* **14,** 127 (1992).
135. Tanabe, K., and Saito K., *J. Catal.* **35,** 247 (1974).
136. Coluccia, S., Segall, R. L., and Tench, A. J., *J. Chem. Soc. Faraday Trans. I* **75,** 289 (1979).
137. Sauer, J., Ugliengo, P., Garrone, E., and Saunders, V. R., *Chem. Rev.* **94,** 2095 (1994).
138. Zecchina, A., Lofthouse, M. G., and Stone, F. S., *J. Chem. Soc. Faraday Trans. I* **71,** 1476 (1975).
139. Pacchioni, G., Ricart, J. M., and Illas, F., *J. Am. Chem. Soc.* **116,** 10152 (1994).
140. Pacchioni, G., and Pescarmona, P., *Surf. Sci.* **412/413,** 657 (1998).
141. Coluccia, S., Barricco, M., Marchese, L., and Zecchina, A., *Spectrochim. Acta* **49A,** 1289 (1993).
142. Marchese, L., Coluccia, S., Matra, G., and Zecchina, A., *Surf. Sci.* **269,** 135 (1992).
143. Snis, A., Strömberg, D., and Panas, I., *Surf. Sci.* **292,** 317 (1993).
144. Snis, A., Panas, I., and Strömberg, D., *Surf. Sci.* **310,** L579 (1994).
145. Pacchioni, G., *Surf. Sci.* **281,** 207 (1993).
146. Ochs, D., Brause, M., Braun, B., Maus-Friedrichs, W., and Kempter, V., *Surf. Sci.* **397,** 101 (1998).
147. Ochs, D., Brause, M., Braun, B., Maus-Friedrichs, W., and Kempter, V., *J. Electron. Spectrosc. Relat. Phenom.* **88,** 757 (1988).
148. Ochs, D., Braun, B., Maus-Friedrichs, W., and Kempter, V., *Surf. Sci.* **417,** 406 (1998).
149. Peng, X. D., and Barteau, M. A., *Catal. Lett.* **7,** 395 (1990).
150. Panella, V., Suzanne, J., and Coulomb, J.-P., *Surf. Sci.* **350,** L211 (1996).
151. Ferry, D., Suzanne, J., Hoang, P. N. M., and Girardet, C., *Surf. Sci.* **375,** 315 (1997).
152. Szymanski, M. A., and Gillan, M. J., *Surf. Sci.* **367,** 135 (1996).
153. Stikaki, M.-A. D., Tsipis, A. C., Tsipis, C. A., and Xanthopoulos, C. E., *J. Chem. Soc. Faraday Trans.* **92,** 2765 (1996).
154. Ferrari, A. F., Huber, S., Knözinger, H., Neyman, K. M., and Rösch, N., *J. Phys. Chem. B* **102,** 4548 (1998).
155. Stubenrauch, J., Brosha, E., and Vohs, J. M., *Catal. Today* **28,** 431 (1996).
156. Kim, K. S., and Barteau, M. A., *J. Catal.* **125,** 353 (1990).

157. Schulz, K. H., and Cox, D. F., *J. Phys. Chem.* **96,** 7394 (1992).
158. Knözinger, H., *Catal. Today* **32,** 71 (1996).
159. Qui, X. O., Wong, N. B., Tin, K. C., and Zhu, Q. M., *J. Chem. Technol. Biotechnol.* **65,** 303 (1996).
160. Stone, F. S., Garrone, E., and Zecchina, A., *Mater. Chem. Phys.* **13,** 331 (1985).
161. Peng, X. D., and Barteau, M. A., *Surf. Sci.* **224,** 327 (1989).
162. Onishi, H., Egawa, C., Aruga, T., and Iwasawa, Y., *Surf. Sci.* **191,** 479 (1987).
163. Nicholas, J. B., Kheir, A. A., Xu, T., Krawietz, T. R., and Haw, J. F., *J. Am. Chem. Soc.* **120,** 10471 (1998).
164. Ellis, W. P., and Schwoebel, R. L., *Surf. Sci.* **11,** 82 (1968).
165. Ellis, W. P., *Surf. Sci.* **45,** 569 (1974).
166. Ellis, W. P., and Taylor, T. N., *Surf. Sci.* **75,** 279 (1978).
167. Taylor, T. N., and Ellis, W. P., *Surf. Sci.* **107,** 249 (1981).
168. Cox, L. E., and Ellis, W. P., *Solid State Commun.* **78,** 1033 (1991).
169. Tasker, P. W., *Surf. Sci.* **78,** 315 (1979).
170. Hubbard, H. V. A., PhD dissertation, Leeds University, 1981.
171. Hubbard, H. V. A., and Griffiths, T. R., *J. Chem. Soc. Faraday Trans.* **2,** 83, 1215 (1987).
172. Griffiths, T. R., Hubbard, H. V. A., Allen, G. C., and Tempest, P. A., *J. Nucl. Mater.* **151,** 307 (1988).
173. Griffiths, T. R., and Hubbard, H. V. A., *J. Nucl. Mater.* **185,** 243 (1991).
174. Sayle, T. X. T., Parker, S. C., and Catlow, C. R. A., *Surf. Sci.* **316,** 329 (1994).
175. Castell, M. R., Muggelberg, C., and Briggs, G. A. D., *J. Vac. Sci. Technol. B* **14,** 966 (1996).
176. Muggelberg, C., Castell, M. R., and Briggs, G. A. D., *Surf. Rev. Lett.* **5,** 315 (1998).
177. Muggelberg, C., Castell, M. R., Briggs, G. A. D., and Goddard, D. T., *Surf. Sci.* **402,** 673 (1998).
178. Muggelberg, C., Castell, M. R., Briggs, G. A. D., and Goddard, D. T., *Appl. Surf. Sci.* **142,** 124 (1999).
179. Nörenberg, H., and Briggs, D. A. D., *Surf. Sci.* **402,** 734 (1998).
180. Rethwisch, D. G., and Dumesic, J. A., *Langmuir* **2,** 73 (1986).
181. Broughton, J. Q., and Bagus, P. S., *J. Electron Spectrosc. Relat. Phenom.* **20,** 261 (1980).
182. Hausner, H., *J. Nucl. Mater.* **15,** 179 (1965).
183. Jollet, F., Petit, T., Gota, S., Thromat, N., Gautier-Soyer, M., and Pasturel, A., *J. Phys. Condens. Matter* **9,** 9393 (1997).
184. Catlow, C. R. A., *J. Chem. Soc. Faraday Trans. II* **74,** 1901 (1978).
185. Pettit, T., Lemaignan, C., Jollet, F., Bogot, B., and Pasturel, A., *Phil. Mag. B.* **77,** 779 (1998).
186. Chen, S. V., Griffith, T., and Idriss, H., *Surf. Sci.* **444,** 187 (2000).
187. Gamble, L., Jung, L. S., and Campbell, C. T., *Surf. Sci.* **348,** 1 (1996).
188. Schulz, K. H., and Cox, D. F., *J. Phys. Chem.* **97,** 643 (1993).
189. Idriss, H., Kim, K. S., and Barteau, M. A., *Surf. Sci.* **262,** 113 (1992).
190. Idriss, H., Kim, K. S., and Barteau, M. A., *J. Catal.* **139,** 119 (1993).
191. Pierce, K. G., and Barteau, M. A., *J. Phys. Chem.* **98,** 3882 (1994).
192. Pierce, K. G., and Barteau, M. A., *Surf. Sci.* **326,** L473 (1995).
193. Idriss, H., Kim, K. S., and Barteau, M. A. *in* "Structure–Activity and Selectivity Relationships in Heterogeneous Catalysis" (R. K. Grasselli and A. W. Sleight, Eds.), p. 327. Elsevier, Amsterdam, 1991.
194. Kim, K. S., and Barteau, M. A., *Surf. Sci.* **223,** 13 (1989).
195. Idriss, H., and Barteau, M. A., *Catal. Lett.* **40,** 147 (1996).
196. Idriss, H., Pierce, K., and Barteau, M. A., *J. Am. Chem. Soc.* **113,** 715 (1991).
197. Idriss, H., Pierce, K. G., and Barteau, M. A., *J. Am. Chem. Soc.* **116,** 3063 (1994).

198. Idriss, H., and Barteau, M. A. *Langmuir* **10,** 3693 (1994).
199. Linsebigler, A. L., Lu, G., and Yates, J. T., Jr., *Chem. Rev.* **95,** 735 (1995).
200. Tompsett, G. A., Bowmaker, G. A., Cooney R. P., Metson J. B., Rodgers, K. A., and Seakins, J. M., *J. Raman Spectosc.* **26,** 57 (1995).
201. Hadjiivanov, K. I., and Klissurski, D. G., *Chem. Soc. Rev.* **25,** 61 (1996).
202. Oliver, P. M., Watson, G. W., Kelsey, E. T., and Parker, S. C., *J. Mater. Chem.* **7,** 563 (1997).
203. Ramamoorthy, M., Vanderbilt, D., and King-Smith, R. D., *Phys. Rev. B* **49,** 16721 (1994).
204. Lide, D. R. (Ed.), "CRC Handbook of Chemistry and Physics," 76th ed. CRC Press, Boca Raton, FL, 1995.
205. Charlton, G., Howes, P. B., Nicklin, C. L., Steadman, P., Taylor, J. S. C., Muryn, C. A., Harte, S. P., Mercer, J., McGrath, R., Norman, D., Turner, T. S., and Thornton, G., *Phys. Rev. Lett.* **78,** 495 (1997).
206. Hird, B., and Armstrong, R. A., *Surf. Sci.* **385,** L1023 (1997).
207. Zhong, Q., Vohs, J. M., and Bonnell, D. A., *Surf. Sci.* **274,** 35 (1992).
208. Cocks, I., Quo, D. Q., and Williams, E. M., *Surf. Sci.* **390,** 119 (1997).
209. Murray, P. W., Condon, N. G., and Thornton, G., *Phys. Rev. B* **51,** 10989 (1995).
210. Onishi, H., and Iwasawa, Y., *Phys. Rev. Lett.* **76,** 791 (1996)
211. Møller P. J., and Wu, M. C., *Surf. Sci.* **224,** 265 (1989).
212. Pang, C. L., Haycock, S. A., Raza, H., Murray, P. W., Thornton, G., Gulseren, O., James, R., and Bullett, D. W., *Phys. Rev. B* **58,** 1586 (1998).
213. Guo, Q., Cocks, I., and Williams, E. M., *Phys. Rev. Lett.* **77,** 3851 (1996).
214. Pang, C. L., Haycock, S. A., Raza, H., Thornton, G., Gulseren, O., James, R., and Bullett, D. W., *Surf. Sci.* **437,** 261 (1999).
215. Tanner, R. E., Castell, M. R., and Briggs, G. A. D., *Surf. Sci.* **437,** 263 (1999).
216. Bates, S. P., Kresse, G., and Gillan, M. J., *Surf. Sci.* **409,** 336 (1998).
217. Persson, P., Stashans, A., Bergstrom, R., and Lunell, S., *Int. J. Quantum Chem.* **70,** 1055 (1998).
218. Lu, G., Linsebigler, A., and Yates, J. T., Jr., *J. Phys. Chem.* **99,** 7626 (1995).
219. Kim, S. H., Stair, P. C., and Weitz, E., *Langmuir* **14,** 4156 (1998).
220. Patthey, L., Rensmo, H., Person, P., Westmark, K., Vayssieres, L., Stahans, A., Petersson, A., Brühwiler, P. A., Siegbahn, H., Lunell, S., and Mårtensson N., *J. Chem. Phys.* **110,** 5913 (1999).
221. Onishi, H., Arugawa, T., and Iwasawa, Y., *Surf. Sci.* **193,** 33 (1988).
222. Cocks, I. D., Guo, Q., Patel, R., Williams, E. M., Roman, E., and de Segovia, J. L., *Surf. Sci.* **377,** 135 (1997).
223. Guo, Q., Cocks, I., and Williams, E. M., *Surf. Sci.* **393,** 1 (1997).
224. Pacchioni, G., Ferrari, A. M., and Bagus, P. S., *Surf. Sci.* **350,** 159 (1996).
225. Casarin, M., Maccato, C., and Vittadini, A., *J. Phys. Chem. B* **102,** 10745 (1998).
226. Sorescu, D. C., and Yates, J. T., Jr., *J. Phys. Chem. B* **102,** 4556 (1998).
227. Anpo, M., Yamashita, H., Ichihashi, Y., and Ehara, S., *J. Electroanal. Chem.* **396,** 21 (1995).
228. Lu, G., Linsebigler, A., and Yates, J. T., Jr., *J. Phys. Chem.* **98,** 11733 (1994).
229. Kurtz, R. L., Stockbauer, R., Madey, T. E., Roman, E., and De Segovia, J. L., *Surf. Sci.* **218,** 178 (1989).
230. Vogtenhuber, D., Podloucky, R., and Redinger, J., *Surf. Sci.* **402–404,** 798 (1998).
231. Goniakowski, J., and Gillan, M. J., *Surf. Sci.* **350,** 145 (1996).
232. Brinkley, D., and Engel, T., *J. Phys. Chem. B* **102,** 7596 (1998).
233. Brinkley, D., and Engel, T., *Surf. Sci.* **415,** L1001 (1998).
234. Suzuki, S., Yamagushi, Y., Onishi, H., Sasaki, T., Fukui, K., and Iwasawa, Y., *J. Chem. Soc. Faraday Trans.* **94,** 161 (1998).

235. Suzuki, S., Onishi, H., Fukui, K.-I., and Iwasawa, Y., *Chem. Phys. Lett.* **304,** 225 (1999).
236. Ferris, K. F., and Wang, L., *J. Vac. Sci. Technol. A* **16,** 956 (1998).
237. Calatayud, M., Andrés, J., and Beltrán, A., *Surf. Sci.* **430,** 213 (1999).
238. Guo, Q., and Williams, E. M., *Surf. Sci.* **433–435,** 322 (1999).
239. Shen, G. L., Casanova, R., Thornton, G., and Colera, I., *J. Phys. Condens. Matter* **3,** S291 (1991).
240. Gercher, V. A., and Cox, D. F., *Surf. Sci.* **312,** 106 (1994).
241. Irwin, J. S. G., Prime, A. F., Hardman, P. J., Wincott, P. L., and Thornton, G., *Surf. Sci.* **352–354,** 480 (1996).
242. Cox, D. F., Fryberger, T. B., and Semancik, S., *Surf. Sci.* **224,** 121 (1989).
243. Kohle, O., Ruile, S., and Grätzel, M., *Inorg. Chem.* **35,** 4779 (1996).
244. Henderson, M. A., *Surf. Sci.* **319,** 315 (1994).
245. Chung, W. Y., Lo, W. J., and Somorjai, G. A., *Surf. Sci.* **64,** 588 (1977).
246. Murray, P. W., Leibsle, F. M., Muryn, C. A., Fisher, H. J., Flipse, C. F. T., and Thornton, G., *Phys. Rev. Lett.* **72,** 689 (1994).
247. Tait, R. H., and Kasowski, R. V., *Phys. Rev. B* **20,** 5178 (1979).
248. Firment, L. E., *Surf. Sci.* **116,** 205 (1982).
249. Nörenberg, H., Dinelli, F., and Briggs, G. A. D., *Surf. Sci.* **436,** L635 (1999).
250. Idriss, H., and Barteau, M. A., *Catal. Lett.* **26,** 123 (1994).
251. Idriss, H., Lusvardi, V. S., and Barteau, M. A., *Surf. Sci.* **348,** 39 (1996).
252. Lusvardi, V. S., Barteau, M. A., Chen, J. G., Eng, J., Jr., Frühberger, B., and Teplyakov, A., *Surf. Sci.* **397,** 237 (1998).
253. Watson, B. A., and Barteau, M. A., *Chem. Mater.* **6,** 771 (1994).
254. Poirier, G. E., Hance, B. K., and White, J. M., *J. Vac. Sci. Technol. B* **10,** 6 (1992).
255. Fan, F. F., and Bard, A. J., *J. Phys. Chem.* **94,** 3761 (1990).
256. Gai-Boyes, P. L., *Catal. Rev. Sci. Eng.* **34,** 1 (1992).
257. Wang, L.-Q., Ferris, F., Winokur, J. P., Schultz, A. N., Baer, D. R., and Engelhard, M. H., *J. Vac. Sci. Technol. A* **16,** 3034 (1998).
258. Schultz, A. N., Jang, W., Hetherington, W. M., Baer, D. R., Wang, L., and Engelhard, M. H., *Surf. Sci.* **339,** 114 (1995).
259. Hoflund, G. B., Yin, H. L., Grogan, A. L., and Asbury, D. A., *Langmuir* **4,** 346 (1988).
260. Rohrer, G. S., Henrich, V. E., and Bonnell, D. W., *Science* **250,** 1239 (1990).
261. Kim, K. S., Barteau, M. A., and Farneth, W. E., *Langmuir* **4,** 533 (1988).
262. Lusvardi, V. S., Barteau, M. A., Dolinger, W. R., and Farneth, W. E., *J. Phys. Chem.* **100,** 18183 (1996).
263. Lusvardi, V. S., Barteau, M. A., and Farneth, W. E., *J. Catal.* **153,** 41 (1995).
264. Brown, R. D., Godfrey, P. D., Champion, R., and McNaughton, D., *J. Am. Chem. Soc.* **103,** 5711 (1981).
265. Kim, K. S., and Barteau, M. A., *Langmuir* **4,** 945 (1988).
266. Kim, K. S., and Barteau, M. A., *Langmuir* **6,** 1485 (1990).
267. Wilson, J. N., Titheridge, D., and Idriss, H., submitted for publication.
268. Xu, C., and Goodman, D. W., *Langmuir* **12,** 1807 (1996).
269. Maier, G., Reisenauer, H. P., Balli, H., Brandt, W., and Janoschek, R., *Angew. Chem. Int. Ed. Engl.* **29,** 905 (1990).
270. McMurry, J. E., *Chem. Rev.* **89,** 1513 (1989).
271. Elschenbroich, Ch., and Salzer, A., "Organometallics: A Concise Introduction," 2nd ed. VCH, Weinheim, 1992.
272. Abdelreheim, I. M., Thornburg, N. A., Sloan, J. T., Caldwell, T. E., and Land, D. P., *J. Am. Chem. Soc.* **117,** 9509 (1995).

273. Tysoe, W. T., Nyberg, G. L., and Lambert, R. M., *J. Chem. Soc. Chem. Commun.,* 623 (1983).
274. Sesselman, W., Woratschek, B., Ertl, G., Küppers, J., and Haberland, H., *Surf. Sci.* **130,** 245 (1983).
275. Lusvardi, V. S., Pierce, K. G., and Barteau, M. A., *J. Vac. Sci. Technol. A* **15,** 1586 (1997).
276. Libby, M. C., Watson, P. C., and Barteau, M. A., *Ind. Eng. Chem. Res.* **33,** 2904 (1994).
277. Watson, P. C., Libby, M. C., and Barteau, M. A., U. S. patent No. 5,475, 144 (1995).
278. Martinez, R., Huff, M. C., and Barteau, M. A., Proc. 12th Int. Congr. Catal., in press.
279. Rajadurai, S., *Catal. Rev. Sci. Eng.* **36,** 385 (1994).
280. Dooley, K. M., and Randery, S. D., *in* "Abstracts, 16th Meeting of the North American Catalysis Society Boston, 1999," p. II-041.
281. Randery, S. D., Warren, J. S., and Dooley, K. M., submitted for publication.
282. Grootendorst, E. J., Pestman, R., Koster, R. M., and Ponec, V., *J. Catal.* **148,** 261 (1994).
283. Pestman, R., Koster, R. M., Pierterse, J. A. Z., and Ponec, V., *J. Catal.* **168,** 255 (1997).
284. Pestman, R., Koster, R. M., Boellaard, E., van der Kraan, A. M., and Ponec, V., *J. Catal.* **174,** 142 (1998).
285. Boffa, A., Lin, C., Bell, A. T., and Somorjai, G. A., *J. Catal.* **149,** 149 (1994).
286. Lebouteiller, A., and P. Courtine, *J. Solid State Chem.* **137,** 94 (1998).
287. Chen, J. G., Früberger, B., Eng, J., Jr., and Bent, B. E., *J. Mol. Catal. A Chem.* **131,** 285 (1988).
288. Duchateau, R., van Wee, C. T., and Teuben, J. H., *Organometallics* **15,** 2291 (1996).
289. Duffy, J. A., "Bonding, Energy Levels, and Bands in Inorganic Solids." Longman, New York, 1990.
290. Allen, L. C., *Int. J. Quantum Chem.* **49,** 253 (1994).
291. Pauling, L., "The Nature of Chemical Bond," 3rd ed. Cornell Univ. Press, Ithaca, NY, 1963.
292. "Thermodynamics Properties of the Elements," Advances in Chemistry Series No. 18, American Chemical Society, Washington, DC, 1956.
293. Matthew, J. A. D., and Parker, S., *J. Electron. Spectrosc. Relat. Phenom.* **85,** 175 (1997).
294. Terashima, K., Hashimoto, T., and Yoko, T., *Phys. Chem. Glasses* **37,** 129 (1996).
295. El-hadi, A. Z., and El-baki, M. A., *Phys. Chem. Glasses* **40,** 90 (1999).
296. Ai, M., *J. Catal.* **83,** 141 (1983).
297. Pacchioni, G., Sousa, C., Illas, F., Parmigiani, F., and Bagus, P. S., *Phys. Rev. B* **48,** 11573 (1993).
298. Yee, A., Morisson, S., and Idriss, H., *J. Catal.* **186,** 279 (1999).
299. Tu, Y.-J., Li, C., and Chen, Y.-W., *J. Chem. Tech. Biotechnol.* **59,** 141 (1994).
300. Idriss, H., and Seebauer, E. G., *J. Mol. Catal.* **152,** 201 (2000).
301. Zhang, W., Desikan, A., and Oyama, S. T., *J. Phys. Chem.* **99,** 14468 (1995).
302. Koga, N., Obara, S., Kitaura, K., and Morokuma, K., *J. Am. Chem. Soc.* **107,** 7109 (1985).
303. Cuberio, M. L., and Fierro, J. L., *J. Catal.* **179,** 150 (1998).
304. The validity of using $E_{(M-O)}$ to correlate reaction rates on oxides has been questioned by Masel. See Masel, R. I., "Principles of Adsorption and Reaction on Solid Surfaces," p. 779. Wiley, New York, 1996.
305. Busca, G., and Lorenzelli, V., *Mater. Chem.* **7,** 89 (1982).
306. Idriss, H., and Seebauer, E. G., submitted for publication.
307. Noller, H., Lercher, J. A., and Vinek, H., *Mater. Chem. Phys.* **18,** 577 (1988).
308. Youssef, A. M., Khalil, L. B., and Girgis, B. S., *Appl. Catal. A* **81,** 1 (1992).
309. Raskò, J., Bontovics, J., and Solymosi, F., *J. Catal.* **146,** 22 (1994).

Catalysis and Surface Science: What Do We Learn from Studies of Oxide-Supported Cluster Model Systems?

H.-J. FREUND, M. BÄUMER, AND H. KUHLENBECK

Fritz-Haber-Institut der Max-Planck-Gesellschaft
Department of Chemical Physics
D-14195 Berlin, Germany

Models of supported metal catalysts have been prepared by deposition of transition metal vapor onto thin, well-ordered oxide layers, particularly alumina. The average sizes of the metal particles and their size distributions are determined by nucleation and growth. Here, we review the morphology and structure of the oxide layers and the metal particles dispersed on them, as measured with a variety of methods including low-energy electron diffraction and scanning tunneling microscopy. Electronic and magnetic structure as a function of particle size, adsorption properties, as well as reactivity with varying particle size are reviewed. The electronic structure of supported palladium particles has been studied and indicates that a nonmetal to metal transition occurs for particles exceeding 70–80 atoms per aggregate. Adsorption of CO has been studied in detail by Fourier transform infrared spectroscopy. Interesting variations in the spectra when very small particles consisting of only a few metal atoms are investigated compared with larger particles and single-crystal surfaces are observed. CO dissociation has been studied on rhodium aggregates and a maximal dissociation rate has been found for aggregates containing several hundred metal atoms. The presence of defects on the particles is deduced to be origin of this behavior. © 2000 Academic Press.

Abbreviations: AFM, atomic force microscopy; EEL, electron energy loss; EELS, electron energy loss spectroscopy; EL, electron loss; FFT, fast Fourier transform; FMR, ferro-magnetic resonance; FT, Fourier-transform; FTIR, Fourier transform infrared; GIXD, grazing incidence X-ray diffraction; HRTEM, high resolution transmission electron microscopy; IR, infrared; LEED, low energy electron diffraction; ML, mono-layer; RAIRS, reflection absorption infrared spectroscopy; SFG, sum frequency generation; SPA-LEED, spot profile analysis-low energy electron diffraction; STM, scanning tunneling microscopy; TDS, thermal desorption spectroscopy; TEM, transmission electron microscopy; UHV, ultrahigh vacuum; XP, X-ray photoelectron; XPS, X-ray photoelectron spectroscopy.

0360-0664/00 $35.00

I. Introduction

From its beginning, surface science has largely been driven by the goal of understanding catalysis at the atomic level, with more recent work being motivated by the goals of understanding microelectronic and sensor materials and phenomena such as corrosion (*1, 2*).

Adsorbates on metal single-crystal surfaces have been characterized with the aim of demonstrating some of the basic mechanisms of catalysis, e.g., how small molecules such as H_2, CO, NO, CO_2, and N_2 interact with structurally well-characterized surfaces, how the electronic and geometric structures of these molecules and the surface are changed as a result of the adsorbate–surface interactions, and how these interactions open pathways for reaction of the adsorbates with coadsorbed species. With the development of an arsenal of surface analytical tools, the past 35 years of surface science have witnessed major progress, although some of the early claims were unrealistic and much remains to be understood. Catalysis remains a strong driving force for surface science (*2*), and it is vital that the catalysis and surface science communities continue to interact with the goal of eventually being united.

The gaps that still exist between catalysis and surface science (*3*) were identified early (4):

1. The materials gap
2. The pressure gap
3. The complexity gap

Surface science has reached a degree of maturity that should allow us to bridge these gaps (*5–17*). The use of model systems (*16*), including model catalysts, is one strategy for bridging the materials and pressure gaps. We believe that these two gaps have to be bridged before the complexity gap can be bridged.

A wide variety of systems serve as models of practical catalysts. Reviews that deal particularly with models of dispersed metal or dispersed metal oxide catalysts include those by Goodman (*11*), Niemantsverdriet *et al.* (*16*), and our group (*10, 13*). Although the approaches to preparation of models of practical supported catalysts have much in common, there are some subtle and important differences among them.

Briefly, the various approaches share the requirement that the substrate onto which the model is built should be electrically conducting to allow application of the appropriate surface science tools, as shown in Fig. 1 (*3, 10, 11, 16, 17*). The catalyst support is prepared on these substrates as a thin layer because most practical supports, such as magnesia, alumina, or silica, are insulators. When they are sufficiently thin, these layers are

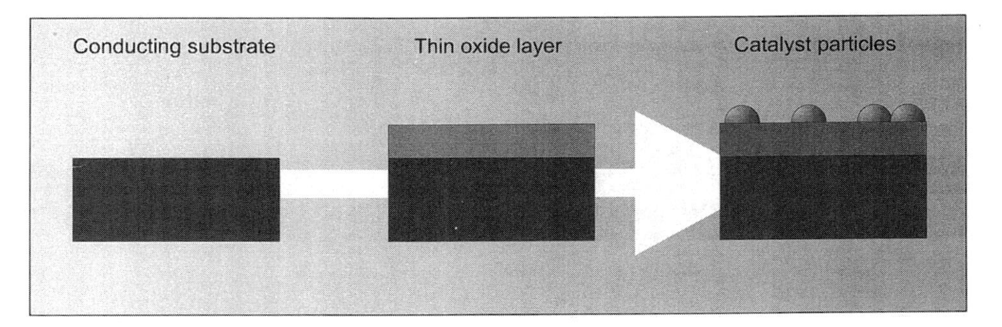

FIG. 1. Schematic diagram of model catalyst systems.

sufficiently conductive to allow application of analytical methods based on sample interrogation with charged particles without severe charging of the sample.

The active catalyst phase is deposited on the support, which is often an oxide. This catalyst phase may consist of aggregates of metal or, alternatively, a second oxide, a carbide, or a nitride, prepared in dispersed form.

Oxide supports are emphasized in this review. The model supports are prepared in various ways. Oxide layers can be grown as native oxides on metals (*3, 9*). In this case, the structure is governed by the natural mismatch of the structures of the metal and its oxide; sometimes the resulting layers are well structured and sometimes they are amorphous. Oxide layers may also be grown by evaporation of the metal onto a second (inert) metal substrate (*11*). The structure of the metal then largely governs the structural quality of the layer. Again, ordered or amorphous layers can be grown. Oxidation can be carried out by evaporation of the metal in an oxygen atmosphere or by exposing the deposited metal layer to oxygen in a second step (*15*). The morphology of the oxide layer may depend critically on the preparation procedure.

Films of carbides (*18*), nitrides (*19*), or sulfides (*20*) can be prepared similarly if a good method for providing the nonmetallic element is known. It is obvious that if the support is conductive (a conductive oxide or graphite), single crystals (*21*) of the respective materials may be used as well.

A variety of methods have been applied for depositing the active catalytic material in a second step. The most popular involves evaporation of the metal. If the deposition rate and the substrate temperature are controlled, then relatively narrow metal particle size distributions can be obtained, and the metal loading can be varied (*10*). Deposition of mass-selected particles may emerge as the preferred method to control particle size for the smallest aggregates (*22*). Chemical vapor deposition of cluster compounds or of organometallic compounds is another possibility (*23*). When these precursors are used, it is usually important to ensure that the ligands do not remain as permanent contaminants of the aggregates.

Additional preparation methods include wet impregnation spin coating (*16*), whereby the precursor of the active component is brought onto the support in solution. The solvent is evaporated and then the precursor is further treated to give the active form.

Although it seems that all the methods yield model systems which are "cum grano solis" the same, there are fundamental differences between the wet impregnation techniques on microcrystalline layers and those involving growth of particles by metal vapor deposition onto single-crystal supports, for example. The first (top-down) approach, although similar to industrially employed catalyst preparation procedures, is characterized by complex processes. The second (bottom-up) approach, on the other hand, starts almost from first principles and tries to be as rigorous as possible in each step of increasing complexity. The bottom-up approach, however, sometimes leads those in the catalysis community to question the relevance of the materials to practical catalysis; one may ask whether the model system has been tested in a catalytic reaction and whether it is active for the reaction it is intended to model. If the answers are negative, then the top-down approach is favored. It can be argued, however, that although the complex model system may do what it is supposed to, its complexity may stymie understanding of the underlying atomic-scale phenomena, which may be more easily isolated by the bottom-up approach.

Microcrystalline powders may also serve as model systems (which are not directly compatible with Fig. 1) that are well-suited for comparison with those made by the bottom-up approach described previously. The groups of Knözinger (*24*) and Zecchina *et al.* (*25*) have made seminal contributions to this area by their infared (IR) and Raman spectroscopic studies of adsorbed probe molecules. If grown and sintered properly, the microcrystals expose single-crystal facets. The adsorption sites on the various facets and possibly on edge and corner sites can be differentiated by the adsorption of probe molecules, the internal mode frequencies of which depend on the adsorption site. Assignment of some of the frequencies by comparison of spectra with those characterizing large single-crystal samples is a first step toward accounting for the complex set of adsorption sites. The complementary nature of the results characterizing microcrystalline and single-crystal materials is helping to close the materials gap (*26*).

Closing the pressure gap can be achieved by application of surface-sensitive techniques that work in the presence of a gas phase, such as thermal desorption spectroscopy (TDS), reflection absorption infrared spectroscopy (RAIRS), sum frequency generation (SFG), scanning tunneling microscopy (STM), and X-ray scattering or X-ray absorption (*27*). Closing the full complexity gap must involve the application of additional methods to characterize the gas phase and mass transport.

Our goal in the following review is to describe the bottom-up approach and provide examples illustrating how far surface science has progressed and what may possibly be accomplished in the near future.

II. Clean Oxide Surfaces: Structure and Adsorption

The preparation of a clean oxide surface is difficult; several strategies have been proposed (*5, 15, 21*). The most straightforward is ultrahigh vacuum is (UHV) *in situ* cleavage, which leads to good results only in certain cases, such as for MgO, NiO, ZnO, and $SrTiO_3$ (*9*). Some catalytically important oxides, such as Al_2O_3, SiO_2, and TiO_2, are difficult to cleave (*21*). A disadvantage encountered in experimental investigations of cleaved bulk single-crystal insulators is the sample charging that results when they are bombarded by charged particles, as in electron spectroscopy experiments. An alternative method of preparing bulk single crystals of oxides is *ex situ* cutting and polishing followed by *in situ* sputtering and subsequent annealing in oxygen. Such a process usually creates a sufficient number of defect sites in the near-surface region and in the bulk to induce conductivity of the material. Thus, electron spectroscopies and STM can be applied (*21*).

Single-crystal oxide surfaces may also be prepared by the growth of thin oxide layers on single-crystal metal supports (*5, 13, 15*). All the surface science tools can be applied to such samples. If the oxide layer is to represent the bulk accurately, care must be taken to provide the proper layer thickness. Furthermore, if adsorption and reactivity investigations are intended, the continuity of the layer must be guaranteed. Several examples illustrate the successful attainment of these goals (*11, 13, 16*).

The best studied clean oxide surfaces are $TiO_2(100)$ and $TiO_2(110)$ (*5, 17, 21*). An STM image of the clean (1×1) $TiO_2(110)$ surface taken by Diebold *et al.* (*28*) is shown in Fig. 2. It is noteworthy that one of the first atomically resolved images of this surface was reported by Murray *et al.* (*29, 30*). The inset shows a ball-and-stick model of the surface. Evidence accumulating from theoretical modeling of the tunneling conditions and from adsorbate studies using molecules assumed to bind to the exposed titanium sites indicates that the bright rows represent titanium atoms. Onishi *et al.* (*31–34*) used formic acid in such an investigation and showed, in line with the theoretical predictions (but in contrast to intuitive expectations

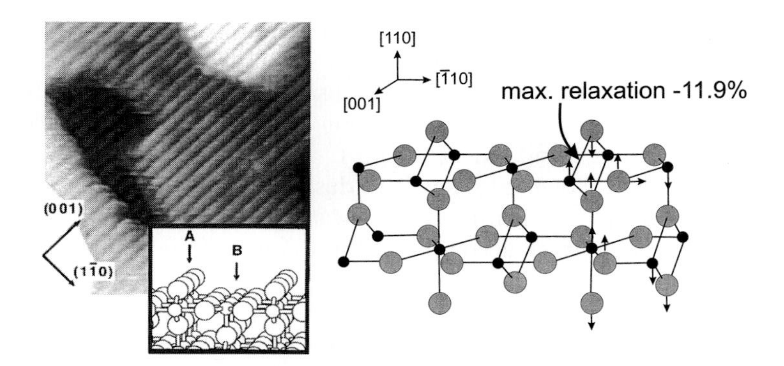

FIG. 2. Structure of the $TiO_2(110)$ surface as determined by STM (left) and grazing incidence X-ray scattering (right). The groups of Bonnell, Diebold, Iwasawa, Thornton, and Vanderbilt contributed to the solution of this problem [reproduced with permission from Diebold *et al.* (*28*) (left) and Charlton *et al.* (*35*) (right)].

based on topological arguments), that the titanium atoms are imaged as bright lines and the oxygen atoms as dark lines. Taking account of the resolvable interatomic distances within the surface layer, the authors inferred that the values correspond to the structure of the neutral truncation of the stoichiometric (110) surface (*35, 36*). Interatomic distances normal to the surface, however, differ substantially from the bulk values, as shown by X-ray scattering experiments. The top-layer, sixfold coordinated titanium atoms move outward and the fivefold coordinated titanium atoms move inward, leading to a surface roughening of 0.3 ± 0.1 Å. The roughening repeats itself in the second layer with an amplitude about half of that in the top layer. Bond length variations range from a contraction of 11.3% to an expansion of 9.3%. These strong relaxations are not atypical for oxide surfaces and have been predicted theoretically (*37–39*).

The relaxations are particularly pronounced for the so-called charge-neutralized polar surfaces (*37–39*). There are several experimental results (*40–43*) that essentially corroborate the theoretical predictions, although the quantitative agreement is not always good (*44–47*). Specifically, the (0001) surfaces of corundum-type materials, such as Al_2O_3 (*44, 45*), Cr_2O_3 (*46*), and Fe_2O_3 (*47*), have been investigated with X-ray diffraction, quantitative low-energy electron diffraction (LEED), STM, and theory.

Figure 3 (*48, 49*) is a reminder that a polar surface [e.g., the (111) orientation of a rock salt structure] exhibits, when the bulk is terminated, a diverging surface potential due to the missing compensation of the interlayer dipole moments, as discussed by Noguera (*39*). Consequently, polar surfaces reconstruct and/or relax substantially, whereas nonpolar surfaces often

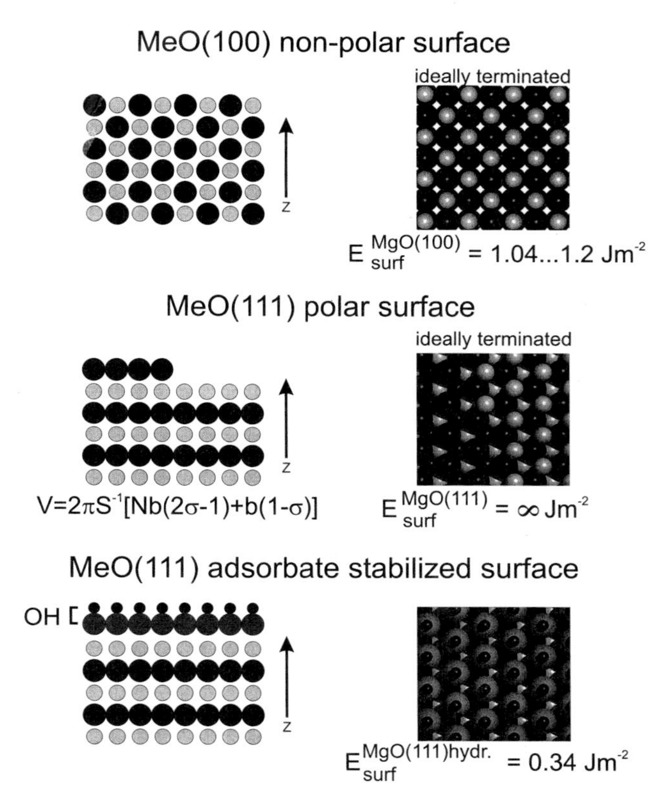

FIG. 3. Schematic representation (side and top views) of the structure of a nonpolar (top) an unreconstructed polar (middle), and a hydroxylated polar (bottom) surface of a rock salt-type crystal. The energies given refer to MgO (*48*).

exhibit much less pronounced relaxations (although, as was shown previously for TiO_2, the degree of relaxation might still be substantial).

Figure 4 shows the results of structural determinations for the three related systems $Al_2O_3(0001)$, $Cr_2O_3(0001)$, and $Fe_2O_3(0001)$. In all three, a stable structure is given by the metal ion-terminated surface retaining only half the number of metal ions in the surface relative to a full buckled layer of metal ions within the bulk. The interlayer distances are strongly relaxed, with the relaxation extending several layers below the surface. The perturbation to the structure caused by the presence of the surface in oxides is considerably more pronounced than in metals, for which the interlayer relaxations are typically on the order of a few percent (*50*). The absence of the screening charge in a dielectric material such as an oxide contributes

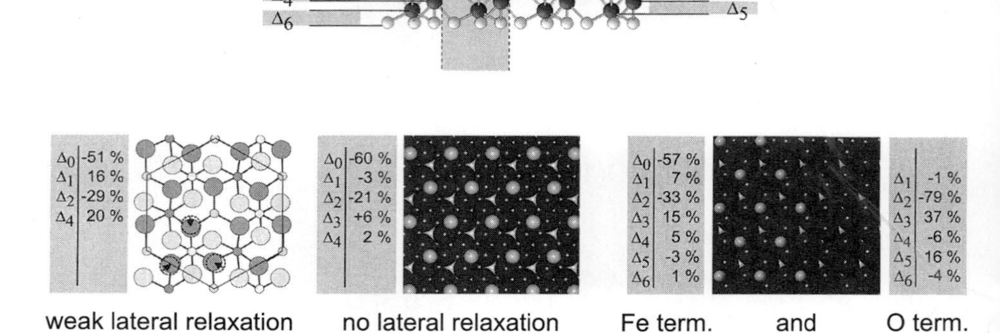

FIG. 4. Experimental results characterizing the structure of corundum-type depolarized (0001) surfaces (side and top views) [reproduced with permission from Renaud (*36*) (left), Rohr *et al.* (*42*) (middle), and Wang *et al.* (*47*) (right)].

significantly to this effect. It has recently been pointed out (*51*) that oxide structures may not be as rigid as has been previously thought on the basis of the relatively stiff phonon spectrum of the bulk. Indeed, at the surface the phonon spectrum may become soft so that the geometric structure becomes flexible and thus significantly dependent on the presence of adsorbed species.

Bulk oxide stoichiometries depend strongly on oxygen pressure (*52*). Both the stoichiometries and the structures of oxide surfaces depend on the oxygen pressure, as illustrated by a recent investigation of the $Fe_2O_3(0001)$ surface (*47*). If a Fe_2O_3 single-crystal layer is grown at a low oxygen pressure, the surface is metal terminated, whereas growth under higher oxygen pressures leads to a complete oxygen termination (*47*). In principle, this surface would be regarded as unstable on the basis of the electrostatic arguments presented previously. However, calculations (*47*) have shown that a strong rearrangement of the electron distribution, as well as relaxation between the layers, leads to a stabilization of the system. STM images (*47*) corroborate the coexistence of oxygen- and iron-terminated layers and thus indicate that stabilization must occur.

Of course, additional structural characterization is needed. The idea of polar and nonpolar surfaces only really holds in its simplest version if the material is highly ionic. Thus, the most extreme cases to investigate are perhaps the polar surfaces of the simple oxides with the rock salt structure

(*53*), such as MgO and NiO [i.e., MgO(111) and NiO(111)]. Barbier and Renaud (*54*) succeeded in preparing a single-crystal NiO(111) surface and in characterizing it by grazing-incidence X-ray diffraction (GIXD). As was shown earlier for thin NiO layers of different crystallographic orientations [i.e., NiO(100) (*55*) and NiO(111) (*56*)], a surface prepared in air or under residual gas pressure exhibits a p(1 × 1) structure, whereas the clean polar (111) surfaces are reconstructed. The p(2 × 2) reconstruction originally reported for the thin layer system has also been found for the bulk single-crystal surfaces (*53, 54*). An initial structural analysis suggested that the structure is not the expected octopolar reconstruction shown in Fig. 5 but instead a more complicated one (*53*). However, recent investigations (*57*) of more carefully prepared bulk single-crystal surfaces revealed that a stoichiometric surface actually reconstructs according to the octopolar scheme (*39, 58*). The small (100)-terminated pyramids are oxygen terminated.

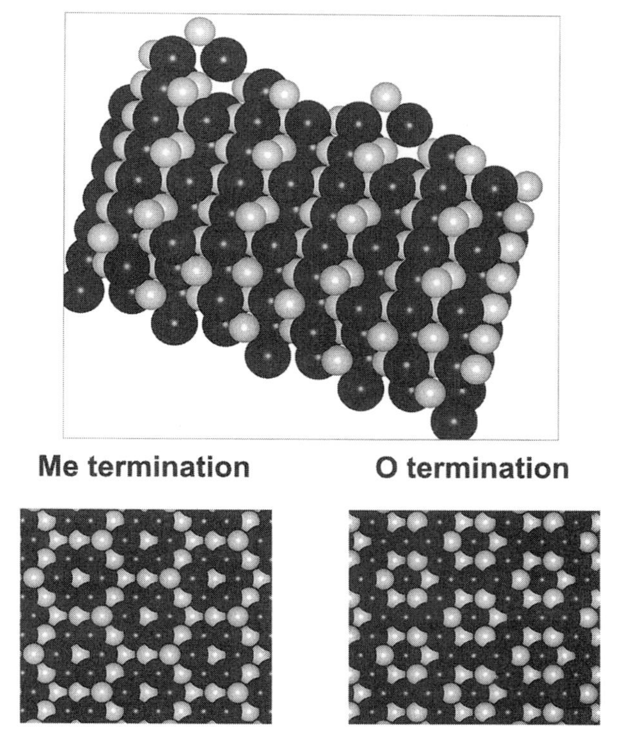

FIG. 5. Schematic representation of the octopolar reconstruction of a polar rock salt (111) surface with oxygen and metal termination [reproduced with permission from Wolf (*53*)].

Furthermore, NiO(111) layers grown on Au(111) which were initially studied by Ventrice *et al.* (*59*) have recently been investigated by GIXD (*60*). The p(2 $\times$ 2) reconstruction was again corroborated, but the structural analyses undertaken to date seem to favor a structure in which oxygen as well as Ni-terminated octopoles, possibly arranged on adjacent terraces, constitute the surface layer. Both the bulk single-crystal surfaces and the NiO(111) layer surfaces grown on Au(111) exhibit a high degree of surface order. This is probably one reason why these surfaces do not quickly restructure upon exposure to water, whereas NiO(111) layers grown on Ni(111) do reconstruct to form a hydroxyl-terminated NiO(111) surface (*56*). A microscopic mechanism would involve massive material transport across the surface, which is less favorable on more ordered surfaces and may therefore be kinetically hindered on well-ordered single crystals.

The interaction of water with polar oxide surfaces is a topic of general interest in geochemical and environmental science (*61*) as well as in catalysis. With respect to the latter, Papp and Egersdörfer (*62*) and Egersdörfer (*63*) found indications that NiO catalysts prepared with preferential (111) crystallographic orientation by topotactical dehydration of $Ni(OH)_2$ show the highest activity for deNO_x reactions after the last monolayer of H_2O has been desorbed. Even in 1977, Fripiat *et al.* (*64*) showed theoretically on the basis of energetic considerations that real crystallites must be terminated partly by polar surfaces the charges of which are reduced by surface OH groups.

Thus, the interactions of molecules with oxide surfaces represent an important field. In the following, we discuss several examples illustrating aspects of the bonding and interaction of molecules with oxide surfaces and, for comparison, metal surfaces.

The bonding of molecules to oxides is different from the bonding of molecules to metals. For example, a CO molecule interacts with metals via chemical bonds of varying strengths involving charge transfer (*65*). Figure 6 schematically illustrates the bonding of CO to a Ni metal atom by σ donation/π back-donation on the basis of a one-electron orbital diagram. The σ and π interactions lead to a shift of those σ and π orbitals involved in the bond with respect to those orbitals not involved. The diagram reflects this shift with the correlation lines.

This bonding of CO to a metal is in contrast to the electrostatically dominated interaction between a CO molecule and a nickel ion in nickel oxide (*66, 67*). There is a noticeable σ repulsion between the CO carbon lone pair and the oxide, leading to a similar shift of the CO 5σ orbital as in the case of the metal atom. However, there is little or no π back-donation so that the CO π orbitals are not modified (*13, 68*). Conceptually, the situation is transparent, and it might be expected that a detailed calculation

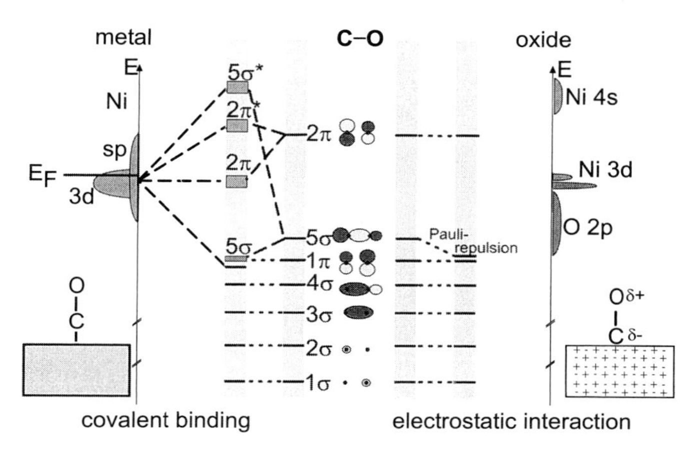

FIG. 6. Orbital diagram for the bonding of CO to nickel metal (left) and to nickel oxide (right).

would reveal the differences quantitatively. However, the description by *ab initio* calculations is rather involved, and a full interpretation cannot be given (*69*). The theoretical prediction is that CO (or NO) binds very weakly to NiO (*69*). The calculated binding energy of CO is on the order of 0.1 eV and expected to be similar to that characterizing CO bonded to MgO(100), i.e., the influence of the nickel d electrons is expected to be negligible (*69*).

To shed light on this problem, thermal desorption measurements were made to characterize cleaved single-crystal surfaces, which are the surfaces with the least number of defects (*70*). In Figs. 7 and 8, TDS data for CO and NO on vacuum-cleaved NiO(100) are compared with data for thin NiO(100) layers grown by oxidation of Ni(100). At temperatures of 30 and 56 K, multilayer desorption of CO and NO, respectively, occurs. The pronounced features at higher temperatures correspond to desorption of the respective adsorbate at (sub)monolayer coverage. In the case of CO, desorption of the second layer occurs at 34 K. The states at 45 and 145 K for CO and NO, respectively, are due to adsorption on defects, as concluded from data obtained with ion-bombarded surfaces (not shown).

For both adsorbates, the thin layer data and the data characterizing the cleaved samples agree well. In particular, for NiO(100) the thin layer data are comparable to those characterizing the more perfect surfaces of the cleaved samples. The higher defect density of the thin layer surfaces leads to small, but clearly visible, additional peaks in the TDS data which, for example, are visible as shoulders near the main peak in the NO spectra.

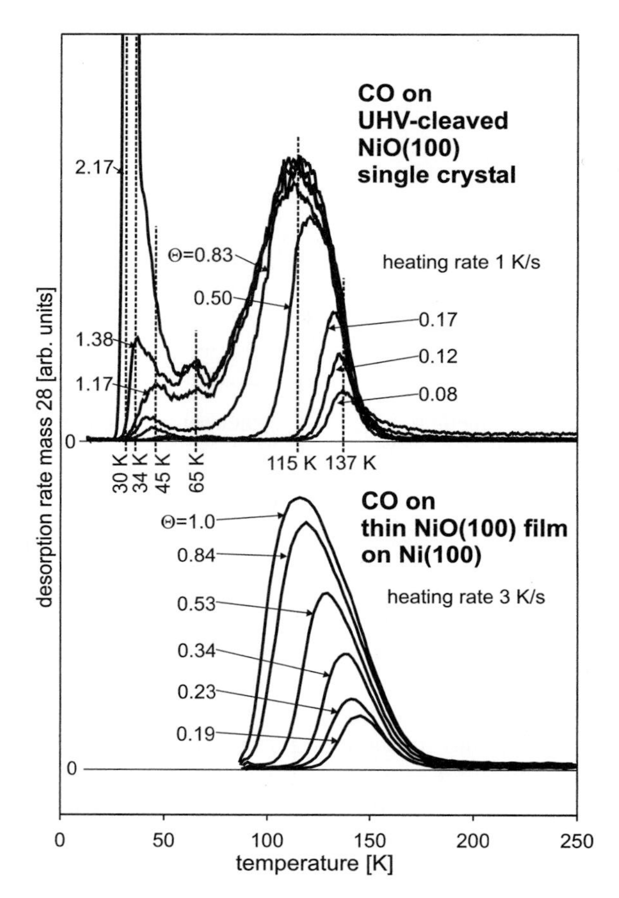

FIG. 7. Thermal desorption spectra of CO on NiO(100) cleaved in vacuum (top) and on a thin NiO(100) layer grown by oxidation of Ni(100) (bottom). The mass spectrometer was set to mass 28 (CO). CO coverages are given relative to full monolayer coverage.

Nevertheless, the overall shapes of the thin layer spectra of both adsorbates are very similar to those of the cleaved samples.

The low-coverage adsorption energies representing CO and NO on NiO(100) and MgO(100) are compiled in Table I. According to theory, the interactions of the adsorbates with MgO(100) and NiO(100) are expected to be similar since the bonding should be mainly electrostatic in nature (*69*) [the electric fields at the surfaces of NiO(100) and MgO(100) are similar]. According to Table I, however, the bonding energies are considerably different, with the higher values being obtained for NiO(100). Covalent

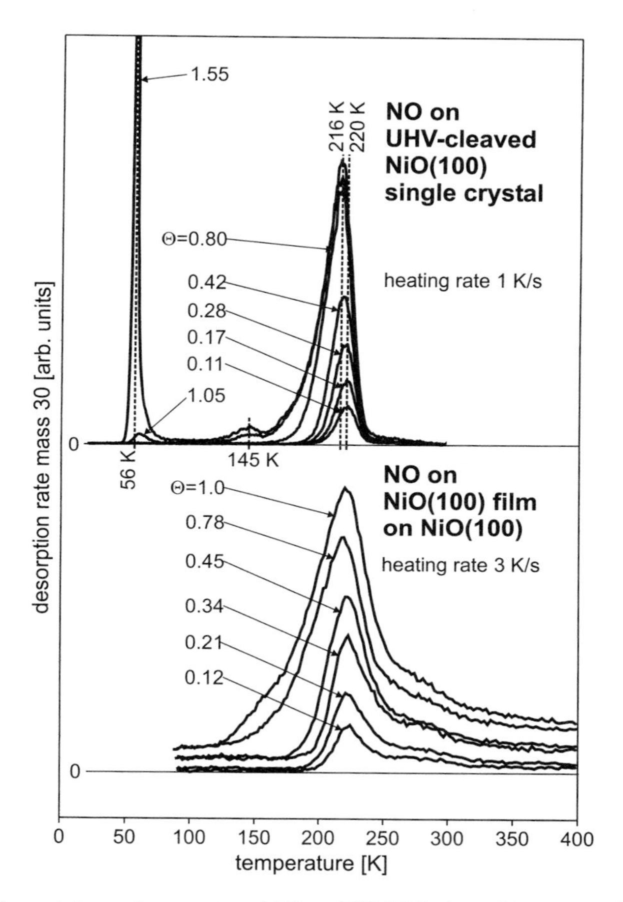

FIG. 8. Thermal desorption spectra of NO on NiO(100) cleaved in vacuum (top) and NO on a thin NiO(100) layer grown by oxidation of Ni(100) (bottom). The mass spectrometer was set to mass 30 (NO). NO coverages are given relative to full monolayer coverage.

interactions involving the nickel 3d electron (which have not become evident in the calculations so far) may play a role in the adsorbate–substrate interaction.

The adsorption of CO on MgO has been investigated thoroughly by Heidberg *et al.* (*71*) with IR spectroscopy and by Weiss *et al.* (*72*) with helium atom scattering. They demonstrated clearly that CO forms ordered phases on the cleavage planes and that order and spectroscopic properties depend on the quality of the prepared surfaces. The influence of the presence of surface defects on adsorption properties is obvious from the data,

TABLE I
Compilation of Low-Coverage Bonding Energies for NO and CO on NiO(100) and MgO(100)[a]

	NiO(100)	MgO(100)
CO	0.30 eV	0.14 eV
NO	0.57 eV	0.22 eV

[a] From Wichtendahl *et al.* (*70*).

but a quantitative evaluation based on the number and the nature of the defects has not been reported.

The quantitative evaluation of defects is a well-defined but difficult problem that awaits future investigations. Water adsorption is an example that lends itself to a study of the influence of defects because, at lower coverages, the (100) cleavage planes of MgO or of NiO do not dissociate water, whereas defects induce water dissociation, as is evident in the TDS spectra of H_2O from (100) rock salt-type surfaces. Figure 9 shows results for H_2O desorption from MgO(100) and NiO(100) (*73*). The most pronounced features in the spectra are due to condensed water layers at the lowest desorption temperature and the conversion of a compact layer [with $c(2 \times 4)$ periodicity in the case of MgO] to the monolayer, which desorbs at 225 K for MgO(100) and at 240 K for NiO (*74, 75*). The difference in desorption temperature between MgO(100) and NiO(100) seems to be characteristic of the H_2O–substrate interaction.

Most of the TDS information is lost when defects are created by sputtering. Thermal desorption is observed up to relatively high temperatures and the features are broad. Which kinds of defects have been created and how many are not known. A combination of techniques to characterize the defects by probe molecule adsorption together with IR, electron spin resonance, and electron spectroscopies may lead to a deeper understanding in the future.

Dissociative adsorption of water on oxide surfaces can also be used in a preparative way, namely, to modify the surface by hydroxylation. We have used this technique for a thin alumina layer to investigate the influence of hydroxyl groups on the nucleation and growth of metal aggregates, as discussed later (*76*). Figure 10 shows the result of such a hydroxylation as measured with vibrational spectroscopies such as high-resolution electron energy loss spectroscopy and Fourier transform infrared (FTIR) spectroscopy (*77*). It is impossible to hydroxylate the thin alumina layer on NiAl(110) just by water dissociation, whereas on a similar layer grown on

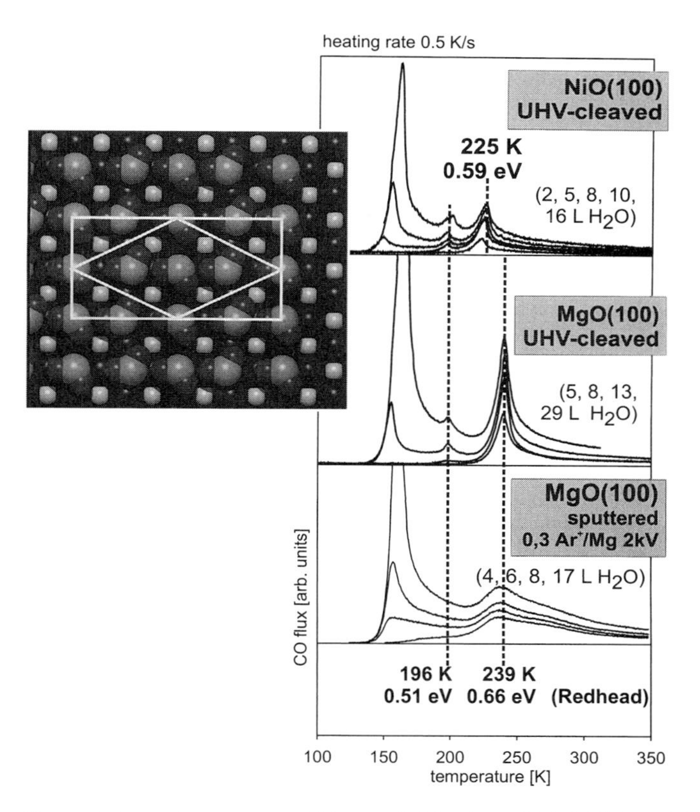

FIG. 9. Thermal desorption spectra of H_2O on UHV-cleaved MgO(100) and NiO(100). A schematic representation of the c(4 × 2) structure is included [reproduced with permission from Heidberg *et al.* (*199*)]. For comparison, a thermal desorption spectrum of MgO(100) taken after creation of defects via sputtering is shown.

NiAl(100) (*78*) formation of OH from dissociative H_2O adsorption occurs. The clean oxide layer surface was exposed to aluminum metal and then the aluminum was hydrolyzed by water adsorption to form a hydroxyl overlayer (*76, 77*). In Fig. 10 (bottom), an electron energy loss spectrum showing the hydroxyl vibration at 465 meV (3750 cm^{-1}) as well as a corresponding spectrum of the clean layer are plotted. The peaks at energies below 120 meV indicate the alumina phonons (*79*), which are broadened by hydroxylation as a result of a change of the surface order. The observed hydroxyl loss coincides clearly with the IR absorption observed for the same sample. In this case, more water was adsorbed so that a broad band from water clusters is seen as well. The sharp extra band at 3705 cm^{-1}

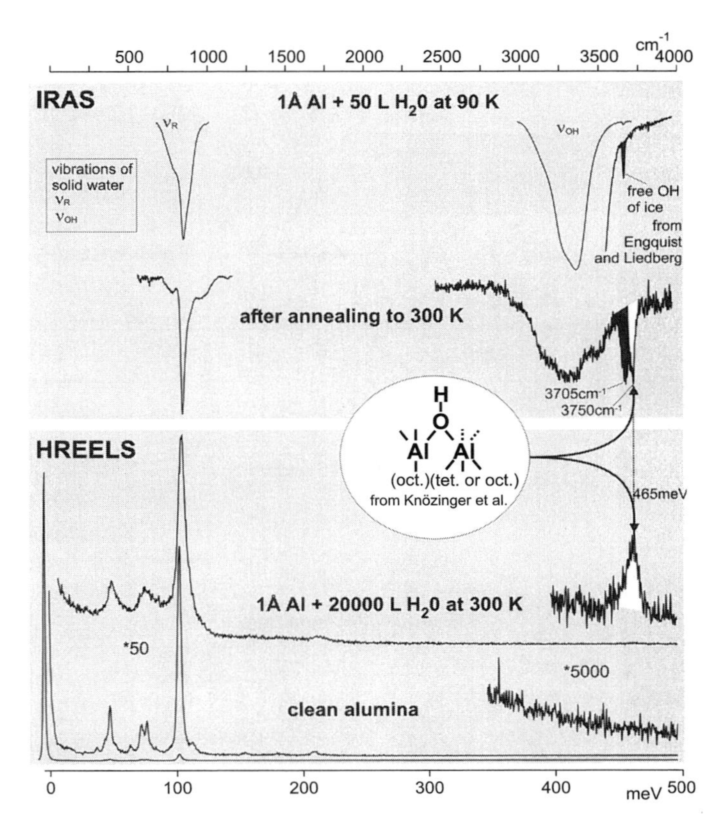

FIG. 10. Fourier transform IR spectra (RAIRS) and electron energy loss spectra (EELS) of a clean and OH (+H_2O)-covered alumina layer.

indicates free OH groups at the surfaces of these water clusters (*80*), as is also known from the surface of ice. Indeed, when a thick ice layer is grown on the alumina layer, this vibration is observed (Fig. 10).

By comparison with literature data (*81*), it is possible to assign the hydroxyl loss on the alumina surface. According to Knözinger (*78*), an OH vibration at 3750 cm^{-1} is characteristic of hydroxyl groups bridging aluminum ions, either both in octahedral sites or one in an octahedral and the other in a tetrahedral site. On alumina layers grown on a different NiAl substrate, other types of OH species may be formed. Therefore, it is conceivable that the influence of the nature of the hydroxyl species on the interaction with additional adsorbates (e.g., metal deposits) could be investigated.

Before we discuss metals on oxides, we describe the adsorption of CO_2 on oxides as an example of a molecular adsorbate system with more degrees of freedom. Figure 11 shows TDS spectra of CO_2 from a clean, flashed $Cr_2O_3(0001)$ surface (*26, 82*). The surface exhibits a structure discussed previously. The TDS spectra indicate that there are more weakly and less weakly bonded CO_2 species on the surfaces. We have investigated the nature of these species by various techniques including IR spectroscopy. Figure 11 shows several sets of IR spectra. The pair of sharp bands at approximately 2300 cm^{-1} is easily assigned to the more weakly bonded CO_2, which is only slightly distorted relative to the gas-phase species. A combination of isotopic labeling of the adsorbed CO_2 (leading to a shift of frequencies) and of the oxide layer (no shift of CO_2 islands) demonstrated that the single band centered at approximately 1400 cm^{-1} is indicative of the presence of a carboxylate species (i.e., a bent anionic CO_2 species) and not, as perhaps would have been expected, a carbonate (*83*).

The bands between 1610 and 1700 cm^{-1} are missing because of surface selection rules which apply to the thin-layer systems. This means that all non-totally symmetric bands are suppressed in intensity. A quick comparison with CO_2 adsorption on chromia microcrystalline material (Fig. 11) reveals the presence of the bands between 1610 and 1700 cm^{-1}, as expected for adsorption on a bulk dielectric material. The similarity between the

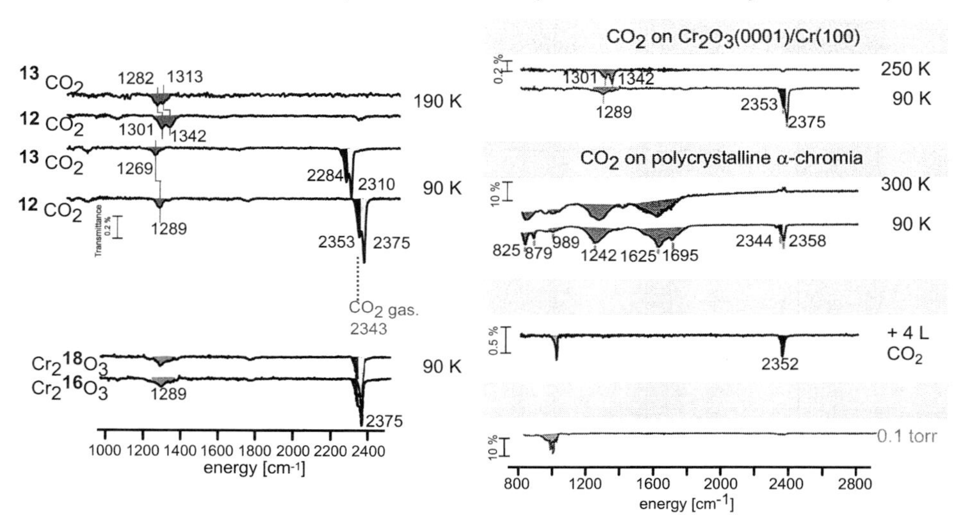

FIG. 11. RAIRS spectra of CO_2 adsorbed on $Cr_2O_3(0001)$ and on polycrystalline chromia. (Left) RAIRS spectra at different surface temperatures with isotopically labeled CO_2 and Cr_2O_3. (Right) Adsorption of CO_2 after preadsorption of oxygen.

thin-layer data and the results for the microcrystalline material are remarkable, as discussed in detail by Seiferth *et al.* (*26*). Furthermore, the responses of the two systems to preadsorption of oxygen are comparable. As shown in Fig. 11, CO_2 adsorption in the form of the less weakly bonded CO_2^- is fully suppressed on the thin layer and very strongly attenuated on the microcrystalline sample. These results indicate that CO_2 occupies the chromium sites because we know that oxygen from the gas phase adsorbs on the chromium ions.

A comment concerning the electronic structure of the $Cr_2O_3(0001)$ surface is appropriate here. Electron loss (*84*) and X-ray photo electron (*85*) spectra have shown that the chromium ions in the surface are in a low oxidation state (i.e., Cr^{2+}), in contrast to chromium ions in the near surface and bulk regions. It is therefore not surprising that such a surface can provide electrons to adsorbed molecules, leading to an electron transfer as shown, for example, by the formation of O_2^- and CO_2^-. The low valence state of the chromium surface ions also has consequences in other reactions, such as the polymerization of ethene [this reaction has been carried out on $Cr_2O_3(0001)$ (*86*)], and in connection with other, more realistic model investigations (*87*).

An area that has not been investigated at all with regard to well-characterized single-crystal oxide surfaces is the photoinduced chemical reaction of large molecules. Photoinduced desorption of small molecules, CO and NO, from oxides, however, has been investigated extensively (*88–91*). Yates and Wovchko (*92*) reported such investigations of powder samples (i.e., rhodium complexes deposited on Al_2O_3 powder), including results characterizing C–H bond activation. We refer to the literature (*92*) for details and note that this should be considered as a new, promising area in connection with single-crystal systems.

III. Metals on Oxides

So far, we have considered clean oxide surfaces and their reactivities. In this section, we deal with the modification of the oxide surface resulting from deposition of metals. This represents a route toward the preparation and characterization of more complex model systems in heterogeneous catalysis that bridge the materials gap.

In the preceding few years, several strategies have been followed along this route (*1*). In early work, small metal particles were deposited onto oxide bulk single-crystal surfaces, particularly MgO, and characterized by transmission electron microscopy (TEM). Poppa (*14*) was the pioneer in this field; important contributions have been reviewed by Henry (*6*), who was involved in the early TEM measurements.

The early efforts were primarily aimed at the preparation of small, well-defined metal particles; another strategy has been followed by Møller *et al.* (*93–96*) and Diebold *et al.* (*12*), who prepared thin metal layers on bulk oxide single crystals such as $TiO_2(110)$. As mentioned previously, the advent of STM has had a substantial influence on the understanding of the structures of clean oxide surfaces. Several groups (*97–99*) have begun investigating metal deposition on TiO_2 surfaces. Interesting initial results have been obtained concerning metal particle migration and oxide migration onto the metal particles (the so-called SMSI effect) (*98, 99*). Particularly well-suited to the application of STM are metal particles deposited onto thin-layer oxide surfaces (*6, 7, 11, 13*). Goodman's group (*11*), for example, has made major contributions to this field. One of the early results of his group was obtained for copper aggregates deposited on an amorphous silica layer (*100*).

Recently, it has been shown that ordered thin silica layers can be grown (*101*). In our laboratory, a silica layer has been grown on Mo(112). Figure 12 shows a LEED pattern of the layer prepared by evaporation of silicon and subsequent oxidation. The surface has hexagonal symmetry, but its

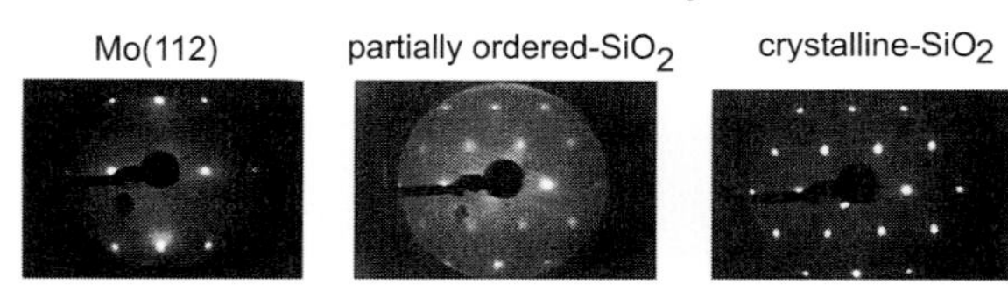

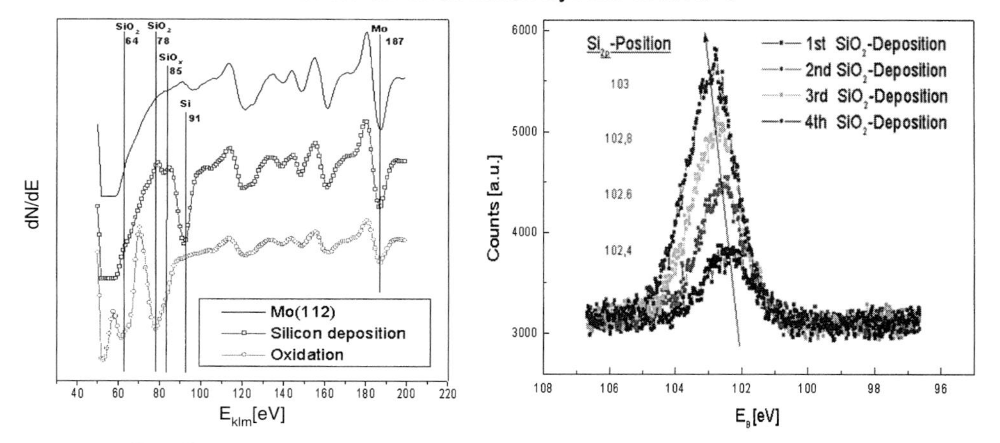

FIG. 12. LEED pattern and Si 2p spectra of a SiO_2 layer grown on Mo(112).

detailed structure is not known. Included in Fig. 12 is a set of Si_{2p} photoelectron spectra shown as a function of layer thickness and preparation conditions. The value observed for the final preparation is practically identical to that observed for thin silica layers on silicon, which are generally known to be amorphous (*102*). These layers will be used in the future as supports for evaporated metal aggregates.

In the past, we primarily used well-ordered alumina layers as substrates; Fig. 13 shows the result of an STM investigation. The left panel shows the clean alumina surface as imaged by a scanning tunneling microscope (*103*). The surface is well ordered, and there are several kinds of surface defects. One constitutes reflection domain boundaries between the two growth directions of $Al_2O_3(0001)$ on the NiAl(110) surface, the substrate on which the layer was grown according to a well-established oxidation recipe (*79*). There are anti-phase domain boundaries within the reflection domains and, in addition, point defects which are not resolved in the images. The image does not change dramatically after hydroxylation of the layer, a procedure which was mentioned previously (*76*). The additional panels show STM images of rhodium deposits on the clean surface at low temperature and at room temperature (*10, 104*) as well as an image recorded after deposition of rhodium at room temperature on a hydroxylated substrate (*105*). The amount deposited onto the hydroxylated surface is equivalent to the amount

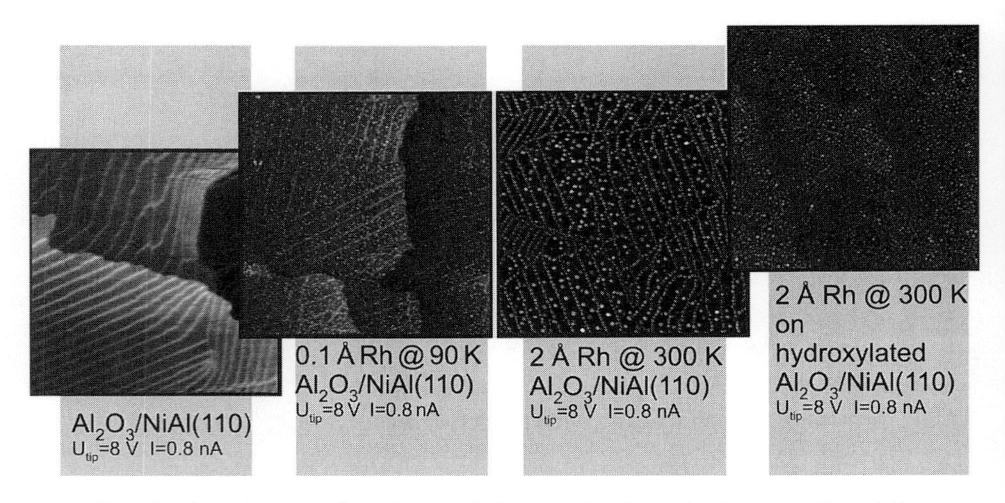

FIG. 13. Scanning tunneling microscopic images of a clean alumina layer (from left to right) after deposition of 0.1 Å of rhodium at 90 K, after deposition of 2 Å of rhodium at 300 K, and after deposition of 2 Å of rhodium at 300 K onto the prehydroxylated alumina layer. All scans: 3000 × 3000 Å.

deposited onto the clean alumina surface at room temperature. Upon vapor deposition of rhodium at low temperature, small particles are observed to nucleate on the point defects of the substrate, giving a narrow distribution of particle sizes. When the deposition of rhodium is carried out at room temperature, the mobility of rhodium atoms is considerably higher than at low temperature so that nucleation at the line defects of the substrate becomes dominant. Consequently, most of the material nucleates on reflection domain and anti-phase domain boundaries. The particles are relatively uniform in size, with the size depending on the amount of material deposited. When the same amount of material is deposited onto a hydroxylated surface, the particles are considerably smaller and distributed across the entire surface, i.e., a much higher metal dispersion is obtained (*76*).

The thermal behavior of the deposits is important to chemical reactivity because the particles may undergo morphological changes to adopt their equilibrium shapes, which could depend on whether or not a reactive gas phase is present. In the present case, detailed experiments have been undertaken to characterize the particles deposited onto the clean substrate, and less detailed experiments have been done to characterize the deposits on the hydroxylated surface (*10*). As a result of these investigations, it is known that the morphology of the ensemble is not altered within a temperature window from 90 to approximately 450–600 K. The window is extended to even higher temperatures when the substrate is hydroxylated. At temperatures above the upper limit, the particles tend to sinter, and they also start to diffuse through the layer into the metal substrate below (*10*).

This sintering process is an interesting subject in its own right, and research on this process is just beginning (*10*). A more basic issue, of course, is the metal atom diffusion on oxide substrates. The obvious technique to use in such an investigation is STM (*106*). However, in contrast to investigations of diffusion on metal surfaces, similar investigations of diffusion on oxide surfaces have not been reported. On the other hand, field ion microscopy characterizations of metal atom diffusion on oxide layers have begun, and a first estimate of activation energies for diffusion has been reported (*107*).

It is obvious that diffusion studies will profit from atomic resolution imaging, once it is obtained routinely for deposited aggregates on oxide surfaces. Whereas for TiO_2 and a few other oxide substrates atomic resolution may be obtained routinely, there are only a few reported investigations of deposited metal particles at atomic resolution (*108*). The first report of an atomically resolved image of a palladium metal cluster on MoS_2 was reported by Piednoir *et al.* (*108*). Atomically resolved images of palladium aggregates deposited on a thin alumina layer have also been obtained (*109*). Figure 14 shows an image of an aggregate of approximately 80 Å in width.

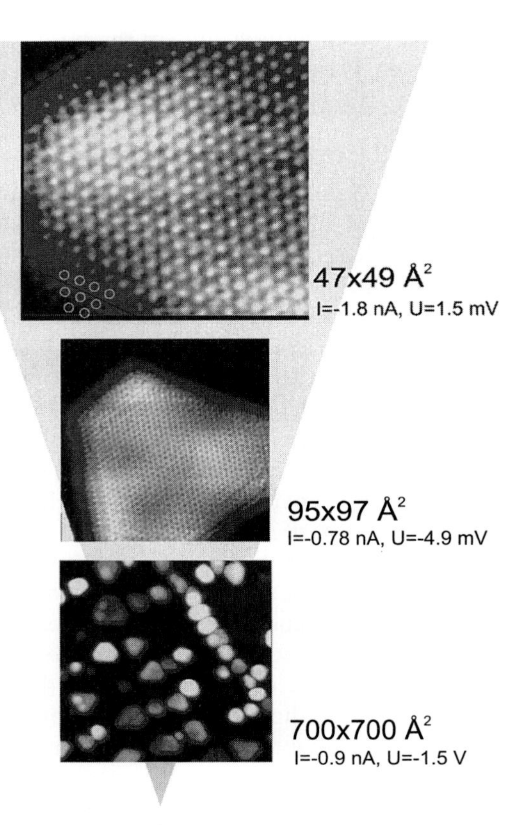

FIG. 14. Atomically resolved scanning tunneling microscopic images of palladium aggregates grown on an alumina layer [reproduced with permission from Hansen *et al.* (*109*)].

The particle is obviously crystalline and exposes on its top a (111) facet. Furthermore, on the side, (111) facets (typical of a cuboctahedral particle) can be discerned. The small (100) facets predicted for the equilibrium shape on the basis of the Wulff construction could not be atomically resolved.

However, if we apply the concept of the Wulff construction, we may deduce the metal surface adhesion energy (*109*). The basic equation is

$$W_{\mathrm{adh}} = \gamma_{\mathrm{oxide}} + \gamma_{\mathrm{metal}} - \gamma_{\mathrm{interface}} \tag{1}$$

Provided the surface energies (γ_{metal}) of the various crystallographic planes of the metal are known (*110*), a relative work of adhesion (W_{adh}) may be defined (*109*). We find a value of $2.9 \pm 0.2\ \mathrm{J/m^2\ eV}$, which is still different from the result of recent calculations by Bogicevic and Jennison (*111*), who reported metal adsorption energies of 1.05 J/m^2 calculated for a thin defect-free alumina layer. It is not unlikely that this discrepancy is related to the complicated nucleation and growth behavior of the aggregates involving defects of the substrate.

Although STM reveals the surface structure of deposited particles, their internal structure, in particular as a function of size, is not easily accessible from STM images. In this regard, TEM images of the same model systems can be of help (*112*). Figure 15 shows a schematic drawing of a sample. After growth of the layer and deposition of the particles, the sample is ion milled from the back so that a small hole is finally formed. In this way, a wedge is obtained which is thin enough for the imaging process.

An added benefit of this procedure is that the unsupported layer next to the edge can also be investigated (*113*). This capability provides the opportunity to assess whether the metal substrate has any structural effect on the deposits. On the basis of numerous high-resolution TEM (HRTEM) images and a subsequent analysis of the Moiré periodicities, it has been possible to calculate the lattice constants as a function of particle size (*112*).

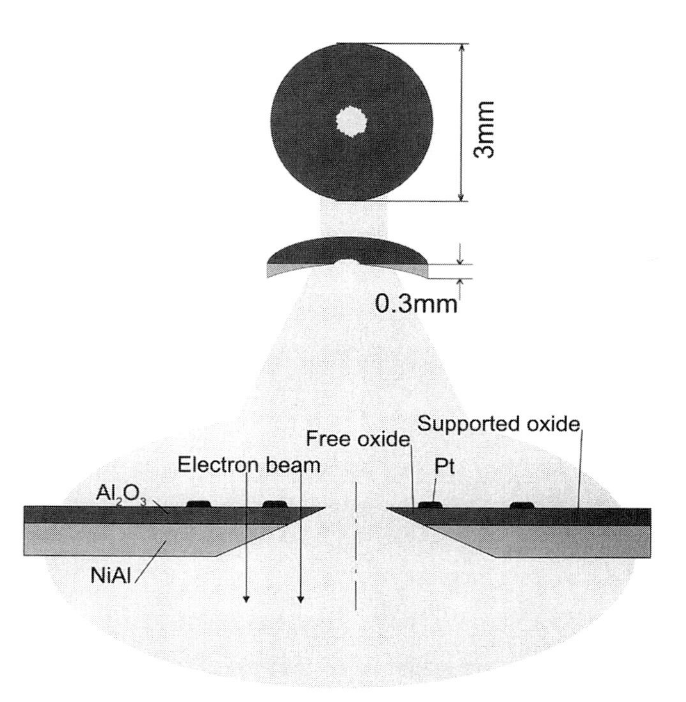

FIG. 15. Schematic drawing of a sample prepared for transmission electron microscopy.

The corresponding plot is depicted in Fig. 16 and indeed demonstrates that the atomic distances decrease continuously to 90% of the bulk value at a cluster size of 10 Å. On the other hand, the lattice constant approaches the platinum bulk value even at a diameter of 30 Å. This effect has also been detected for tantalum and for palladium clusters on thin alumina layers, but it seems to be less pronounced in these cases (*114, 115*).

The deposits discussed so far were prepared with the intention of maintaining the size distribution narrow. The spacing of aggregates on the surface, however, has not been an issue. If we consider reacting systems, diffusion of species between the particles (i.e., spillover processes) may become important. Therefore, it may be desirable to control not only the particle size and morphology but also the distances between particles. Several methods have been used, but we refer only to those based on electron beam lithography. Rupprechter *et al.* (*116, 117*) reported the preparation of two-dimensional arrays of platinum particles deposited onto amorphous SiO_2 layers. Particles of 25- to 40-nm average size were produced, as shown in Fig. 17. The atomic force microscopy (AFM) image reveals an average height of 20 nm for these particles, which were obtained after several reaction-cleaning cycles. Other, similar images have been reported (*118–122*). The average metal particle size reported in these investigations is still an order of magnitude larger than that of the particles nucleated and grown under UHV conditions.

There are other methods to prepare model systems which are not classi-

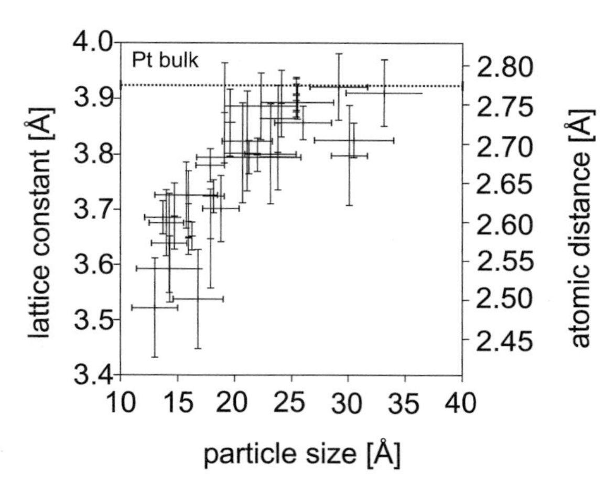

FIG. 16. Lattice constants and interatomic distance of platinum particles grown on Al_2O_3/NiAl(110) as a function of their size. (The ends of the horizontal bars represent the width and the length of the particular particles, respectively, and the vertical bars are error bars.)

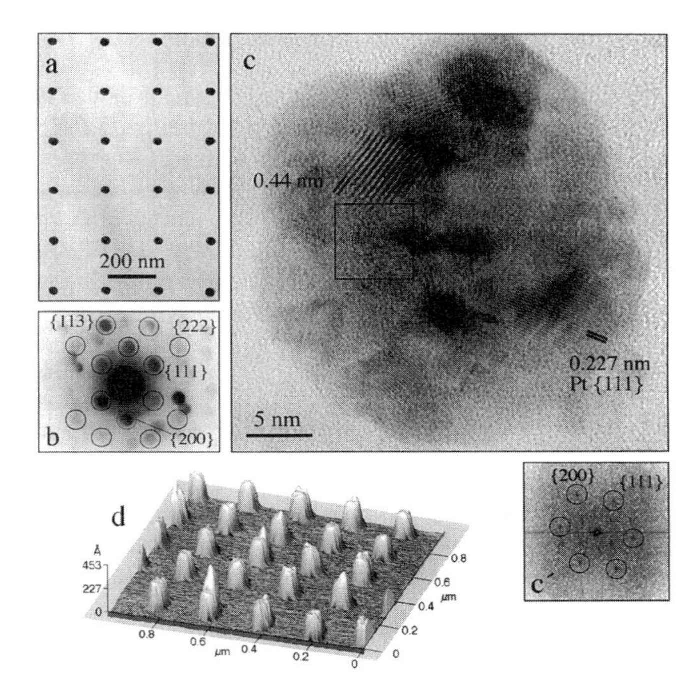

FIG. 17. (a) Transmission electron micrograph of a platinum nanoparticle array on SiO_2 (mean particle diameter, 40 nm; interparticle distance, 230 nm). (b) Microdiffraction pattern of an individual platinum particle showing its polycrystallinity [spots originating from a (110)-oriented crystalline grain within the polycrystalline platinum particle are marked by circles]. (c) HRTEM micrograph and (c′) fast Fourier transform of a 25-nm platinum model catalyst particle. (d) AFM image of a platinum nanocluster array after several reaction-cleaning cycles.

fied according to Fig. 1. One is the preparation of "inverse" model systems consisting of oxide islands grown on metallic supports. The idea is to leave parts of the metal support uncovered so that the interface between meal and oxide is exposed. Several groups (*123–126*) followed such routes. We refer to the literature for details and restrict this discussion to the deposited metal aggregates on oxides.

Of course, the electronic structure of deposited metal aggregates reflects the geometric structure to a degree and vice versa. The electronic structure, which is discussed next, has been investigated using various methods, including photoemission, X-ray absorption, and scanning tunneling spectroscopy. One particularly interesting aspect is the size dependence of the electronic structure in relation to adsorption and reactivity.

Starting from an atomic-level diagram, Fig. 18 shows the development when increasingly more atoms are agglomerated to form an aggregate and finally a solid with a periodic lattice. Upon formation of an aggregate from equivalent atoms, the atomic levels are split into molecular orbitals, many of which are degenerate if the symmetry of the system is high. The splittings are characteristic of the interatomic interactions. Depending on the interaction strength, the split levels derived from a given atomic orbital start to overlap energetically with levels derived from other atomic orbitals. As long as the system has molecular character, there is an energy gap left between occupied and unoccupied levels. This situation is in contrast to that encountered for an infinite periodic metallic solid, as presented on the right-hand side of Fig. 18: where there is no longer a gap between occupied and unoccupied levels. It is not difficult to envision that, as the number of atoms in an agglomerate is slowly increased, the gap between occupied

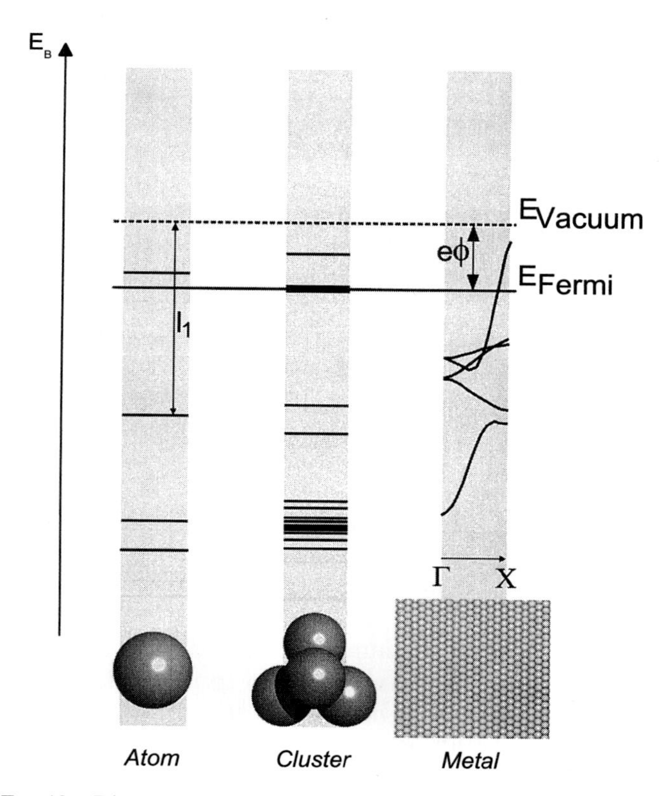

FIG. 18. Diagram illustrating the transition from an atom to a metal.

and unoccupied orbitals effectively vanishes. This is the case if the gap decreases to a value close to kT. In this situation, the changes in the electronic structure would be responsible for an insulator (molecule)–metal transition.

The question arises as to how many atoms are necessary to induce such a transition. There are several reports claiming numbers ranging from 20 to several hundred atoms in this respect (*115, 127–140*). An interesting extrapolation was deduced from spectroscopic measurements of the gap of inorganic carbonyl cluster compounds as a function of the cluster size. The result is shown in Fig. 19 (*129*). The extrapolation suggests that 70 atoms are sufficient to close the gap. The extrapolation yields a vanishing gap just below 100 metal atoms.

On the other hand, we have investigated deposited clusters of varying size with a combination of photoelectron spectroscopy (*132, 141*) and

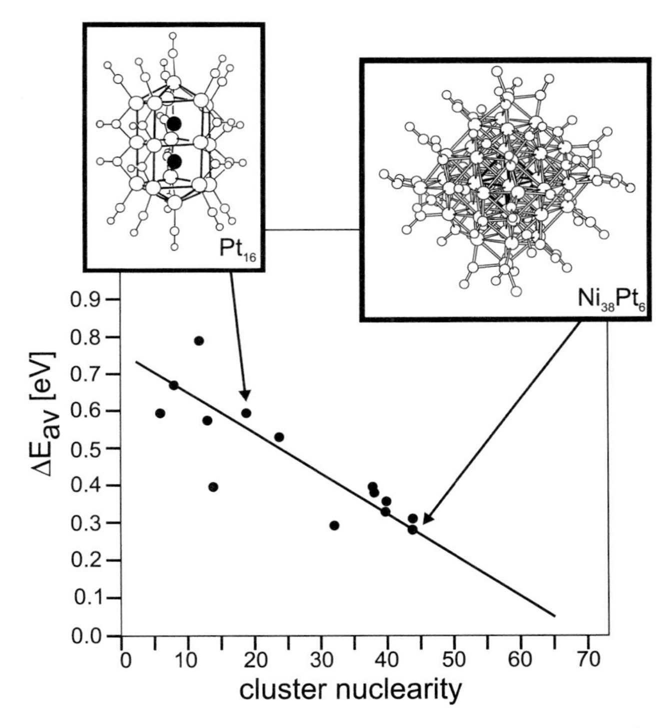

FIG. 19. Electronic excitation of lowest energy for several cluster compounds as a function of the number of metal atoms in the cluster [reproduced with permission from de Biani *et al.* (*129*)].

X-ray absorption (*140*). Figure 20 shows the chemical shifts measured for aggregates of palladium and of rhodium of varying size and plotted as a function of the inverse diameter (*130, 140, 142–148*). There is a linear correlation characterizing the clean as well as the CO-covered aggregates. It has generally been accepted (*130, 140, 142–148*) that this behavior is typical of the response of metallic spheres on the creation of a core hole, which in turn is screened by the metal electrons of the sphere. The different slopes for the clean and the CO-covered clusters can be explained by the electron-withdrawing effect of the adsorbed CO molecules. The withdrawal of charge from the clusters changes the effective electron density in the cluster and the screening properties. The plots are fully compatible with a metallic behavior, and there is no clear indication for a metal-to-non-metal transition.

According to the calculation characterizing gold clusters (*149*), this is not surprising. The screening properties are governed by the dielectric function. Its value is infinite for a metal, and it is still very large for a finite cluster, even though the cluster may have developed a small gap between occupied and unoccupied levels. This makes it hard to use the results as

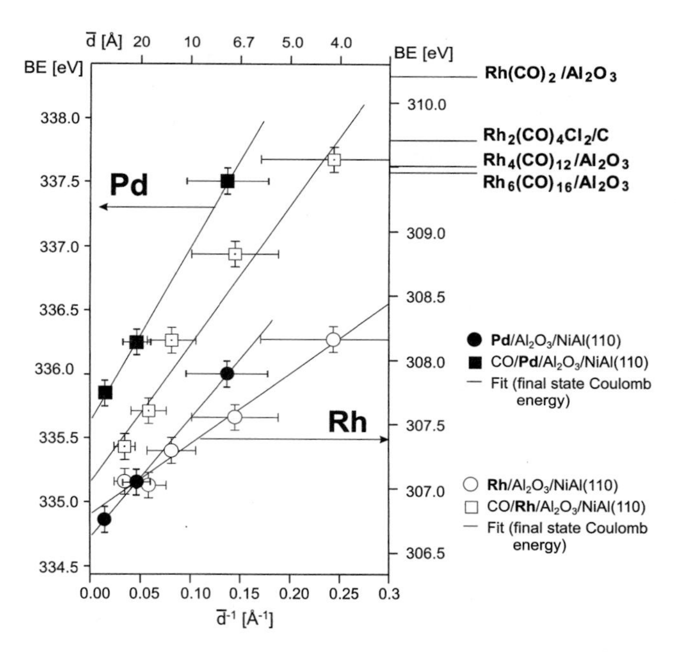

FIG. 20. Shift of XPS binding energies as a function of aggregate size for palladium aggregates and rhodium aggregates.

test cases. Therefore a second set of experiments (*140*) has been undertaken for CO-covered palladium clusters exclusively (Fig. 21).

The C 1s ionization potentials of CO molecules adsorbed on palladium aggregates of different size have been measured. The ionization potential is then compared with the onset of X-ray absorption of light in the wavelength region where the C 1s–2π absorption is observed. The lowest wavelength where this adsorption can be induced corresponds to a state of the system in which screening is optimal and the system can attain the state of lowest energy. In a photoelectron spectrum, on the other hand, where a C 1s core hole is created and charge is transferred from the metal particle to the 2π unoccupied states to screen the core hole, the same screened state of the system is obtained. In other words, the C 1s ionization potential is pinned to the onset of the X-ray absorption Pd 4d–CO 2π band (*150, 151*).

Again, this is the situation if the aggregate is a metal. However, if the aggregate is not metallic, then this relationship does not hold, and thus the ionization potential floats with respect to the onset of the X-ray absorption spectrum. Consequently, in the case of a nonmetal-to-metal transition, we

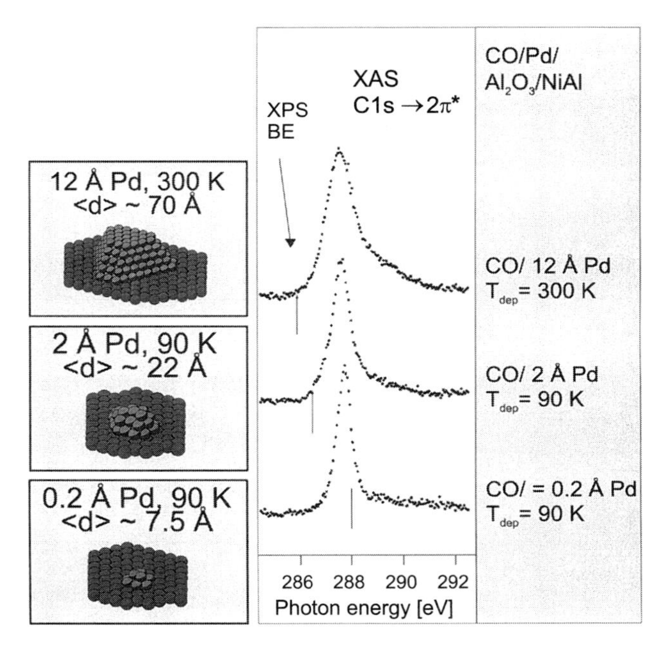

FIG. 21. X-ray absorption spectra of CO adsorbed on palladium aggregates of varying size. The ionization potentials as derived from photoelectron spectra are marked in the spectra.

observe a switch from a situation with floating Fermi levels to a pinning of the X-ray onset to the ionization potential.

Figure 21 shows a set of X-ray adsorption spectra and also the positions of the corresponding ionization potentials for various cluster sizes. Whereas for the larger cluster sizes there is a clear pinning of the Fermi energy indicating a metallic state of the system, the onset of the X-ray absorption spectrum does not align with the ionization energy for the small cluster sizes. It appears that such a situation is reached when the diameter of the aggregate decreases to values below 25 Å diameter and a height of 15–20 Å. The aggregate of this size contains 75–100 atoms, and the size correlates with the extrapolation of the spectroscopic data of metal carbonyl cluster compounds discussed in connection with Fig. 19 as well as with our results. We consider this comparison as a strong indication that, at least for the carbon monoxide-covered clusters, a nonmetal-to-metal transition occurs in the vicinity of such a size.

Goodman and his group used scanning tunneling spectroscopy to investigate the electronic structure of aggregates deposited on oxides (*152*). Figure 22 shows typical current–voltage curves for some aggregate sizes, i.e., gold on $TiO_2(110)$ (*77*). Although the large particles do not exhibit a plateau near $I = V = 0$, the smaller clusters do show the behavior expected for a system with a gap. However, the discrete structures observed for other systems (i.e., nanoparticles on graphite and related substrates) are not found (*109, 153, 154*). The authors (*77*) report indications that it is particularly the second layer in the gold aggregates that is responsible for the nonmetal-

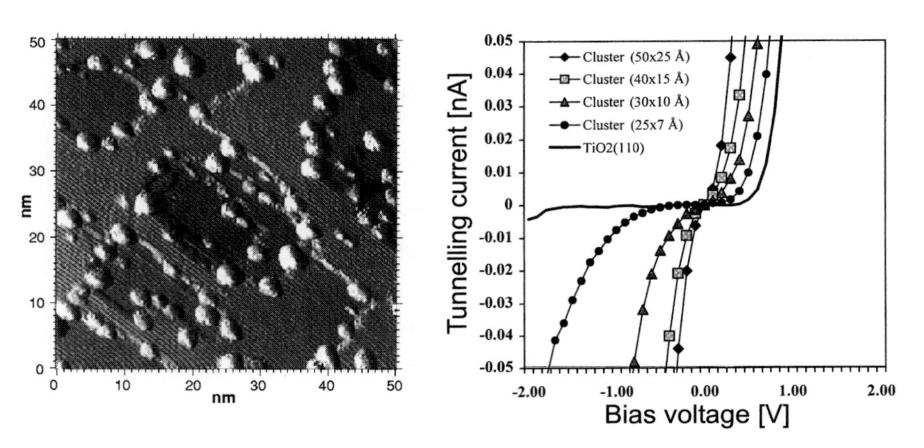

FIG. 22. Current–voltage curves determined for clusters of various sizes in the Au/$TiO_2(110)$ system as depicted in the STM image [reproduced with permission from Valden and Goodman (*77*)].

to-metal transition. Gold is an interesting low-temperature CO oxidation catalyst and we return to this when we discuss chemical reactivities.

Another particular aspect of the electronic structure is the magnetic behavior of small particles (*155*). Although much is known about thin metal layers in UHV as well as about supported powder catalysts, studies of magnetic properties of deposited particles under well-defined conditions are very rare (*156–159*). We briefly mention here studies using the technique of ferromagnetic resonance spectroscopy (Fig. 23), a modification of electron spin resonance.

Aggregates of iron and cobalt have been studied after deposition onto

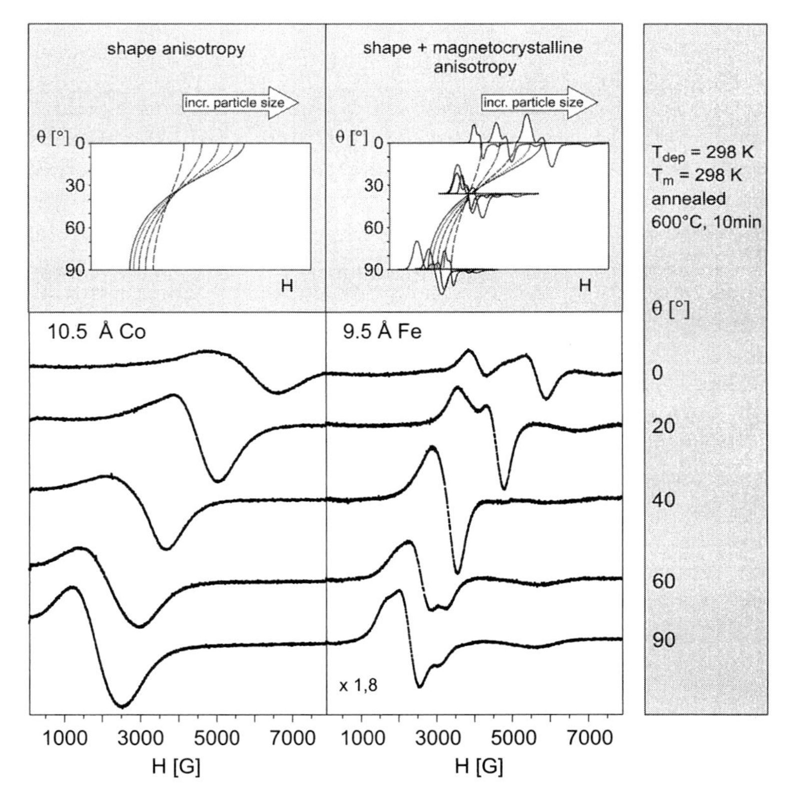

FIG. 23. Ferromagnetic resonance spectra taken for cobalt aggregates deposited on sapphire (left) and iron aggregates on sapphire (right). The angle of the magnetic field was varied with respect to the surface normal. The insets show the theoretically expected angular dependencies as a function of the deposited amount (left) for a layer with particles exhibiting uniaxial anisotropies (right, with particles with triaxial anisotropies).

alumina, prepared both as a thin layer and as bulk alumina, i.e., sapphire substrate. The dispersion of the particles has been investigated by STM and their magnetism by angle-dependent ferromagnetic resonance (FMR) measurements of the kind shown in Fig. 23 for two particle size distributions. The particular data have been taken for iron particles of 30–40 Å. The observed line shape and resonance frequency changes as a function of the angle of the external magnetic field can be fit by assuming a particle size distribution of the log normal type and a dependence of the magnetic properties of the particles as a function of their size. A schematic diagram taking the various contributions into account is included in Fig. 23. The resonance position of the absorption changes as the angles between external field and surface normal are varied. The magnetic anisotropy is the reason for this behavior. For the different particle sizes, characteristic variations in the anisotropy are expected ranging from a strong dependence for the large particles to a weak dependence for the small particles. The weak dependence is characteristic of superparamagnetic particles. For each size, a line shape has to be chosen. The superposition then results in the variations observed in Fig. 23. The presented data and the analysis are only preliminary and more detailed measurements and simulations have to be performed.

It is yet to be determined to what degree these magnetic properties can be used in applications, such as sensors and information storage. The FMR signal can be used, of course, to monitor the magnetic properties of the particles even under ambient conditions, when X-ray photoelectron spectroscopy (XPS) and other surface-sensitive tools fail, because the measurement relies on photons.

Before further discussing the reactivities of deposited particles, we briefly discuss adsorption properties using CO as a probe molecule. An advantageous technique to study CO adsorption is FTIR spectroscopy because it provides the resolution to differentiate between various adsorbed species. Again, the thin-layer-based systems are particularly well suited since the metallic support of the oxide layers acts as a mirror at IR frequencies. However, it is also possible to perform such experiments on surfaces of bulk dielectrics as shown by the Hayden group (*160, 161*).

Rainer and Goodman (*152*) published a study of CO adsorption on palladium aggregates on Al_2O_3 layers. The results have been interpreted as characteristic of the adsorption of CO on different facets of the small crystalline aggregates. Although this interpretation does not take into account adsorption on the various defect sites of the aggregates (*132*), as pointed out recently (*162*) the data are indicative of the potential of this technique for the characterization of size-dependent absorption phenomena.

We recently prepared metal deposits on well-ordered alumina layers at low temperatures in the range of 50 to 90 K (*163, 164*) to determine the IR characteristics of specific sites. The IR spectrum characterizing a rhodium deposit prepared and saturated with CO at 90 K (the average cluster contained nine atoms) is displayed in Fig. 24a (top). The most prominent feature in the stretching region of terminally bonded CO molecules is a sharp, intense band at 2117 cm^{-1}. This signal has previously been shown to arise from isolated rhodium atoms trapped at oxide defects (*162*). Both the number of adsorbed CO molecules and the nature of the defect site remained unclear. Features at lower frequencies, on the other hand, have been assigned to molecules on rhodium aggregates.

To gain insight into the stoichiometry of the rhodium carbonyl species giving rise to the band at 2117 cm^{-1}, isotopic mixture experiments were performed. If such a signal were indicative of Rh–CO, an equimolar mixture of ^{12}CO and ^{13}CO should result in the appearance of a second IR band of equal intensity about 47 cm^{-1} lower. In the case of $Rh(CO)_2$, three species of different isotopic composition would be formed with relative abundances 1:2:1, giving rise to three IR bands with intensities reflecting this ratio.

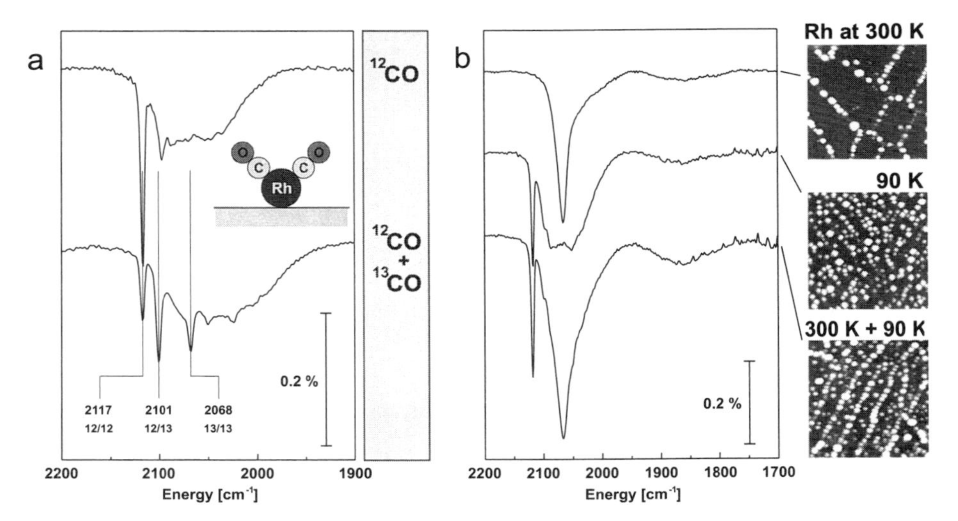

FIG. 24. (a) IR spectra taken after deposition of 0.028 monolayer (ML) of rhodium and subsequent saturation with ^{12}CO (top) and an approximately equimolar mixture of ^{12}CO and ^{13}CO (bottom) at 90 K. The isotopic compositions giving rise to the three dicarbonyl bands are indicated below the corresponding wavenumbers. The average particle contained nine atoms. (b) IR spectra recorded after CO saturation of rhodium deposits at 90 K, along with corresponding room-temperature STM images (500 × 500 Å). Top, 0.057 ML of rhodium deposited at 300 K; middle, 0.057 ML of rhodium deposited at 90 K; bottom, 0.057 ML of rhodium deposited at 300 K, followed by the same exposure at 90 K.

The latter observation was indeed made experimentally (Fig. 24a, bottom), thereby showing that $Rh(CO)_2$ is responsible for the IR band under investigation. Such rhodium gem-dicarbonyl species, with a rhodium oxidation state of 1, result from the disruption of metal crystallites on Rh/Al_2O_3 catalysts (symmetric stretch, ~2100 cm^{-1}; antisymmetric stretch, ~2035 cm^{-1}) (*165–167*). These bands have also been observed by the group of Solymosi, who characterized rhodium on $TiO_2(110)$ single crystals (~2110 cm^{-1}/~2030 cm^{-1}) (*160, 161*), and followed a similar disruption process by STM (*168*).

No signal is discerned in our data that is indicative of the antisymmetric stretch (Fig. 24a). Since for thin oxide layers on metal substrates the surface selection rule applies, its dynamic dipole moment must be oriented parallel to the oxide surface.

We now discuss the type of rhodium nucleation site responsible for the IR band at 2117 cm^{-1}. Rhodium deposition at room temperature results in the formation of metal aggregates located at oxide domain boundaries. Upon CO saturation at 90 K, no $Rh(CO)_2$ is formed (Fig. 24b, top). This is in clear contrast to what is observed for a low-temperature deposit. Nucleation inside the domains gives rise to the dicarbonyl band (Fig. 24b, middle). STM measurements showed that point defects are the primary nucleation sites under these conditions (*169*). Rhodium decoration of the line defects at room temperature preceding deposition at 90 K does not suppress rhodium dicarbonyl formation (Fig. 24b, bottom). Consequently, we may conclude that the $Rh(CO)_2$ species are not associated with the domain boundaries but rather with oxide point defects, which are the dominant rhodium nucleation sites at 90 K.

To reduce the particle sizes further, growth and adsorption experiments were performed at 60 K. In the IR spectrum of ^{12}CO adsorbed on such a rhodium deposit (the average cluster at 300 K incorporated five atoms), many distinct signals are observed at frequencies between 2172 and 1961 cm^{-1}, along with a broad absorption band in the range of 2120 to 1950 cm^{-1} (Fig. 25, bottom) and a very weak feature due to multiply coordinated CO at 1856 cm^{-1}.

As in the case of the deposits formed at 90 K, the broad band and the sharp feature at 2116 cm^{-1} are attributed to terminally bonded CO on rhodium aggregates of various size and to $Rh(CO)_2$ species at oxide point defects, respectively. Physisorption on the Al_2O_3 (*170*) is reflected by the band at 2172 cm^{-1}, which is also observed upon exposure of the pristine oxide layer to CO.

One or more new types of rhodium nucleation sites clearly come into play at 60 K, in contrast to the situation at 90 K, as indicated by the observation of a smaller mean particle size, by the lower intensity of the IR band due to rhodium nucleated at the point defects, and by the appearance of new signals at 2087, 2037, 1999, and 1961 cm^{-1}. The small half-

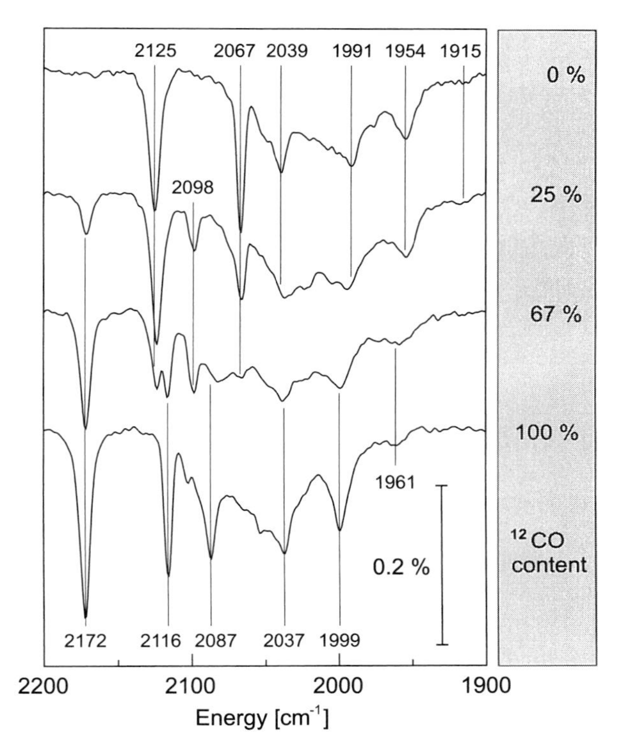

FIG. 25. Series of IR spectra acquired at 60 K after deposition of 0.020 ML of rhodium at 60 K and subsequent saturation with isotopic mixtures of ^{12}CO and ^{13}CO at the same temperature. The percentage of ^{12}CO is given next to the spectra. The average particle contained five atoms.

widths of the bands point to the presence of uniform, isolated $Rh_n(CO)_m$ species. Similar features are obtained for deposits with even lower mean particle sizes. Therefore, it is suggested that they most likely originate from atomic, dimeric, or possibly trimeric Rh–CO species.

Again, the use of isotopic mixtures provides additional information regarding the identities of the surface species. As expected, saturation with ^{13}CO results in a downward shift of the entire spectrum by 42–49 cm^{-1} (Fig. 25, top). Slight changes in relative band intensities are due to small variations in sample temperature. Following the spectral evolution with isotopic composition, differences in behavior of the individual bands are observed (Fig. 25):

1. As expected for single, isolated CO molecules, the intensities of the bands at 2172 and 2125 cm^{-1} vary linearly with the concentration of the corresponding isotopes.

2. In line with the statistics of mixtures, the rhodium dicarbonyl bands at 2116 and 2067 cm^{-1} are more strongly attenuated upon addition of the other isotope, and a band due to the mixed dicarbonyl is observed at 2098 cm^{-1}.

3. Between the bands at 1999 and 1954 cm^{-1}, no such signal at an intermediate frequency is found, implying that they do not originate from rhodium dicarbonyl species at a different surface site. Their slow intensity variation may indicate Rh–CO.

4. In clear contrast to this, the 2087-cm^{-1} band vanishes completely upon admixture of 33% ^{13}CO. This change is conceivable only if the surface complex contains three or more CO molecules.

The information gained from the other signals is less specific. In summary, we conclude that several different types of rhodium particles are responsible for the observed IR features. Currently, density functional calculations characterizing small rhodium carbonyls are in progress (*171*). Calculated vibrational frequencies of such systems may help to identify the species present on the alumina layer.

Studies characterizing small rhodium particles have been extended to neighboring elements in the periodic table. IR spectra recorded after deposition of comparable amounts of palladium, rhodium, and iridium and subsequent CO saturation at 90 K are displayed in Fig. 26. We note differences in the low-wavenumber region, where vibrational frequencies of molecules in multiple coordinated sites are found. As on single crystals, the population of such sites is highest on palladium (*172, 173*), whereas no such CO is observed on iridium (*174, 175*).

The differences in the region of terminally bonded CO, however, are much more pronounced. In the case of iridium, several distinct features are observed. In analogy to the $Rh(CO)_2$ band at 2117 cm^{-1}, the sharp signal at 2107 cm^{-1} may be attributed to $Ir(CO)_2$ species via isotopic mixture experiments (not shown). Bands with similar frequencies have been assigned to the symmetric stretch of $Ir^{+}(CO)_2$ on Ir/Al_2O_3 catalysts (2107–2090 cm^{-1}) (*176*) and on the iridium-loaded zeolite HZSM-5 (2104 cm^{-1}) (*177*). The appearance of many bands at lower wavenumbers is reminiscent of the rhodium deposits formed at 60 K (Fig. 26), indicating a comparable nucleation behavior.

In contrast to these results, no signs of atomically dispersed palladium or structurally well-defined aggregates are observed. Indeed, the IR spectrum is similar to that observed for much larger, disordered palladium aggregates (*162*). At the same metal exposure, the palladium particles are found to be larger than the rhodium aggregates by room-temperature STM.

Our observations show that IR spectra of adsorbed CO provide valuable

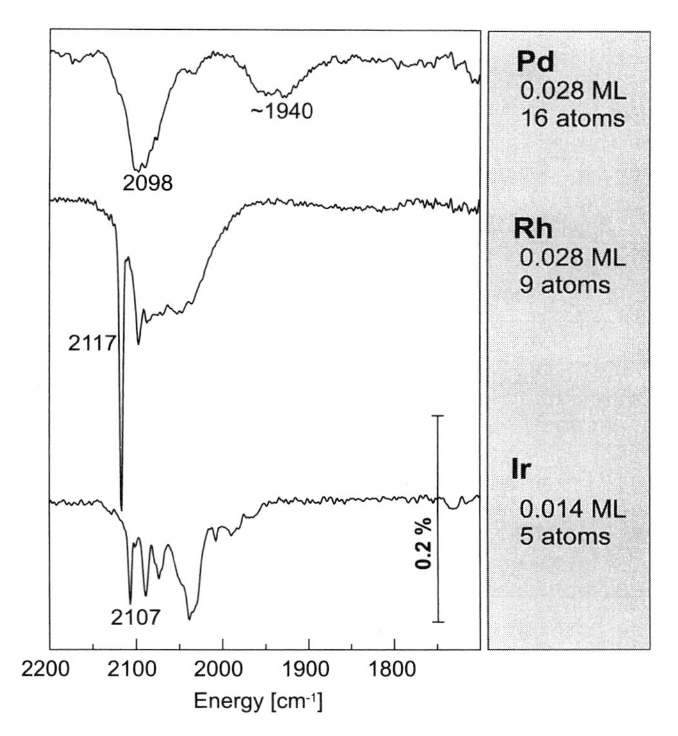

FIG. 26. IR spectra of palladium, iridium, and rhodium deposited at 90 K and saturated with CO at the same temperature.

information on the size of metal nanoparticles, as has long been recognized in the catalysis-related literature. The literature contains several adsorption studies (*178*) employing other probe molecules such as hydrocarbons, but in this case reaction also comes into play and renders the situation even more complicated.

Next, several simple chemical reactions of O_2, CO, and hydrocarbons on small aggregates are addressed. A simple reaction is the dissociative adsorption of oxygen on small particles. The palladium aggregate shown in Fig. 14 can be imaged at atomic resolution after a dosage to saturation with molecular oxygen from the gas phase (*179*). On the side facets the corrugation due to the presence of adsorption of oxygen can be identified. A doubled periodicity corresponding to a p(2 $\times$ 2) structure can be identified. This structure is very similar to the p(2 $\times$ 2) structure observed after dissociative oxygen adsorption on Pd(111) (*180*). We therefore conclude that a similar situation is encountered in the case of the deposited aggregates. The p(2 × 2) structure interestingly appears on the different facets at different tunneling conditions. When the oxygen-covered palladium aggregates are exposed to carbon monoxide, the reactivity of the different facets appears to be different in the sense that the oxygen adsorbate structure is lost on the various facets at various temperatures and exposures. It will be interesting to investigate these effects in more detail.

CO oxidation has been characterized for gold aggregates supported on TiO_2 at low temperatures and as a function of particle size (*77*). Gold clusters ranging in diameter from 10 to 60 Å were prepared on titania single-crystal surfaces and exposed to O_2 and CO. It was deduced that the structure sensitivity of this reaction in this system is related to a quantum size effect with respect to the thickness of the gold islands on the $TiO_2(110)$-(1 × 1) surface. The result of these reaction experiments is shown in Fig. 27 (top). The authors found a marked size effect of the catalytic activity, which correlates with the original observations (*181*) for gold on high-area, titania-supported catalysts. The aggregates near 35 Å in diameter show the maximum activity.

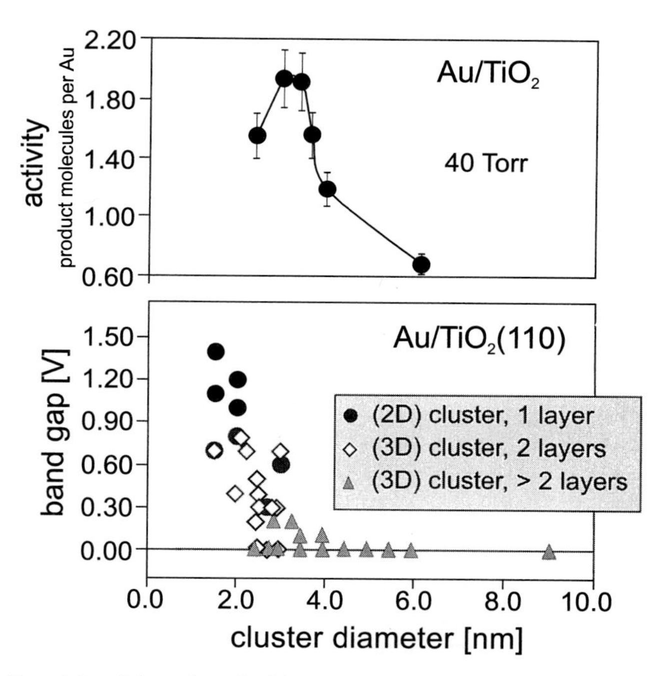

FIG. 27. Reactivity of size-selected gold aggregates deposited on $TiO_2(110)$ (top) correlated with the band gap of the aggregate as deduced from STM current–voltage curves (bottom) [reproduced with permission from Valden and Goodman (*77*)].

In the future, it will be important to perform kinetics measurements for such reactions under well-defined conditions, in both UHV and ambient environments. A good example was recently reported (*182*)—a molecular-beam investigation of CO oxidation on palladium aggregates deposited on a MgO single crystal. The results shed light on how the rates are affected by the role of adsorbate diffusion on the MgO substrate, spillover to the metal deposits, and the role of different CO and O adsorption sites on the various facets of the aggregates. Kasemo *et al.* (*183, 184*) reported model calculations which consider the dependence of the kinetic phase diagrams (reaction rate versus reactant pressure) on the details of diffusion and sticking parameters on the crystallite facets. The experimental results from Henry's group (*182*) are compatible with such considerations.

In the work reported up to now, the aggregate distributions have not been monodisperse. However, Heiz *et al.* (*22, 185, 186*) reported adsorption and reaction experiments characterizing deposits of platinum and of gold aggregates on MgO layers; these were prepared from impinging size-selected, gas-phase clusters of the metal and inferred to be monodisperse, but they have not been imaged. The adsorbates were characterized by TDS

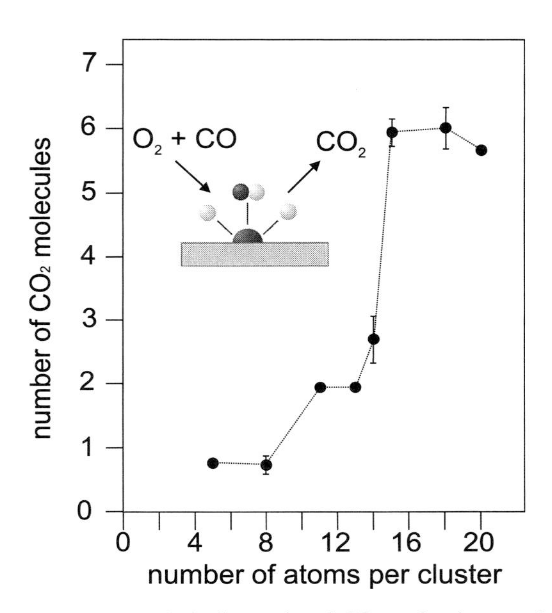

FIG. 28. Total number of catalytically produced CO_2 molecules as a function of cluster size. The clusters have been mass selected in the gas phase and deposited on a MgO film [reproduced with permission from Heiz *et al.* (*186*)].

and IR spectroscopy. The main reaction was CO oxidation; a typical result is shown in Fig. 28. It seems that with respect to the rate of the reaction, every atom of the aggregate counts in the regime of very small clusters.

Oxygen adsorption on palladium and on gold does not lead to oxide formation under the reported conditions, but it is expected that more reactive metals form oxides. We recently tested this assumption for tantalum aggregates and compared them with continuous tantalum layers (*187*). The continuous layer was oxidized to Ta_2O_5, as monitored with XPS, whereas the deposited aggregates adsorbed oxygen without being fully oxidized. On the other hand, cobalt and iron aggregates are easily oxidized under comparable conditions (*158*). This work needs to be extended to determine the effects of aggregate size.

Similar to experiments concerned with CO oxidation on gold aggregates summarized previously, we have undertaken an investigation of CO dissociation on rhodium particles (*188–190*). C 1s photoelectron spectra were recorded as a function of sample temperature and rhodium particle size. An example is shown in Fig. 29 for a temperature range in which the morphology of the ensemble of aggregates did not change. At low temperature, the signal typical of molecular CO was observed. At a temperature near 400 K a second signal appeared, indicating the dissociation of CO into carbon and oxygen atoms. At 500 K all molecular CO had been either dissociated or desorbed. The dissociation probability is then given by the ratio of the molecular to the atomic C 1s signal, which is plotted in Fig. 30 as a function of the aggregate size. Also included in Fig. 30 are data for very small aggregates, for which it has been shown that CO dissociation is negligible (*191*), as it is for close-packed, single-crystal surfaces (*192*), which correspond to the limiting case of infinitely large aggregate size. In contrast, the data obtained for stepped rhodium surfaces indicates that there is a probability for CO dissociation (*193*). The dissociation probability is highest for intermediate-sized aggregates consisting of 100–200 atoms.

Although electronic effects cannot be completely excluded as a reason for the onset of the dissociation phenomenon for small particles, an explanation on the basis of structural properties seems more likely. Since the rhodium deposits are basically disordered, it is easily imaginable that aggregates of medium size exhibit a maximum density of defects, such as steps, kinks, and other low-coordinated surface atoms. Smaller units are expected to contain fewer defects, especially if they are two-dimensional. Furthermore, spatial constraints may play a role [accommodation of carbon and oxygen atoms on adjacent sites (Fig. 30)]. At high exposures, the step density is reduced as a consequence of coalescence processes. For deposition at 300 K, the observed tendency to form crystalline aggregates in the high-coverage regime also contributes to a lower defect density, consistent with

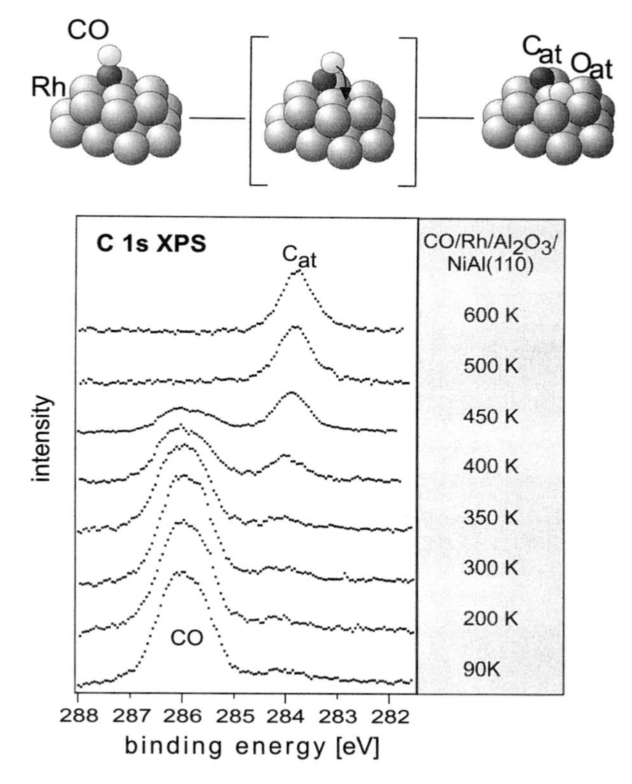

FIG. 29. CO dissociation on $Rh/Al_2O_3/NiAl(110)$: representative series of C 1s spectra taken after CO saturation at 90 K and heating to the indicated temperatures (data acquisition at 90 K).

the observation that the dissociation activity declines much faster in this case (Fig. 30).

A detail of the dissociation process is evident from a close inspection of the C 1s emission of the molecularly adsorbed CO (*189*). As shown in Fig. 31a, the peak is resolved into two components, denoted A and B. If the fraction of the total intensity found for component B after heating to 300 K is compared with the fraction of CO finally dissociating (Fig. 30), the species giving rise to B can be regarded as a kind of dissociation precursor. The evolution of each of these two quantities as a function of the particle size is identical, i.e., each pass through a maximum at the same particle size (*189*). At 90 K, however, the relative intensity is constant: It is the heating step which causes a shift of intensity from component A to component B (i.e., an increase of the B species which is most pronounced

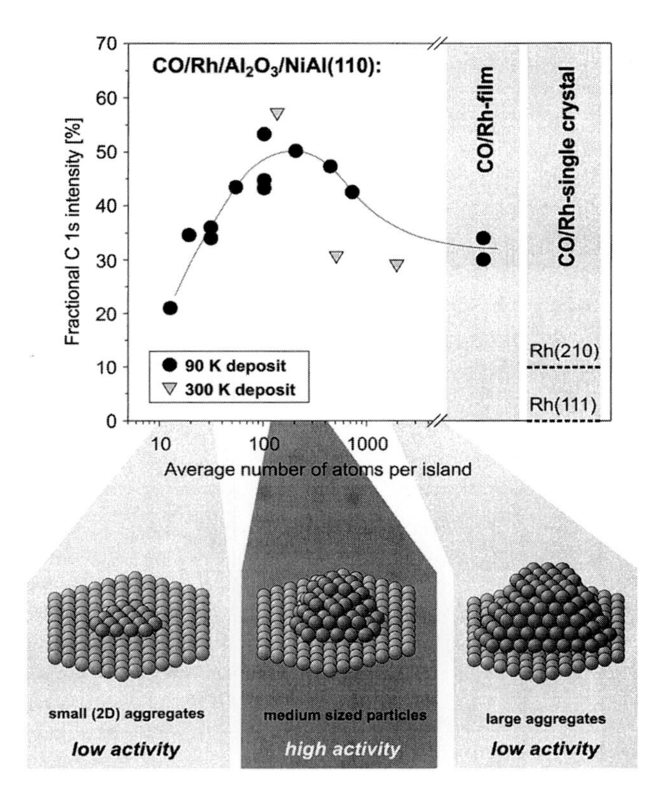

FIG. 30. CO dissociation activity on rhodium particles deposited on Al_2O_3/NiAl(110) as determined by XPS. According to Frank *et al.* (*188, 189*), the dissociation activity also passes through a maximum for the 300-K deposits, i.e., the activity decreases in the regime of small particle sizes reflecting the behavior of the 90-K deposits.

for the medium-sized particles). This conversion is irreversible; cooling does not lead to an intensity redistribution.

The conclusion that B is indeed a dissociation precursor is additionally corroborated by the results of Fig. 31b showing the intensity changes for the A and B peaks as well as the losses which result from either desorption or dissociation (*190*). Unambiguously, the desorption curve follows the curve for component A, whereas the dissociation curve mimics the development of the component B. Unfortunately, the results allow no further statement as to the nature of the A and B species. It can be assumed, however, that the B species is associated with CO adsorbed on defects. On the basis of the fact that more highly coordinated CO species give rise to

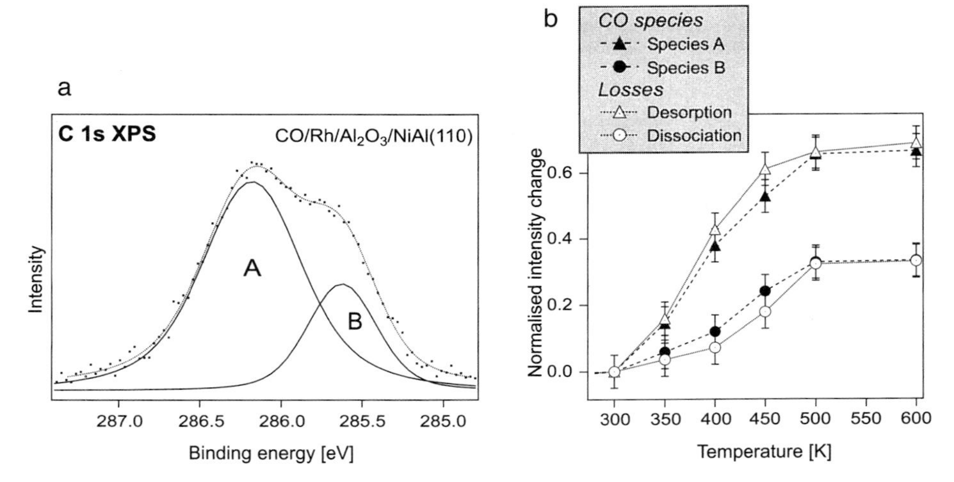

FIG. 31. (a) C 1s spectrum of CO adsorbed on rhodium particles after saturation at 90 K; (b) intensity changes for the components A and B as well as the intensity losses due to dissociation and desorption as a function of the annealing temperature (the average particle contained $\sim 10^4$ atoms).

lower C 1s binding energies, it may be further speculated whether B is associated with CO in a higher coordination than the A species.

As the final example of the size-dependent properties of well-defined model systems, we review recent data for the photoinduced reaction of methane interacting with palladium aggregates supported on alumina (*194*). Figure 32 shows TDS data for CD_4 desorbing from palladium aggregates of various sizes without having been exposed to light. The variation of the maximum desorption temperature with particle size represents differences of several tens of meV in interaction energy between CH_4 and the palladium aggregates. Since we know from photoemission measurements that the density of metal states at the Fermi energy varies and that energy gaps exist for very small aggregates, it is conceivable that the dispersive molecule–surface interactions vary as the density of states at the Fermi energy varies.

In the dark, CH_4 does not dissociate in the current situation. If the system is exposed to ultraviolet light with a wavelength in the range of 5.2–6.4 eV, however, CH_4 dissociation into adsorbed CH_3 and H fragments is observed (*194*). In a characteristic way, recombinative desorption occurs at temperatures exceeding 110 K, and this desorption is clearly separated from the CH_4 molecular desorption peak. By evaluating the relative areas

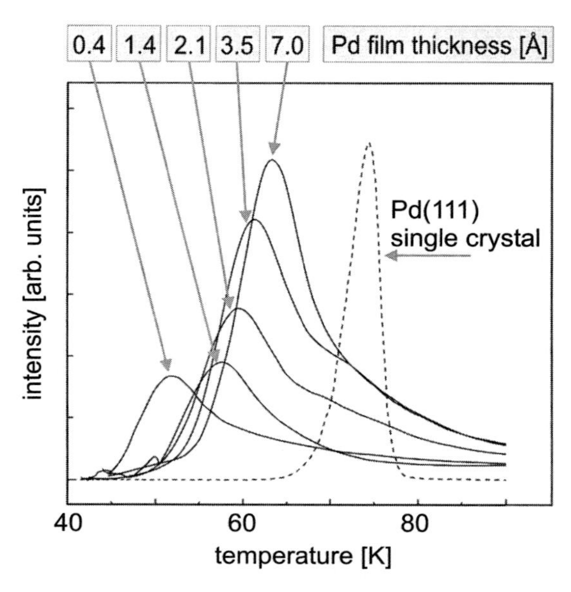

FIG. 32. Series of CD_4 TDS spectra (m/e = 20; solid curves) of palladium clusters of various sizes deposited on a thin Al_2O_3 layer epitaxially grown on NiAl(110). The sample was exposed to 0.5 liters of CD_4 at 40 K. The desorption peaks are indicative of molecular desorption, CD_4 (ads) $\rightarrow$ CD_4(g). The numbers denote the total palladium coverages as a measure of the cluster size. The dashed curve corresponds to a TDS spectrum of a Pd(111) single-crystal surface after exposure to 0.6 liters of CD_4 and is plotted with a different scale. The heating rate for the clusters was 0.5 K s^{-1} and that for the Pd(111) single crystal was 0.4 K s^{-1}.

in the TDS spectra characterizing direct and recombinative desorption, we can determine the ratio of photon-stimulated molecular desorption to photoinduced dissociation. In Fig. 33, these quantities are plotted as a function of aggregate size. The data show qualitatively that small aggregates support photodesorption, whereas photodissociation sets in for larger aggregates. In the limit of large aggregates, the behavior on Pd(111) single-crystal surfaces is approached (*195*).

Watanabe *et al.* (*195*) found that CH_4 may be photodissociated on Pd(111) at a photon energy of 6.4 eV, whereas in the gas phase this process cannot be induced at a photon energy lower than 8.4 eV (*196*). A model has been proposed in which the relevant excited state (of Rydberg character) of the CH_4 molecule is stabilized by the interaction with unoccupied states of the substrate (*197*), thus leading to a shift of the dissociation threshold to lower energy. The size dependence results from the change of the unoccupied

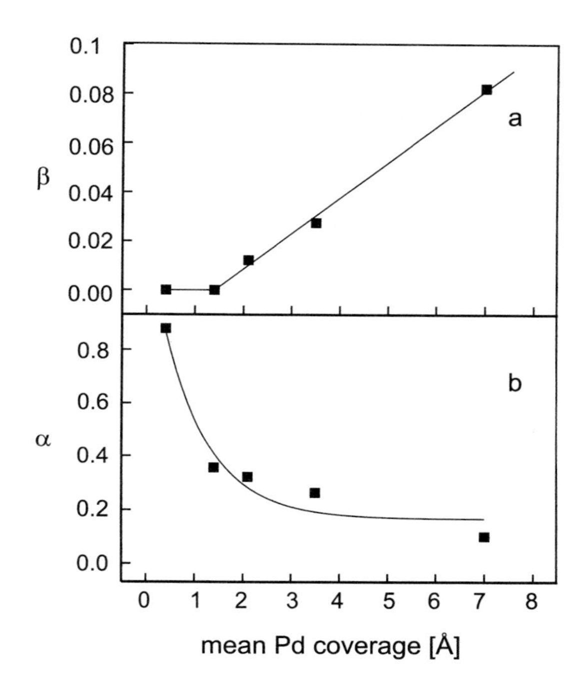

FIG. 33. (a) Plot of the CD_3H formation ratio β ($\beta = B/A_0$, where B is the integrated TDS peak area of CD_3H after irradiation and A_0 is the integrated TDS peak area of initially adsorbed CD_4 before irradiation) as a function of the total palladium coverage. (b) Plot of the CD_4 depletion ratio α ($\alpha = 1 - A/A_0$, where A_0 is the integrated peak area of CD_4 before irradiation and A is the integrated peak area of CD_4 after irradiation with 1.5×10^{19} photons/cm^2) as a function of the total palladium coverage. The total palladium coverages of 1.4, 2.1, 3.5, and 7 Å correspond to average cluster sizes of 37, 49, 65, and 73 atoms, respectively, as determined by a spot profile LEED analysis.

density of states in the aggregate with size, which for the smallest aggregates does not allow effective coupling to the CH_4 excited state. An alternative model has been proposed (*198*) which is based on the idea that the photon induces charge transfer from the CH_4 molecule to the aggregate. The energy necessary for such a process is then governed by the electron affinity of the aggregate, which varies with size and the ionization potential of CH_4. An important additional ingredient is the stabilization of the charge transfer by an image charge in the aggregate, which also varies with size. A qualitative estimate yields a threshold energy of 5 eV for dissociation, which is consistent with the currently available experimental results.

The field of investigations of chemical reactivity as a function of aggregate size is still emerging, and there are more exciting results on the horizon.

IV. Conclusions

After 30 years of surface science, which have seen an enormous development of methods and instrumentation, the field is now in a position to tackle questions of a complex nature. Naturally, metal surfaces have been the focus of attention in surface science to date, and this trend will continue. However, the complexity of problems is constantly increasing, particularly if molecular adsorbates and self-organized systems are considered. Metal oxide surfaces have received some attention in the recent past, and the study of such systems as well as more complex metal/metal-oxide composite systems has already defined a direction in surface science that promises interesting results of fundamental interest as well as of potential value for catalytic applications. It is encouraging that surface science is still a very active and productive field. Without a doubt, further impact on catalysis research can be expected in the future.

Acknowledgments

We are grateful to those colleagues whose names appear in the references. We also thank many colleagues for stimulating discussions and collaboration, especially Ralph Wichtendahl for his help in preparing the figures. The following have provided support for our work: Deutsche Forschungsgemeinschaft, Bundesministerium für Bildung und Forschung, Ministerium für Wissenschaft und Forschung des Landes Nordrhein-Westfalen, Fonds der Chemischen Industrie, German–Israeli Foundation, European Union, NEDO International Joint Research Grant on Photon and Electron Controlled Surface Processes, Hoechst Celanese, and Synetix, a member of the ICI group, through their Strategy Research Fund.

References

1. Duke, C. B. (Ed.), "Surface Science: The First Thirty Years." Elsevier, Amsterdam, 1994.
2. Ertl, G., and Freund, H.-J., *Phys. Today* **52,** 32 (1999).
3. Freund, H.-J., *Ber. Bunsenges. Phys. Chem.* **99,** 1261 (1995).
4. Bonzel, H. P., *Surf. Sci.* **68,** 236 (1977).
5. Freund, H.-J., and Umbach, E. (Eds.), "Adsorption on Ordered Surfaces of Ionic Solids and Thin Films." Springer, Heidelberg, 1993.
6. Henry, C. R., *Surf. Sci. Rep.* **31,** 231 (1998).
7. Campbell, C. T., *Surf. Sci. Rep.* **27,** 1 (1997).
8. Campbell, C. T., *Curr. Opin. Solid State Mater. Sci.* **3,** 439 (1998).
9. Freund, H.-J., Kuhlenbeck, H., and Staemmler, V., *Rep. Progr. Phys.* **59,** 283 (1996).
10. Bäumer, M., and Freund, H.-J., *Progr. Surf. Sci.* **61,** 127 (1999).
11. Goodman, D. W., *Surf. Rev. Lett.* **2,** 9 (1995).
12. Diebold, U., Pan, J.-M., and Madey, T. E., *Surf. Sci.* **331–333,** 845 (1995).
13. Freund, H.-J., *Angew. Chem. Int. Ed. Engl.* **36,** 452 (1997).
14. Poppa, H., *Catal. Rev. Sci. Eng.* **35,** 359 (1993).
15. Vurens, G. H., Salmeron, M., and Somorjai, G. A., *Progr. Surf. Sci.* **32,** 333 (1989).
16. Gunter, P. L. J., Niemantsverdriet, J. W. H., Ribeiro, F. H., and Somorjai, G. A., *Catal. Rev. Sci. Eng.* **39,** 77 (1997).

17. Bonnell, D. A., *Progr. Surf. Sci.* **57,** 187 (1998).
18. Chen, J. G., *Surf. Sci. Rep.* **30,** 1 (1997).
19. Müller, C., Uebing, C., Kottcke, M., Rath, C., Hammer, L., and Heinz, K., *Surf. Sci.* **400,** 87 (1998).
20. Wiegenstein, C. G., and Schulz, K. H., *Surf. Sci.* **396,** 284 (1998).
21. Henrich, V. E., and Cox, P. A., "The Surface Science of Metal Oxides." Cambridge Univ. Press, Cambridge, UK, 1994.
22. Heiz, U., and Schneider, W.-D., *in* "Cluster–Solid Surface Interaction" (K.-H. Meiwes-Broer, Ed.). Springer, Berlin, 1999.
23. Gates, B. C., Guczi, L., and Knözinger, H., (Eds.), "Metal Clusters in Catalysis." Elsevier, Amsterdam, 1986.
24. Knözinger, H., *Catal. Today* **32,** 71 (1996).
25. Zecchina, A., Scarano, D., Bordiga, S., Ricchiardi, G., Spoto, G., and Geobaldo, F., *Catal. Today* **27,** 403 (1996).
26. Seiferth, O., Wolter, K., Dillmann, B., Klivenyi, G., Freund, H.-J., Scarano, D., and Zecchina, A., *Surf. Sci.* **421,** 176 (1999).
27. Hävecker, M., Knop-Gericke, A., Schedel-Niedrig, T., and Schlögl, R., *Angew. Chem. Int. Ed. Engl.* **37,** 1939 (1998).
28. Diebold, U., Anderson, J. F., Ng, K.-O., and Vanderbilt, D., *Phys. Rev. Lett.* **77,** 1322 (1996).
29. Murray, P. W., Leibsle, F. M., Muryn, C. A., Fisher, H. J., Flipse, C. F. J., and Thornton, G., *Phys. Rev. Lett.* **72,** 689 (1994).
30. Murray, P. W., Condon, N. G., and Thornton, G., *Phys. Rev. B* **51,** 10989 (1995).
31. Onishi, H., and Iwasawa, Y., *Surf. Sci.* **313,** L783 (1994).
32. Onishi, H., and Iwasawa, Y., *Chem. Phys. Lett.* **226,** 111 (1994).
33. Onishi, H., Fukui, K., and Iwasawa, Y., *Bull. Chem. Soc. Jpn.* **68,** 2447 (1995).
34. Onishi, H., and Iwasawa, Y., *Phys. Rev. Lett.* **76,** 791 (1996).
35. Charlton, G., Howes, P. B., Nicklin, C. L., Steadman, P., Taylor, J. S. G., Muryn, C. A., Harte, S. P., Mercer, J., McGrath, R., Norman, D., Turner, T. S., and Thornton, G., *Phys. Rev. Lett.* **78,** 495 (1997).
36. Renaud, G., *Surf. Sci. Rep.* **32,** 1 (1998).
37. Tasker, P. W., *Adv. Ceram.* **10,** 176 (1984).
38. Mackrodt, W. C., Davey, R. J., Black, I. N., and Docherty, R., *J. Cryst. Growth* **82,** 441 (1987).
39. Noguera, C., "Physics and Chemistry at Oxide Surfaces." University Press, Cambridge, UK, 1996.
40. Guénard, P., Renaud, G., Barbier, A., and Gantier-Soyer, M., *Surf. Rev. Lett.* **5,** 321 (1998).
41. Rohr, F., Bäumer, M., Freund, H.-J., Mejias, J. A., Staemmler, V., Müller, S., Hammer, L., and Heinz, K., *Surf. Sci.* **372,** L291 (1997).
42. Rohr, R., Bäumer, M., Freund, H.-J., Mejias, J. A., Staemmler, V., Muller, S., Hammer, L., and Heinz, K., *Surf. Sci.* **389,** 391 (1997).
43. Weiss, W., *Surf. Sci.* **377–379,** 943 (1997).
44. Manassidis, L., and Gillan, M. J., *Surf. Sci.* **285,** L517 (1993).
45. Manassidis, L., and Gillan, M. J., *J. Am. Ceram. Soc.* **77,** 335 (1994).
46. Rebhein, C., Harrison, N. M., and Wander, A., *Phys. Rev. B* **54,** 14066 (1996).
47. Wang, X.-G., Weiss, W., Shaikhutdinov, S. K., Ritter, M., Petersen, M., Wagner, F., Schloegl, R., and Scheffler, M., *Phys. Rev. Lett.* **81,** 1038 (1998).
48. Refson, K., Wogelius, R. A., Fraser, D. G., Payne, M. C., Lee, M. H., and Milman, V., *Phys. Rev. B* **52,** 10823 (1995).

49. Birkenheuer, U., Boettger, J. C., and Rösch, N., *J. Chem. Phys.* **100,** 6826 (1994).
50. Heinz, K., *Surf. Sci.* **299,** 433 (1984).
51. Harrison, N. M., Wang, X.-G., Muscat, M., and Scheffler, M., *Faraday Disc.* **114,** 305 (1999).
52. Schmalzried, H., "Chemical Kinetics of Solids." VCH, Weinheim, 1995.
53. Wolf, D., *Phys. Rev. Lett.* **68,** 3315 (1992).
54. Barbier, A., and Renaud, G., *Surf. Sci.* **392,** L15 (1997).
55. Cappus, D., Xu, C., Ehrlich, D., Dillmann, B., Ventrice Jr., C. A., Al-Shamery, K., Kuhlenbeck, H., and Freund, H.-J., *Chem. Phys.* **177,** 533 (1993).
56. Rohr, F., Wirth, K., Libuda, J., Cappus, D., Bäumer, M., and Freund, H.-J., *Surf. Sci.* **315,** L977 (1994).
57. Barbier, A., Proceedings of the 1st International Workshop on Oxide Surfaces, Elmau, Germany, 1999.
58. Lacman, R., *Colloq. Int. CNRS* **152,** 195 (1965).
59. Ventrice, C. A., Jr., Bertrams, T., Hannemann, H., Brodde, A., and Neddermeyer, H., *Phys. Rev. B* **49,** 1773 (1994).
60. Barbier, A., Mocuta, C., Kuhlenbeck, H., Peters, K., Richter, B., and Renaud, G., to be published.
61. Casey, W. H., Westrich, H. R., and Arnold, G. W., *Geochim. Cosmochim. Acta* **53,** 2795 (1988).
62. Papp, H., and Egersdörfer, B., personal communication.
63. Egersdörfer, B., PhD thesis, Ruhr-Universität, Bochum, Germany, 1993.
64. Fripiat, J. G., Lucas, A. A., André, J. M., and Derouane, E. G., *Chem. Phys.* **21,** 101 (1977).
65. Henzler, M., and Göpel, W., "Oberflächenphysik des Festkörpers." Teubner-Verlag, Stuttgart, 1991.
66. Neyman, K. M., Pacchioni, G., and Rösch, N., *in* "Recent Developments and Applications of Modern Density Functional Theory and Computational Chemistry" (J. M. Seminario, Ed.), p. 569. Elsevier, Amsterdam, 1996.
67. Pöhlchen, M., and Staemmler, V., *J. Chem. Phys.* **97,** 2583 (1992).
68. Pacchioni, G., and Bagus, P. S., *in* "Adsorption on Ordered Surfaces of Ionic Solids and Thin Films" (H.-J. Freund and E. Umbach, Eds.), Vol. 33, p. 180. Springer-Verlag, Berlin, 1993.
69. Nygren, M. A., and Pettersson, L. G. M., *J. Chem. Phys.* **105,** 9339 (1996).
70. Wichtendahl, R., Rodriguez-Rodrigo, M., Härtel, U., Kuhlenbeck, H., and Freund, H.-J., *Phys. Stat. Sol. A* **173,** 93 (1999).
71. Heidberg, J., Kandel, M., Meine, D., and Wildt, U., *Surf. Sci.* **333,** 1467 (1995).
72. Gerlach, R., Glebov, A., Lange, G., Toennies, J. P., and Weiss, W., *Surf. Sci.* **331–333,** 1490 (1995).
73. Wichtendahl, R., PhD thesis, Freie Universität, Berlin, 1999.
74. Stirniman, M. J., Huang, C., Smith, R. C., Joyce, J. A., and Kay, B. D., *J. Chem. Phys.* **105,** 1295 (1996).
75. Xu, C., and Goodman, D. W., *Chem. Phys. Lett.* **L65,** 341 (1997).
76. Libuda, J., Frank, M., Sandell, A., Andersson, S., Brühwiler, P. A., Bäumer, M., Mårtensson, N., and Freund, H.-J., *Surf. Sci.* **384,** 106 (1997).
77. Valden, M., and Goodman, D. W., *Science* **281,** 1647 (1998).
78. Hemminger, J., personal communication.
79. Jaeger, R. M., Kuhlenbeck, H., Freund, H.-J., Wuttig, M., Hoffmann, W., Franchy, R., and Ibach, H., *Surf. Sci.* **259,** 235 (1991).
80. Engquist, I., and Liedberg, B., *J. Phys. Chem.* **100,** 20089 (1996).
81. Knözinger, H., and Ratnasaniy, P., *Catal. Rev. Sci. Eng.* **17,** 31 (1978).

82. Ehrlich, D., PhD. Ruhr-Universität, thesis, Bochum, Germany, 1995.
83. Kuhlenbeck, H., Xu, C., Dillmann, B., Haßel, M., Adam, B., Ehrlich, D., Wohlrab, S., Freund, H.-J., Ditzinger, U. A., Neddermeyer, H., Neuber, M., and Neumann, M., *Ber. Bunsenges. Phys. Chem.* **96,** 15 (1992).
84. Xu, C., Dillmann, B., Kuhlenbeck, H., and Freund, H.-J., *Phys. Rev. Lett.* **67,** 3551 (1991).
85. Xu, C., Haßel, M., Kuhlenbeck, H., and Freund, H.-J., *Surf. Sci.* **258,** 23 (1991).
86. Hemmerich, I., Rohr, F., Seiferth, O., Dillmann, B., and Freund, H.-J., *Z. Phys. Chem.* **202,** 31 (1997).
87. Thüne, P. C., and Niemantsverdriet, J. W., *Israel. J. Chem.* **38,** 385 (1998).
88. Menges, M., Baumeister, B., Al-Shamery, K., Freund, H.-J., Fischer, C., and Andresen, P., *J. Chem. Phys.* **101**(4), 3318 (1994).
89. Beauport, I., Al-Shamery, K., and Freund, H.-J., *Chem. Phys. Lett.* **256,** 641 (1996).
90. Klüner, T., Freund, H.-J., Staemmler, V., and Kosloff, R., *Phys. Rev. Lett.* **80,** 5208 (1998).
91. Al-Shamery, K., *Appl. Phys. A* **63,** 509 (1996).
92. Wovchko, E. A., and Yates, J. T., Jr., *Langmuir* **15,** 3506 (1999).
93. Wu, M. C., and Møller, P. J., *Surf. Sci.* **221,** 250 (1989).
94. Möller, P. J., and Wu, M. C., *Surf. Sci.* **224,** 265 (1989).
95. Wu, M. C., and Møller, P. J., *Surf. Sci.* **235,** 228 (1990).
96. Møller, P. J., and Nerlov, J., *Surf. Sci.* **307–309,** 591 (1993).
97. Murray, P. W., Shen, J., Condon, N. G., Peng, S. J., and Thornton, G., *Surf. Sci.* **380,** L455 (1997).
98. Stone, P., Bennett, R. A., and Bowker, M., *New J. Phys.* **1,** 8 (1989-1999).
99. Diebold, U., Proceedings of the 1st International Workshop on Oxide Surfaces, Elmau, Germany, 1999.
100. Xu, X., Vesecky, S. M., and Goodman, D. W., *Science* **258,** 788 (1992).
101. Schröder, T., Adelt, M., Richter, B., Naschitzki, M., Bäumer, M., and Freund, H.-J., submitted for publication.
102. Gunthaner, F. J., Gunthaner, P. J., Vasquez, R. P., Lewis, B. F., Maserjian, J., and Madhukar, A., *J. Vac. Sci. Technol.* **16,** 1443 (1979).
103. Libuda, J., Winkelmann, F., Bäumer, M., Freund, H.-J., Bertrams, T., Neddermeyer, H., and Müller, K., *Surf. Sci.* **318,** 61 (1994).
104. Stempel, S., Bäumer, M., and Freund, H.-J., *Surf. Sci.* **402–404,** 424 (1998).
105. Heemeier, M., PhD thesis, Freie Universität, Berlin, manuscript in preparation.
106. Linderoth, T. R., Horch, S., Lænsgaard, E., Stensgaard, I., and Besenbacher, F., *Surf. Sci.* **404,** 308 (1998).
107. Ernst, N., Duncombe, B., Bozdech, G., Naschitzki, M., and Freund, H.-J., *Ultramicroscopy* **79,** 231 (1999).
108. Piednoir, A., Pernot, E., Granjeand, S., Humbert, A., Chapon, C., and Henry, C. R., *Surf. Sci.* **391,** 19 (1997).
109. Hansen, K. H., Worren, T., Stempel, S., Lægsgaard, E., Bäumer, M., Freund, H.-J., Besenbacher, F., and Stensgaard, I., submitted for publication.
110. Methfessel, M., Hennig, D., and Scheffler, M., *Phys. Rev. B* **46,** 4816 (1992).
111. Bogicevic, A., and Jennison, D. R., *Phys. Rev. Lett.* **82,** 4050 (1999).
112. Klimenkov, M., Nepijko, S., Kuhlenbeck, H., Bäumer, M., Schlögl, R., and Freund, H.-J., *Surf. Sci.* **391,** 27 (1997).
113. Klimenkov, M., Nepijko, S., Kuhlenbeck, H., and Freund, H.-J., *Surf. Sci.* **385,** 66 (1997).
114. Nepijko, S., Klimenkov, M., Kuhlenbeck, H., Zemlyanov, D., Herein, D., Schlögl, R., and Freund, H.-J., *Surf. Sci.* **413,** 192 (1998).
115. Nepijko, S. A., Klimenkov, M., Adelt, M., Kuhlenbeck, H., Schlögl, R., and Freund, H.-J., *Langmuir* **15,** 5309 (1999).

116. Rupprechter, G., Eppler, A. S., and Somorjai, G. A., *in* "Electron Microscopy 1998" (H. A. C. Benavides and M. J. Yacaman, Eds.), Vol. 2, p. 369. Institute of Physics Publishing, Philadelphia, 1998.
117. Eppler, A. S., Rupprechter, G., Guczi, L., and Somorjai, G. A., *J. Phys. Chem. B* **101,** 9973 (1997).
118. Wong, K., Johansson, S., and Kasemo, B., *Faraday Disc.* **105,** 237 (1996).
119. Jacobs, P. W., Ribero, F. H., Somorjai, G. A., and Wind, S. J., *Catal. Lett.* **37,** 131 (1996).
120. Gotschy, W., Vonmetz, K., Leitner, A., and Aussenegg, F. R., *Appl. Phys. B* **63,** 381 (1996).
121. Cerrina, F., and Marrian, C., *MRS Bull.* **21,** 56 (1996).
122. Fischer, P. B., and Chou, S. Y., *Appl. Phys. Lett.* **62,** 2989 (1993).
123. Queeney, K. T., Pang, S., and Friend, C. M., *J. Chem. Phys.* **109,** 8058 (1998).
124. Queeney, K. T., and Friend, C. M., to be published.
125. Leisenberger, F. P., Surnev, S., Vitali, L., Ramsey, M. G., and Netzer, F. P., *J. Vac. Sci. Technol. A* **17,** 1743 (1999).
126. Grant, J. L., Fryberger, T. B., and Stair, P. C., *Surf. Sci.* **239,** 127 (1990).
127. Schmid, G., (Ed.), "Clusters and Colloids: From Theory to Applications." VCH, Weinheim, 1994.
128. Ekardt, W. (Ed.), "Metal Clusters." Wiley, Chichester, UK, 1999.
129. de Biani, F. F., Femoni, C., Iapalucci, M. C., Longoni, G., Zanello, P., and Ceriotti, A., *Inorg. Chem.* **38,** 3721 (1999).
130. Johnson, B. F. G. (Ed.), "Transition Metal Clusters." Wiley, Chichester, UK, 1980.
131. Wertheim, G. K., DiCenzo, S. B., and Buchanan, D. N. E., *Phys. Rev. B* **33,** 5384 (1986).
132. Sandell, A., Libuda, J., Brühwiler, P., Andersson, S., Maxwell, A., Bäumer, M., Mårtensson, N., and Freund, H.-J., *J. Electron Spectrosc. Relat. Phenom.* **76,** 301 (1995).
133. Unwin, R., and Bradshaw, A. M., *Chem. Phys. Lett.* **58,** 58 (1978).
134. Wertheim, G. K., *Z. Phys. D* **12,** 319 (1989).
135. Lee, S.-T., Apai, G., Mason, M. G., Benbow, R., and Hurych, Z., *Phys. Rev. B* **23,** 505 (1981).
136. de Gouveia, V., Bellamy, B., Hadj Romdhane, Y., Mason, A., and Che, M., *Z. Phys. D* **12,** 587 (1989).
137. Kuhrt, C., and Harsdorff, M., *Surf. Sci.* **245,** 173 (1991).
138. Sandell, A., Libuda, J., Brühwiler, P. A., Andersson, S., Maxwell, A. J., Bäumer, M., Mårtensson, N., and Freund, H.-J., *J. Vac. Sci. Technol. A* **14,** 1546 (1996).
139. Goodman, D. W., *J. Vac. Sci. Technol. A* **14,** 1526 (1996).
140. Sandell, A., Libuda, J., Brühwiler, P. A., Andersson, S., Bäumer, M., Maxwell, A. J., Mårtensson, N., and Freund, H.-J., *Phys. Rev. B* **55,** 7233 (1997).
141. Frederick, B. B., Apai, G., and Rhodin, T. N., *J. Am. Chem. Soc.* **109,** 4797 (1987).
142. Wertheim, G. K., *Z. Phys. B* **66,** 53 (1987).
143. Mason, M. G., *Phys. Rev. B* **27,** 748 (1983).
144. Cini, M., de Crescenzi, M., Patella, F., Motta, N., Sastry, M., Rochet, F., Pasquali, R., Balzarotti, A., and Verdozzi, C., *Phys. Rev. B* **41,** 5685 (1990).
145. Vijayakrishnan, V., and Rao, C. N. R., *Surf. Sci. Lett.* **255,** L516 (1991).
146. Jirka, I., *Surf. Sci.* **232,** 307 (1990).
147. Wertheim, G. K., CiCenzo, S. B., and Youngquist, S. E., *Phys. Rev. Lett.* **51,** 2310 (1983).
148. DiCenzo, S. B., and Wertheim, G. K., *Comments Solid State Phys.* **11,** 203 (1985).
149. Häberlen, O. D., Chung, S.-C., Stener, M., and Rösch, N., *J. Chem. Phys.* **106,** 5189 (1997).
150. Björneholm, O., Nilsson, A., Zdansky, E. O. F., Sandell, A., Hernnäs, B., Tillborg, H., Anderson, J. N., and Mårtensson, N., *Phys. Rev. B* **46,** 10353 (1992).
151. Nilsson, A., Björneholm, O., Zdansky, E. O. F., Tillborg, H., Mårtensson, N., Andersen, J. N., and Nyholm, R., *Chem. Phys. Lett.* **197,** 1892 (1992).

152. Rainer, D. R., and Goodman, D. W., *in* "NATO ASI" (G. Pacchioni and R. M. Lambert, Eds.), Vol. 331, p. 27. Kluwer, Dordrecht, 1997.
153. Meiwes-Broer, K. H., *Phys. B1.* **55,** 21 (1999).
154. Adelt, M., Nepijko, S., Drachsel, W., and Freund, H.-J., *Chem. Phys. Lett.* **291,** 425 (1998).
155. Dormann, J. L., and Fiorani, D. (Eds.), "Magnetic Properties of Fine Particles." North-Holland, Amsterdam, 1992.
156. Hill, T., Mozaffari-Afshar, M., Schmidt, J., Risse, T., Stempel, S., Heemeier, M., and Freund, H.-J., *Chem. Phys. Lett.* **292,** 524 (1998).
157. Hill, T., Stempel, S., Risse, T., Bäumer, M., and Freund, H.-J., *J. Magn. Magn. Mater.* **198/199,** 354 (1999).
158. Hill, T., Mozaffari-Afshar, M., Schmidt, J., Risse, T., and Freund, H.-J., *Surf. Sci.* **429,** 246 (1999).
159. Risse, T., Hill, T., Mozaffari-Afshar, M., and Freund, H.-J., submitted for publication.
160. Evans, J., Hayden, B., Mosselman, F., and Murray, A., *Surf. Sci.* **279,** L159 (1992).
161. Evans, J., Hayden, B., Mosselman, F., and Murray, A., *Surf. Sci.* **301,** 61 (1994).
162. Wolter, K., Seiferth, O., Kuhlenbeck, H., Bäumer, M., and Freund, H.-J., *Surf. Sci.* **399,** 190 (1998).
163. Frank, M., Kühnemuth, R., Bäumer, M., and Freund, H.-J., *Surf. Sci.* **427/428,** 288 (1999).
164. Frank, M., Kühnemuth, R., Bäumer, M., and Freund, H.-J. submitted for publication.
165. Yates, J. T., Jr., Duncan, T. M., Worley, S. D., and Vaughan, R. W., *J. Chem. Phys.* **70,** 1219 (1979).
166. Basu, P., Panayotov, D., and Yates, J. T., Jr., *J. Am. Chem. Soc.* **110,** 2074 (1988).
167. Solymosi, F., and Knözinger, H., *J. Chem. Soc. Faraday Trans.* **86,** 389 (1990).
168. Berkó, A., and Solymosi, F., *J. Catal.* **183,** 91 (1999).
169. Bäumer, M., Frank, M., Heemeier, M., Kühnemuth, R., Stempel, S., and Freund, H.-J., to be published.
170. Jaeger, R. M., Libuda, J., Bäumer, M., Homann, K., Kuhlenbeck, H., and Freund, H.-J., *J. Electron Spectrosc. Relat. Phenom.* **64/65,** 217 (1993).
171. Mineva, T., and Russo, N., personal communication.
172. Uvdal, P., Karlsson, P.-A., Nyberg, C., Andersson, S., and Richardson, N. V., *Surf. Sci.* **202,** 167 (1988).
173. Giessel, T., Schaff, O., Hirschmugl, C. J., Fernandez, V., Schindler, K. M., Theobald, A., Bao, S., Lindsay, R., Berndt, W., Bradshaw, A. M., Baddeley, C., Lee, A. F., Lambert, R. M., and Woodruff, D. P., *Surf. Sci.* **406,** 90 (1998).
174. Kisters, G., Chen, J. G., Lehwald, S., and Ibach, H., *Surf. Sci.* **245,** 65 (1991).
175. Lauterbach, J., Boyle, R. W., Schick, M., Mitchell, W. J., Meng, B., and Weinberg, W. H., *Surf. Sci.* **350,** 32 (1996).
176. Solymosi, F., Novák, É., and Molnár, A., *J. Phys. Chem.* **94,** 7250 (1990).
177. Voskobojnikov, T. W., Shpiro, E. S., Landmesser, H., Jaeger, N. I., and Schulz-Ekloff, G., *J. Mol. Catal. A Chem.* **104,** 299 (1996).
178. De La Cruz, C., and Sheppard, N., *J. Chem. Soc. Faraday Trans.* **93,** 3569 (1997).
179. Hansen, K. H., Stempel, S., Lægsgaard, E., Besenbacher, F., and Stensgaard, I., personal communication.
180. Conrad, H., Ertl, G., and Küppers, J., *Surf. Sci.* **76,** 323 (1978).
181. Haruta, M., *Catal. Today* **36,** 153 (1997).
182. Piccolo, L., Becker, C., and Henry, C. R., *Eur. Phys. J.* **8** (1999).
183. Zhdanov, V. P., and Kasemo, B., *Phys. Rev. B* **55,** 4105 (1997).
184. Persson, H., Thormählen, P., Zhdanov, V. P., and Kasemo, B., *Catal. Today* **53,** 273 (1999).
185. Vanolli, F., Heiz, U., and Schneider, W.-D., *Chem. Phys. Lett.* **277,** 527 (1997).

186. Heiz, U., Sanchez, A., Abbet, S., and Schneider, W.-D., *J. Am. Chem. Soc.* **121,** 3214 (1999).
187. Richter, B., Wilkes, J., Wichtendahl, R., Kuhlenbeck, H., and Freund, H.-J., unpublished manuscript.
188. Frank, M., Andersson, S., Libuda, J., Stempel, S., Sandell, A., Brena, B., Giertz, A., Bruhwiler, P. A., Bäumer, M., Mårtensson, N., and Freund, H.-J., *Chem. Phys. Lett.* **279,** 92 (1997).
189. Andersson, S., Frank, M., Sandell, A., Giertz, A., Brena, B., Brühwiler, P. A., Mårtensson, N., Libuda, J., Bäumer, M., and Freund, H.-J., *J. Chem. Phys.* **108,** 2967 (1998).
190. Andersson, S., Frank, M., Sandell, A., Libuda, J., Brena, B., Giertz, A., Brühwiler, P. A., Bäumer, M., Mårtensson, N., and Freund, H.-J., *Vacuum* **49,** 167 (1998).
191. Irion, M. P., personal communication.
192. Yates, J. T., Williams, E. D., and Weinberg, W.H., *Surf. Sci.* **91,** 562 (1980).
193. Rebholz, M., Prins, R., and Kruse, N., *Surf. Sci.* **259,** L791 (1991).
194. Watanabe, K., Matsumoo, Y., Kampling, M., Al-Shamery, K., and Freund, H.-J., *Angew. Chem. Int. Ed.* **38,** 2192 (1999).
195. Watanabe, K., and Matsumoto, Y., *Surf. Sci.* **390,** 250 (1997).
196. Herzberg, G., "Molecular Spectra and Molecular Structure." Van Nostrand, New York, 1966.
197. Akinaga, Y., Taketsugu, T., and Hirao, T., *J. Chem. Phys.* **107,** 415 (1997).
198. Jennison, D. R., personal communication.
199. Heidberg, J., Redlich, B., and Wetter, D., *Ber. Bunsenges. Phys. Chem.* **99,** 1333 (1995).

Sum Frequency Generation: Surface Vibrational Spectroscopy Studies of Catalytic Reactions on Metal Single-Crystal Surfaces

GABOR A. SOMORJAI AND KEITH R. McCREA

Department of Chemistry
University of California at Berkeley
Berkeley, CA 94720

Sum frequency generation (SFG) has been used for molecular-level investigations of adsorbates under both ultrahigh vacuum and high-pressure catalytic reaction conditions to characterize several important catalytic reactions on platinum single crystals. These reactions include ethylene hydrogenation, propylene hydrogenation and dehydrogenation, cyclohexene hydrogenation and dehydrogenation, and carbon monoxide oxidation. From SFG spectra, important catalytic reaction intermediates were determined, and possible reaction spectator species were also identified. For ethylene and propylene hydrogenation, important reaction intermediates were determined to be π-bonded ethylene and propylene, respectively. During cyclohexene hydrogenation, 1,3-cyclohexadiene was observed to be the important reaction intermediate on both Pt(111) and Pt(100). By comparing SFG spectra characterizing cyclohexene dehydrogenation on Pt(111) and on Pt(100), it was determined that 1,3-cyclohexadiene is also an important reaction intermediate. Another intermediate, 1,4-cyclohexadiene, was also observed under dehydrogenation conditions on Pt(111) but not on Pt(100). This species was determined to be a spectator species. During carbon monoxide oxidation, incommensurate CO species were determined to be important reaction intermediates.

Abbreviations: AES, Auger electron spectroscopy; AGS, $AgGaS_2$; BBO, BaB_2O_4; CHD, cyclohexadiene; DFG, difference frequency generation; EELS, electron energy loss spectroscopy; HREELS, high resolution electron energy loss spectroscopy; IRAS, infrared adsorption spectroscopy; KD*P, KD_2PO_4; LEED, low energy electron diffraction; ML, monolayer; OPA, optical parametric amplification; OPG, optical parametric generation; RFA, retarding field analyzer; SFG, sum frequency generation; TOR, turn over rate; UHV, ultra high vacuum; UPS, ultraviolet photoemission spectroscopy; XPS, X-ray photoelectron spectroscopy.

0360-0564/00 $35.00

I. Introduction

Since the advent of modern surface science, ultrahigh vacuum (UHV) techniques have been used to probe the identity of molecules adsorbed on the surfaces of single crystals. Information about structure and bonding was obtained with low-energy electron diffraction (LEED), Auger electron spectroscopy, X-ray photoelectron spectroscopy, high-resolution electron energy loss spectroscopy (HREELS), and other techniques (*1–5*). These techniques utilize electrons to probe and characterize surfaces; however, they are limited to low pressures because of the large mean free path required for the electrons to reach either the sample or the detector.

Obviously, these techniques are not practical for *in situ* characterization of surfaces during high-pressure catalytic reactions. To understand the surface composition during reactions with these electron-utilizing techniques, the single crystal was characterized in UHV prior to application of catalytic reaction conditions. The catalytic reaction was then monitored by gas chromatographic analysis of the gas-phase composition during the reaction to provide data for the determination of kinetics. Once the reaction was completed, the sample was again exposed to a low pressure to allow the use of these electron techniques to once again characterize the surface(*6–9*). These investigations were very insightful in showing that many reactions cause surface restructuring.

Additional characterizations allowed the identification of reaction intermediates on catalysts by performing reactions in UHV and using techniques such as HREELS. The catalyst was exposed to a few Langmuirs of reactants, usually at low temperatures. The sample was then heated, and surface species were observed to change; reaction intermediates were then proposed. However, it is difficult to extrapolate these results to actual high-pressure reactions because the surface compositions under UHV and reaction conditions are most likely substantially different from each other.

Because of the limitations described previously, it became clear that molecular-level investigations of surfaces under actual high-pressure catalytic reaction conditions are necessary to obtain a better understanding of catalytic reactions. Photons, in contrast to electrons, are not limited to UHV conditions and can readily probe surfaces under high pressure. The difficulty in using techniques such as infrared (IR) spectroscopy is that both the bulk material of the catalyst and gas-phase reactants can interfere with surface measurements, necessitating the subtraction of both gas-phase and bulk spectra from IR spectra to obtain surface information.

Thus, a technique utilizing photons with surface specificity was required. Recently, sum frequency generation (SFG) has been developed to study surfaces and interfaces (*10–14*). SFG has the advantage that its selection rules necessitate that a vibrational mode must have a change in dipole and a change in polarizability in order for the process to occur. Only at surfaces and interfaces are these selection rules met. Another advantage of SFG is that it is sensitive to a submonolayer coverage of adsorbates.

With this new surface-specific technique, several high-pressure catalytic reactions have been characterized *in situ,* and they are discussed here. The reactions include ethylene hydrogenation, propylene hydrogenation, cyclohexene hydrogenation and dehydrogenation, and carbon monoxide oxidation. Along with surface characterization by SFG, kinetics data acquired by gas chromatographic analysis of the gas phase were obtained to allow correlation of the kinetics with reaction intermediates identified in the SFG spectra.

II. Theory Underlying SFG: Surface-Specific Vibrational Spectroscopy

Sum frequency generation is a surface-specific vibrational spectroscopy with submonolayer sensitivity. A visible laser beam at a fixed wavelength is overlapped with a tunable IR laser beam. Vibrational spectra can be acquired by scanning the IR beam over the vibrational region of interest.

SFG is powerful for investigating catalytic reactions on smooth surfaces such as those of a single crystal or an ordered material. Unfortunately, porous materials scatter the input beams so that only a weak SFG beam can be detected, and therefore the applications of SFG to most typical solid catalysts are limited. Another disadvantage of SFG is that a spectrum may take from 10 min to 1 h to obtain, depending on the number of laser shots averaged and width of frequency range investigated. Currently, the technique of broadband SFG is under development to allow the acquisition of spectra several hundred wavenumbers in width simultaneously with each laser shot, which will greatly reduce the acquisition time.

A brief discussion of SFG is presented here, and more detailed descriptions can be found in the literature (*10, 11, 14–17*). The principle of SFG is governed by second-order nonlinear optics, and the technique is made possible by the use of high-energy pulsed lasers. Under weak electric fields, the polarization of a material is described by

$$\vec{P} = \vec{P}^{(0)} + \vec{P}^{(1)} \tag{1}$$

$$\vec{P}^{(1)} = \varepsilon_0 \chi^{(1)} \vec{E}(r) \cos(\omega t) \tag{2}$$

where $\vec{P}^{(0)}$ is the static polarization, $\vec{P}^{(1)}$ is the first-order linear polarization, $\chi^{(1)}$ is the linear susceptibility, ε_0 is the permittivity of free space, t is time, and $\vec{E}(r)\cos(\omega t)$ describes the electric field. In linear optics, as this equation indicates, the frequency of light is invariant as it passes through a medium.

Under strong electric fields, such as those produced by lasers, second and higher order polarization terms must be added to Eq. (1):

$$\vec{P} = \vec{P}^{(0)} + \vec{P}^{(1)} + \vec{P}^{(2)} + \vec{P}^{(3)} + \ldots \tag{3}$$

$$P_i^{(2)} = \varepsilon_0 \sum_{j,k} \chi_{ijk}^{(2)} \vec{E}_j(r) \cos(\omega_1 t) \vec{E}_k(r) \cos(\omega_2 t) \tag{4}$$

where $\chi^{(2)}$ is the second-order nonlinear susceptibility. The subscripts i, j, and k refer to the axes of the coordinate system. If the frequencies ω_1 and ω_2 are the same ($\omega_1 = \omega_2 = \omega$), $P_i^{(2)}$ can be rearranged by simple trigonometry to give the form

$$P_i^{(2)} = \tfrac{1}{2}\,\varepsilon_0 \sum_{j,k} \chi_{ijk}^{(2)} \vec{E}_j(r)\vec{E}_k(r)(1 + \cos 2\omega t) \tag{5}$$

which indicates that a second frequency of light oscillating at 2ω can be generated from an oscillating dipole in the medium. This process is called second harmonic generation. If ω_1 and ω_2 are oscillating at different frequencies, then Eq. (4) can be rearranged to the form

$$P_i^{(2)} = \tfrac{1}{2}\,\varepsilon_0 \sum_{j,k} \chi_{ijk}^{(2)} \vec{E}_j(r)\vec{E}_k(r)[\cos(\omega_1 + \omega_2)t + \cos(\omega_1 - \omega_2)t] \tag{6}$$

From the term in brackets in Eq. (6), it can be seen that there is now the possibility of generation by an oscillating dipole of frequencies at the sum and difference of ω_1 and ω_2. These two processes are known as SFG and difference frequency generation (DFG), respectively. In this article, we focus on SFG, although DFG also occurs.

The magnitude of the SFG signal is proportional to the absolute square of $\chi^{(2)}(|\chi^{(2)}|)$ which is made up of both a nonresonant susceptibility term ($\chi_{\mathrm{NR}}^{(2)}$) and a resonant susceptibility ($\chi_{\mathrm{R}}^{(2)}$) term:

$$\chi^{(2)} = \chi_{\mathrm{NR}}^{(2)} + \chi_{\mathrm{R}}^{(2)} \tag{7}$$

The nonresonant susceptibility term originates from the substrate surface and is typically invariant as the IR beam is scanned; it is called nonresonant background. The resonant susceptibility, which originates from vibrational modes on the surface, is described by

$$\chi_{\mathrm{R}}^{(2)} = \chi_{i,j,k}^{2} \sum_{l,m,n} \langle(\hat{\imath}\cdot\hat{l})(\hat{\jmath}\cdot\hat{m})(\hat{k}\cdot\hat{n})\rangle \frac{A_q}{\omega_{\mathrm{IR}} - \omega_q + i\Gamma_q} \tag{8}$$

where A_q is the strength of the qth vibrational mode, ω_{IR} is the frequency of the infrared laser beam, ω_q is the frequency of the qth vibrational mode, and Γ_q is the damping constant of the qth vibrational mode. The subscripts l, m, and n refer to the axes for the molecular coordinate system. From Eq. (8), it is observed that $\chi_R^{(2)}$ is at a maximum when $\omega_{IR} = \omega_q$, and hence a vibrational spectrum is acquired by scanning the IR frequency. The selection rules for the SFG process are inferred from the following equation:

$$A_q = \frac{1}{2\omega_q} \frac{\partial \mu_n}{\partial q} \frac{\partial \alpha_{lm}^{(1)}}{\partial q} \tag{9}$$

where μ_n is the dipole moment and $\alpha_{lm}^{(1)}$ is the linear polarizability. Hence, in order for $\chi_R^{(2)}$ to be nonzero, the vibrational mode of interest must obey both IR and Raman selection rules; there must be both a change in the dipole and a change in the polarizability. Because SFG is a second-order nonlinear process, only a medium without inversion symmetry can generate SFG signals under the electric-dipole approximation. Surfaces or interfaces lack inversion symmetry; therefore, modes at surfaces and interfaces are SFG allowed. Centrosymmetric bulk materials and isotropic gas phases have centrosymmetry and are therefore SFG forbidden. Therefore, SFG is a surface-specific process. Because gas phases are isotropic and not SFG active, experiments can easily be performed from pressures of 1×10^{-10} Torr up to 10^3 Torr without the concern of gas-phase interference.

The intensity of an SFG signal is also dependent on the ordering of the surface or interface. If molecules on the surface are randomly oriented, then $\chi^{(2)}$ becomes zero and SFG spectra will not be generated. Therefore, strong SFG features indicate that the surface adsorbates are well ordered.

By using different polarization combinations for the IR, visible, and detected SFG light, information about molecular orientation on the surface may be obtained for nonmetal surfaces (*16*, *18*). For each polarization combination used during a SFG experiment, different susceptibility components are measured. By modeling these susceptibility components, it is then possible to determine the orientation of surface molecules. However, this technique is not possible for characterization of metal surfaces. Because of the metal surface-selection rules (MSSRs), an s-polarized IR beam is canceled out due to the image field of the electrons in the metal. Therefore, an s-polarized IR beam cannot excite a dipole along a metal surface; therefore, in the performance of SFG experiments characterizing a metal surface, the IR beam is always p polarized. The visible beam may be either s or p polarized, but the signal is almost 40 times weaker for s-polarized visible light. Therefore, in the experiments reported here, both the IR and visible light were p polarized, resulting in p-polarized SFG output.

III. Experimental Considerations

A. Laser System

A schematic diagram of the experimental setup is shown in Fig. 1. The SFG experiments were carried out by using a Nd:YAG laser. The laser operated at 20 Hz and provided a 20-ps pulse at a fundamental frequency of 1064 nm with 35 mJ of energy per pulse. Just beyond the Nd:YAG laser, the beam is split into two separate beams. The first passed through a KD_2PO_4 (KD*P) nonlinear crystal which doubled the frequency of the fundamental beam to 532 nm. This beam was used for the visible portion of the SFG experiment and had an output energy of 400 μJ/pulse.

The second part of the fundamental beam passed through one of two angle-tuned optical parametric generation/optical parametric amplification (OPG/OPA) stages for the generation of the tunable IR source (*19*). A $LiNbO_3$ OPG/OPA stage was used to generate a tunable IR beam with frequencies between 2700 and 3600 cm^{-1}, which allowed investigation of the vibrational C–H stretching modes of adsorbed hydrocarbons. The maximum output of this stage was at 2850 cm^{-1} with 200 μJ of energy and a full width at half-maximum (FWHM) of 12 cm^{-1}. The second OPG/OPA stage consisted of BaB_2O_4 and $AgGaS_2$ nonlinear crystals to produce a tunable IR beam with frequencies between 1400 and 2300 cm^{-1} to probe vibrational modes associated with CO adsorbed in various coordinations. The maximum output of this stage occurred at 2200 cm^{-1} with 120 μJ of energy and a FWHM of 4 cm^{-1}.

Both the IR and visible beams were p polarized and both were spatially and temporally overlapped on a single crystal mounted in a UHV chamber

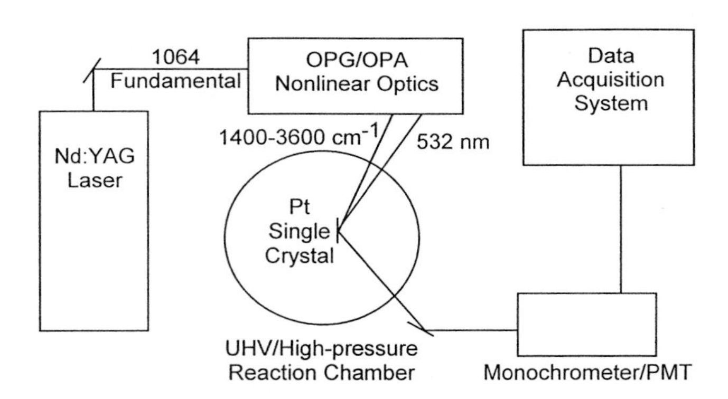

FIG. 1. Schematic of the experimental setup including a YAG laser, the OPG/OPA optics, the UHV/high-pressure reaction chamber, and the detection system.

(described later). The visible beam made an angle of 50° with respect to the surface normal, whereas the IR beam was at an angle of 55° with respect to the surface normal. The generated SFG beam was then sent through a monochrometer, and the signal intensity was detected by a photomultiplier tube and integrated by a gated integrator. During experiments, 200–300 laser shots at each frequency were collected and averaged.

B. UHV/High-Pressure Reaction Chamber

Single-crystal samples were mounted in a UHV/reaction chamber pumped by a turbomolecular pump and an ion pump to a base pressure of 5×10^{-10} Torr. The experimental apparatus has been described in detail elsewhere (*20*). The chamber was equipped with a retarding field analyzer for Auger electron spectroscopy (to check sample cleanliness) and with the capability for LEED (to check sample surface structure). By use of resistive heating, the sample could be heated up to a temperature of 1200 K and cooled under vacuum with liquid nitrogen to 115 K. The samples were cleaned in UHV by repeated cycles of Ar^+ bombardment and annealing at temperatures up to 1200 K. After the sample was clean, gases were introduced through a gas manifold system. During high-pressure catalytic reactions, the sample was isolated from the vacuum pumps by a gate valve. A recirculation pump was used to mix the gases in the chamber, and a septum in the recirculation line was used to sample the gas for analysis by gas chromatography. For SFG experiments, the chamber had input and output windows made of CaF_2 to allow the IR light to enter the chamber.

To identify vibrational peaks in the high-pressure catalytic reaction experiments, low-pressure experiments were first performed in which possible intermediate species were introduced into the chamber. Not only do the low-pressure experiments help in the assignment of high-pressure intermediate species but also they allow the correlation of the SFG spectra with results of previous low-pressure vibrational studies of these systems.

IV. Ethylene Hydrogenation on Pt(111)

The first mechanism proposed for the hydrogenation of the simplest olefin, ethylene, on a platinum surface was that of Horiuti and Polanyi, reported in the 1930s (*21*). According to their model, ethylene adsorbs on a clean platinum surface by breaking one of the C=C double bonds and then forming two σ bonds with the metal surface. This intermediate is known as di-σ-bonded ethylene. One of the metal–carbon bonds was then presumed to be hydrogenated with adsorbed hydrogen, creating an ethyl intermediate, allowing ethane production through the final hydrogenation of the remaining metal–carbon bond.

Surface techniques such as ultraviolet photoemission spectroscopy were later used under UHV conditions to investigate the mechanism of ethylene hydrogenation on platinum single crystals. It was shown that at temperatures lower than 52 K, ethylene physisorbs through the π bond, giving π-bonded ethylene (*22*). As the temperature is heated higher than 52 K, the π bond is broken and di-σ-bonded ethylene is formed (*23*). Ethylidyne ($M{\equiv}CCH_3$) is formed as di-σ-bonded ethylene is dehydrogenated and transfers a hydrogen atom from one carbon atom to the other (*24*). As the surface is heated further, ethylidyne decomposes into graphitic precursors (*25*).

Ethylidyne is not believed to be a reaction intermediate in ethylene hydrogenation. Davis *et al.* (*26*) showed that ethylidyne hydrogenation was several orders of magnitude slower than the overall hydrogenation of ethylene to ethane. Furthermore, Beebe and Yates (*27*), using *in situ* IR transmission spectroscopy, showed that the reaction rate was the same on a supported Pd/Al_2O_3 catalyst, whether or not the surface was covered with ethylidyne. The results of these investigations indicate that ethylidyne is a spectator species that is not directly involved in ethylene hydrogenation.

To determine the importance of π-bonded ethylene and di-σ-bonded ethylene, Mohsin *et al.* (*28*), using transmission IR spectroscopy, showed that both species are hydrogenated on a Pt/Al_2O_3 catalyst as hydrogen flows over the surface. In addition, Mohsin *et al.* showed that only di-σ-bonded ethylene is converted to ethylidyne in the absence of hydrogen when the catalyst is heated. These studies, however, were not performed under actual high-pressure reaction conditions because gas-phase ethylene interferes with IR experiments. Since gas-phase ethylene does not generate a SFG signal, SFG is an ideal technique to characterize ethylene hydrogenation *in situ* under high-pressure reaction conditions.

A. Low-Pressure SFG Experiments Characterizing Ethylene Adsorption

To aid in the identification of high-pressure reaction intermediates adsorbed on the surface under reaction conditions, experiments were first performed under UHV conditions in which di-σ-bonded ethylene, ethyl, ethylidyne, and π-bonded ethylene were prepared independently on a Pt(111) single crystal (*29*). Figure 2 shows the results of these UHV experiments. By exposing the surface to 4 Langmuirs (L) of ethylene at 200 K, a saturation coverage of di-σ-bonded ethylene was formed on the surface; the resultant SFG spectrum is shown in Fig. 2a. A single peak at 2904 cm^{-1}

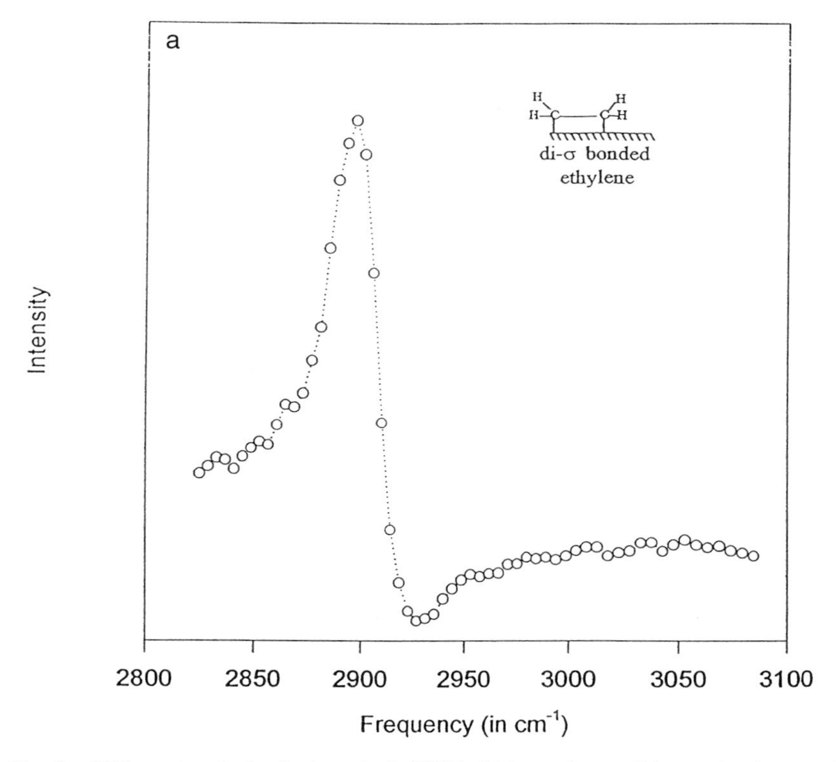

FIG. 2. SFG spectra of adsorbed species in UHV which may be possible reaction intermediates during high-pressure ethylene hydrogenation on Pt(111): (a) di-σ-bonded ethylene at 202 K, (b) ethylidyne at 300 K, (c) ethyl groups at 193 K, and (d) coadsorption of di-σ- and π-bonded ethylene at 120 K.

is observed, which is assigned to the $\nu_s(CH_2)$ peaks of the methylene groups. The surface was then heated to 300 K, a temperature at which di-σ-bonded ethylene is dehydrogenated to form ethylidyne. The SFG spectrum of ethylidyne on Pt(111) (Fig. 2b) shows a single peak at 2884 cm^{-1}, which is assigned to $\nu_s(CH_3)$.

By exposure of clean Pt(111) to ethyl iodide and annealing to temperatures above 170 K, ethyl groups can be formed on the surface (*30*). The SFG spectrum of ethyl groups on Pt(111) is shown in Fig. 2c. The peaks at 2860 and 2920 cm^{-1} are assigned to a Fermi resonance ($\nu_s + 2\nu_{def}$) and to $\nu_s(CH_3)$, respectively. There is only one Fermi resonance band because of both the orientation of ethyl on the surface and the polarization of the laser beams used during the experiment. A spectrum (Fig. 2d) of both π-

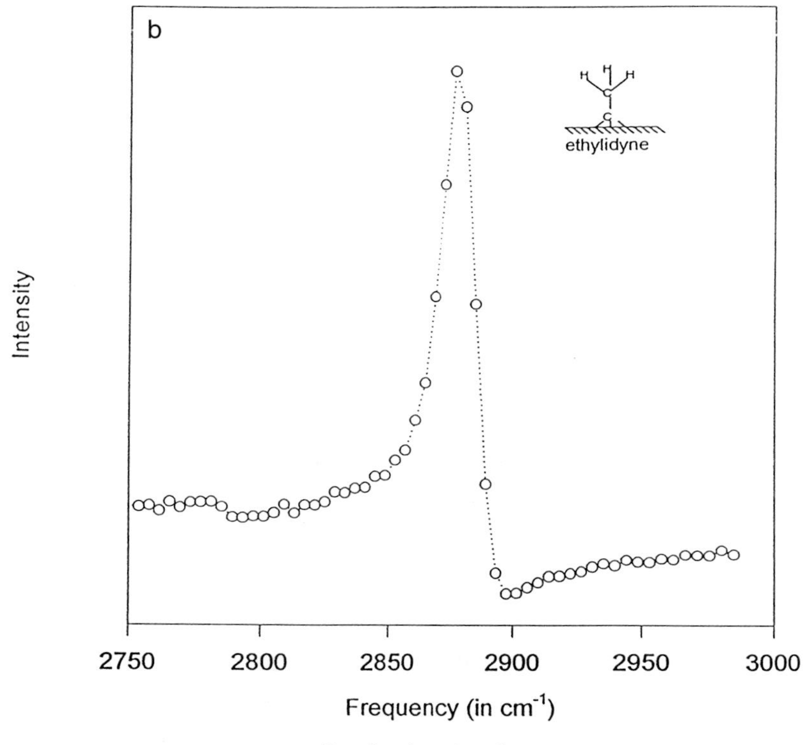

FIG. 2. (*continued*)

bonded ethylene and di-σ-bonded ethylene was obtained by exposing a clean Pt(111) crystal to a saturation coverage of O_2 at 300 K, followed by exposure to ethylene at 120 K (*31, 32*). The resultant peak at 2915 cm^{-1} is di-σ-bonded ethylene, and a smaller peak at 2995 cm^{-1} is associated with the $\nu_s(CH_2)$ of π-bonded ethylene. The 2995 cm^{-1} peak is much smaller than that representing the di-σ-bonded ethylene because the dynamic dipole of the molecule's $\nu_s(CH_2)$ is nearly in plane with the metal surface.

B. High-Pressure Catalytic SFG Experiments Characterizing Ethylene Hydrogenation

High-pressure ethylene hydrogenation experiments were monitored *in situ* with SFG under hydrogen pressures between 2 and 700 Torr (*29*). In each experiment, the ethylene pressure was kept below the hydrogen pressure, and the ethylene pressures were kept in the regime in which the

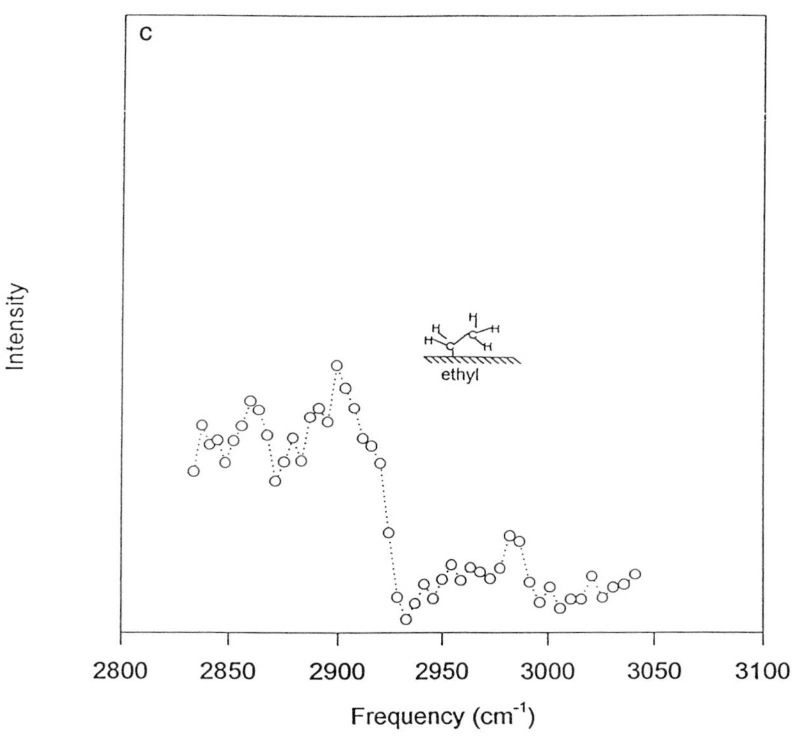

FIG. 2. *(continued)*

reaction order was zero with respect to ethylene. Results shown here are for experiments in which 35 Torr of ethylene, 100 Torr of hydrogen, and 625 Torr of He were introduced at 295 K. Figure 3a shows an SFG spectrum in which signals of several surface species are observed. Ethylidyne was observed at 2878 cm^{-1} and di-σ-bonded ethylene at 2910 cm^{-1}; π-bonded ethylene appeared as two broad features near 3000 cm^{-1}. The intensities of the peaks correspond to 0.15, 0.08, and 0.04 monolayers (ML) for ethylidyne, di-σ-bonded ethylene, and π-bonded ethylene, respectively. The spectrum shown in Fig. 3a remained unchanged for hours, indicating that the composition of adsorbates remained essentially the same on the surface over the lifetime of the experiment.

As SFG spectra were being recorded, the gas-phase composition was monitored by gas chromatography. The chromatographic data showed a turnover rate (TOR) of 11 $\pm$ 1 ethylene molecules converted per surface platinum atom per second.

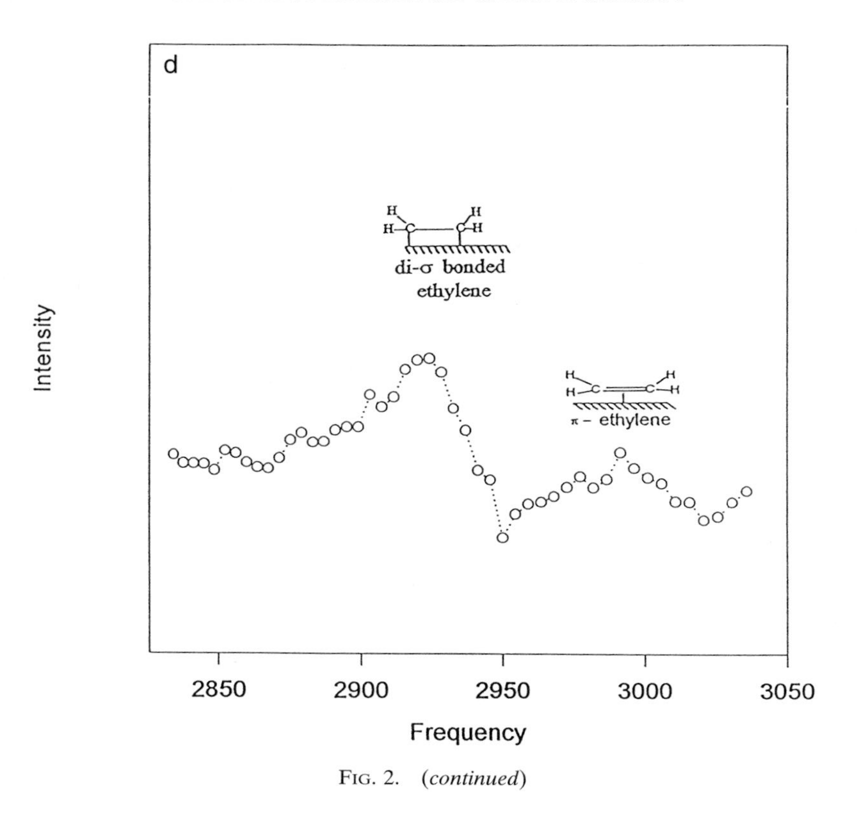

FIG. 2. *(continued)*

After the batch reactor was evacuated, another SFG spectrum was recorded (Fig. 3b), and it shows that only ethylidyne remained on the surface. The peak intensity also indicates that the amount of ethylidyne adsorbed increased to saturation coverage (0.25 ML). This change in surface coverage was not unexpected because the UHV experiments showed that di-σ-bonded ethylene is hydrogenated to ethylidyne at 300 K when hydrogen is not present. This result confirms that di-σ-bonded ethylene and π-bonded ethylene can exist at 300 K only when hydrogen is present.

Figure 3c was acquired during a similar high-pressure experiment, although the surface was pretreated in UHV with 4 L of ethylene at 295 K. This experiment was performed to determine the effect of a saturation coverage of ethylidyne during a reaction. After the pretreatment with ethylene, 35 Torr of ethylene, 100 Torr of hydrogen, and 625 Torr of He were introduced into the batch reactor. The vibrational spectrum is considerably different from that of Fig. 3a. Ethylidyne appears at a higher coverage than in the previous experiment, whereas the feature representing the di-σ-

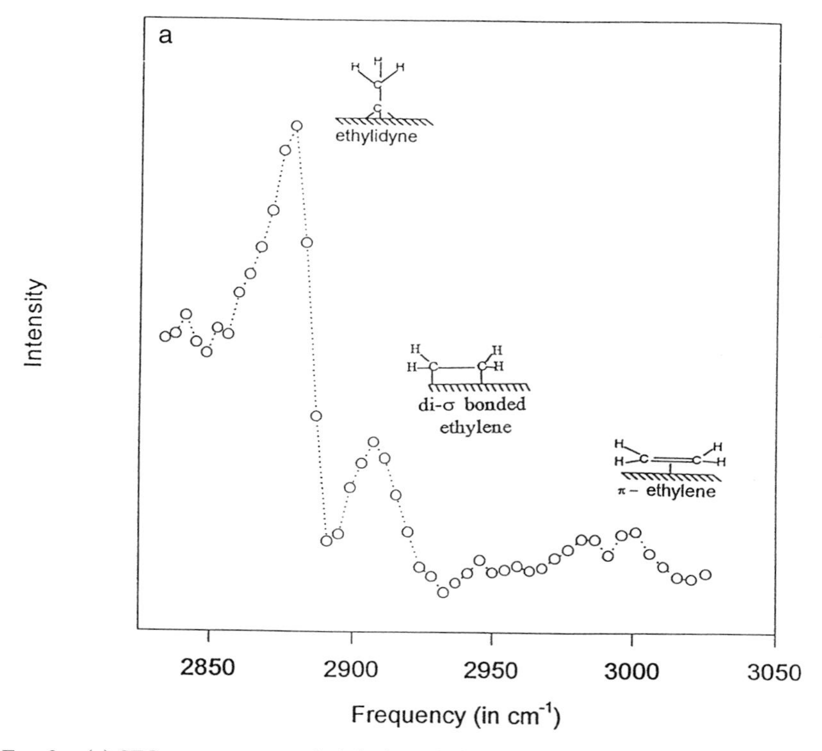

FIG. 3. (a) SFG spectrum recorded during ethylene hydrogenation at 295 K on Pt(111): 100 Torr H_2, 35 Torr of ethylene, and 615 Torr of He. Species observed under reaction conditions include ethylidyne, di-σ-bonded ethylene, and π-bonded ethylene. (b) SFG spectrum after the reaction cell was evacuated. Only strongly bound ethylidyne is observed. (c) SFG spectrum under the same conditions as shown in a on Pt(111) preadsorbed with 0.25 ML of ethylidyne.

bonded species is much smaller (0.02 ML compared to 0.08 ML previously). In addition, the π-bonded ethylene appears at a similar coverage, as shown in Fig. 3a. Gas chromatographic data revealed a TOR of 12 $\pm$ 1 molecules per platinum site per second, which is almost identical to that observed in the previous experiment.

Because of the fact that the rate of the reaction remained the same while the amount of surface di-σ-bonded species decreased, it is apparent that di-σ-bonded ethylene is not an important reaction intermediate in ethylene hydrogenation. Furthermore, it appears that both ethylidyne and di-σ-bonded ethylene compete for sites. Once ethylidyne forms on the surface, di-σ-bonded ethylene is blocked from adsorbing to the surface. From previ-

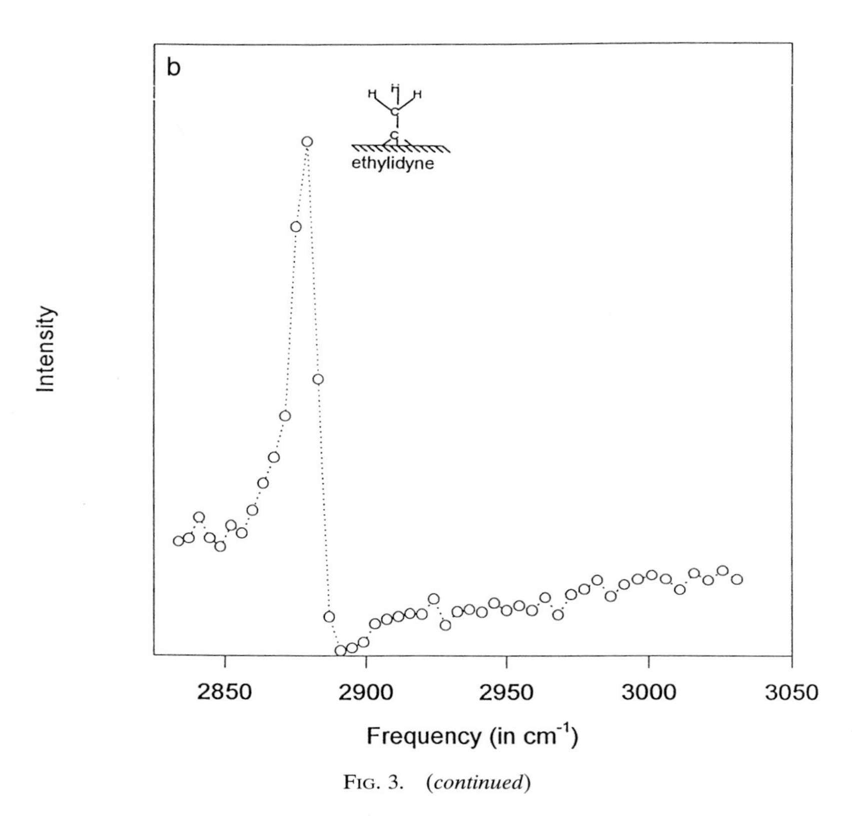

FIG. 3. *(continued)*

ous work, it is known that ethylidyne is not a reaction intermediate. This study indicates that di-σ-bonded ethylene is also not a reaction intermediate.

Because the concentration of π-bonded ethylene was the same in both cases and because the reaction rate remained the same, it appears that π-bonded ethylene is likely the key reaction intermediate in ethylene hydrogenation on Pt(111).

Additional high-pressure ethylene hydrogenation reactions were performed in which the hydrogen pressure and ethylene pressure were increased to values exceeding those of the previous case. A clean Pt(111) crystal was exposed to 723 Torr of hydrogen and 60 Torr of ethylene at 295 K. In this case (Fig. 4), the SFG spectrum shows that the surface was not saturated with ethylidyne as in the previous experiment with 100 Torr of hydrogen. There is little change in the size of the π-bonded ethylene peak; however, there is a decrease in the intensities of the peaks associated

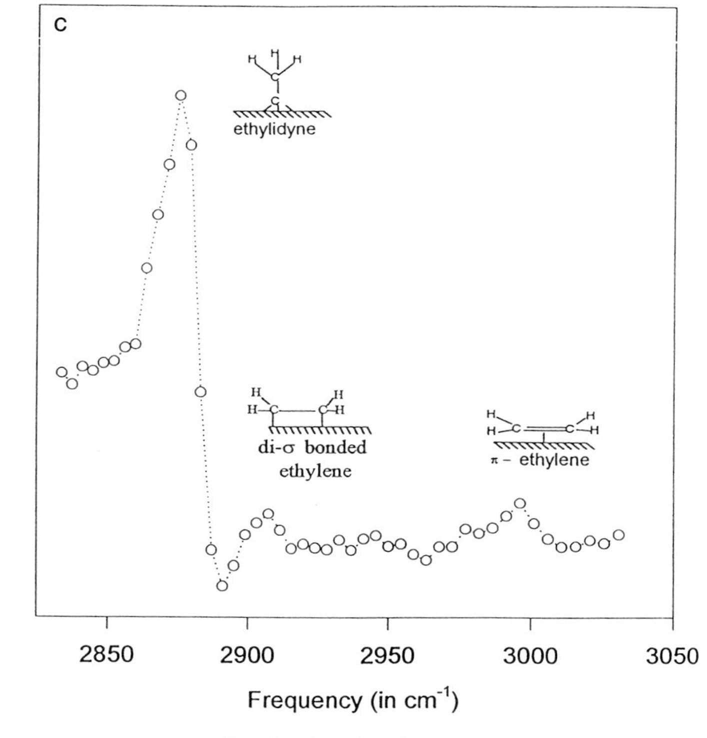

FIG. 3. *(continued)*

with di-σ-bonded ethylene relative to those of Figs. 3a and 3c. Two new peaks, at 2850 and 2925 cm^{-1}, were also observed and correspond to an ethyl species; these peaks are similar to the peaks observed in the UHV experiment (Fig. 2c). The gas chromatographic data revealed a TOR of 61 $\pm$ 3 molecules per platinum site per second.

With lower pressures of hydrogen and ethylene (1.75 Torr of hydrogen, 0.25 Torr of ethylene, and 758 Torr of argon), the spectrum (Fig. 5) showed that the surface was saturated with ethylidyne; only a small amount of di-σ-bonded ethylene was observed. As the reaction progressed, the di-σ-bonded ethylene peak decreased until it was no longer observable after 200 min. Then ethylidyne was the only observable surface, and at no time during the reaction was π-bonded ethylene observed. The TOR throughout the reaction was 1.6 molecules per platinum site per second.

From these high-pressure ethylene hydrogenation experiments in which the gas-phase composition was monitored by gas chromatography at the

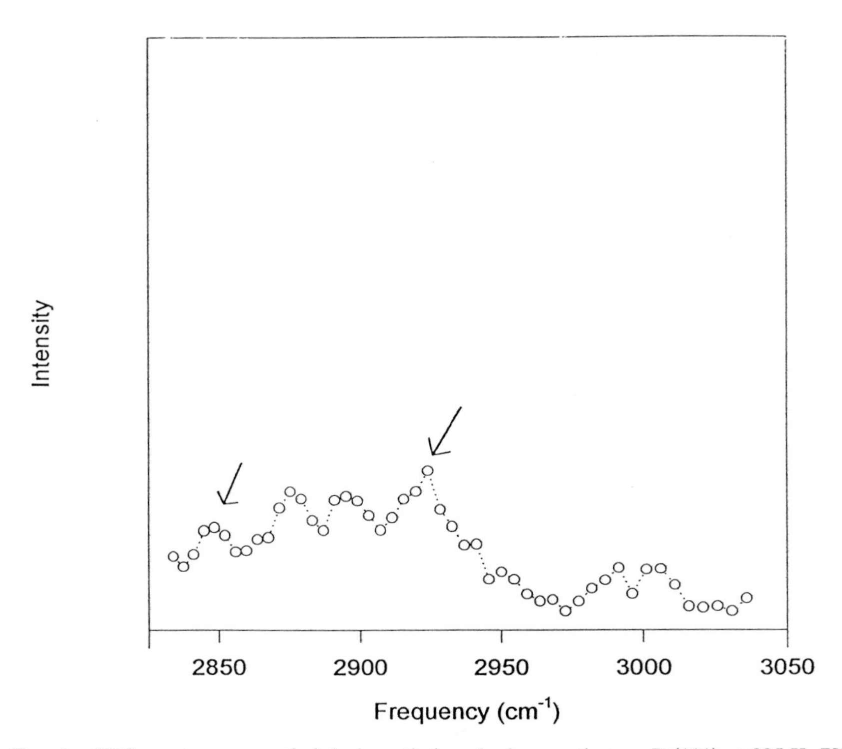

FIG. 4. SFG spectrum recorded during ethylene hydrogenation on Pt(111) at 295 K: 727 Torr H_2 and 60 Torr ethylene. Ethyl species are observed and noted with the arrows.

same time SFG spectra were being acquired, it was observed that the ethylene hydrogenation TOR was independent of the surface concentration of di-σ-bonded ethylene. It is thus suggested that ethylidyne and di-σ-bonded ethylene are spectator species for the reaction. The most likely reaction intermediate is inferred to be π-bonded ethylene. The coverage of the surface with π-bonded ethylene was determined to be approximately 4% of a monolayer, suggesting the TOR is actually 25 times higher if reported per surface intermediate rather than per platinum atom. These results indicate that not all platinum atoms on the surface are active at any given time. Studies have shown that ethylene hydrogenation is a structure-insensitive reaction, and therefore the specific location at which key reaction steps take place must be sites that are available on all crystallographic planes of platinum (*33*).

Thus, from the data presented, the following reaction pathway for ethylene hydrogenation is proposed (Fig. 6): hydrogen dissociatively chemisorbs

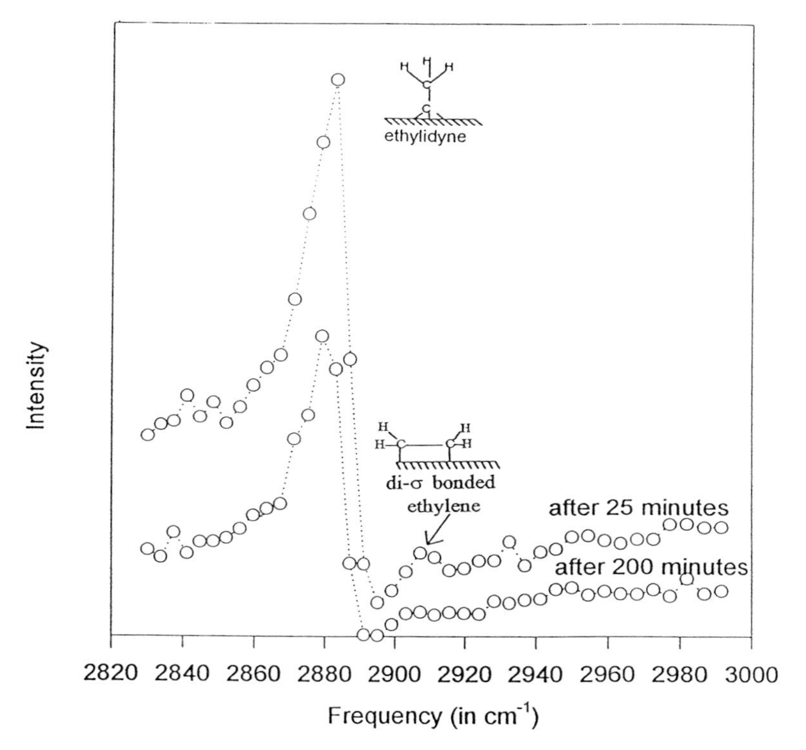

FIG. 5. SFG spectra recorded during ethylene hydrogenation on Pt(111) at 295 K after 25 and 200 min: 1.75 Torr H_2, 0.25 Torr ethylene, and 758 Torr of argon. Ethylidyne and di-σ-bonded ethylene are observed after 25 min, but only ethylidyne is observed after 200 min.

slow pathway

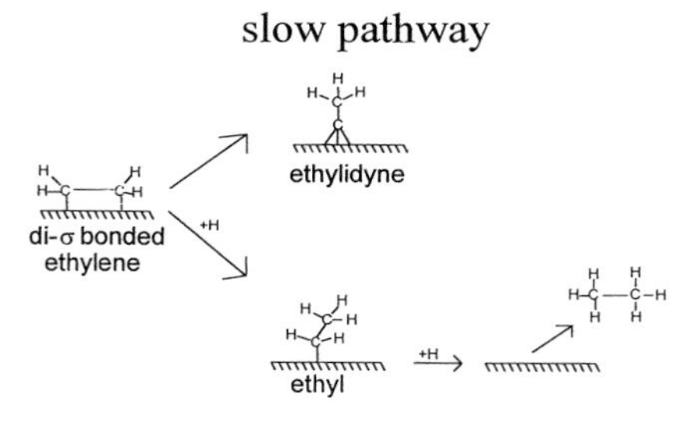

fast pathway

FIG. 6. Proposed reaction pathway for ethylene hydrogenation. The slow pathway proceeds through an ethylidyne species and the fast pathway proceeds through an ethyl intermediate.

on a clean or ethylidyne-covered platinum surface. Ethylene then physisorbs to form π-bonded ethylene, which is then hydrogenated stepwise to form ethyl and then ethane.

V. Propylene Hydrogenation and Dehydrogenation on Pt(111)

In the few investigations of propylene hydrogenation and dehydrogenation on Pt(111), it has been shown that propylene adsorbs as a di-σ-bonded species in UHV at temperatures lower than 220 K (*34*). The chemistry is similar to that of ethylene; once the surface is heated, the di-σ-bonded propylene species is converted to propylidyne at a temperature near 300 K while hydrogen is evolved (*35*). At temperatures higher than 300 K, propylidyne decomposes into graphitic precursors.

Kinetics data show that propylene hydrogenation at 300 K and atmospheric pressures on platinum is 0.5 order in hydrogen and zero order in propylene (*36, 37*). An investigation of Pt/SiO_2 after exposure to propylene and hydrogen showed the presence of several species: propylidyne, di-σ-bonded propylene, and π-bonded propylene (*38*). Two pathways were proposed for the stepwise hydrogenation of propylene. The reaction may proceed through either a 2-propyl intermediate or a 1-propyl intermediate, depending on whether the first hydrogen is added to the terminal carbon atom or the internal carbon atom of the olefin.

A. Low-Pressure SFG Experiments Characterizing Propylene Adsorption

Experiments under UHV conditions were first performed to help in the identification of high-pressure reaction intermediates observed under catalytic reaction conditions (*39*). On a clean Pt(111) crystal, di-σ-bonded propylene was first studied by exposure of the surface to 4 L of propylene at 187 K. The vibrational spectrum of the resulting di-σ-bonded propylene species is shown in Fig. 7a. Two large peaks are observed at 2825 and 2880 cm^{-1}, assigned to a Fermi resonance ($\nu_s + 2\nu_{def}$) and to the $\nu_s(CH_3)$ of the terminal methyl group, respectively. A small shoulder at 2880 cm^{-1} indicates either a $\nu_s(CH_2)$ or $\nu(CH)$, as determined by the results of an otherwise identical experiment performed with CD_3CHCH_2 (Fig. 7b).

When the Pt(111) was exposed to 4 L of propylene at 310 K, a propylidyne species was observed in the SFG spectrum (Fig. 7c). There are three strong features at 2855, 2920, and 2960 cm^{-1}. The feature at 2855 cm^{-1} indicates a Fermi resonance ($\nu_s + 2\nu_{def}$), and the 2920 cm^{-1} peak indicates $\nu_s(CH_3)$. The strong feature at 2960 cm^{-1} is due to $\nu_a(CH_3)$. The SFG spectrum of

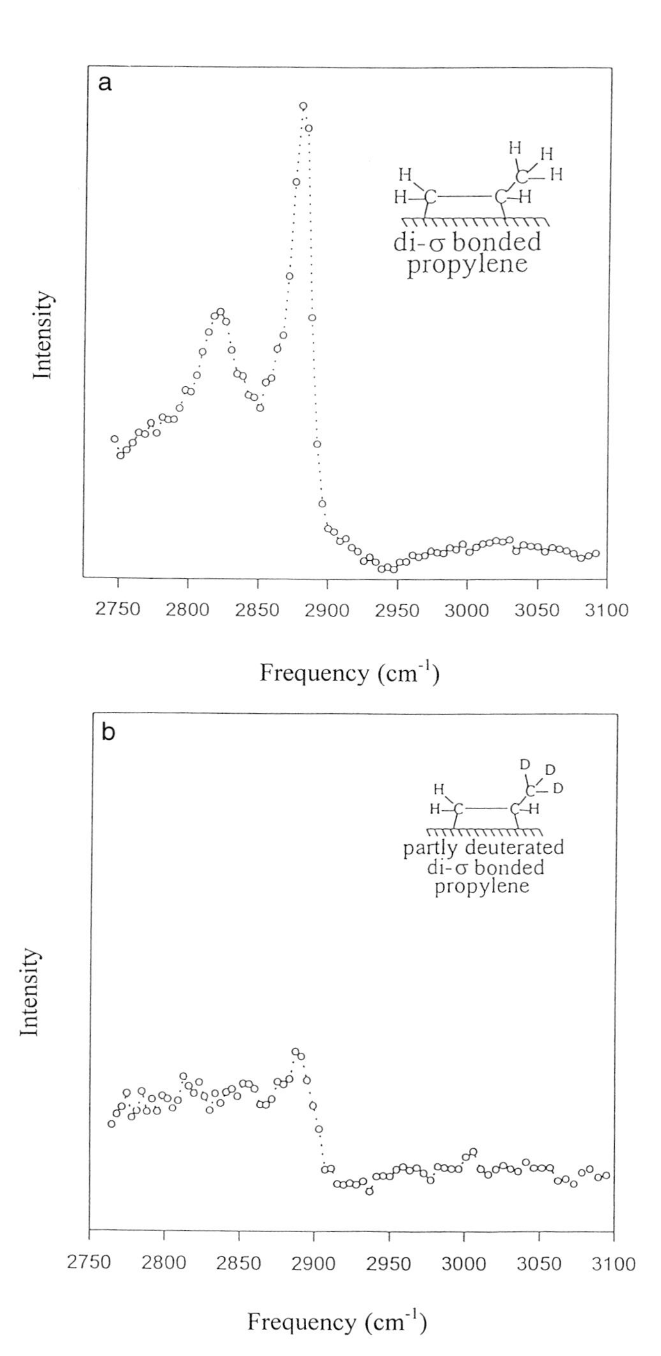

FIG. 7. SFG spectra of adsorbed species in UHV which may be possible reaction intermediates during high-pressure propylene hydrogenation on Pt(111). (a) Di-σ-bonded propylene at 187 K, (b) partially deuterated di-σ-bonded propylene at 187 K, (c) propylidyne at 310 K, and (d) partially deuterated propylidyne at 310 K.

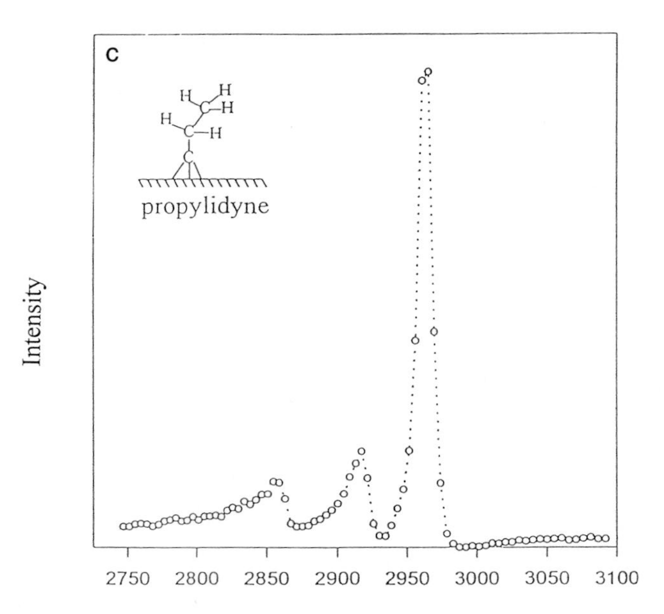

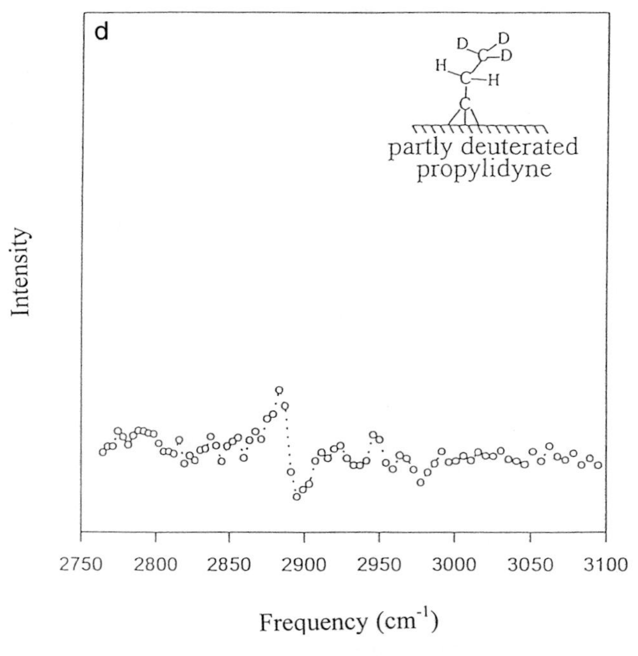

FIG. 7. (*continued*)

CD_3CHCH_2 (Fig. 7d) reveals a feature at 2880 cm^{-1}, most likely representing CH_2 modes.

To follow the conversion of di-σ-bonded propylene to propylidyne, a clean Pt(111) crystal was exposed to 4 L of propylene at 187 K and annealed to successively higher temperatures for 30-s periods. After the sample had been heated, it was allowed to cool to 187 K, whereupon the SFG spectra were taken. The results (Fig. 8a) show that di-σ-bonded propylene is stable up to a temperature of 231 K, at which the di-σ bonded species begins to

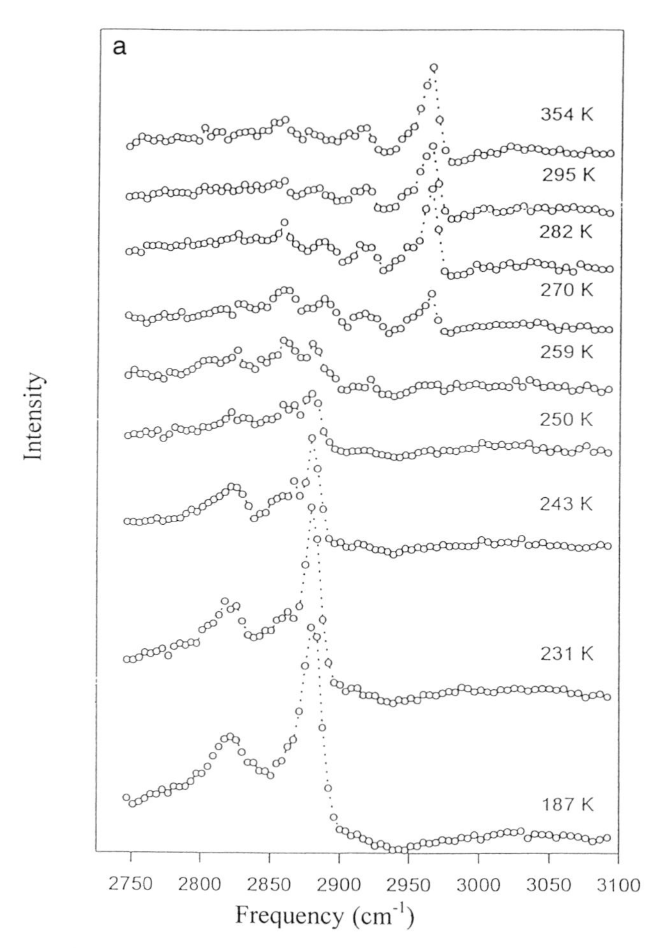

FIG. 8. (a) SFG spectra indicating the thermal evolution of a saturated coverage of di-σ-bonded propylene under UHV on Pt(111). (b) SFG spectra of the thermal evolution of a saturated coverage of propylidyne on Pt(111) under UHV conditions.

be dehydrogenated. When the temperature reached 295 K, all the di-σ-bonded species had been converted to propylidyne.

With further heating, propylidyne began to undergo dehydrogenation, and at temperatures higher than 390 K new spectral features were observed near 2900 and 3000 cm^{-1} (Fig. 8b). The feature near 3000 cm^{-1} is most likely indicative of an olefinic species converted as a result of propylidyne dehydrogenation. As the sample was heated to temperatures higher than 420 K, the features became less intense.

Other possible reaction intermediates could be 1-propyl and 2-propyl surface moieties, which can be formed on Pt(111) by adsorption of the respective iodides and heating the surface to break the carbon–iodine bond (*30*). A clean Pt(111) crystal was exposed to 10 L of 1-propyl iodide at 140 K and then annealed to 190 K for 10 s. In the SFG spectrum of 1-

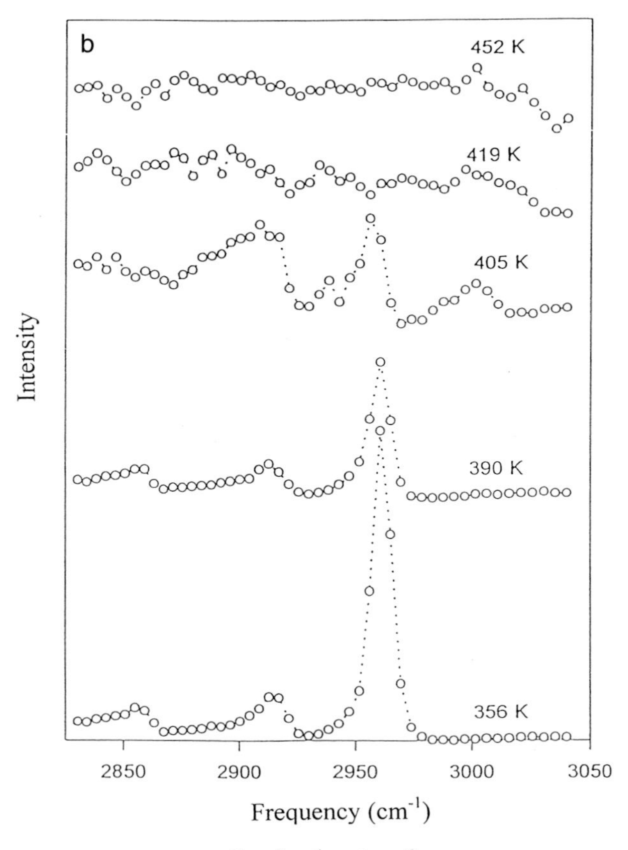

FIG. 8. (*continued*)

propyl (Fig. 9a), several features are observed. The largest three peaks, at 2875, 2934, and 2974 cm^{-1}, can be tentatively assigned to a Fermi resonance ($\nu_s + 2\nu_{def}$), $\nu_s(CH_3)$, and $\nu_a(CH_3)$, respectively. A lower coverage of 1-propyl was also obtained by exposing the Pt(111) to less than 4 L. The feature in the SFG spectrum (Fig. 9b) is considerably weaker than before and may be a result of a different orientation of the surface 1-propyl moieties. The most notable change was the appearance of the largest feature at 2875 cm^{-1}, with the feature at 2974 cm^{-1} being considerably weaker. The peak at 2934 cm^{-1} appears to have the same intensity as in Fig. 9a.

Hydrogenation of 1-propyl moieties was also carried out on an initially saturated overlayer of 1-propyl iodide by exposing the crystal to 1.4×10^{-5} Torr of hydrogen while annealing the sample to 190 K for 100 s (Fig. 9c). After the crystal had been cooled to 140 K, an SFG spectrum was acquired. For comparison, the spectrum in Fig. 9a was superimposed on this result, indicating a 20% reduction in the spectral intensity with only minor changes

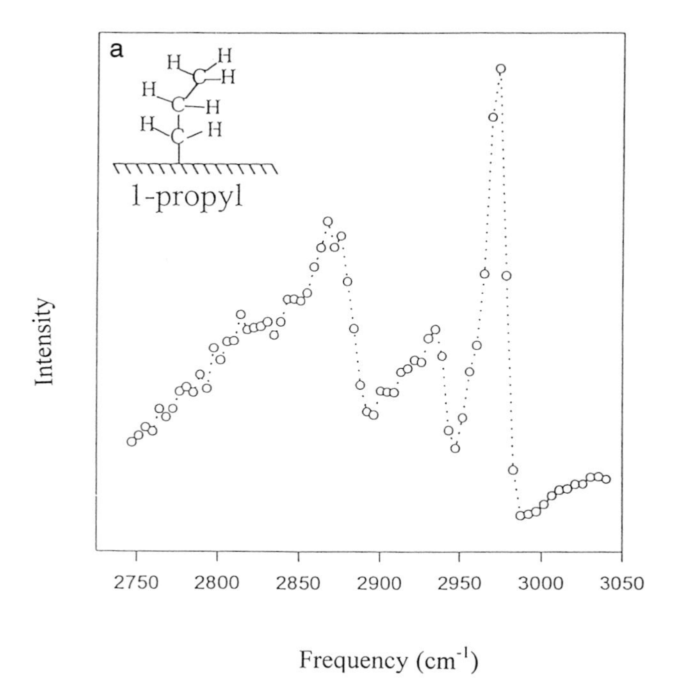

FIG. 9. (a) SFG spectrum of a saturated exposure of 1-propyl iodide on Pt(111) at 140 K. (b) SFG spectrum of a low-coverage exposure of 1-propyl iodide on Pt(111) at 140 K. (c) SFG spectra of a saturated exposure of 1-propyl iodide before and after a 1400 L exposure of H_2 at 190 K.

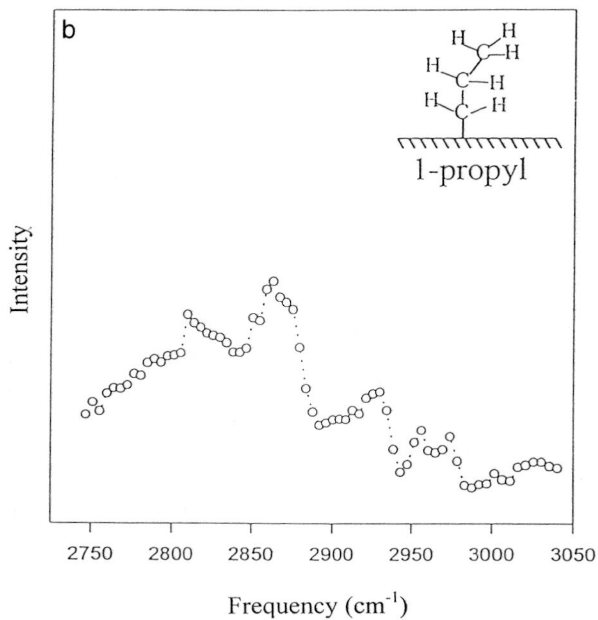

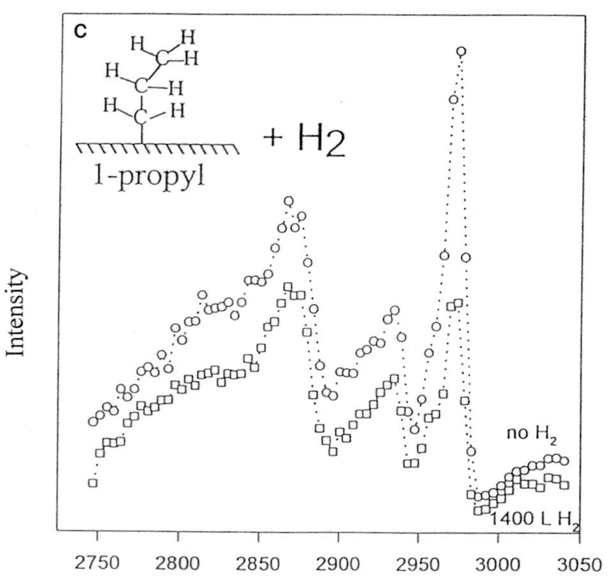

FIG. 9. *(continued)*

in relative peak heights. These results suggest that 1-propyl species were hydrogenated to propane, accounting for the decrease in SFG spectral intensity.

Experiments similar to those performed for 1-propyl were also carried out for the 2-propyl moiety formed from 2-propyl iodide. The results are shown in Fig. 10. Notably, the SFG spectrum representing the saturated 2-propyl moiety is considerably weaker than its analog with the 1-propyl moiety. Three features are seen at 2838, 2863, and 2913 cm^{-1}. The peak at 2863 cm^{-1} can be assigned to $\nu_s(CH_3)$, and the 2913 cm^{-1} feature possibly indicates $\nu(CH)$. In the low-coverage spectrum (Fig. 10b), it is unclear whether the 2863 or the 2838 cm^{-1} peak represents the $\nu_s(CH_3)$ peak.

Comparable to what was observed in the hydrogenation of the 1-propyl species, the SFG spectral intensity of the 2-propyl moiety was also attenuated by approximately 40% when the sample was exposed to 1400 L of hydrogen at 190 K, and otherwise there were no changes in the qualitative appearance of the spectrum.

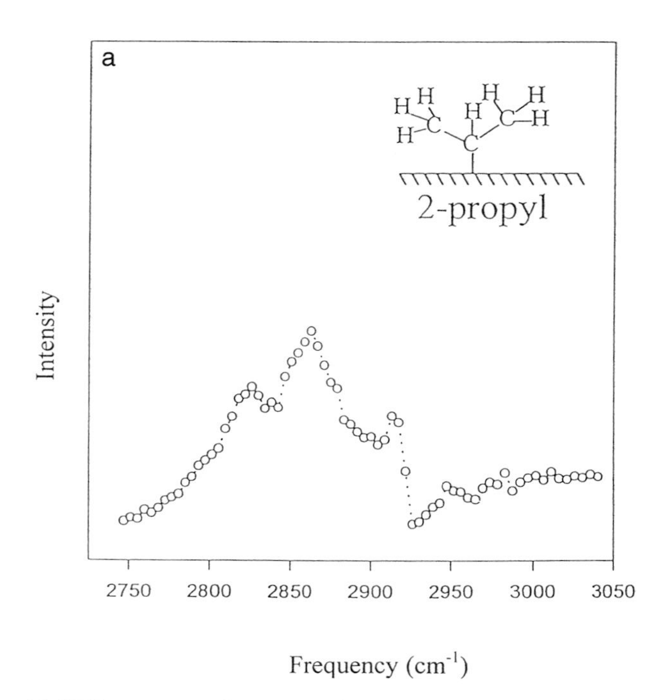

FIG. 10. (a) SFG spectrum of a saturated exposure of 2-propyl iodide on Pt(111) at 140 K. (b) SFG spectrum of a low-coverage exposure of 2-propyl iodide on Pt(111) at 140 K. (c) SFG spectra of a saturated exposure of 2-propyl iodide before and after a 1400 L exposure of H_2 at 190 K.

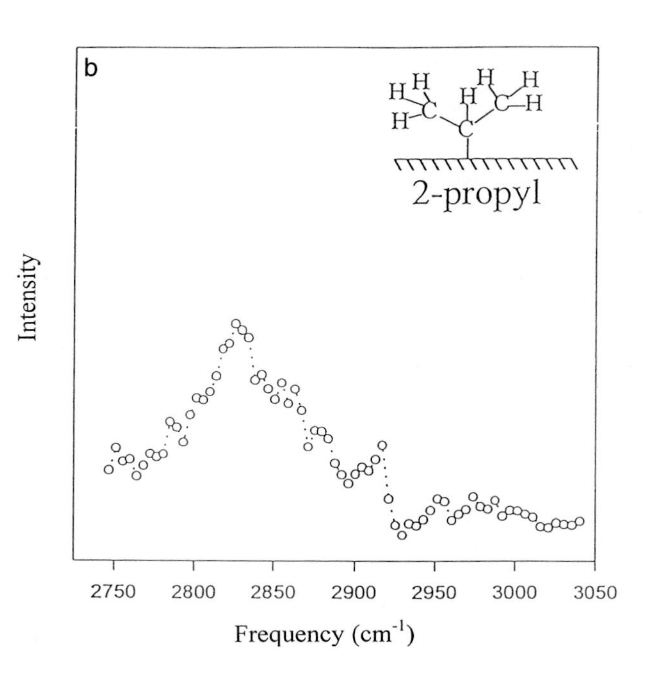

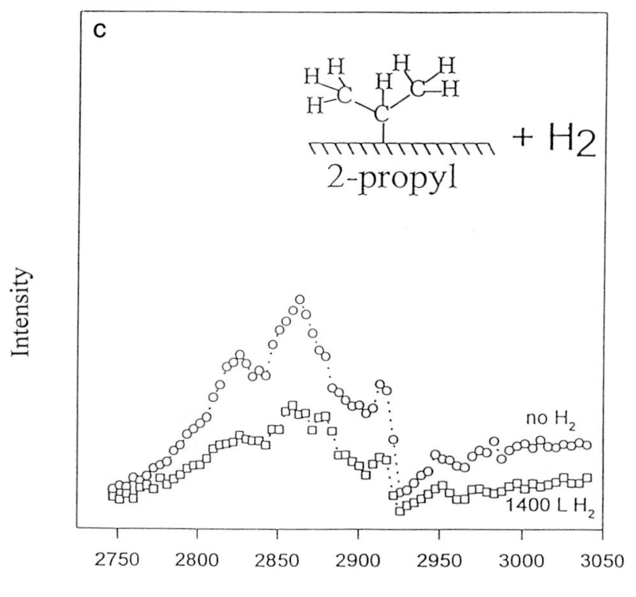

FIG. 10. (*continued*)

B. High-Pressure Catalytic SFG Experiments

High-pressure catalytic reactions were performed during acquisition of SFG spectra, and the gas-phase composition in the batch reactor was monitored by gas chromatography. Under the highest-pressure conditions, 723 Torr of hydrogen and 40 Torr of propylene were introduced into the reactor at 295 K. The SFG spectrum (Fig. 11a) revealed seven different peaks. The largest peak, at 2830 cm^{-1}, is assigned to a 2-propyl group. The peaks at 3050 and 2960 cm^{-1} correspond to π-bonded propylene and propylidyne, respectively. The assignment of the peaks between 2860 and 2940 cm^{-1} is more difficult, but these peaks may indicate 1-propyl, 2-propyl, and/or di-σ-bonded propylene. Under these reaction conditions, the TOR was 29 propane molecules formed per exposed platinum atom per second. Surprisingly, considering that the intensity of the di-σ-bonded propylene species should be smaller than that of the π-bonded propylene species because of the metal dipole selection rules and the SFG selection rules (*22*), the π-bonded propylene peak is actually larger than that indicating the di-σ-bonded propylene species. This evidence suggests that the concentration of the π-bonded species was higher than that of the di-σ-bonded propylene species.

Another interesting point is that the intensity of the peak associated with 2-propyl species is at least three times greater than the intensity of the peak associated with the 1-propyl species. This result also indicates that the 2-propyl species was present at a higher concentration than the 1-propyl species because the UHV experiments revealed that the 1-propyl species had the higher SFG cross section. Furthermore, the peaks associated with the 2-propyl species resemble those of the low-coverage spectrum more than they do those of the saturation spectra in Figs. 10a and 10c.

When the reaction was performed at 100 Torr of hydrogen, 40 Torr of propylene, and 617 Torr of He at 295 K, both the SFG spectrum (Fig. 11b) and the TOR were dramatically different from those of the previously mentioned high-pressure case. The TOR decreased to 8.6 molecules per platinum site per second, and the peaks at 2960 and 2920 cm^{-1} can be assigned to propylidyne. Both the SFG features at 3050 and 2830 cm^{-1} associated with π-bonded propylene and 2-propyl, respectively, were considerably weaker than in the high-pressure experiment. However, the feature at 2863 cm^{-1} was more intense, indicating a possible net reorientation of the 2-propyl species. The feature at a frequency just below 2900 cm^{-1} increased in intensity; it is assigned to di-σ-bonded propylene. The increase in the propylidyne and di-σ-bonded propylene features along with a decrease in the TOR help rule out these species as important reaction intermediates in propylene hydrogenation.

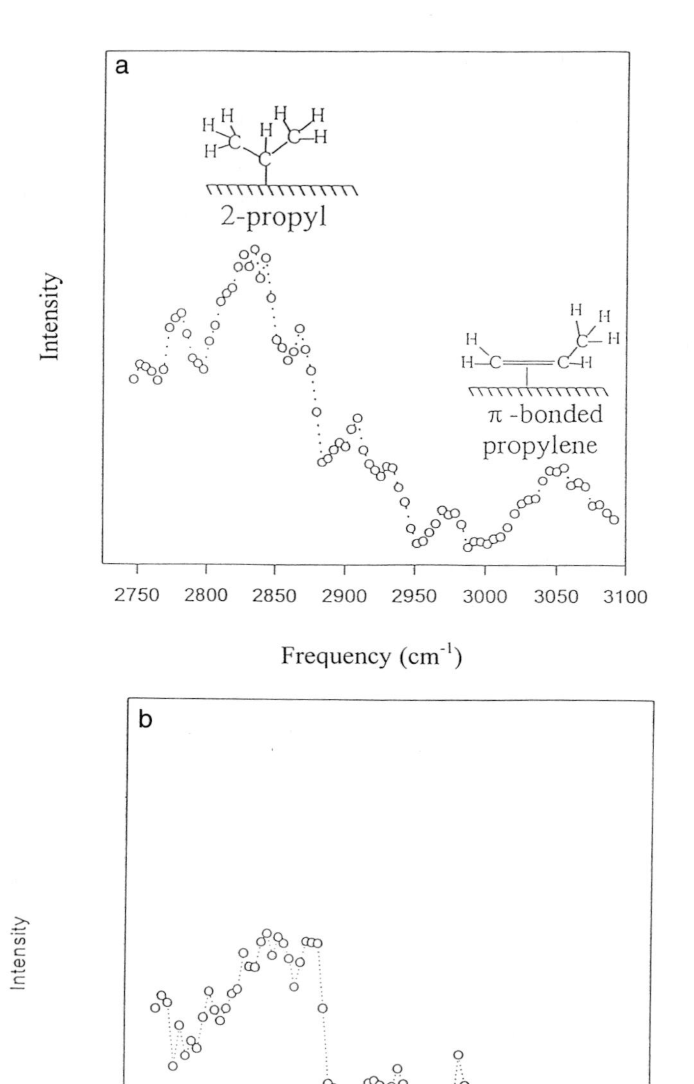

FIG. 11. (a) SFG spectrum recorded during propylene hydrogenation on Pt(111) at 295 K with 723 Torr of H_2 and 40 Torr of propylene. (b) SFG spectrum recorded during propylene hydrogenation on Pt(111) at 295 K with 100 Torr of H_2, 40 Torr of propylene, and 617 Torr of He.

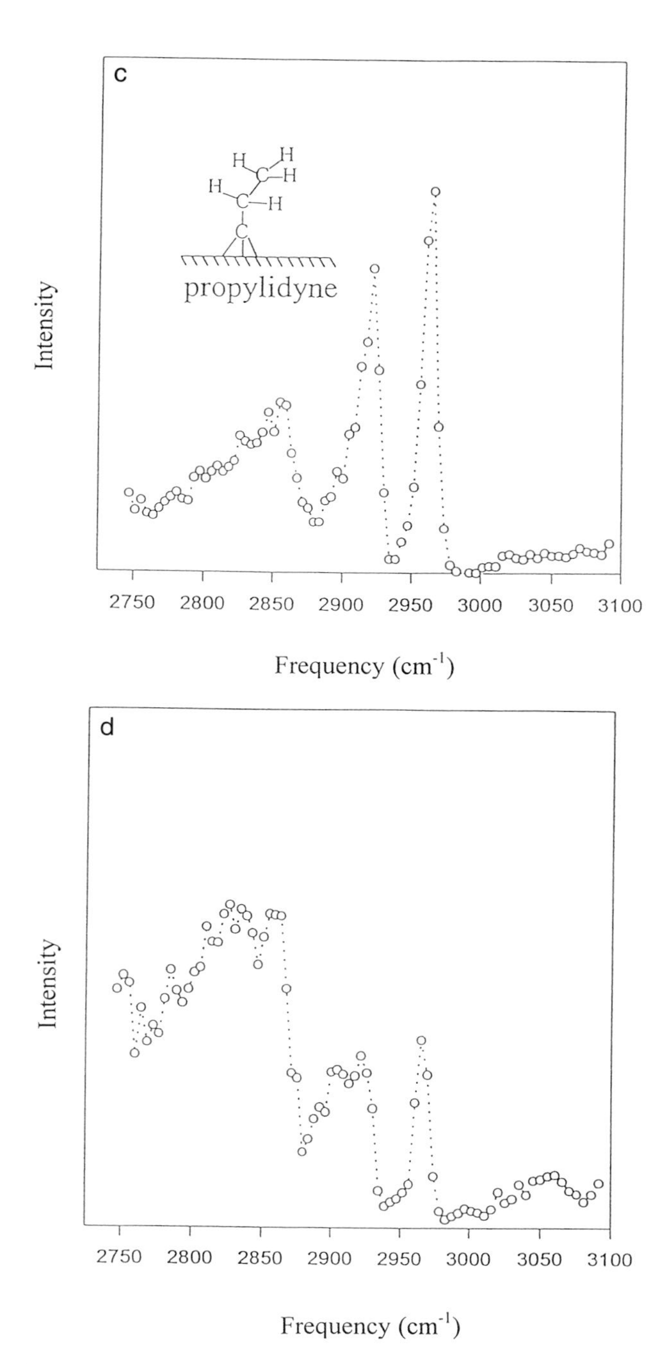

FIG. 11. (*continued*) (c) SFG spectra recorded during propylene hydrogenation on Pt(111) at 295 K with 13.5 Torr of H_2, 5 Torr of propylene, and 746 Torr of He. (d) SFG spectra recorded during propylene hydrogenation on Pt(111) with preadsorbed propylidyne at 295 K with 100 Torr of H_2, 40 Torr of propylene, and 617 Torr of He.

As the reactant pressures were lowered further (to 13.5 Torr of hydrogen, 5 Torr of propylene, and 746 Torr He), the TOR decreased to 3.3 molecules per platinum site per second. This rate is nearly an order of magnitude lower than the high-pressure value discussed previously, and the SFG spectrum was dominated by large features associated with propylidyne at 2850, 2915, and 2960 cm^{-1} (Fig. 11c). The broad intensity near 2850 cm^{-1} probably indicates a low concentration of 2-propyl.

To determine the effect of propylidyne on the surface under reaction conditions, a clean Pt(111) sample was exposed to 4 L of propylene at 295 K, forming a saturation coverage of propylidyne. The sample was then exposed to 40 Torr of propylene, 100 Torr of hydrogen, and 760 Torr of He. The TOR was 8.4 molecules per platinum site per second, which is nearly identical to the value associated with Fig. 11b. The SFG spectrum (Fig. 11d) was also very similar to that of Fig. 10b. This comparison demonstrates that the amount of adsorbed propylidyne quickly comes to equilibrium during the hydrogenation process, regardless of how much propylidyne is initially adsorbed.

From the SFG experiments summarized previously, it is clear that there are two important reaction intermediates in propylene hydrogenation. Analogous to ethylene hydrogenation, propylene hydrogenation involves adsorption of the olefin as a π-bonded species on platinum. This π-bonded species can be hydrogenated to either 1-propyl or 2-propyl moieties. The SFG spectra indicate that the 2-propyl species is present at a higher concentration than the 1-propyl species, and the UHV data show that 2-propyl incorporates hydrogen faster than 1-propyl in the presence of iodine. This inference is reasonable because the bond between an internal carbon atom and the underlying metal should be slightly weaker than the bond between a terminal carbon atom and the metal. With these data in mind, it is believed that propylene hydrogenation proceeds faster through a 2-propyl intermediate than a 1-propyl intermediate. The resulting proposed reaction pathway is shown in Fig. 12.

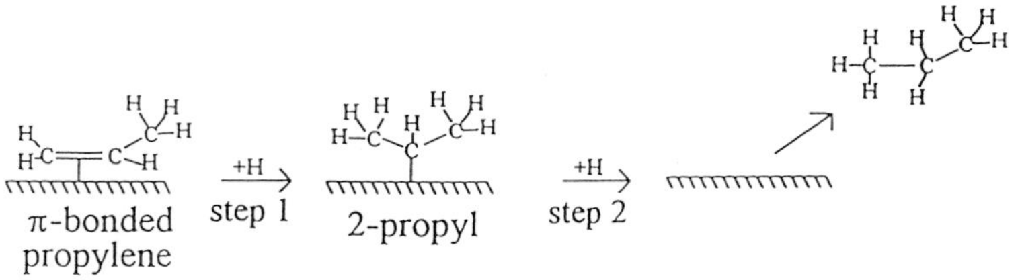

π-bonded propylene — +H step 1 → 2-propyl — +H step 2 →

FIG. 12. Proposed reaction pathway for propylene hydrogenation on Pt(111). The important reaction intermediates are π-bonded propylene and 2-propyl.

VI. Cyclohexene Hydrogenation and Dehydrogenation on Pt(111) and Pt(100): A Structure Sensitivity Study

The literature reports the application of several spectroscopic techniques to characterize cyclohexene adsorbed under UHV conditions on Pt(111) and Pt(100). These techniques include HREELS, EELS, and infrared adsorption spectroscopy (*40–44*). The results show that cyclohexene is dehydrogenated to a C_6H_9 intermediate en route toward benzene formation on platinum surfaces. These techniques are limited to low-pressure studies, and high-pressure experiments are needed to provide insight into the reaction pathways prevailing under catalytic reaction conditions.

Using SFG, a comparison between cyclohexene hydrogenation and dehydrogenation on Pt(111) and Pt(100)(5 × 20) single crystals at high pressures and temperatures was made to determine how the surface structure influences these reactions. Cyclohexene was also investigated under UHV conditions to allow comparison of results with the reported observations. In addition to cyclohexene, 1,3-cyclohexadiene and 1,4-cyclohexadiene were also investigated under both high-pressure and low-pressure reaction conditions. The resultant data help in the assignment of possible high-pressure reaction intermediates present during cyclohexene hydrogenation and dehydrogenation.

A. Low-Pressure Investigations of Cyclohexene Adsorption on Pt(111) and Pt(100)

A clean Pt(111) crystal was exposed to 4 L of cyclohexene at 130 K, and a SFG spectrum was acquired. The sample was then annealed to consecutively higher temperatures and allowed to cool to 130 K, after which SFG spectra were taken. The results are shown in Fig. 13 (*45*). At 130 K, the SFG spectrum is similar to the IR/Raman spectra of liquid-phase cyclohexene (*46*). The peaks at 2875 and 2958 cm^{-1} are assigned to C–H stretch modes, and the peak at 2918 cm^{-1} is assigned to a Fermi resonance ($\nu_s + 2\nu_{def}$) of the bending mode of C–H groups.

As a free molecule, cyclohexene has a double bond and a half-chair conformation with C_2 symmetry (*46, 47*). Cyclohexene adsorbs by donating its π electron density to the metal as it bonds to the surface. As a result of the different electronic coupling with the carbon ring, the equatorial C–H appears at a higher stretching frequency than the axial C–H (*48, 49*). Except for a small shift in frequencies, this spectrum is also nearly identical to the Raman spectrum of deuterated cyclohexene 3,3,6,6-d_4 (*46*). In the deuterated molecule, the vibrational features arise from CH_2 groups at the C_4 and C_5 positions, and therefore the peaks observed in the SFG spectra

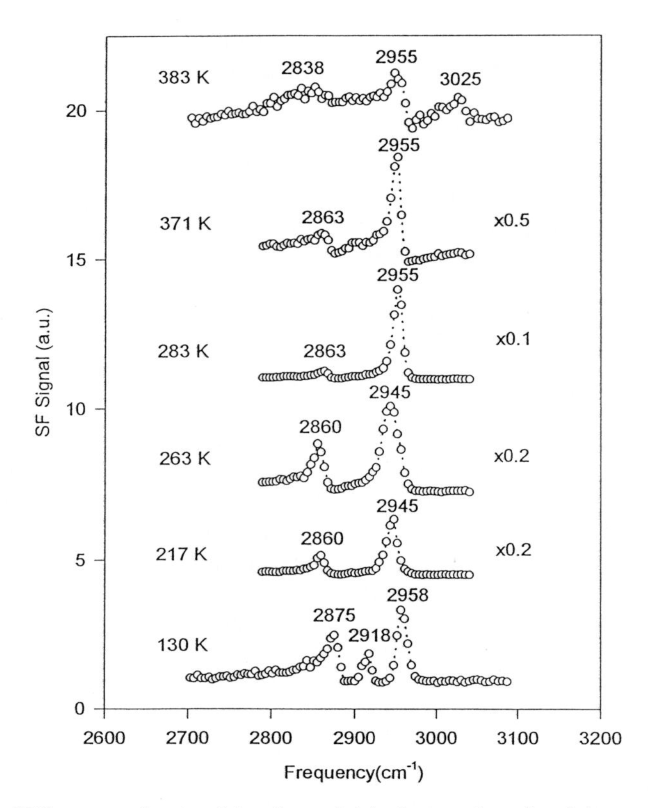

FIG. 13. SFG spectra characterizing thermal dehydrogenation of cyclohexene on Pt(111) at various temperatures.

are assigned to C–H bonds at C_4 and C_5. Because features corresponding to C_3 and C_6 C–H bonds were not observed, it is believed that these bonds are parallel to the surface and not observed because of the MSSR. Furthermore, because the SFG spectrum appears to be that of a molecular species, it is believed that cyclohexene is π/σ hybridized as it adsorbs.

As the sample was heated, the π/σ-hybridized cyclohexene was present up to a temperature of 200 K, whereupon a change in the SFG spectrum was observed. At 217 K, two new peaks with slightly higher intensity were observed at 2860 and 2945 cm^{-1}. This change in the spectrum, similar to what has been observed by others (*40*), is explained as the conversion of the π/σ-bonded species to a di-σ-bonded cyclohexene. The shift in the two species is probably due to the electronic effect as the hybridization of the double bond has changed from sp^2 to sp^3. The higher intensities of these two peaks relative to those of the previous example probably result because the di-σ-bonded species is more nearly perpendicular to the metal surface than its π/σ-bonded predecessor.

The SFG spectra recorded at temperatures higher than 217 K remained relatively unchanged until a temperature of about 283 K was reached, whereupon the two spectral features shifted to 2955 and 2863 cm^{-1}. Dehydrogenation of the di-σ-bonded cyclohexene species is inferred to have occurred, and the new peaks are believed to be associated with a π-allyl c-C_6H_9 species, as observed by others (*40, 43, 44*). A proposed structure of this C_6H_9 intermediate is shown in Fig. 14. The strong feature at 2955 cm^{-1} is attributed to the equatorial C–H stretching in the c-C_6H_9 species. The proposed C_6H_9 structure shows the equatorial C–H on C_4 and C_6 as perpendicular to the surface, and this structure explains the high intensity of this peak.

The C_6H_9 intermediate was observed on the surface up to a temperature of 371 K, and at higher temperatures more dehydrogenation occurred, as evidenced by the greatly attenuated spectral intensity and the appearance of a feature at 3025 cm^{-1}. This new feature is indicative of a sp^2-hybridized CH stretch and may arise from benzene or other decomposition products on the surface. The dehydrogenation temperature in this investigation was slightly higher than those reported previously, and the difference suggests site blocking effects associated with the relatively high coverage of cyclohexene used in this investigation. Only at higher temperatures will desorption and decomposition occur, freeing up active dehydrogenation sites.

A similar experiment was performed with a Pt(100)(5 × 20) single crystal exposed to 4 L of cyclohexene at 130 K (Fig. 15). The SFG spectrum observed at this temperature is practically identical to those observed with the Pt(111) crystal; cyclohexene adsorbs through the double bond in a

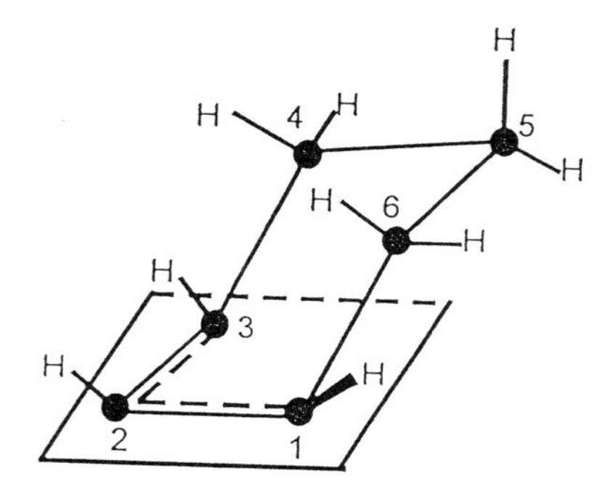

FIG. 14. Possible structure of the c-C_6H_9 intermediate species of cyclohexene hydrogenation under UHV.

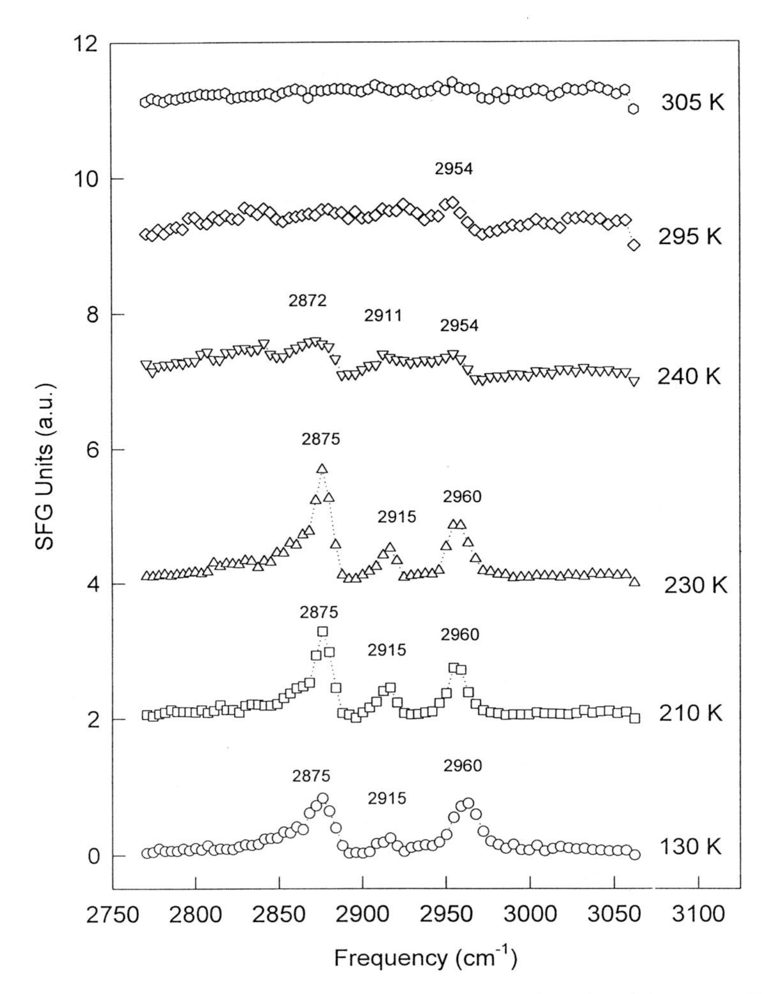

FIG. 15. SFG spectra characterizing thermal dehydrogenation of cyclohexene on Pt(100) at various temperatures.

π/σ hybridization. Again, the peaks at 2875 and 2960 cm^{-1} correspond to ν(CH) at C_4 and C_5, and the feature at 2915 cm^{-1} may indicate a Fermi resonance ($\nu_s + 2\nu_{def}$). As the sample was heated to temperatures higher than 200 K, differences were seen between the SFG spectra of the adsorbates on Pt(111) and Pt(100). In contrast to the peak shift observed for the species on Pt(111), the peaks at 2875 and 2960 cm^{-1} representing the species on Pt(100) did not shift their positions. Furthermore, on the Pt(111) surface, the feature at 2918 cm^{-1} disappeared, whereas the feature at 2915 cm^{-1} in the spectrum of the species on Pt(100) actually became more intense. Bradshaw *et al.* (*42*) used IR spectroscopy to observe the same effect on a Pt(100)(5 × 20) crystal. On the basis of EELS data representing the same surface, Bradshaw *et al.* believe that cyclohexene undergoes a conformational change from the half-chair to the half-boat structure. With this conformational change, the axial hydrogen atoms at C_3 and C_6 become perpendicular to the metal surface, and therefore the corresponding peaks in the vibrational spectra become more intense. As the Pt(100) crystal was heated to approximately 300 K, there were no more features observed in the SFG spectrum, and the surface was probably covered with benzene.

On the basis of the difference in the results obtained for the Pt(111) and Pt(100)(5 × 20) crystals for cyclohexene adsorption under UHV conditions, it is expected that these two surfaces will exhibit differences under catalytic reaction conditions.

B. Low- and High-Pressure Experiments Characterizing 1,3-Cyclohexadiene and 1,4-Cyclohexadiene on Pt(111)

Possible important intermediates in high-pressure catalytic reactions of cyclohexene may be 1,3-cyclohexadiene and 1,4-cyclohexadiene. Thus, it is important to obtain SFG spectra of these species on platinum under both UHV and high-pressure conditions (*50*). A clean Pt(111) crystal was exposed to 1 L of 1,3-cyclohexadiene at 130 K. The SFG spectrum (Fig. 16a) has four peaks, at 2830, 2875, 2900, and 3020 cm^{-1}, and a shoulder at 2770 cm^{-1}. On the basis of the liquid IR spectrum of 1,3-cyclohexadiene (Fig. 16b), the features at 2830, 2875, and 2900 cm^{-1} are assigned to the C–H stretch modes of CH_2, and the peak at 3020 cm^{-1} is assigned to the C–H stretch mode of the C=C–H group. For reasons similar to those invoked to explain cyclohexene adsorption at this temperature, it is believed that 1,3-cyclohexadiene is also adsorbed intact. The nonplanar conformation of the 1,3-cyclohexadiene molecule with two CH_2 groups pointing in opposite directions probably hinders the interaction between its π electrons and the d orbitals of surface platinum atoms and could result in a tilted configuration of the adsorbed 1,3-cyclohexadiene molecule. This geometry has a significant consequence on the chemistry of 1,3-cyclohexadiene upon annealing.

A similar UHV experiment was also performed with 1,4-cyclohexadiene on platinum. At 130 K, a clean surface was exposed to 1 L of 1,4-cyclohexadiene. The SFG spectrum (Fig. 17a) reveals only one strong feature, at 2770 cm^{-1}. Because 1,4-cyclohexadiene is nearly planar in the gas phase,

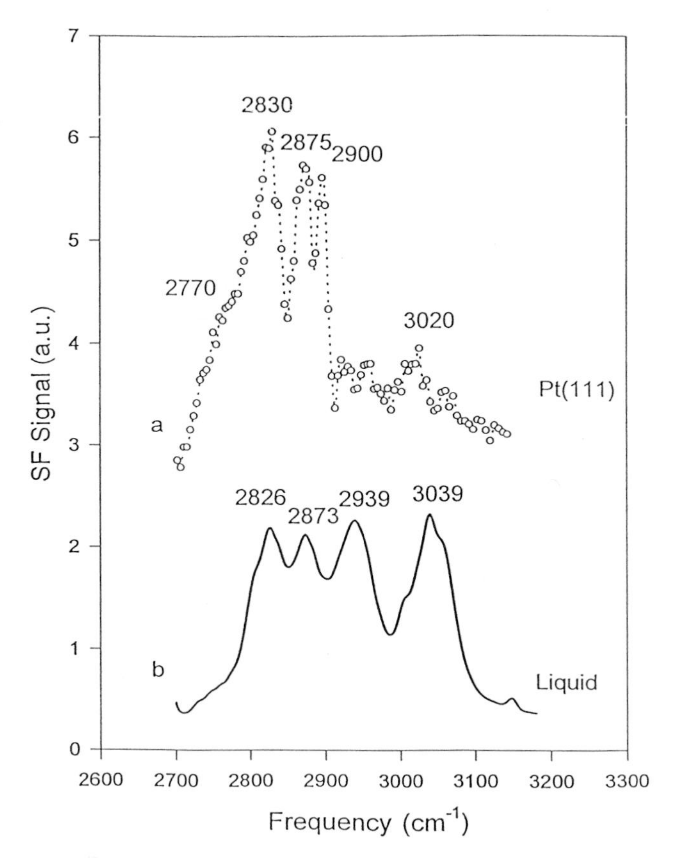

FIG. 16. (a) SFG spectrum of a monolayer of 1,3-cyclohexadiene on Pt(111) at 130 K. (b) Liquid IR spectrum of 1,3-cyclohexadiene.

it should adsorb flat on the platinum to allow the interaction between both sets of π electrons and the d orbitals of the metal. Hugenschmidt *et al.* (*51*) showed that 1,4-cyclohexadiene adsorbs molecularly at temperatures lower than 230 K, but the SFG spectrum is quite different from a liquid IR spectrum (Fig. 17b). The difference in the spectra can be explained by a flat, "quatra-σ-bonding" adsorption geometry. The C–H bonds in the C=C–H group are in the molecular plane and parallel to the surface, and therefore SFG intensity is not expected from these modes. The single feature at 2770 cm^{-1} is thus assigned to the C–H stretch mode in the CH_2 groups, and the frequency shift of this feature is probably due to the strong electron-withdrawing effect from the surface platinum atoms. Conse-

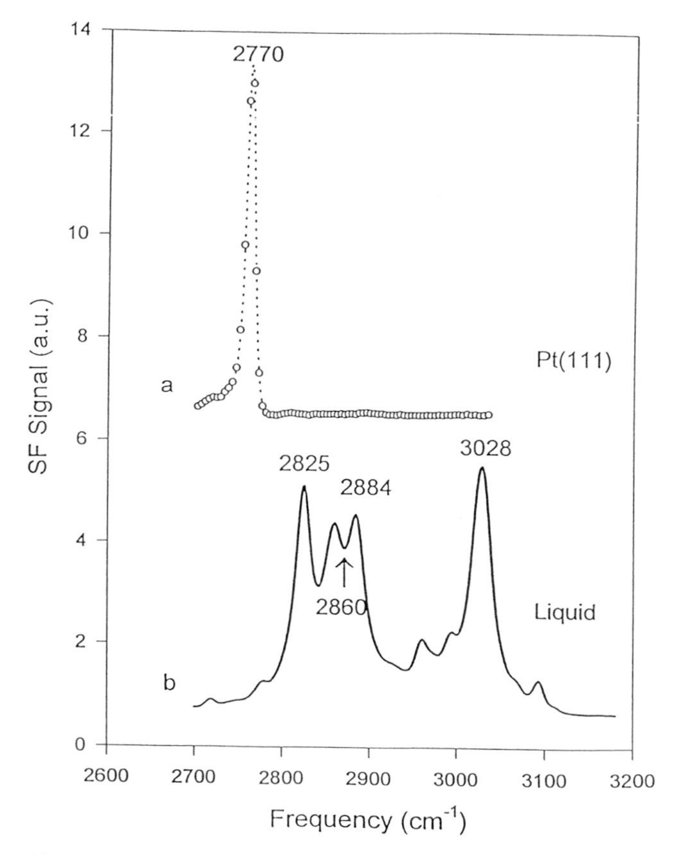

FIG. 17. (a) SFG spectrum of a monolayer of 1,4-cyclohexadiene on Pt(111) at 130 K. (b) Liquid IR spectrum of 1,4-cyclohexadiene.

quently, the electron density in the two C–H bonds flows to the surface platinum atoms and weakens the C–H bond.

As the adsorbed 1,4-cyclohexadiene was heated, there was no change in the SFG spectra up to a temperature of 283 K (Fig. 18). At temperatures higher than 300 K, the feature at 2770 cm^{-1} was no longer observed because of the dehydrogenation of 1,4-cyclohexadiene to form benzene, as was observed by others. The weak feature at approximately 3030 cm^{-1} could arise from benzene.

As the adsorbed 1,3-cyclohexadiene was heated to 244 K, a substantial change in the SFG spectrum ensued. As shown in Fig. 19, a peak associated with 1,4-cyclohexadiene appeared at 2770 cm^{-1}, whereas the associated 1,3-

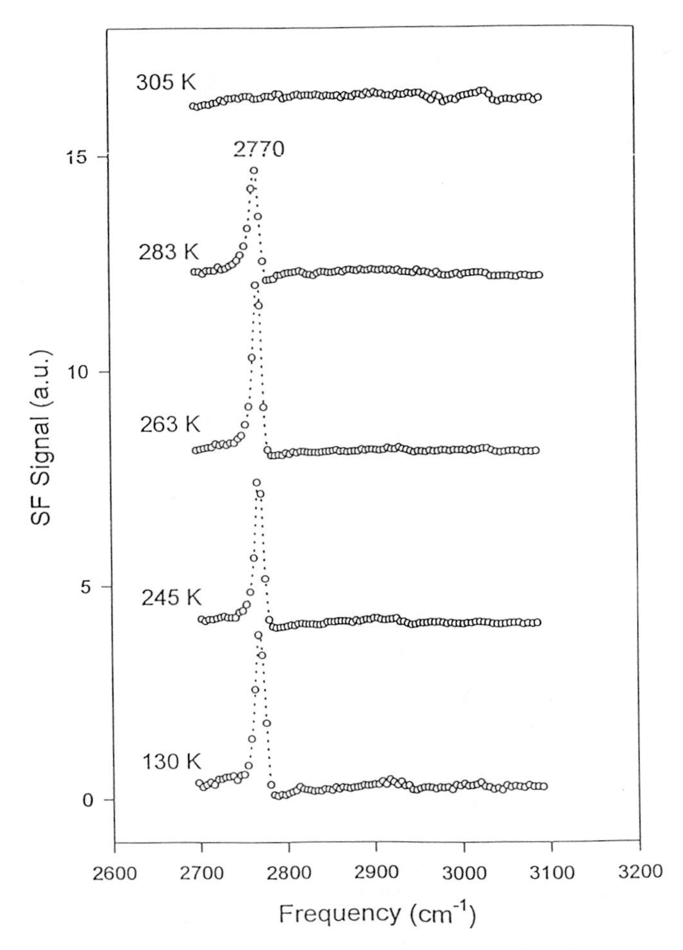

FIG. 18. SFG spectra characterizing the thermal evolution of 1,4-cyclohexadiene on Pt(111) under UHV conditions.

cyclohexadiene features became weaker. As the sample was further heated to 283 K, the strongest SFG feature was now due to 1,4-cyclohexadiene. At temperatures higher than 300 K, this feature was absent from the spectrum and only a weak feature was left near 3030 cm^{-1}, again indicating benzene as a dehydrogenation product. These data show that 1,3-cyclohexadiene can be dehydrogenated directly to give benzene, or it is first isomerized to 1,4-cyclohexadiene followed by dehydrogenation to give benzene.

Two different high-pressure experiments characterizing catalytic hydro-

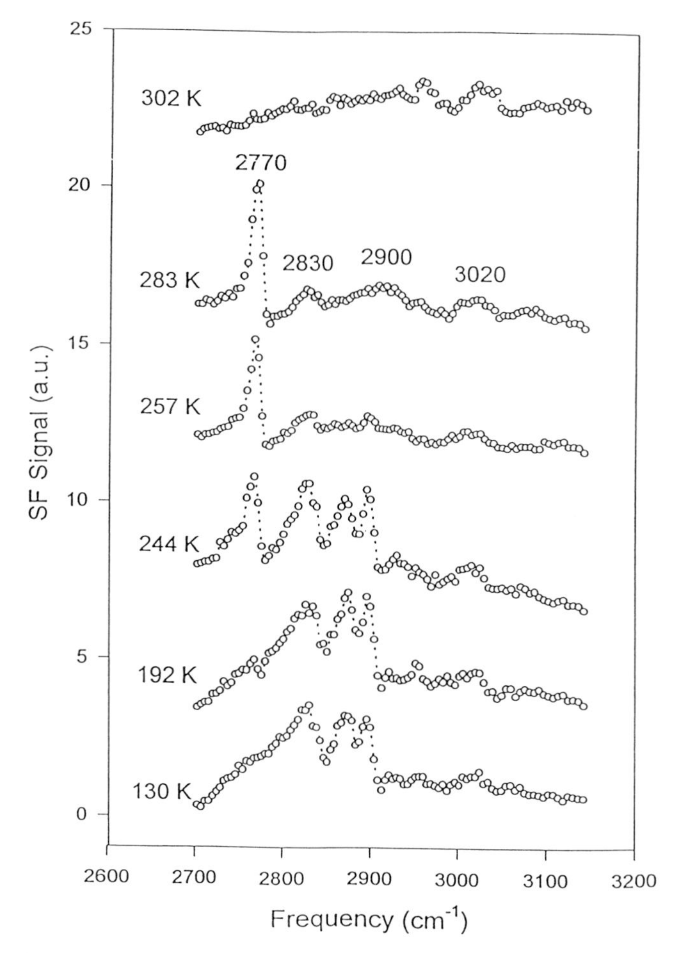

FIG. 19. SFG spectra characterizing the thermal evolution of 1,3-cyclohexadiene on Pt(111) under UHV conditions.

genation of 1,3-cyclohexadiene and of 1,4-cyclohexadiene were carried out on Pt(111) at 295 K. The TOR for 1,3-cyclohexadiene was very high (100 molecules per platinum site per second), whereas the hydrogenation of 1,4-cyclohexadiene was much slower, with a TOR of only 5 molecules per platinum site per second. The SFG spectra characterizing the two reactions are shown in Figs. 20 and 21. These data are important for determining the principal reaction intermediates during cyclohexene hydrogenation and dehydrogenation on Pt(111) and on Pt(100).

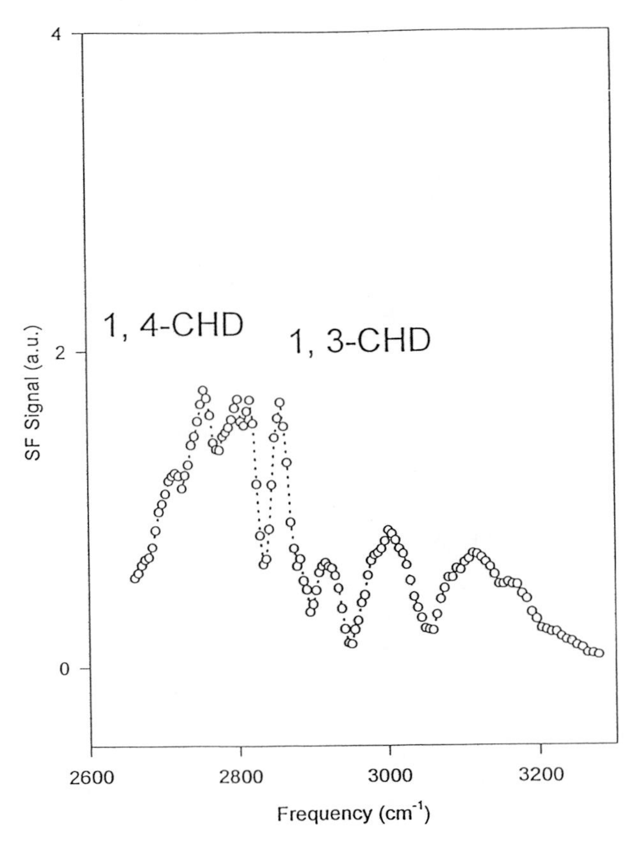

FIG. 20. SFG spectrum recorded during 1,3-cyclohexadiene (CHD) hydrogenation on Pt(111). Both 1,4- and 1,3-CHD are observed.

C. High-Pressure Catalytic Cyclohexene Hydrogenation and Dehydrogenation on Pt(111) and Pt(100) Monitored by SFG and Gas Chromatography

To mimic a high-pressure industrial reaction, a platinum single crystal was placed into a UHV chamber/batch reactor and cleaned by conventional methods. Into the reactor, 10 Torr of cyclohexene, 100 Torr of hydrogen, and 650 Torr of He were introduced (*45*). The sample was heated to temperatures between 300 and 600 K while changes in the spectra of surface species were monitored by using SFG, and the gas composition was monitored by gas chromatography. Turnover rates were determined from the chromatographic data at each temperature, as shown in Figs. 22a and 22b

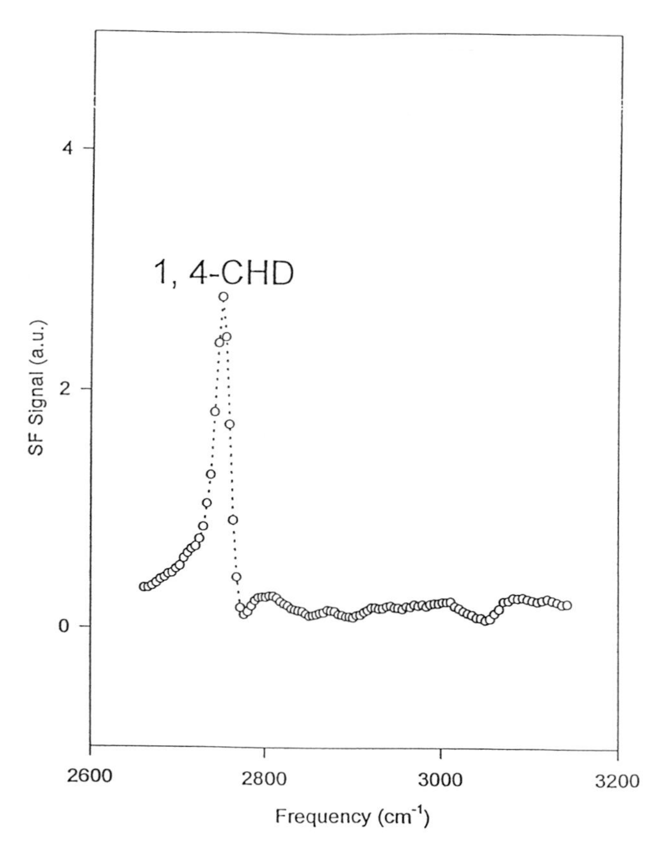

FIG. 21. SFG spectrum recorded during 1,4-cyclohexadiene(CHD) hydrogenation on Pt(111). Only 1,4-CHD is observed.

for cyclohexene on Pt(111) and Pt(100), respectively. At 300 K, the hydrogenation and dehydrogenation TORs were found to be negligible. Upon heating of the crystal sample, the hydrogenation TOR increased and reached a maximum of 78 molecules per platinum site per second at 400 K for Pt(111) and a maximum of 38 molecules per platinum site per second at 425 K for Pt(100). Near the temperature of the maximum hydrogenation TOR on either crystal, the dehydrogenation TOR began to increase while the hydrogenation rate began to decrease with increasing temperature. On Pt(111), the maximum dehydrogenation TOR was 58 molecules per platinum site per second at 475 K, and on Pt(100) the maximum dehydrogenation TOR was 75 molecules per platinum site per second at 500 K.

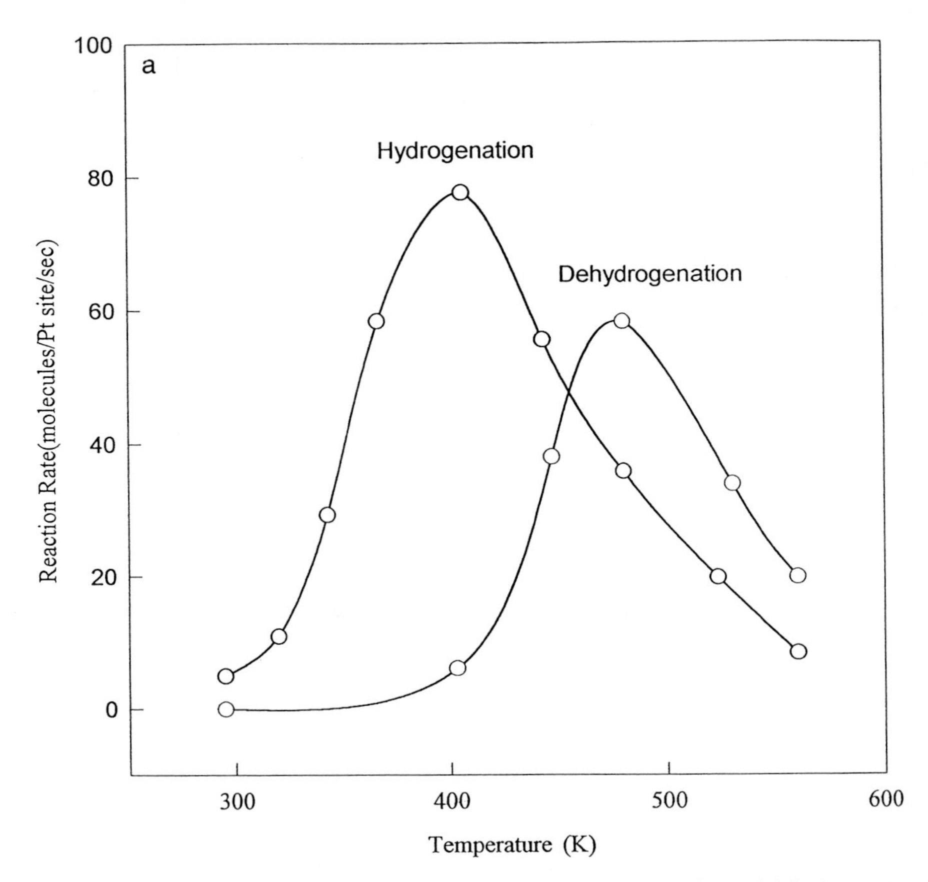

FIG. 22. (a) The temperature dependence of the TOR of hydrogenation and dehydrogenation reactions of cyclohexene on Pt(111): 10 Torr cyclohexene, 100 Torr of H_2, and 650 Torr of He. (b) The temperature dependence of the TOR of hydrogenation and dehydrogenation reactions of cyclohexene on Pt(100).

Assuming that the dehydrogenation TOR is negligible in the region of increasing hydrogenation TOR, the activation energies for cyclohexene hydrogenation were calculated to be 8.9 and 15.5 kcal/mol for Pt(111) and Pt(100), respectively. As the temperature was increased above the cyclohexene hydrogenation rate maximum, the hydrogenation rate decreased as the dehydrogenation rate increased. This result may indicate that the numbers of active surface sites for the two processes change. Otherwise, the hydrogenation reaction rate would obey the Arrhenius law over the entire temperature range.

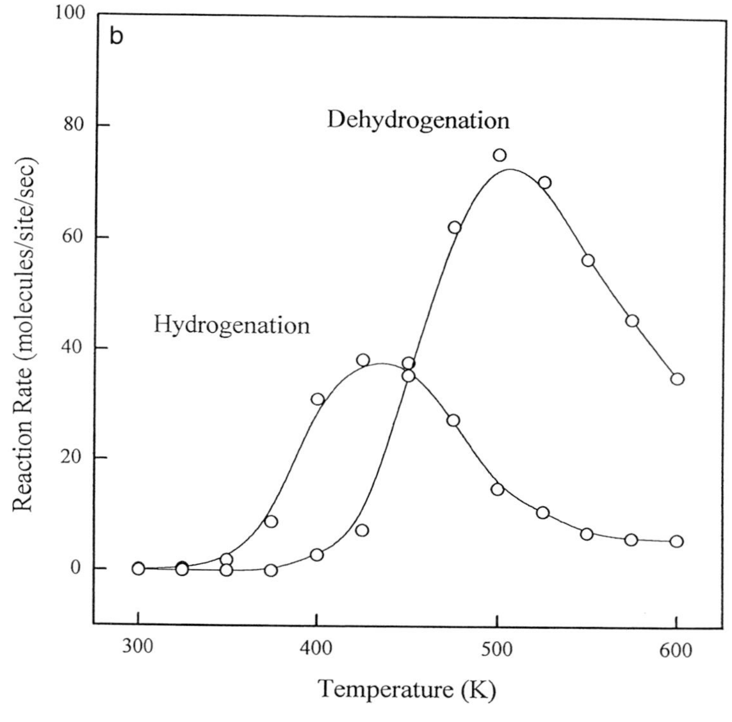

FIG. 22. *(continued)*

While the catalytic reactions were being performed, SFG spectra were collected. Figure 23 shows the SFG spectra for the temperature at which the gases were introduced into the chamber, the spectra for the maximum hydrogenation TOR, and the spectra for the maximum dehydrogenation TOR. In the spectrum of the adsorbates on Pt(111) (Fig. 23a) at 295 K, a sharp feature is evident at 2755 cm^{-1}. The SFG spectrum (Fig. 21) representing the high-pressure reaction of 1,4-cyclohexadiene allows the assignment of the 2755 cm^{-1} feature to 1,4-cyclohexadiene. On the basis of the high-pressure data for hydrogenation of 1,3-cyclohexadiene (Fig. 20), the weaker features at higher frequencies in Fig. 23a are attributed to 1,3-cyclohexadiene. We stress that there is no evidence of a c-C_6H_9 intermediate species, as was observed under UHV conditions. The major features of the species on Pt(100) at 300 K (Fig. 23b) are assigned to 1,3-cyclohexadiene; a weak accompanying feature at approximately 2780 cm^{-1} is assigned to 1,4-cyclohexadiene.

At the maximum hydrogenation TOR—at 403 K and 425 K for Pt(111)

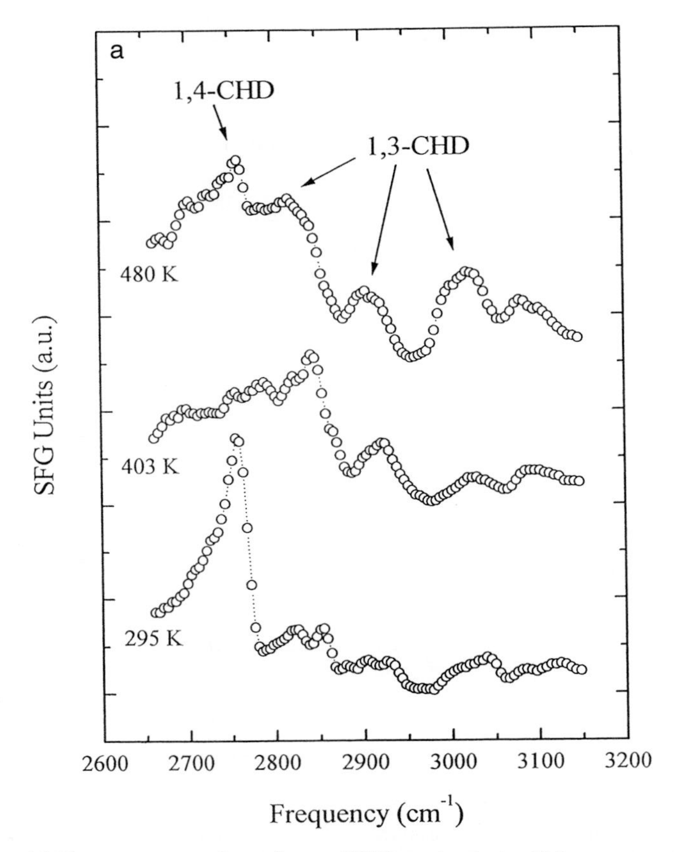

FIG. 23. (a) The temperature dependence of SFG spectra during high-pressure cyclohexene hydrogenation and dehydrogenation on Pt(111): 10 Torr cyclohexene, 100 Torr H_2, and 650 Torr He. At the maximum dehydrogenation temperature, both 1,3- and 1,4-CHD species are observed. (b) The temperature dependence of SFG spectra during high-pressure cyclohexene hydrogenation and dehydrogenation on Pt(100). Only 1,3-CHD is observed.

and Pt(100), respectively—the two similar SFG spectra indicate the presence of 1,3-cyclohexadiene as the major surface species. This spectral evidence indicates that cyclohexene hydrogenation proceeds through a 1,3-cyclohexadiene intermediate. As the temperature was increased to the maximum of the dehydrogenation TOR, the SFG spectra of adsorbates on the two crystal surfaces again became different from each other. The spectrum of the species on Pt(111) indicates the presence of both 1,3- and 1,4-cyclohexadiene, whereas only 1,3-cyclohexadiene was observed on Pt(100). Considering the differences in the two SFG spectra and the difference

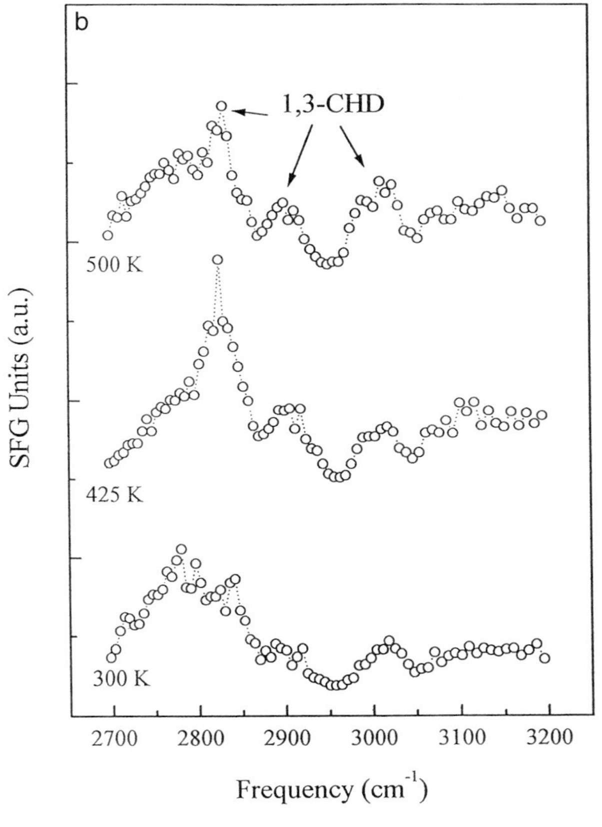

FIG. 23. *(continued)*

of the TORs for the maximum dehydrogenation, reaction pathways are proposed for dehydrogenation on each platinum surface. On Pt(111), cyclohexene dehydrogenation can proceed through either a 1,3- or 1,4-cyclohexadiene intermediate, whereas on Pt(100) dehydrogenation occurs only through a 1,3-cyclohexadiene intermediate. Figure 24 shows the difference of the reaction pathways for the two surfaces.

Because cyclohexene dehydrogenation occurs faster on Pt(100) than on Pt(111), and because 1,4-cyclohexadiene is absent from the Pt(100) surface, it is believed that 1,4-cyclohexadiene inhibits dehydrogenation on the Pt(111) surface. This inference is reasonable considering that 1,4-cyclohexadiene conceivably must isomerize to 1,3-cyclohexadiene before it is completely dehydrogenated to form benzene.

Reaction Pathway on Pt(100)

400 - 600 K

400 - 600 K

300 - 600 K

300 - 600 K

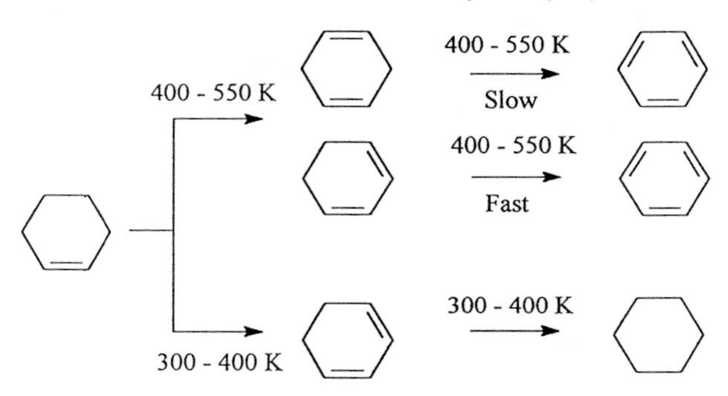

FIG. 24. Proposed reaction pathways for cyclohexene hydrogenation and dehydrogenation on Pt(111) and Pt(100). The absence of 1,4-CHD during dehydrogenation conditions on Pt(100) explains the structure sensitivity of the reaction.

VII. CO Oxidation on Pt(111)

One of the most thoroughly investigated surface-catalyzed reactions is the oxidation of carbon monoxide to carbon dioxide on transition metals. UHV studies have shown that the thermal excitation of adsorbed oxygen initiates the reaction, and IR spectra studies show that atop CO is the major carbon monoxide species on the surface (*52*). The reaction proceeds via a Langmuir–Hinshelwood mechanism (reaction on the surface following adsorption of both reactants) (*53*). Under certain conditions, the reaction shows oscillatory behavior, hysteresis, and formation of CO_2 with excess translational and vibrational energy (*54–57*). It has also been shown that the metal catalyst surface exhibits structural rearrangements in some range of temperature and reactant pressure (*58*).

Under UHV conditions the adsorbed oxygen must first be thermally excited for the reaction to be initiated (*59*). Furthermore, the desorption of CO_2 is dependent on the adsorption state of the oxygen species. The mobile oxygen atoms on the surface that approach and react with CO molecules are possibly responsible for CO_2 production. IR measurements made under UHV conditions have shown that atop CO is the major CO species on the surface under oxidative conditions.

The majority of what is known about CO oxidation has come from UHV studies, and because of the limitations of UHV experiments the investigation of CO oxidation under high pressures of CO and O_2 with techniques such as SFG is of practical importance.

A high-pressure catalytic reaction was performed as 100 Torr of CO, 40 Torr of O_2, and 600 Torr of He were introduced into a batch reactor (*60*). Figure 25 shows the SFG spectra at various temperatures. At temperatures lower than 740 K, only atop-bonded CO (with a feature at 2090 cm^{-1}) and an incommensurate CO overlayer formation represented by a low-frequency shoulder were observed (*61*). As the temperature was increased to 1100 K, the intensity of the feature indicating atop CO species decreased, which indicates that the concentration of atop CO decreased. Simultaneously, the intensity of the feature indicating the incommensurate CO overlayer increased. The ignition temperature of the reaction under these conditions was 760 K; at this temperature, the reaction becomes self-sustaining, proceeding at a high constant temperature because of the high exothermicity of the reaction. When a comparison was made between the SFG spectra recorded at temperatures higher and lower than the ignition temperature, it became obvious that the two are completely different. A low-frequency broad band and peak at 2045 cm^{-1}, which dominate the spectrum recorded at 1100 K, are assigned to incommensurate CO and CO adsorbed at defect sites. The reaction rate was found to increase with an increase of the intensity of this band.

Another high-pressure CO oxidation experiment was performed, but instead of CO present in excess, O_2 was in excess. In this experiment, 100 Torr of O_2, 40 Torr of CO, and 600 Torr of He were introduced into the chamber. The SFG spectra recorded under these conditions at various temperatures are shown in Fig. 26. At low temperatures, the reaction rate was found to be low, and the SFG spectrum was dominated by the atop CO species, similar to what was observed when CO was present in excess. At 540 K, a TOR of 28 molecules of CO_2 per platinum surface site per second was determined. With increasing temperature, the atop CO peak intensity decreased, but the reaction rate increased. The ignition temperature was near 600 K, and at higher temperatures the system reached the high-reactivity regime, with the reaction becoming self-sustaining. As in

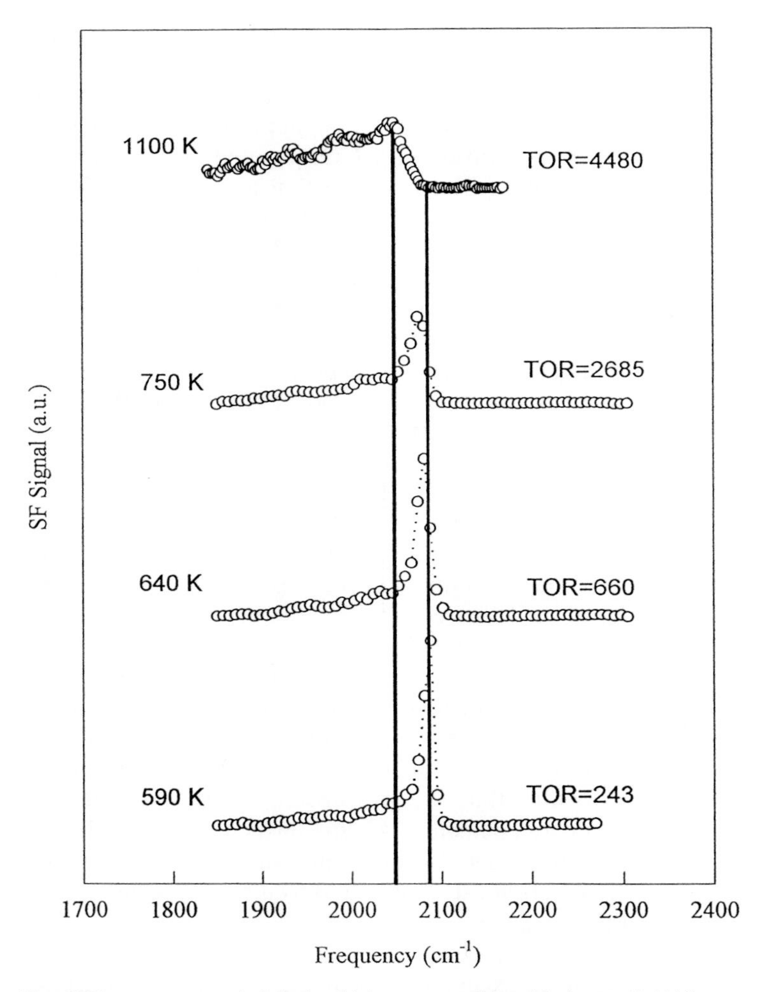

FIG. 25. SFG spectra recorded during high-pressure CO oxidation on Pt(111) at temperatures above and below the ignition temperature: 40 Torr O_2, 100 Torr CO, and 600 Torr He (TOR units: molecules converted per platinum atom per second).

the experiment with excess CO, the SFG spectra recorded at temperatures higher and lower than the ignition temperature were considerably different from each other. The atop CO peak, which was the dominant feature at temperatures lower than the ignition temperature, was not observed at all at higher temperatures. The only new feature to note was again at 2045 cm^{-1}, as observed in the experiment with excess CO. Two other peaks, at

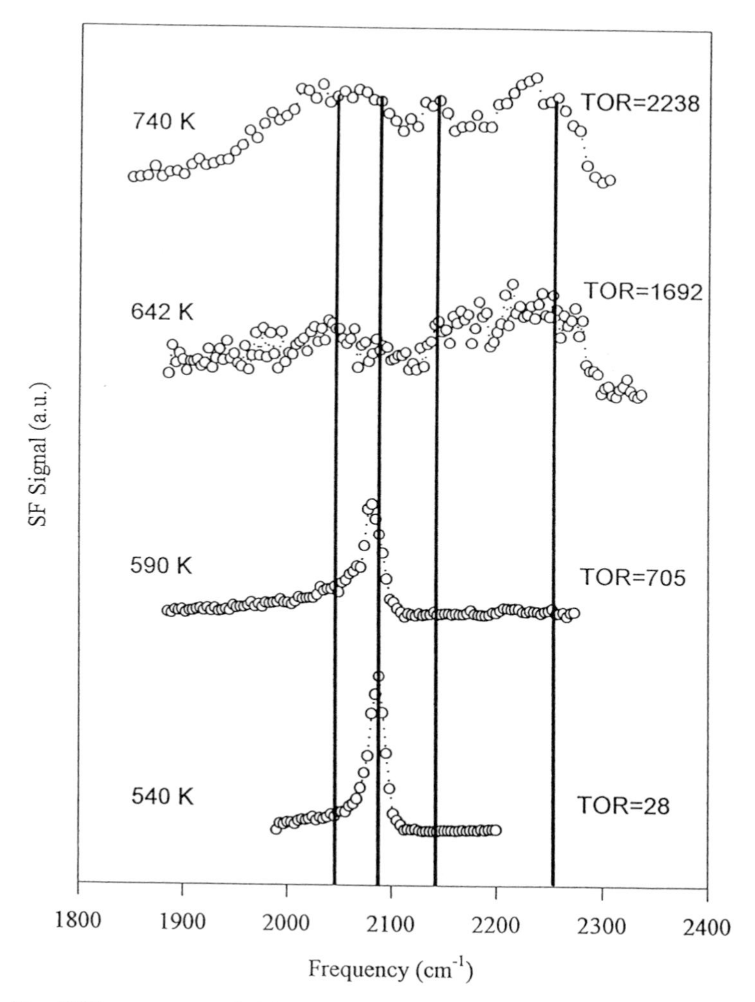

FIG. 26. SFG spectra recorded during high-pressure CO oxidation on Pt(111) at temperatures above and below the ignition temperature: 100 Torr O_2, 40 Torr CO, and 600 Torr He (TOR units: molecules converted per platinum atom per second).

2130 and 2240 cm^{-1}, were also observed, but these peaks were not real; in the high-temperature regime, the nonlinear background was higher than that in the low-temperature regime, and the observed peaks at 2130 and 2240 cm^{-1} were the results of a decrease in the IR beam intensity. The decrease in the IR power was caused by gas-phase CO in the chamber and by atmospheric CO_2 outside the chamber.

Figure 27 shows the correlation between the reaction rate and the relative CO coverage of various surface species. As Fig. 27a indicates, when the reaction rate increases, intensity of the atop CO species decreases. This result clearly indicates that atop CO is not the reactive participant in CO oxidation. Its presence on the surface may actually inhibit the reaction.

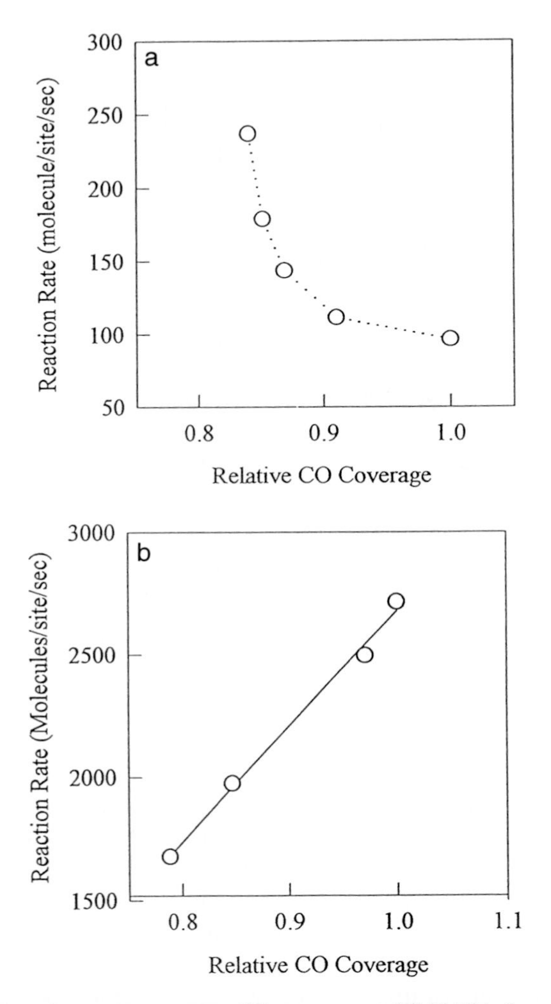

FIG. 27. (a) Reaction rate versus atop CO coverage at 590 K. This figure shows atop CO is not the reactive species in CO oxidation. (b) Reaction rate versus surface coverage of incommensurate CO at 720 K. (c) Reaction rate versus surface coverage of incommensurate CO at 590 K.

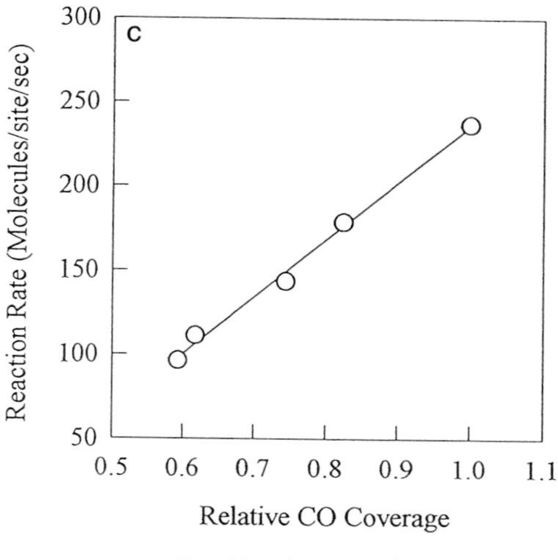

FIG. 27. *(continued)*

Figures 27b and 27c show that the reaction rate is proportional to the concentration of the incommensurate CO species, at temperatures both higher and lower than the ignition temperature. These data show that this incommensurate surface species must be directly responsible for CO oxidation at all the investigated temperatures.

VIII. Summary and Conclusions

The data presented here show clearly that SFG is a flexible tool that allows molecular-level characterization of surfaces at ambient pressures between 10^{-11} and 10^{3} Torr, a range of 14 orders of magnitude. Low-pressure experiments served two purposes; they allowed correlation of results with those of previous experiments carried out with electron spectroscopies and ion scattering techniques, and they provided information about possible reaction intermediates under high pressures. By correlating high-pressure SFG spectra recorded under reaction conditions and kinetics data, it has been possible to determine the important reaction intermediates of several surface catalytic reactions.

The ethylene hydrogenation study showed that although ethylidyne and di-σ-bonded ethylene may be present on the platinum surface at high concentrations, the important reaction intermediate is actually π-bonded ethylene. After ethylene adsorbs through its π electrons, it is hydrogenated stepwise through an ethyl intermediate to form ethane. This work also showed that strongly bound species such as ethylidyne may not turn over, but weakly bound species such as π-bonded ethylene that desorb under UHV condition do turn over.

Similar results characterize propylene hydrogenation. Propylene also adsorbs through its π bond, giving a more important reaction intermediate than propylidyne and di-σ-bonded propylene. After the π-bonded surface species is formed, it can be hydrogenated to propane through either a 1-propyl or a 2-propyl moiety, and the results show that the 2-propyl species is the more important reaction intermediate.

Under UHV conditions, cyclohexene is dehydrogenated to benzene through a C_6H_9 intermediate, but under high-pressure conditions this reaction intermediate is not observed. By allowing a comparison of the structure sensitivity of cyclohexene hydrogenation and dehydrogenation on Pt(111) and Pt(100) at high pressure, SFG data and corresponding kinetics data showed that hydrogenation occurs on both surfaces through a 1,3-cyclohexadiene intermediate. Dehydrogenation occurs faster on Pt(100) than on Pt(111), and from an examination of the SFG spectra it became clear that both 1,3-cyclohexadiene and 1,4-cylcohexadiene exist on Pt(111), whereas only 1,3-cyclohexadiene exists on Pt(100). The evidence shows that dehydrogenation occurs faster through the 1,3-cyclohexadiene species than through the 1,4-cyclohexadiene species, and it may explain why the Pt(100) surface gives a higher TOR. No active sites are blocked by 1,4-cyclohexadiene on Pt(100), but some are on Pt(111). The structure sensitivity can be explained by the predominance of one of the two reaction intermediates, which provides a more rapid reaction pathway than the other reaction intermediate.

The CO oxidation work showed that there are two different reaction regimes for CO oxidation on Pt(111). An atop CO species dominates the SFG spectra at low temperatures, and an incommensurate CO species dominates the spectra at temperatures higher than the ignition temperature, at which the reaction becomes self-sustaining and proceeds faster. On the basis of the relationship of the surface concentration of the incommensurate CO species to the kinetics, it was shown that this species is more important during oxidation than atop CO.

The biggest limitation of SFG is that it cannot be used on supported powder catalysts. The surface must be ordered and reflective to detect an SFG signal. Also, the frequency range (1650–3500 cm^{-1}) in which tunable IR beams can be generated limits the vibrational modes which can be investigated. Currently, a $AgGaSe_2$ nonlinear crystal is being used to generate a tunable IR beam between 1200 and 2000 cm^{-1}. This frequency range will allow deformation and other low-frequency vibrational modes to be studied.

The application of SFG is also being extended in our laboratory. Currently, broadband SFG is being developed to shorten the time required to acquire spectra and to monitor surface kinetics on the femto-second timescale. Also, it has been found recently by this group that SFG spectra can be acquired for molecules adsorbed on structured nanoparticle arrays supported on oxide surfaces (*62*). It appears that the SFG signal generated on these nanoparticle arrays is enhanced for reasons similar to those of surface-enhanced Raman spectroscopy. SFG studies of these nanoparticle arrays will allow the investigation of model supported catalysts.

Thus, SFG is a unique and excellent technique for investigating single-crystal surfaces under high-pressure catalytic reaction conditions. SFG spectra combined with kinetics data provide much insight into the important reaction intermediates present on a surface during a catalytic reaction.

References

1. Somorjai, G. A., "Introduction to Surface Chemistry and Catalysis." Wiley, New York, 1994.
2. Somorjai, G. A., *Surf. Sci.* **299/300,** 849 (1994).
3. Goodman, D. W., *Chem. Rev.* **95,** 523 (1995).
4. Ertl, G. *Surf. Sci.* **299/300,** 742 (1994).
5. Woodruff, D. P., and Delchar, T. A., "Modern Techniques of Surface Science." Cambridge Univ. Press, New York, 1986.
6. Blakely, D. W., Kozak, E., Sexton, B. A., and Sormorjai, G. A., *J. Vac. Sci. Technol.* **13,** 1091 (1976).
7. Cabrera, A. L., Spencer, N. D., Kozak, E., Davies, P. W., and Somorjai, G. A., *Rev. Sci. Instrum.* **53,** 1888 (1982).
8. Rupprechter, G., and Somorjai, G. A., *Catal. Lett.* **48,** 17 (1997).
9. Davis, S. M., Zaera, F., and Somorjai, G. A., *J. Am. Chem. Soc.* **104,** 7453 (1982).
10. Shen, Y. R., *Surf. Sci.* **299/300,** 551 (1994).
11. Du, Q., Superfine, R., Freysz, E., and Shen, Y. R., *Phys. Rev. Lett.* **70,** 2313 (1993).
12. Johal, M. S., Ward, R. N., and Davies, P. B., *J. Phys. Chem.* **100,** 274 (1996).
13. Conboy, J. C., Messmer, M. C., and Richmond, G. L., *J. Phys. Chem.* **100,** 7617 (1996).
14. Shen, Y. R., *Nature* **337,** 519 (1989).
15. Shen, Y. R., "The Principles of Nonlinear Optics." Wiley, New York, 1984.
16. Guyot-Sionnest, P., Hunt, J. H., and Shen, Y. R., *Phys. Rev. Lett.* **59,** 1597 (1987).
17. Bain, C. D., *J. Chem. Soc. Faraday Trans.* **91,** 1281 (1995).
18. Hirose, C., Yamamoto, H., Akamatsu, N., and Domen, K., *J. Chem. Phys.* **97,** 10064 (1993).
19. Zhang, D., Huang, J., Shen, Y. R., and Shen, C., *J. Opt. Soc. Am.* **10,** 1758 (1993).
20. Cremer, P. S., Su, X., Shen, Y. R., and Somorjai, G. A., *J. Am. Chem. Soc.* **118,** 2942 (1996).
21. Horiuti, I., and Polanyi, M., *Trans. Faraday Soc.* **30,** 1164 (1934).
22. Cassuto, A., Kiss, J., and White, J., *Surf. Sci.* **255,** 289 (1991).
23. Ibach, H., and Lehwald, S., *Surf. Sci.* **117,** 685 (1982).
24. Cremer, P., Stanners, C., Niemantsverdriet, J., Shen, Y., and Somorjai, G., *Surf. Sci.* **328,** 111 (1993).
25. Land, T., Michely, T., Behm, R., Hemminger, J., and Comsa, G., *J. Chem. Phys.* **97**(9), 6774 (1992).
26. Davis, S., Zaera, F., Gordon, B., and Somorjai, G., *J. Catal.* **92,** 250 (1985).
27. Beebe, T., and Yates, J., *J. Am. Chem. Soc.* **108,** 663 (1986).

28. Mohsin, S., Trenary, M., and Robota, H., *J. Phys. Chem.* **92,** 5229 (1988).
29. Cremer, P. S., Su, X., Shen, Y. R., and Somorjai, G. A., *J. Am. Chem. Soc.* **118,** 2942 (1996).
30. Zaera, F., *Surf. Sci.* **219,** 453 (1989).
31. Steiniger, H., Ibach, H., and Lehwald, S., *Surf. Sci.* **117,** 685 (1982).
32. Cassuto, A., Mane, M., Hugenschmidt, M., Dolle, P., and Jupille, J., *Surf. Sci.* **237,** 63 (1990).
33. Schlatter, J., and Boudart, M., *J. Catal.* **24,** 482 (1972).
34. Avery, N., and Sheppard, N., *Proc. R. Soc. London A* **405,** 1 (1986).
35. Koestner, R., Frost, J., Stair, P., Van Hove, M., and Somorjai, G., *Surf. Sci.* **116,** 85 (1982).
36. Lok, L., Gaidai, N., and Kiperman, S., *Kinet. Katal.* **32,** 6.1406 (1991).
37. Otero-Schipper, P., Wachter, W., Butt, J., Burwell, R., and Cohen, J., *J. Catal.* **50,** 494 (1977).
38. Shahid, G., and Sheppard, N., *Spectrochim. Acta* **46A**(6), 999 (1990).
39. Cremer, P. S., Su, X., Shen, Y. R., and Somorjai, G. A., *J. Phys. Chem.* **100**(40), 16302 (1996).
40. Bussell, M. E., Henn, F. C., and Campbell, C. T., *J. Phys. Chem.* **96,** 5965 (1992).
41. Rodriguez, J. A., and Campbell, C. T., *J. Phys. Chem.* **93,** 826 (1989).
42. Lamont, C. L. A., Borbach, M., Martin, R., Gardner, P., Jones, T. S., Conrad, H., and Bradshaw, A. M., *Surf. Sci.* **374,** 215 (1997).
43. Land, D. P., Erley, W., and Ibach, H., *Surf. Sci.* **289,** 773 (1993).
44. Martin, R., Gardner, P., Tushaus, M., Bonev, C. H., Bradshaw, A. M., and Jones, T. S., *J. Electron Spectrosc. Relat. Phenom.* **54/55,** 773 (1990).
45. Su, X., Kung, K., Lahtinen, J., Shen, Y. R., and Somorjai, G. A., *Catal. Lett.* **54,** 9 (1998).
46. Lespade, L., Rodin, S., Cavagnat, D., and Abbate, S., *J. Phys. Chem.* **97,** 6134 (1993).
47. Neto, N., *Spectrochim. Acta* **23A,** 1763 (1967).
48. Haines, J., and Gilson, D. F. R., *J. Phys. Chem.* **94,** 4712 (1990).
49. Caillod, J., Saur, O., and Lavalley, J. C., *Spectrochim. Acta* **36A,** 185 (1980).
50. Su, X., Shen, Y. R., Kung, K. Y., Lahtinen, J., and Somorjai, G. A., *J. Mol. Catal. A* **141,** 9 (1999).
51. Hugenschmidt, M. B., Diaz, A. L., and Campbell, C. T., *J. Phys. Chem.* **96,** 5974.
52. Yoshinobu, J., and Kawai, M., *J. Chem. Phys.* **103,** 3220 (1995).
53. Campbell, C. T., Ertl, G., Kuipers, H., and Segner, J., *J. Chem. Phys.* **73**(11), 5862 (1980).
54. Ertl, G., *Science* **254,** 1750 (1991).
54. Hong, S., and Richardson, H., *J. Vac. Sci. Technol. A* **11**(4), 1951 (1993).
55. Allers, K.-H., Pfnur, H., Feulner, P., and Menzel, D., *J. Chem. Phys.* **100,** 3985 (1994).
57. Bald, D. J., Kunkel, R., and Bernasek, S. L., *J. Chem. Phys.* **104,** 7719 (1996).
58. Falta, J., Imbihl, R., Sander, M., and Henzler, M., *Phys. Rev. B* **45,** 6858 (1992).
59. Yoshinobu, J., and Kawai, M., *J. Chem. Phys.* **103,** 3220 (1995).
60. Su, X., Cremer, P. S., Shen, Y. R., and Somorjai, G. A., *J. Am. Chem. Soc.* **119,** 3994 (1997).
61. Su, X., Cremer, P. S., Shen, Y. R., and Somorjai, G. A., *Phys. Rev. Lett.* **77**(18), 3858 (1996).
62. Baldelli, S., Eppler, A., Somorjai, G. A., manuscript in preparation.

Index

E

F

G

H

R

S

T

U

Z

ISBN 0-12-007845-7